ADVANCE MICROPROCESSOR

DR. A.K. GAUTAM

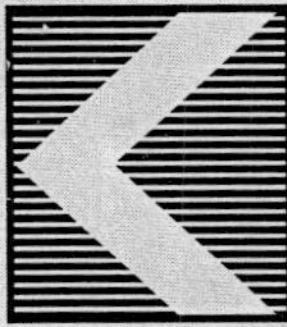

KHANNA BOOK PUBLISHING CO. (P) LTD.

Publisher of Engineering and Computer Books

4C/4344, Ansari Road, Darya Ganj, New Delhi-110002

Phone : 011-23244447-48 **Mobile:** +91-9910909320

E-mail : contact@khannabooks.com

Website : www.khannabooks.com

Dear Readers,

To prevent the piracy, this book is secured with HIGH SECURITY HOLOGRAM on the front title cover. In case you don't find the hologram on the front cover title, please write us to at contact@khannabooks.com or whatsapp us at +91-99109 09320 and avail special gift voucher for yourself.

Specimen of Hologram on front Cover title:

Moreover, there is a SPECIAL DISCOUNT COUPON for you with EVERY HOLOGRAM.

How to avail this SPECIAL DISCOUNT:

Step 1: Scratch the hologram
Step 2: Under the scratch area, your "coupon code" is available
Step 3: Logon to www.khannabooks.com
Step 4: Use your "coupon code" in the shopping cart and get your copy at a special discount
Step 5: Enjoy your reading!

Thanks,
Khanna Publishing House

ADVANCE MICROPROCESSOR
by **Dr. A.K. Gautam**

ISBN: 978-93-81068-92-2

To view complete list of books, please scan the QR Code:

Published by:
Khanna Book Publishing Co. (P) Ltd.
CIN: U22110DL1998PTC095547

Visit us at: **www.khannabooks.com**
Write us at: **contact@khannabooks.com**

Printed in India.

PREFACE

Microprocessors have become an integral part of human's daily life. Yet there exists a requirement for a text that provides deep insight into the working of these very complex digital devices. The purpose of this book is to present and emphasis on the working, interfacing and programming of the x86 family of microprocessor.

Chapter 1 provides the basics and evolution of computer, the 80186/80188 and 80286 microprocessors and also different digital devices that may be interfaced to the microprocessor.

Chapter 2 discussed the architect, pin diagram and assembly language programming of the 8085 microprocessor. Chapter 3 provides the architecture of the 8086, memory segmentation and deep insight into the different addressing modes used by the 8086 as well as the basic instruction template for forming the 8086 machine codes. Chapter 4 discusses thoroughly the pin out as well as the IC module of the 8086 and 8088 microprocessor.

Chapter 5,6,7,8 discusses assembly language programming, as well as provides an understanding of the 8086/8088 instruction set allowing the reader to form working programs for the 8086/8088 by use of common assemblers available. Chapter 9 thoroughly discusses the 8087 coprocessor, as well as provides a comprehensive coverage of its instruction set.

Chapter 10 deals with the different available support chips and provides a complete coverage of these chips. Chapter 11 completes the 8086/8088 microprocessor system design by providing knowledge of the 8086 memory module and interface techniques. Interrupts have also been discussed thoroughly in chapter 11.

Chapter 12 deals with advanced microcontrollers such as the 8051 microcontroller. The analog to digital and digital to analog converters are played a wide role when digital systems have to interact with analog world. Shift registers design. Thus, A/D and D/A converters are covered in chapter 13.

Considerable efforts have been to present the material at such a level that there is consistent progression from theoretical concepts to design considerations without getting mired in too many complex theoretical concepts. Efforts have been made throughout to provide working programs which should aid the student who are attempting to program the 8085 and 8086. The book shall serve as a text book for several universities. The book will also be useful for practicing engineers, who may like to keep up to date their knowledge in this very important field. We will be grateful to any comments which should provide the help to improve the book in the following edition. Students can interact with us in this very vast topic by emailing their questions and comments to talk drakgautam@ieee.org.

Any effort of this magnitude is not possible without the assistance of number people. In particular, I would like to thank Mr. Papendra Kumar, AP, CSED, GBPEC, Pauri for providing numerous suggestions for improving the technical contents of this book. I also thank my M. Tech. and B. Tech students for the untold hours spend by them in thoroughly reviewing every chapter for language and contents. For publishing and typing the text with great diligence I would like to thank Mr. Punit Khanna for Khanna publication, New Delhi. I also wish to acknowledge many others who have contributed in various ways to this work and regret that space will not allow individual mention. Last but not lease, all the patience and encouragement consistently shown by my family can only be acknowledged to some degree by dedicating this book to them.

Overview

Contents

Chapter-1

INTRODUCTION TO MICROPROCESSOR

INTRODUCTION

A microprocessor is the chip containing some control and logic circuits that is capable of making arithmetic and logical decisions based on input data and produces, the corresponding Arithmetic or logical output. The word 'processor' is taken from the word 'process' that means to carry out systematic operations on data. The computer that is being used to write this page of the manuscript uses a microprocessor to do its work. The microprocessor is the heart of any computer, whether it is a desktop machine, a server or a laptop. The microprocessor may be a Pentium, a K6, a PC, a Sparc or any of other brands and types but it does approximately the same thing in the same way. All logically enabled devices use the microprocessor without that they can't do anything. The microprocessor is not only used in computers, but also used in many other devices such as cell phones, satellites, and many other hand held devices. They are also present in modern day cars in the form of microcontrollers. A microprocessor is also known as a CPU or central processing unit, which is a complete computational engine that is fabricated on a single chip. The historical background of different microprocessors will help us to better understand the design compromises they made as well as to understand the legacy issues surrounding the CPU's design.

1.1 THE COMPUTER

Before the launch of microprocessors a computer was designed as shown in figure 1.1

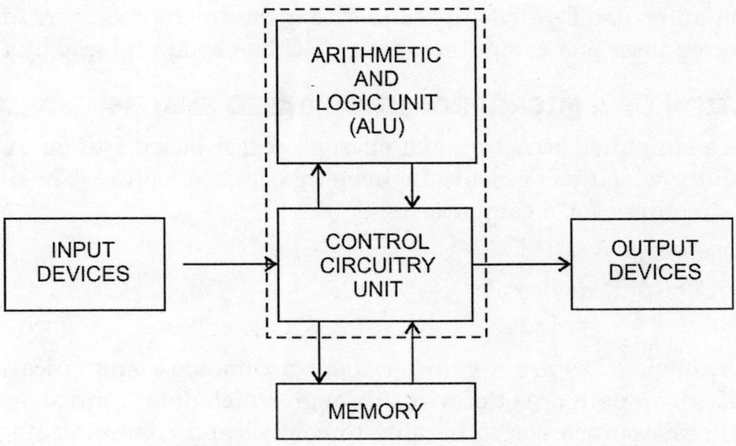

Figure 1.1 Block diagram of a Computer

The computer is divided into four components namely: Memory, Input devices, output devices and the CPU (Central Processing Unit). The CPU contains registers to store data, the ALU which performs arithmetic and logic operations on data, the instruction decoders, counters, circuits and control lines. During the days of SSI fabrication (Small Scale Integration) all these different components were contained on separate IC's and wired together so as to work collectively as the CPU. The CPU's job is to read instructions from memory and perform the operation specified, Its job also includes communication with input and output devices so as to accept or send data. Since earlier, dicrete components (i.e separate IC's) were used to implement each stage of the CPU (i.e., ALU, decoder, counter etc.) the size of CPU was extremely large, The large size as well as delays of each component created CPU's which were very slow.

However, with the advent of LSI (Large Scale Integration), VLSI (Very Large Scale Integration), and ULSI (Ultra Large Scale Integration) the complete CPU was fabricated on a single wafer of Silicon. This resulted in an IC which was compact, cheap and capable of doing all the functions of the CPU, This integrated circuit came to be known as the microprocessor. The name itself signifies

micro + processing unit = microprocessor

The traditional block diagram of figure 1.15 of a computer can thus be replaced by that of figure 1.2.

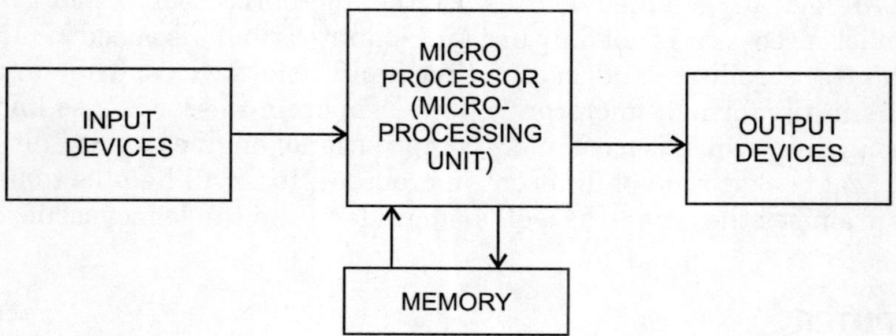

Figure 1.2. The Computer with the microprocessor as the CPU

The microcomputer itself derives its name from the microprocessor. All computer have a CPU but a microcomputer is a computer whose CPU has been replaced by the microprocessor.

1.2 ORGANIZATION OF A MICROPROCESSOR BASED SYSTEM

Figure l.3 Shows a simplified structure of a microprocessor based system. A typical application is the microcomputer which will essentially have the same structure. The diagram includes all vital components required for a computer :

1. Microprocessor,
2. I/O (Input/Output) devices,
3. Memory.

All the three components are organized about a common communication path called the bus. A bus is basically a path or set of wires through which data, control and other signals can flow so as for the microprocessor to be able to control and communicate with the complete system. Each component is referred to as a subsystem. The group of components collectively constitute the microcomputer or microprocessor based system.

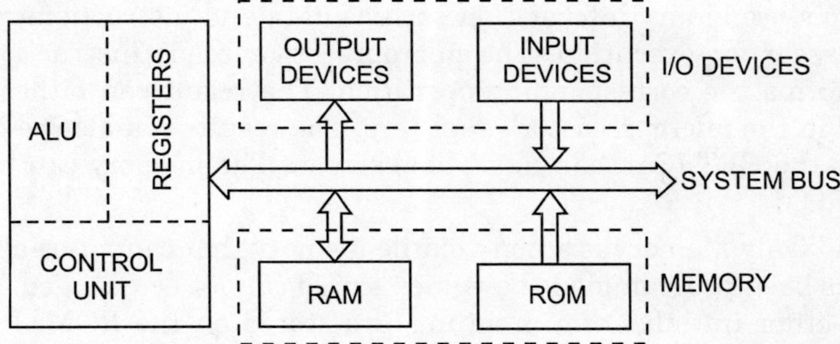

Figure 1.3. A microprocessor based microcomputer using bus architecture

Thus the microprocessor is one of the many components of a microcomputer and the microcomputer is a computer whose CPU operations are performed by the microprocessor. Similarly the input and output devices (I/O) are called peripherals. Now let us discuss this system in detail.

1.2.1. Microprocessor

The microprocessor is a programmable integrated circuit consisting of a high density of electronic logic circuits and is fabricated by either LSI, VLSI, or ULSI techniques. Like most sequential digital devices, it is driven by a clock signal so as to synchronize all the components present in the microprocessor. The microprocessor is capable of performing arithmetic and logical operations on data and making logical decisions for the change of program execution sequence. This complete capability is available to the user through its instruction set. The user can in any way manipulate these capabilities so as to achieve a desired result. The microprocessor is similar to the CPU except for the fact that the microprocessor contains all the arithmetic and logic circuitry, registers and control unit on a single chip or wafer of silicon. The microprocessor can be divided into three main units as shown in figure 1.17: ALU, Registers and Control Unit.

(i) Arithmetic and Logic Unit (ALU) : This is the area on the chip where all the circuitry for the various computing functions are fabricated. The circuitry collectively is called the ALU. The ALU performs operations such as addition, subtraction, multiplication division and logical operations such as AND, OR, XOR, NOT etc. Each of these operations can be done by using specific instructions.

(ii) Registers: This is the area of the microprocessor chip which can be used. to store data temporarily. These registers are accessible to the user through instructions. Some registers are also present in the microprocessor for storage of addresses of the memory locations that must be accessed by the microprocessor.

(iii) Control Unit: The control unit contains many parts such as the instruction decoder, and the control and synchronization circuits. The instructions are decoded by the control unit and the corresponding control signals are generated for the instruction execution. The control unit also synchronizes all components of the microprocessor. The control unit also controls the flow of data between the microprocessor and memory (and peripherals).

1.2.2. MEMORY

Memory is used to store binary information such as data and instructions which are required by the microprocessor for execution. The microprocessor reads instructions and data from memory and performs the corresponding operation. The result can either be transferred to memory or stored in the microprocessor's registers. The results can also/be outputted on the I/O devices such as the CRT (Cathode Ray Tube) screen. The memory can be divided into two parts: RAM and ROM

- ROM or Read Only Memory is a non-volatile memory that cannot be altered. That is once the ROM has been programmed the values stored cannot be changed. The boot up loader (BIOS) and other initialization programs are stored on the ROM. Programs and data stored in ROM cannot be altered.
- RAM or Random Access Memory is volatile alterable·memory. It is also called the user memory as the user can read or write wherever and whenever in the RAM. The operating system etc. are loaded into RAM and then executed by the microprocessor.

1.2.3. I/O Devices

The third component of a microprocessor is the input and output devices. These can be the keyboard (input device), monitor (output device), hard disk drive (input and output device) etc. The I/O devices are also called the peripherals.

Input devices are used to transfer binary data to the microprocessor. The microprocessor reads this data and performs the corresponding operation required. There are many input devices such as keyboards, mouse etc.

The microprocessor transfers the results or data to output devices for display or other purposes. The output devices include CRT monitors, printer, XY recorders etc.

1.2.4. SYSTEM BUS

The system bus is a common communication path between the microprocessor, input/output (I/O) devices and the memory. The system bus is nothing but a set of wires to carry bits. All equipments connected to the microprocessor share the same system bus, but the microprocessor interacts with one peripheral at a time. The system bus can be subdivided into three parts namely :

(*i*) Control bus,

(*ii*) Data bus,

(*iii*) Address bus.

The data bus is a set of wires that are used to transmit data from the microprocessor to peripherals or vice-versa. Similarly the address bus is a set of wires that is used to transmit address from the microprocessor to the peripherals. This address is used to select one peripheral and to enable the required device. After the device has been enabled, the data can be transferred from the microprocessor to the device or vice-versa. Since the address is always sent from the microprocessor to a device for selection it is a unidirectional bus. The data bus however is used to transfer data from the microprocessor to a device or vice-versa and is thus a bidirectional bus. Most microprocessors have a multiplexed address and data bus so as to reduce the number of external pins of the microprocessor IC. This means that these microprocessors use the same bus for address and data on a time shared basis.

The control bus is used to transfer control signal to peripheral for timing and control. Sometimes some peripherals have to tell the microprocessor about certain conditions requiring control lines that must be input to the microprocessor. Thus the control bus is also bidirectional. The complete timing for the synchronization of the system is provided by the microprocessor control bus.

1.3 WORKING OF A MICROPROCESSOR

At this point a question that may arise is how the microprocessor works. Here we shall briefly explain the working ofa general microprocessor. The detailed working of the 8086 microprocessor is explained in the following chapters. We at this stage assume that a program has been entered into the RAM (user memory).

We are not concerned how the program is stored in memory because that shall be explained in later chapters. The program we assume includes instructions and data on which the microprocessor must perform arithmetic or logic operations and then display the result on the seven segment LED output. On providing the microprocessor with a command to execute this program, it reads and executes one instruction from a specific address in the RAM. For this it would have to send an address and the corresponding signal to enable the RAM. The RAM on receiving this signal obliges and sends out the instructions and data to the microprocessor. The microprocessor then decodes these instructions and performs the operations required. Similarly, all other instructions are read, decoded and executed by the processor. After all instructions have executed it finally sends the address and control signal to the seven segment LED display. This is followed by the result which is obligingly displayed by the LED display.

This process can be correlated to an electronics engineer designing an electronic circuit. The electronics engineer would first have to read all the basics of electronics. After reading all basics he will have to interpret these and then sit in a laboratory and execute the necessary designing steps. Similarly, the microprocessor reads the first instruction from memory. It then decodes the instruction and follows its by executing the instruction. The sequence of fetch, decode and execute is performed until the microprocessor encounters an instruction that tells it to perform a halt operation. During this entire process the microprocessor uses the system bus to fetch instructions and data from memory (or I/O). It will use the registers as a temporary storage device for the data or result and the ALU to perform arithmetic or logic operations. After all this it will send data using the same system bus to I/O devices (7 segment LED) or memory.

1.4. EVOLUTION OF MICROPROCESSORS

The first general purpose programmable computer system was developed in 1946 at the university of Pennslyvania. This computer was called the ENIAC or Electronic Numerical Integrator and Calculator. Though this was preceded by years of research by many other scientists including Charles Babbage (Analytical Computer), Konard Zuse (Z3 computer) etc, yet this was the first successful computer to be designed and used. However, ENIAC was a massive machine having over 17000 vacuum tubes. The ENIAC too, had its sets of problems. Programming the ENIAC was cumbersome as it required rewiring of the circuitry which took lots of time. Also the use of vacuum tubes required lot of maintenance as the vacuum tube components have short life.

However breakthroughs in the field of electronics such as the development of the transistor in 1948 followed by the invention of the Integrated circuit in the late fifties laid the foundation of development of smaller and more reliable computers. Integrated circuits led to the development of digital IC in the 60's and was followed by the launch of the microprocessor in 1971 by Intel corporation.

The world's first microprocessor debuted in 1971 and was developed by Intel Corporation. This microprocessor was named 4004. 4004 was a 4 bit microprocessor that could only address 4K of 4 bit memory (4096 x 4 bits). It was fabricated by P-channel MOSFET technology and was limited to an execution speed of 50 KIPS (50 Kilo Instructions per second). The 4004 was a very slow processor compared to the ENIAC which was able to execute 100,000 instructions per second. The 4004 however was extremely small in size and weight in comparison to the 30 ton ENIAC. The 4004 was used in videogame consoles, calculators and small control systems such as microwave ovens. However, its slow speed, low memory size and data width limited its widespread use.

As demand for faster processors with larger data widths grew, the Intel corporation launched the 8008 microprocessor which was the extended version of the 4004 microprocessor. The 8008 could access 16 K bytes of memory and had a total of 48 instructions that enabled its use in more advanced control systems. However, as the industry developed more demanding uses for the 8008 microprocessor, they discovered that its small memory size, slow speed and small instruction set was a big limitation. The Industry's increased demand paved the launch of the Intel 8080 microprocessor in 1973. The 8080 is considered to be the first modern 8 bit processor. Soon other companies too launched their own microprocessor such as Motorola launched the MC6800 microprocessor. The 8080 was at that time one of the most advanced microprocessors. The 8080 was almost ten times faster than the 8008 and could access 64 K bytes of memory. Following the 8080, Intel launched the 8085 microprocessor, which was an upgraded version of the 8080. The 8085 microprocessor boasted of faster speed and lower cost. The key features that were present in the 8080 and 8085 that led to their widespread use were :

- high speed,
- low cost
- directly compatible with TTL IC's leading to low cost of interfacing the complete microprocessor based system.

These were the reasons why the microprocessor revolution is attributed to the 8080 and 8085 microprocessors. The 8085 was very widely used as it had better speed as well as in built clock generator, system controller circuits which reduced the overall cost of the 8085 based system in comparison to the 8080 based system.

The modern microprocessor age was ushered in with the launch of the 8086 microprocessor by Intel Corporation in 1978. This was followed soon by the 8088 microprocessor. Both these devices were 16 bit microprocessors. (The 8086 however is a true 16 bit processor as it has a 16 bit data bus. The 8088 however is internally a 16 bit processor but externally has an 8 bit data bus).

These microprocessors brought in what is known as the PC (Personal Computer) revolution. These were implemented with the then state of the art n-channel NMOS fabrication technology and capable of operating at 2.5 MPS (Million instruction per second). These were definitely faster than the preceding 8080 and 8085 microprocessors. They also used advanced techniques for faster execution such as pipelining and were capable of addressing 1 M byte of memory. The presence of small caches* in 8088 and 8086 also helped in increase of execution speeds. These microprocessors formed the base for the development of more advanced micro processors such as 80186,80286,80386,80486, Pentium microprocessor.

The 8086/8088 microprocessors were widely used in PC's by the IBM corporation and led to further advances in microprocessor fabrication techniques. These microprocessors (8086, 80186, 80286,80386,80486, Pentium) are collectively called the X86 family of microprocessors. They belong to a class of microprocessors called ClSC or Complex Instruction set computer, because of the large number and complexity of the instructions that their instruction set contains.

During the mid 80's another class of microprocessors were launched. These microprocessors are called RISC (Reduced instruction set computer) processors. These microprocessors use a very small but effective instruction set and execute instructions in one or less than one clock periods. These processors were designed for use in high speed computers such as Servers, Maniframes etc. They employ very modern architecture techniques such as pipelining, parallel processing, superscalar architecture and harvard architecture so as to be able to execute instructions at very high speeds. The first RISC processor to be launched was the MIPS 4000. This was followed by the intel 80860, Motorola Power PC, DEC Alpha, Sun Microsystems Ultra Spare processors etc. The latest RISC processor from intel the i960 processor is capable of execution speeds upwards of 500-600 MIPS (Million Instruction per second).

1.5 THE 80186/80188 ARCHITECTURE

The 80186 and 80188 microprocessor are similar to each other just like the 8086 and 8088 microprocessors. These processors too, differ only due to the width of their data buses. The registers that are present internally in the 80186/80186/80188 too are virtually identical to the 8086. The 80186 and 80188 microprocessor however contain additional reseuged interrupt vectors and in built circuits such as DMA circuitry, PIC, block generators, Timer Circuits and Power Management circuitry. The presence of such circuits reduces the number of external support devices required for an 80186/80188 microprocessor based system. Due to these reasons the 80186 has found extensive use in systems where it functions as a controller. The 80186/ 80188 is therefore also referred to as an embedded controller. The 80186/80188 is found in LAN Cards, Routers, Gateways, Control Cards for industrial systems where it is used as a controller.

1.5.1 Internal Architecture

Figure 1.6.1 shows the basic block diagram of the 80186/80188 microprocessor except for features that are outlined in figure for the different versions of the 80186 and 80188 microprocessors. All other features that are shown in the block diagram remain same except for the prefetch queue. The 80188 contains a 4 byte prefetch queue while the 810186 contains a 6 byte prefetch

* The 8086/8088 microprocessors donot have a true cache. However, the 6 byte instruction queue does work like one though in size it is very small to be classified as a cache memory.

queue. Like the 8086/8088, the 80186/80188 microprocessors can be divided into two functional units namely the EU (Execution unit) and the BIU (Bus Interface unit).

In addition to these two functional units the 80186/80188 microprocessor contains the following units :

1. Clock Generator unit.
2. Programmable Interrupt Controller.
3. Programmable Timers.
4. DMA unit.
5. Programmable Chip Selection unit.
6. A Power management unit.

Due the presence of the in built circuits the 80186/80188 has found use in a wide variety of systems due to overall reduction in peripheral component required to implement a system. The 80186 can be found in many subsystems interfaced to a personal computers such as in LAN Cards, disk controllers etc.

Features	80C186 XL 80C188 XL	80C186EC 80C188EC	80C186 EB 80C188EB	80C186EA 80C188EA
1. Power down mode	N.A.	Present	present	present
2. Coprocessor interface	present	present	presetn	present
3. PIC	present	present	present	present
4. Timer Unit	present	present	present	present
5. CSU	present (enhanced)	present (enhanced)	present	present
6. DMA Controller	present 2 channel	present 4 channel	N.A.	present 2 channel
7. Refresh Controller	present	present	present	
8. SCU	N.A.	present	present	N.A.
9. I/O Ports	N.A 22 bit	present 16 bit	present	N.A
10. Power save Mode	present	present	N.A.	present

N.A = Not available,PIC = Programmable Interrupt Controller
SCU = Serial Communications unitCSU = Chip selection unit

Figure 1.4 Different circuit present on the different versions of 80188/80186 embedded controller

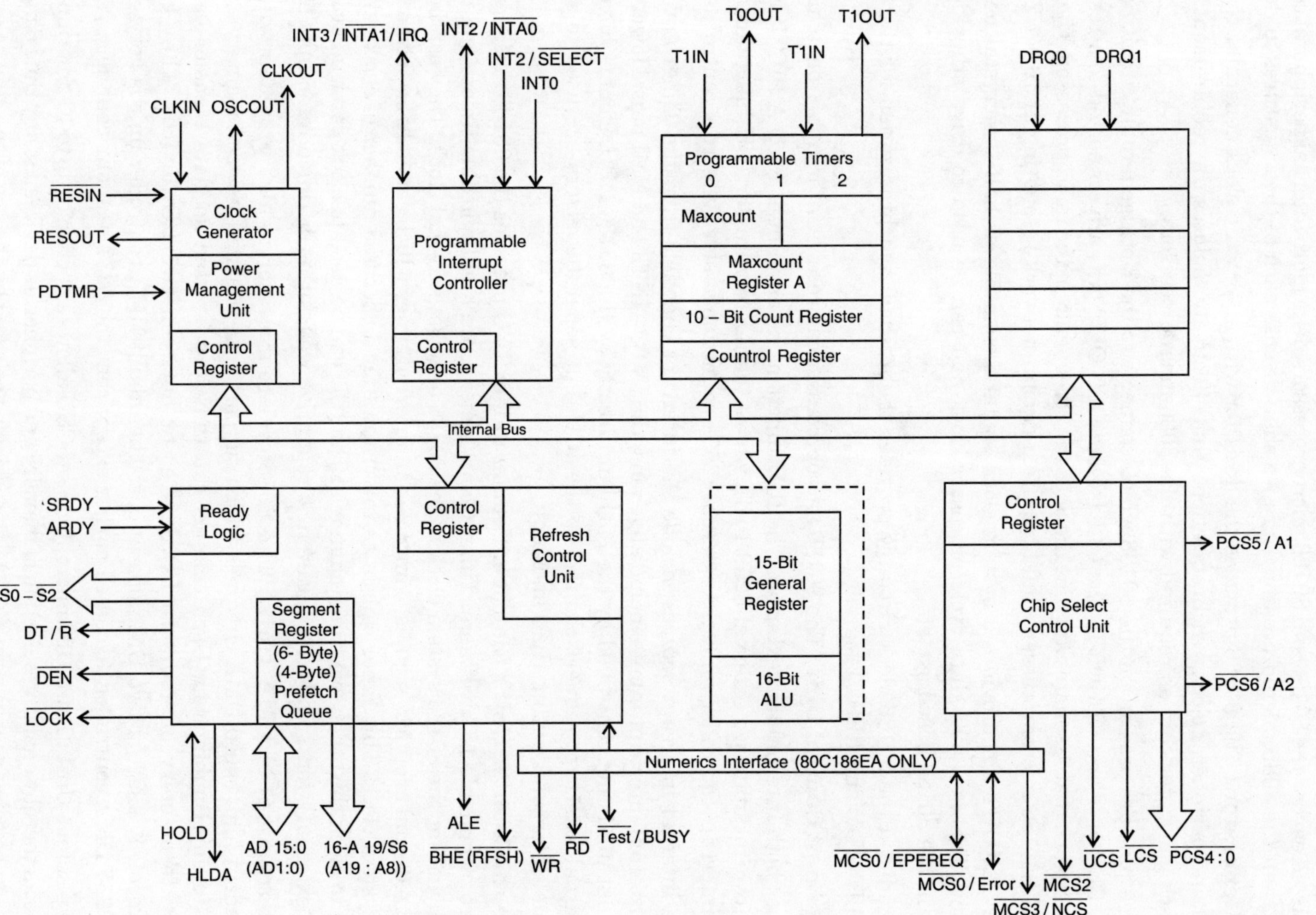

Figure 1.5 The 80816/80'88 block diagram

1.5.2. 80186/80188 Basic Internal Features

As discussed earlier the 80186 and 80188 microprocessors have more internal peripheral control circuits than the 8086 microprocessor. Here we shall discuss each of these features in detail.

Clock Generator : The 80186 provides the designer with an on chip clock generator for both internal and peripheral clock synchronization. This effectively replaces the requirement of the 8284 clock generator chip as required with the 8086 microprocessors.

Three pins are present on the 80186 which connects to the clock generator circuit namely CLKIN, OSCOUT and CLKOUT. The CLKIN and OSC OUT pins are connected to a resonant crystal network that resonates at twice the operating frequency of the microprocessor. Therefore an 8 MHZ 80186 microprocesser would require a crystal network resonating at 16 MHZ.

The CLKOUT pin provides a system clock signal resonating at half the crystal network frequency. This is a 50% duty cycle processor clock and can be used to drive other devices present in the 80186 based system.

In addition the 80186 clock circuitry provides the internal synchronization for the 80186 internal READY and RESET signals.

Interrupt Control Unit : The 80186 microprocesser can receive interrupts from a number of sources both internal and external. The ICU functions as an in built 8259 A PIC. It can be used to control upto two external 8259 PIC's where the microprocesser ICU functions as the master interrupt controller while the external 8259A's function as slave interrupt control devices.

The Internal interrupt sources include the interrupt channels from the Timer and DMA units. External interrupt inputs include five pins, the INTO - INT3 and NMI pins. However the NMI input pin is not controlled by the ICU but directly by the 80186 CPU. Timers have only one internal interrupt request channel to the ICU; but separate vectors are used to service individual interrupts within the Timer unit.

Timers/Counter Unit : The Timer/Counter unit or TCU provides the 80186 designer with three programmable 16 bit timers. Timer 0 and Timer 1 are highly flexible timers and can be driven by either the 80186 system clock of the 80186 or by an external clock. They can be used to count external events or signals. Timer 2 is an internal timer which can only be driven by the 80186 system clock. This timer causes an interrupt after a specified number of clocks and can be used to clock the other two timer channels. The ICU can be used to clock external events or to time external events and to generate new repetitive waveforms or timed interrupts etc.

Programmable DMA Unit : The 80186 microprocessor is also provided with two independent high speed DMA channels which support data transfer between memory and I/O devices in any combination. Each channel can therefore be used for data transfer between memory to memory, memory to I/O devices or between two I/O devices. They can be simultaneously used for two DMA operations. The inbuilt DMA controller unit is very similar to the 8237 DMA controller discussed earlier except that the 8237 has four channels.

Chip select Unit : The 80186 microprocesser CSU allows selection of up to 13 programmable chip selects thereby enabling selection of memory and peripherals devices. At any time 6 memory devices can be selected while seven I/O devices can be selected by the use of the chip select unit.

Refresh Control Unit (RCU) : The 80186 has integrated refesh control unit which automatically generates a periodic memory read bus cycle to keep the DRAM interfaced constantly refreshed.

Power Management Unit : The 80816 has three operational modes to control power consumption of the microprocesser. These are Power Save Mode, Idle mode and Power down mode.

The power save mode allows the processor clock to be divided by a programmable value (either 4,8 or 16) therefore allowing the processor power consumption to be reduced (current is proportional frequency). Power save mode is entered by software routines. It is however exited by a hardware event such as an interrupt.

The idle mode places the EU and BIU internal clocks at zero state while all peripherals operate normally.

Power down mode freezes all internal clocks thereby limiting current consumption only to transistor leakage currents. The HLT instruction causes the 80186 to enter power down mode while it is exited by a preset interrupt type.

1.5.3. The 80186 Pin Out

Figure 1.6 shows the pin out of the 80186 microprocesser. The 80186 is available in two package: a 68 pin lead less chip carrier (PLCC) package or a 68 pin PGA (pin Grid drray) package. The figure 1.6 shows the 80186 68 pin PLCC package.

Pin definitions: The following list defines all the 80186 pins and their functions.

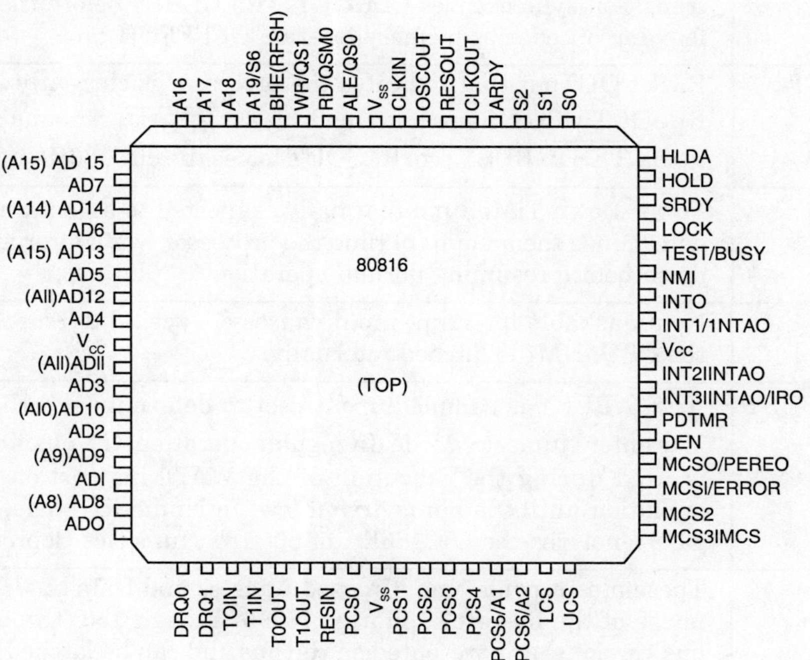

Figure 1.6 Pin out of the 80816

Pin definitions

Pin	Name Description
Vcc	POWER connections consist of six pins which must be shorted externally to a V_{CC} Supply.
VSS	GROUND connections consist of five pins which must be shorted externally to a V_{SS}.
CLKIN	CLOCK INput is an input for an external clock. An external oscillator operating at two times the required processor operating frequency can be connected to CLKIN. For crystal operation CLKIN (along with OSCOUT) are the crystal connections to an internal oscillator.
OSCOUT	Oscillator Output is only used while using a crystal to generate the external clock. OSCOUT (along with CLKIN) are the crystal R(Q) connections to an internal Pierce type oscillator. OSCOUT does not float in ONCE mode.
CLKOUT	CLocK OUTput provides a timing reference for inputs and outputs of the processor, and is one-half the input clock (CLKIN) frequency. CLKOUT has a 50% duty cycle and transitions every falling edge of CLKIN.
$\overline{\text{RESIN}}$	RESet IN causes the processor to immediately terminate any bus cycle in progress and assume an initialized state. All pins will be driven to a known state, and RESOUT will also be driven active. The rising edge (low-to-high) transition synchronizes CLKOUT with CLKIN before the processor begins fetching opcodes at memory location OFFFFOH.
RESOUT	RESet OUTput that indicates the processor is currently in the reset state. RESOUT will remain active as long as $\overline{\text{RESIN}}$ remains active. When tied to the $\overline{\text{TEST}}$ / BUSY pin, RESOUT forces the 80C186EA into Numeric Mode.
PDTMR	Power Down TiMeR pin (normally connected to an external capacitor) that determines the amount of time the processor waits after an exit from power down before resuming normal operation.
NMI	Non-Maskable interrupt input causes a Type 2 interrupt to be serviced by the CPU. NMI is latched internally.
$\overline{\text{TEST}}$/BUSY	$\overline{\text{TEST}}$ /BUSY is sampled upon reset to determine whether the 80C186EA is to enter Numerics Mode. In regular operation, the pin is active low. $\overline{\text{TEST}}$ is used during the execution of the WAIT instruction to suspend CPU operation until the pin is driven low. In Numerics Mode, the pin is BUSY. BUSY notifies the 80C186EA of 80C187 Numerics Coprocessor activity.
AD15;0 (AD7;0)	These pins provide a multiplexed Address and Data bus. During the addess phase of the bus cycle, address bits 0 through 15(0 through 7 on the 8-bit bus versions) are presented on the bus and can be latched using ALE. 8 – or 16 bit data information is transferred during the data phase of the bus cycle.

Pin	Description
A19/S6-A16 (Al9-A8)	These pins provide multiplexed Address during the address phase of the bus cycle. Address bits 16 through 19 are presented on these pins and can be latched using ALE. A_{18}-A_{16} are driven to a logic 0 during the data phase of the bus cycle. On the 8 bit bus versions, A15 - A8 provide valid address information for the entire bus cycle. Also during the data phase, S6 is driven to a logic 0 to indicate a CPU-initiated bus cycle or logic 1 to indicate a DMA-initiated bus cycle or a refresh cycle.
$\overline{S2}-\overline{S0}$	Bus Cycle Status are encoded on these pins to provide bus transaction information. $\overline{S2}-\overline{S0}$ are encoded as follows:

$\overline{S2}$	$\overline{S1}$	$\overline{S0}$	Bus Cycle Initiated
0	0	0	Interrupt Acknowledge
0	0	1	Read 110
0	1	0	Write 110
0	1	1	Processor HALT
1	0	0	Queue Instruction Fetch
1	0	1	Read Memory
1	1	0	Write Memory
1	1	1	Passive (no bus activity)

Pin	Description
ALE/QSO	Address Latch Enable output is used to strobe address information into a transparent type latch during the address phase of the bus cycle. In Queue Status Mode, QSO provides queue status information along with QSl.
$\overline{BHE}$/ $\overline{(RFSH)}$	Byte High Enable output to indicate that the bus cycle in progress is transfering data over the upper half of the data bus. $\overline{BHE}$ and AO have the following logical encoding:

A0	$\overline{BHE}$	Encoding (for 80C186EA/80L186EA Only)
0	0	Word Transfer
0	1	Even Byte Transfer
1	0	Odd Byte Transfer
1	1	Refresh Operation

On the 80C188EA/80L188EA, $\overline{RFSH}$ is asserted low to indicate a Refresh bus cycle.

Pin	Description
$\overline{RD}$ / $\overline{QSMD}$	Read output signals that the accessed memory or I/O device must drive data information onto the data bus. As $\overline{QSMD}$,it enables Queue Status Mode when grounded. In Queue Status Mode, the ALE/QSO and $\overline{WR}$/QSl pins provide the following information about processor instruction queue interaction;

QSl	QSO	Queue Operation
0	0	No Queue Operation
0	1	First Opcode Byte Fetched from the Queue
1	1	Subsequent Byte Fetched from the Queue
1	0	Empty the Queue

$\overline{\text{WR}}$/QSl	WRite outputs signals that the data available on the data bus are to be written into the accessed memory or I/O device. In Queue Status Mode, QSl provides queue status information along with QSO.
ARDY	Asynchronous ReaDY is an input to signal for the end of a bus cycle. ARDY is asynchronous on rising CLKOUT and synchronous on falling CLKOUt. ARDY or SRDY must be active to terminate any processor bus cycle, unless they are ignored due to correct programming of the Chip Select Unit.
$\overline{\text{DEN}}$	Data ENable output to control the enable of bidirectional transceivers when buffering a system. $\overline{\text{DEN}}$ is active only when data is to be transferred on the bus.
$\overline{\text{LOCK}}$	LOCK output indicates that the bus cycle in progress is not to be interrupted. The processor will not service other bus requests (such as HOLD) while $\overline{\text{LOCK}}$ is active. This pin is configured as a held high input while $\overline{\text{RESIN}}$ is active and must not be driven low.
HOLD	HOLD request input to signal that an external bus master desires to gain control of the local bus. The processor will relinquish control of the local bus between instruction and execution provided that the instruction is not prefixed by a LOCK prefix.
HLDA	HOLD Acknowledge output to indicate that the processor has relin-quished control of the local bus. When HLDA is asserted, the processor will have (or has) floated its data bus and control signals allowing another bus master to drive the signals directly.
$\overline{\text{UCS}}$	Upper Chip Select will go active whenever the address of a memory or I/O bus cycle is within the address limitations programmed by the user. After reset, $\overline{\text{UCS}}$ is configured to be active for memory accesses between OFFCOOH and 0FFFFFH. During a processor reset, $\overline{\text{UCS}}$ and $\overline{\text{LCS}}$ are used to enable ONCE Mode.
$\overline{\text{LCS}}$	Lower Chip Select will go active whenever the address of a memory bus cycle is within the address limitations programmed by the user. R(1) $\overline{\text{LCS}}$ is inactive after a reset. During a processor reset, $\overline{\text{UCS}}$ and $\overline{\text{LCS}}$ are used to enable ONCE Mode.

$\overline{\text{MCS0}}$/PEREQ $\overline{\text{MCS1}}$/$\overline{\text{ERROR}}$ $\overline{\text{MCS2}}$ $\overline{\text{MCS3}}$/$\overline{\text{NCS}}$	These pins provide a multiplexed function. If enabled, these pins normally comprise a block of Mid-Range Chip Select outputs which will go active whenever the address of a memory bus cycle is within the address limitations programmed by the user. In Numerics Mode (80C186EA only), three of the pins become handshaking pins for the 80C187. The CoProcessor Request input signals that a data transfer is pending. $\overline{\text{ERROR}}$ is an input which indicates that the previous numerics coprocessor operation resulted in an exception condition. An interrupt Type 16 is generated when $\overline{\text{ERROR}}$ is sampled active at the beginning of am" series operation. Numerics Coprocessor Select is an output signal generated when the processor accesses the 80C187.
$\overline{\text{PCS4}}$ – $\overline{\text{PCS0}}$	Peripheral Chip Selects go active whenever the address of a memory or I/O bus cycle is within the address limitations by the user. They may also be programmed to provide latched Address A2: 1 signals.
$\overline{\text{PCS5}}$ / A1 $\overline{\text{PCS6}}$ / A2	These pins provide a multiplexed function. As additional Peripheral Chip Selects, they go active whenever the address of a memory or I/O bus cycle is within the address limitations by the user. They may also be programmed to provide latched Address A2 : 1 signals.
TOOUT TI0UT TOIN	Timer OUTput pins can be programmed to provide a single clock or continuous. waveform generation, depending on the timer mode selected. Timer INput is used either as clock or control signals, depending on the timer mode selected T1IN A(E).
DRQO DRQl	DMA ReQuest is asserted by an external request when it is prepared for a DMA transfer.
INTO	Maskable INTerrupt pin will cause a vector to a specific Type interrupt routine to allow interrupt expansion. INTO andlor INTl can be used with $\overline{\text{INTA0}}$ and $\overline{\text{INTA1}}$ to interface with an external slave controller. INTI becomes $\overline{\text{SELECT}}$ when the ICU is configured for Slave Mode.
INT2/$\overline{\text{INTA0}}$ INT3/$\overline{\text{INTA1}}$ / IRQ	These pins provide multiplexed functions. As inputs, they provide a maskable INTerrupt that will cause the CPU to vector to a specific Type interrupt routine. As outputs, each is programmatically controlled to provide an InTerrupt Acknowledge handshake signal to allow interrupt expansion. INT3/$\overline{\text{INTA1}}$ becomes IRQ when the ICU is configured for Slave Mode.

N.C.	No Connect. For compatibility with future products, do not connect to these pins.

NOTE: Pin names in parentheses apply to the 8OC188EA and 80L188EA.

1.6 THE 80286 MICROPROCESSOR

The 80286 is an advanced high performance microprocessor with specially optimized capabilities for multiuser and multitasking systems. The 80286 has built in memory protection that supports operating systems and task isolation as well as program and data privacy within tasks. A 25 MHz 80286 microprocessor is capable of providing upto twenty times the maximum throughput that can be achived on a 5 MHz 8'086 processor based system. The 80286 microprocessor however doesnot have a multiplexed address data bus as in 8086. You may observe in figure 13,.that 80286 has 16 data lines and 24 address lines. The 80286 can thus directly address 40 MByte of physical memory. However due to the enhanced memory management unit (MMU) 80286 has the capability of address 1 G byte of memory.

1.6.1 The 80286 Architecture

Figure 13.4 shows the internal block diagram of the 80286 microprocessor. However unlike the 80186, the 16 bit 80286 microprocessor doesnot contain any additional peripheral control circuits such as DMA controller etc. Physically the 80286 can be divided into four functional units.

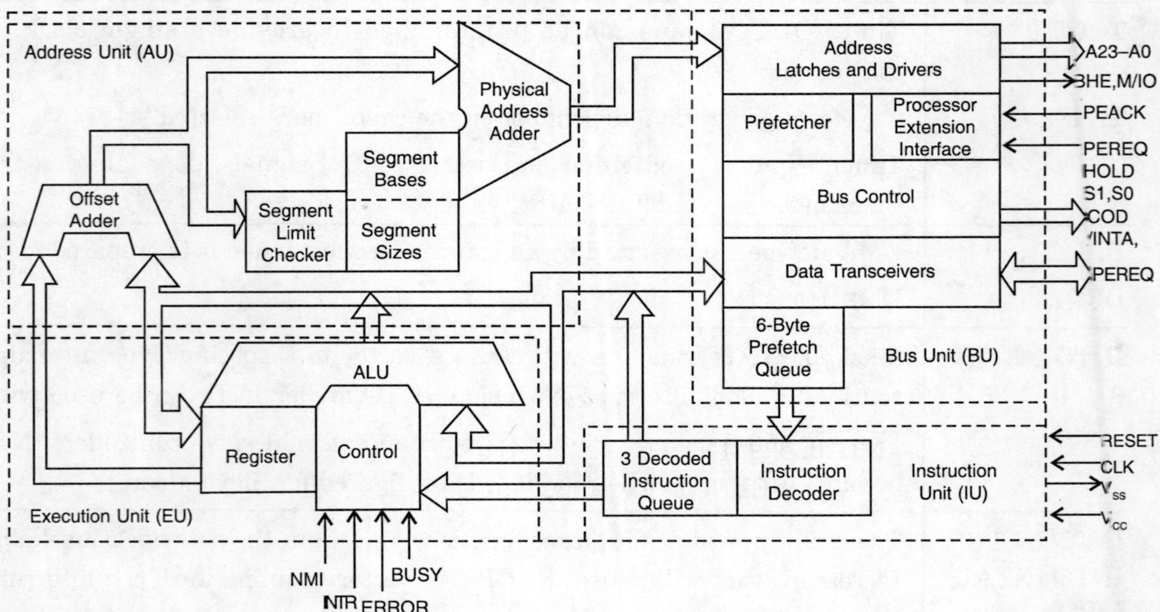

Figure 1.7 The 80286 Architecture

- Adress unit (AU)
- Instruction (IU)
- Bus Unit (BU)
- Execution unit (EU)

The 80286 can be operated in two modes: the real address mode and the protected virtual address mode.

In the real address mode the 80286 operates identically to the 8086. It can address only 1 Mbyte of memory. However in the protected mode the 80286 can address its full memory space of 16 Mbytes.

1.6.2 The 80286 Pin Out

Figure 1.8 (a) and 1.8 (b) shows the pin out for the two package available for the 80286 microprocessor. The 68 pin PLCC and the 68 pin PGA packages.

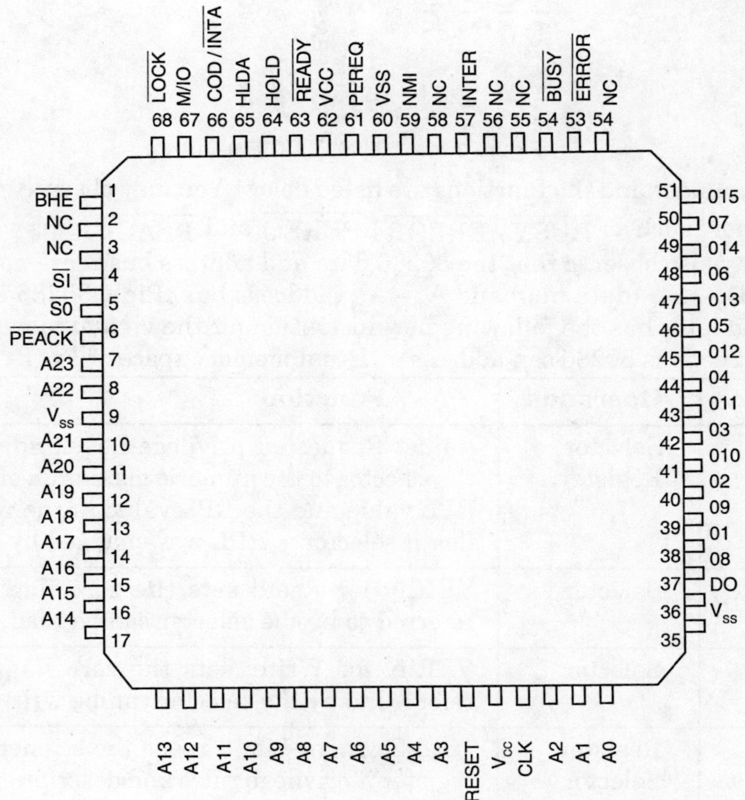

Figure 1.8(a) 80286 PLCC Pin Out

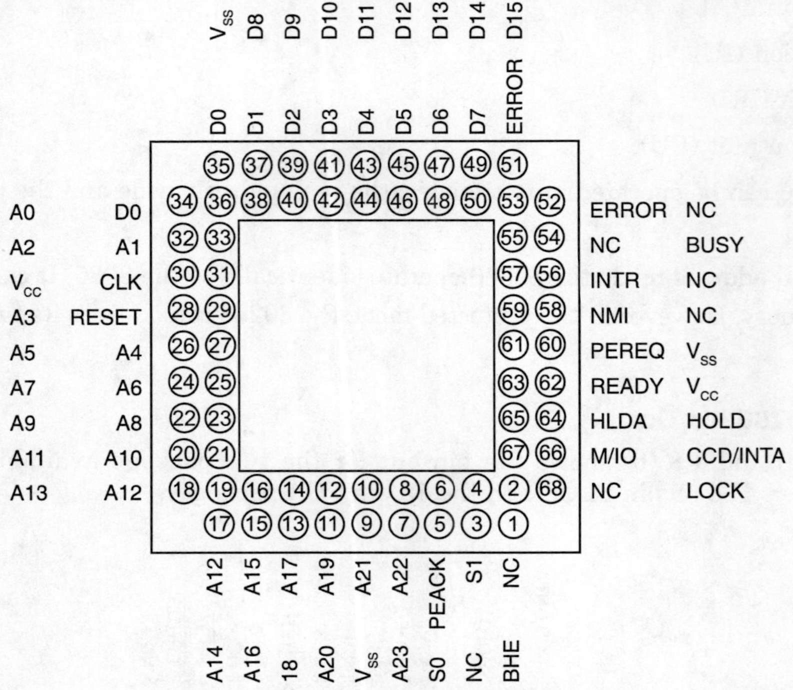

Figure 1.8(b) 80286 PLCC Pin Out

The pin assignment and the functions are listed below, You may observe here that there are many new pins here such as $\overline{\text{BUSY}}$, $\overline{\text{ERROR}}$, $\overline{\text{PEREQ}}$ and $\overline{\text{PEACK}}$ pins which donot appear on 8086, You may also observe that the 80286 data and address buses are not multiplexed but new appear as $D_0 - D_{15}$ (data bus) and $A_0 - A_{23}$ (address bus) Pins, 80286 in addition to the instruction listed later has the following new instrution for the virtual memory management. Using these instructions 80286 can address a virtual memory space of 1 G Byte.

Instruction	Operands	Function
ARPL	Selector, Register	Adjust Requested privilege Level: adjusts the RPL of the selector to the numeric maximum of current selector RPL value and the RPL value in the register. Set zero flag if selector RPL was changed by ARPL,
VERR	Selector	VERify for Read: sets the zero flag if the segment referred to by the selector can be read,
VERW	Selector	VERify for Write: sets the zero flag if the segment referred to by the selector can be written,
LSL	Register Selector	Load Segment Limit: reads the segment limit into the register if privilege rules and description type allow, Set zero flag if successful.
LAR	Register, Selector	Load Access Rights: reads the description access rights byte into the register if privilege rules allow, Set zero flag if successful.

TABLE: 80C286 Pointer Test Instructions

Pin Descriptions The following pin function descriptions are for the 80C286 microprocessor

SYMBOL	Description
CLK	SYSTEM CLOCK: provides the fundamental timing for the 80C286 system. It is divided by two inside 80C286 to generate the processor Clock. The internal divide-by-two circuitry can be synchronized to an external clock generator by a LOW to HIGH transition on the RESET input.
$D_{15} - D_0$	DATA BUS: inputs data during memory, I/O, and interrupt acknowledge read cycles; outputs data during memory and I/O write cycles. The data bus is active HIGH and is held at high impedance to the last valid logic level during bus hold acknowledge.
$A_{23} - A_0$	ADDRESS BUS: outputs physical memory and I/O port addresses. $A_{23} - A_{16}$ are LOW during I/O transfers. Ao is LOW when data is to be transferred on pins $D_7 - D_0$ (see table below). The address bus is active High and floats to three-state off during bus hold acknowledge.
$\overline{BHE}$	BUS HIGH ENABLE: indicates transfer of data on the upper byte of the data bus, $D_{15} - D_8$ Eight-bit oriented devices assigned to the upper byte of the data bus would normally use BHE to condition chip select functions. BHE is active LOW and floats to three-state OFF during bus hold acknowledge.

$\overline{BHE}$ ANDAOTRUTHTABLE

$\overline{BHE}$	A_0	FUNCTION
0	0	Word transfer
0	1	Byte transfer on upper half of data bus ($D_{15} - D_8$)
1	0	Byte transfer on lower half of data bus ($D_7 - D_0$)
1	1	Reserved

SYMBOL	Description
$\overline{S1}, \overline{S0}$	BUS CYCLE STATUS: indicates initiation of a bus cycle and along with $M/\overline{IO}$ and $COD/\overline{INTA}$, defines the type of bus cycle. The bus is in a T_S state whenever one or both are LOW. $\overline{SI}$ and $\overline{S0}$ are active LOW and are held at a high impedance logic one during bus hold acknowledge.

80C286 BUS CYCLE STATUS DEFINITION

COD/$\overline{INTA}$	M/$\overline{IO}$	$\overline{S0}$	$\overline{SI}$	BUS CYCLE INITIATED
O(LOW)	0	0	0	Interrupt acknowledge
0	0	0	1	Reserved
0	0	1	0	Reserved
0	0	1	1	None; not a status cycle

0	1	0	0	If A_1, = 1 then halt: else shutdown
0	1	0	1	Memory data read
0	1	1	0	Memory data write
0	1	1	1	None; not a status cycle
l(HIGH)	0	0	0	Reserved
1	0	0	1	I/O read
1	0	1	0	I/O write
1	0	1	1	None: not a status cycle
1	1	0	0	Reserved
1	1	0	1	Memory instruction read
1		1	1	Reserved
1	1	1	0	Reserved
1	1	1	1	None; not a status cycle

M/$\overline{IO}$	Memory I/O SELECT: distinguishes memory access from I/O access. If HIGH during Ts' a memory cycle or a halt/shutdown cycle is in progress. If LOW, an I/O cycle or an interrupt acknowledge cycle is in progress. M/$\overline{IO}$ is held at high impedance to the last valid logic state during bus hold acknowledge.
COD/$\overline{INTA}$	CODE/INTERRUPT ACKNOWLEDGE: distinguishes instruction fetch cycles from memory data read cycles. Also distinguishes interrupt acknowledge cycles from I/O cycles. COD/$\overline{INTA}$ is held at high impedance to the last valid logic state during bus hold acknowledge. Its timing is same as M/$\overline{IO}$.
$\overline{LOCK}$	BUS LOCK: indicates that other system bus masters are not to gain control of the system bus for the current and following bus cycles. The $\overline{LOCK}$ signal may-be activated explicitly by the "LOCK" instruction prefix or automatically by 80C286 hardware during memory XCHG instructions, interrupt acknowledge, or description table access $\overline{LOCK}$ is active LOW and is held at a high impedance logic one(s) during bus hold acknowledge.
$\overline{READY}$	BUS READY: terminates a bus cycle. Bus cycles are extended without limit until terminated by $\overline{READY}$ LOW. $\overline{READY}$ is an active LOW synchronous input requiring setup and hold times relative to the system clock be met for correct operation. $\overline{READY}$ is ignored during bus hold acknowledge. (See Note 1)

Signal	Description
HOLD	BUS HOLD REQUEST AND HOLD ACKNOWLEDGE: control ownership of the 80C286 local bus. The HOLD input allows another local bus master to request control of the local bus. When control is granted, the 80C286 will float its bus drivers and then activate HLDA, thus entering the bus hold acknowledge condition. The local bus will remain granted to the requesting master until HOLD becomes inactive which results in the 80C2'86 deactivating HLDA and regaining control of the local bus. This terminates the bus hold acknowledge condition. HOLD may be asynchronous to the system clock. These signals are active HIGH. Note that HLDA never floats.
INTR	INTERRUPT REQUEST: requires the 80C286 to suspend its current program execution and service a pending external request. Interrupt requests are masked whenever the interrupt enable bit in the flag word is cleared. When the 80C286 responds to an interrupt to an interrupt request, it performs two interrupts. To ensure program interruption, INTR must remain active until an interrupt acknowledge bus cycle is initiated. INTR is sampled at the beginning of each processor cycle and must be active HIGH at least two processor cycles before the current instruction ends in order to interrupt before the next instruction. INTR is level sensitive, active HIGH and may be asynchronous to the system clock.
NMI	NON-MASKABLE INTERRUPT REQUEST: interrupts the 80C286 with an internally supplied vector value of two. No interrupt acknowledge cycles are performed. The interrupt enable bit in the 80C286 flag word does not affect this input. The NMI input is active HIGH, may be asynchronous to the system clock, and is edge triggered after internal synchronization. For proper recognition, the input must have been previously LOW for at least four system clock cycles and remain HIGH for at least four system clock cycles.
PEREQ $\overline{\text{PEACK}}$	PROCESSOR EXTENSION OPERAND REQUEST AND ACKNOWLEDGE: extend the memory management and protection capabilities of the 80C286 to processor extensions. The PEREQ input requests the 80C286 to perform a data operand transfer for a processor extension. The PEREQ input requests the 80C286 to perform a data operand transfer for a processor extension. The $\overline{\text{PEACK}}$: output signals the processor extension when the requested operand is being transferred. PEREQ is active HIGH. $\overline{\text{PEACK}}$: is active LOW and is held at a high impedance logic one during bus hold acknowledge. PEREQ may be asynchronous to the system clock.

$\overline{BUSY}$
$\overline{ERROR}$

PROCESSOR EXTENSION BUSY AND ERROR: indicates the operation of a processor extension to the 80C286. An active $\overline{BUSY}$ input stops 80C286 program execution on WATT and some ESC instructions until $\overline{BUSY}$ becomes inactive (HIGH). The 80C286 may be interrupted while waiting for $\overline{BUSY}$ to become inactive. An active $\overline{ERROR}$ input causes the 80C286 to perform a processor extension interrupt when executing WAIT or some ESC instructions. These inputs are active LOW and may be asynchronous to the system clock.

RESET

SYSTEM RESET: clears the internal logic of the 80C286 and is active HIGH. The 80C286 may be reinitialize at any time with a LOW to HIGH transition, on RESET which remains active for more than 16 system clock cycles. During RESET active, the output pins of the 80C286 enter the state shown below.

80C286 PIN STATE DURING RESET

PIN VALUE	PIN NAMES
1(HIGH)	$\overline{SO}$, $\overline{S1}$, $\overline{PEACK}$ $A_{23} - A_0$, $\overline{BHE}$, $\overline{LOCK}$
0(LOW)	$M/\overline{IO}$, $COD/\overline{INTA}$, HLDA (Note 2)
HIGH IMPEDANCE	$D_{15} - D_0$

Operation of the 80C286 begins after a HIGH to LOW transition on RESET . The HIGH to LOW transition of RESET must be synchronous to the system clock. Approximately 50 system clock cycles are required by the 80C286 for internal initializations before the first bus cycle to fetch code from the power on execution address is performed. A LOW to HIGH transition of RESET synchronous to the system clock will end a processor cycle at the second HIGH to LOW transition of the system clock. The LOW to HIGH transition of RESET may be asynchronous to the system clock; however, in this case it cannot be predetermined which phase of the processor clock will occur during the next system clock period. Synchronous LOW to HIGH transitions of RESET are required only for systems where the processor clock must be phase synchronous to another clock.

V_{SS}

SYSTEM GROUND: are the ground pins (all must be connected to system ground).

VCC

SYSTEM POWER: + 5V power supply pins.

1.7 THE 80386 MICROPROCESSOR

The successor of the 80286 was the Intel 80386, a 16-bit processor with a segment-based memory management and protection system. The 80386 added a 32-bit architecture and a paging translation unit, which made it much easier to implement operating systems that used virtual memory. It also had support for hardware debugging.

The 80386 has three operating modes: real mode, protected mode and virtual mode. The protected mode which introduced in the 286 was extended to allow the 386 to address up to 4 GB of memory. The all new virtual 8086 mode (or VM86) made it possible to run one or more real mode programs in a protected environment, although some programs were not compatible. Figure 1.9 shows a simplified arcitecture of 80386 microprocessor.

1.7.1 Datatypes of 80386

The 80386 supports the following 17 data types each of which is discussed here in brief.

I. Bit.

II. Bit field-A group of at most 32 bits i.e., 4 bytes.

III. Bit string- A string of continuous bits of maximum 4Gbytes length.

IV. Signed Byte- Signed byte data. Sign of the operand depends upon its most significant bit.

If it is 0, then the number is positive. else it is negative. Range is from -128 to 127.

V. Unsigned Byte-Unsigned byte data. Range from 0 to 255.

VI. Integer word- Signed 16-bit data. Range from -32,768 to 32,767.

Vll. Long Integer-32-bit signed data that is represented in 2's complement form. Range is from - 2.147*109 to 2.1479.

VIII. Unsigned Integer Word- Unsigned 16-bit data. Range from 0 to 65,535.

IX. Unsigned Long Integer-Unsigned 32 bit data. Range from 0 to 4,294,967,295.

X. Signed quad word- A signed 64-bit or four word data.

Xl. Unsigned Quad Word-An unsigned 64-bit data.

XII. Offset-A 16 or 32- bit displacement that references a memory location using any of the addressing modes.

XIII. Pointer-16-bit selector and 16 or 32 bit offset.

XIV. Character-An ASCII equavelent to any of the alphanumeric or control characters.

XV. Strings-These are the sequences of bytes, words or double words. A string may contain up to maximum 4 Gbytes.

XVI. BCD- Decimal digits from 0-9 represented by unpacked bytes.

XVII. Packed BCD- This represents two packed BCD digits using a byte from 00 to 99.

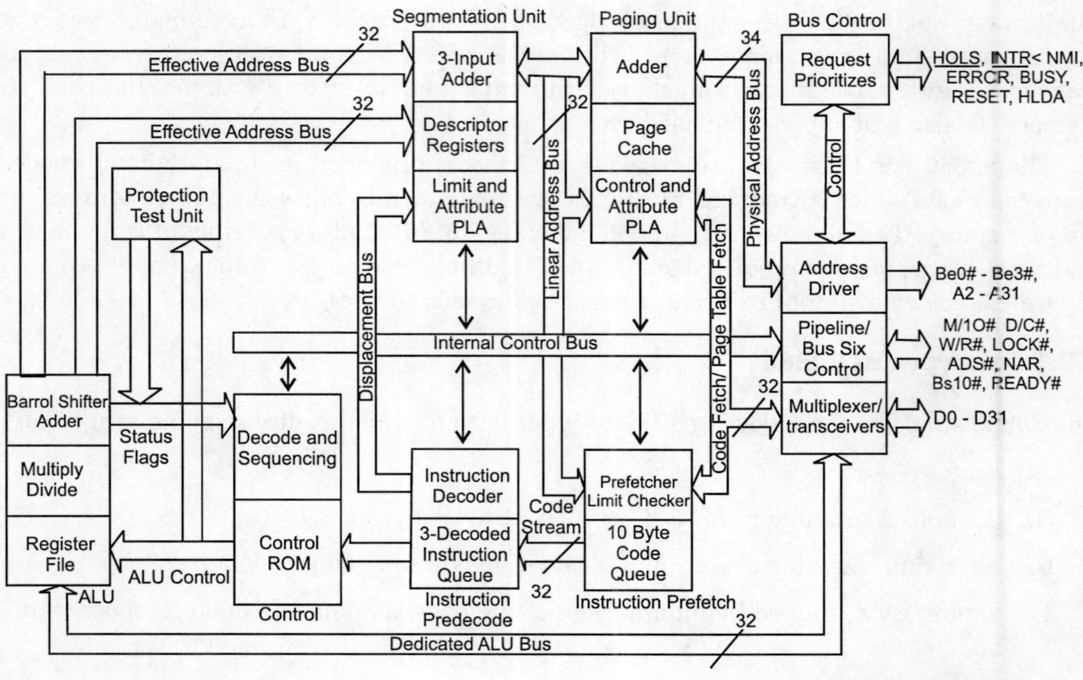

Fig 1.9 Arctitecture of the 80386 microprocessor.

1.8 ARCHITECTURE OF THE 80486 MICROPROCESSOR

The Intel 80486 microprocessor (alias i486 or Inte1486) was higher performances follow up on the Intel 80386. Introduced in 1989, it was the first tightly pipelined x86 design as well as the first x86 chip to use more than a million transistors, due to a large on-chip cache and an integrated floating point unit. It represents a fourth generation of binary compatible CPUs since the original 8086 of 1978. A 50 MHz 80486 executed around 40 million instructions per second on average and was able to reach 50 MIPS peak performance. Thp. i486 was without the usual 80-prefix because of a court ruling that prohibited trademarking numbers (such as 80486). Later, with the introduction of the Pentium brand, Intel began branding its chips with words rather than numbers. Figure 1.10 shows a simplified arcitecture of 80386 microprocessor.

1.9 DIFFERENCE BETWEEN i386 AND i486;

1. An 8 KB on-chip (level 1) SRAM cache stores the most recently used instruction and data (16 KB and/or write-back on some later models). The 386 had no such internal cache but like the i486, supported a slower off-chip cache (level 2 cache).
2. Tightly coupled pipelining allows the 486 to complete a simple instruction like ALU reg, or reg in every clock cycle. The 386 needed two clock cycles for this.
3. Integrated FPU (disabled or absent in SX models) with a dedicated local bus; together with faster algorithms on more extensive hardware than in the i387, this gives faster floating point calculations compared to the i386 + i387 combination.

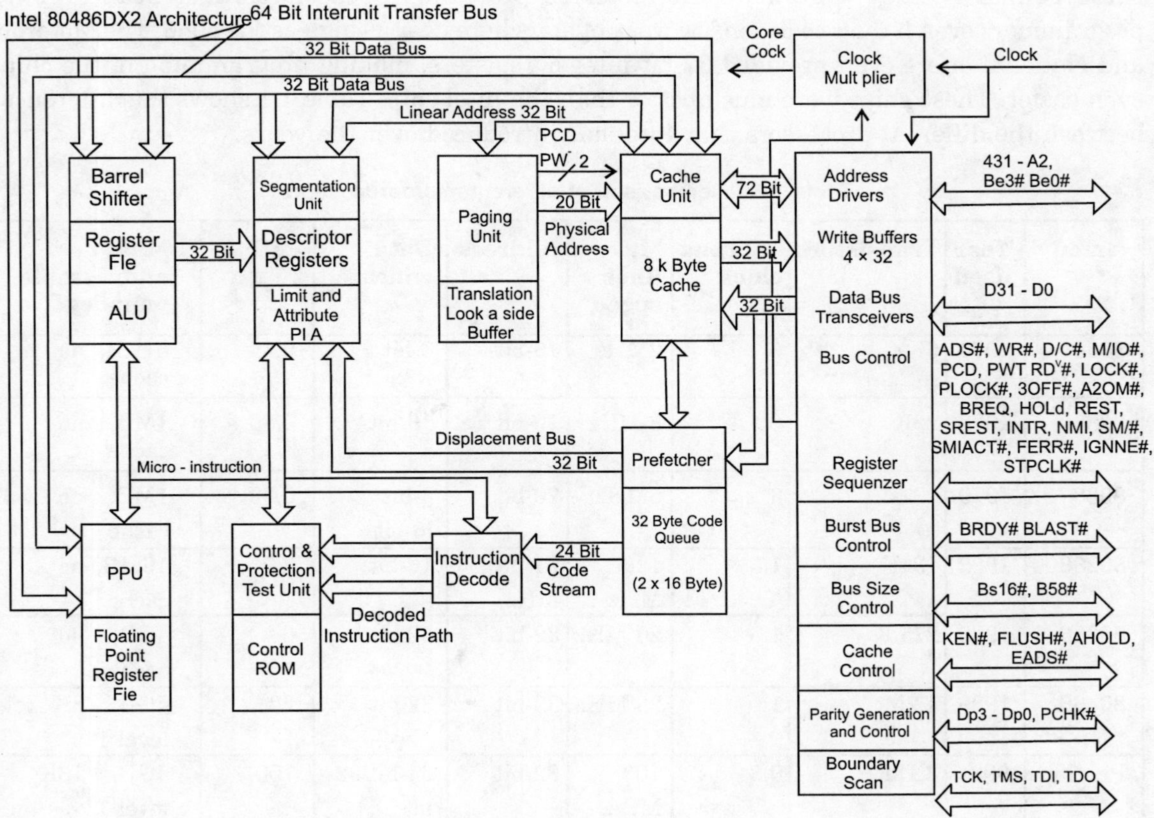

Fig 1.10 Arctitecture of the 80486 microprocessor.

4. Improved MMU perormance. The 486 has a 32-bit data bus and a 32-bit address bus. This required either four matched 30-pin (8-bit) SIMMs or one 72-pin (32-bit) SIMM on a typical PC motherboard. Just like the 80386, the 32-bit address bus of the 80486 enabled up to 4 gigabytes of memory to be directly addressed using a flat memory model with 32-bit linear address in protected mode. Just as with the 80386, the ability to use memory directly without segmentation helped performance in compliant operating system and applications.

In the early 1970's, Intel developed and delivered the first commercially viable microprocessor, namely the 4004 and 4040 device. The 4004 was not very powerful and all it could do was add and subtract with 4-bit data only at a time. But it was amazing in those days that everything was on one chip. Prior to the 4004, engineers built computers either from collections of chips or from discrete components (Transistor wired one at a time). The early machines were not portable and were very bulky, and power consuming. The 4004 powered one of the first portable electronic calculators known as 'Bascom'.

Later on Zilog created their own 16-bit processor that they called Z8000, Motorola created the 68000, their own 16-bit processor, and National Semiconductor introduced the 16032 device (later to be renamed the 32016). The designers of these chips had different design goals than Intel. Primarily, they were more interested in providing a reasonable instruction set for programmers even if their code density was not anywhere near as high as the 8086. The Motorola and National offers even provided 32-bit integer registers, making programming of the chips even easier. These chips were much better than the Intel chip. Table 1.1 shows the differences between the different processors that Intel has introduced over the years.

Table 1.2.1 Comparison of different microprocessors

Name	Year used	Transistors	Microns clock	Max lines speed	Address	Data width	MIPS	Max addressable memory
8080	1974	6K	6	2MHz	16-bit	8-bit	0.64	64KB, no cache
8086	1978	29K	6	8 MHz	20-bit	16-bit	0.80	1MB, no cache
8088	1979	29 K	3	5MHz	20-bit	8-bit 16-bit	0.33	1MB, no cache
80286	1982	134K	1.5	12.5 MHz	24-bit MHz	16-bit	2.7	16MB, no cache
80386	1985	275 K	1.5	20 MHz	32-bit	32 bit	6	4GB, no cache
80486	1989	1.2M	1	25 MHz	32-bit	32 bit	20	4GB, 8K level 1
Pentium	1993	3.1M	0.8	100 MHz	32-bit	64-bit/32-bit	100	4GB, 16K level 1
Pentium Pro	1995	5.5M	0.8	440 MHz	32-bit	64-bit/32-bit	440	64GB, 16K level 1
Pentium II	1997	7.5M	0.35	266 MHz	32-bit	64-bit/32-bit	446	64GB, 32K level 1, 256K/512 level 2
Pentium III	1999	9.5M	0.25	500 MHz	32-bit	64-bit/32-bit	1000	64GB, 32K level 1, 256K/512 level 2
Pentium IV	2000	42M	0.18	1.5 GHz	32-bit	64-bitl/32-bit	1,700	64GB, 32K level 1, 256K/512 level 2
Pentium IV Prescott	2004	125M	0.09	3.6 GHz	32-bit	64-bitl32-bit	7,000	64GB, 32K level 1, 256K/512 level 2

1.11 Differences between 8085 and 8086 microprocessor are

So No.	Microprocessors 8085	Microprocessors 8086
1	developed in 1977	develo ed in 1978
2	8 bit microprocessor	16 bit microprocessor.
3	It contains 16 bit address bus	It contains 20 bit address bus.
4	contains 8 bit data bus	It contains 16 bit data bus.
5	The clock speed of is 3 MHZ	The clock speed can vary between 5, 8 and 10 MHz for three different 8086 microprocessor.
6	does not have memory segmentation	Features a special concept of memory segmentation which allows parallel processing.
7	It has 6,500 transistors in the circuit board	It has 29,000 transistors in the circuit board.
8	No pipelining is used in 8085 microprocessor	whereas in 8086 pipeling is used
9	a accumulator based microprocessor	a General purpose register based microprocessor.
10	Only program counter and address counter are on 16 bit otherwise all otherwise all other counters are 8 bits	All flags and counters are of 16 bits.
11	no minimum or maximum	Both minimum and maximum modes are present.
12	only 1 processor is used	more than I processor is used
13	It has only 64 kb memory is used together	Whereas in 8086 1MB is used by dividing into segment registers.
14	no control flag is us	It uses a control flag.

1.10 KEY POINTS TO REMEMBER

1. The first microprocessor to be introduced was the intel 4004. This was followed by 8008. The 4004 microprocessor was a 4 bit microprocessor which could address 4K of 4 bit memory which was extremely low compared to today's micriprocessors.
2. The first modern 8-bit microprocessor is considered to be the intel 8080. This was followed by the very widely used intel 8085 microprocessor.

3. The modern microprocessor family designed for applications in microcomputers started with the launch of the 16-bit 8086 micriprocessor and 8-bit 8088 microprocessor.
4. A general computer block diagram is given below

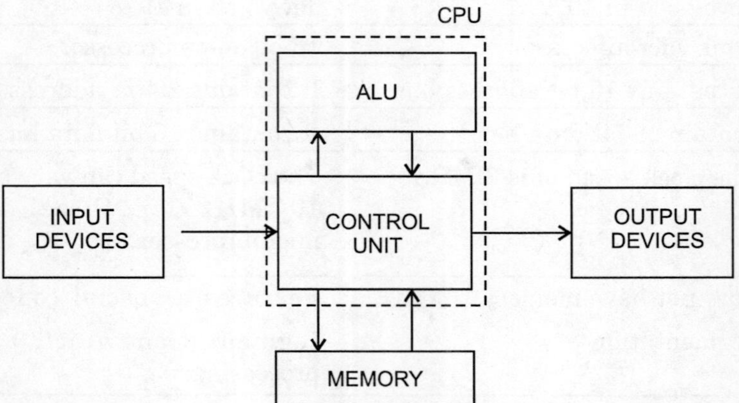

5. The computer can be divided into four main components namely: CPU, memory, Input devices, and output devices.
6. An electronic IC package designed to handle all the functions of a traditional CPU is known as a microprocessor. Microprocessors are implemented by modern IC fabrication techniques such as VLSI and ULSI Microprocessor's contain all the functioning circuitry (that function as a complete CPU) on one wafer of silicon.
7. A microprocessor based microcomputer can be shown by the simplified block diagram given below:

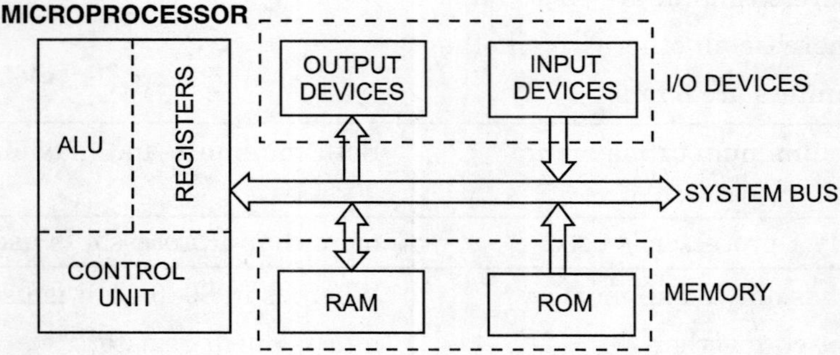

8. A microprocessor itself can be divided into three components mainly registers, ALU and control unit.
9. I/O Devices constitute the input and output devices interfaced to a microprocessor based system. These include keyboard and Mouse (Input devices) and printers and CRT display monitors (Output devices).
10. The system memory is composed of two types of memory
    - RAM: Random Access Memory (Volatile memory)
    - ROM: Read Only Memory (Non-volatile memory)

11. The system bus is a common communication path provided between the micprocessor, I/O devices and memory. It itself is nothing more than a set of wires required to carry different bits. The system bus can be divided in three components
    - Control bus
    - Address bus
    - Data bus
12. Digital devices such as buffer, flip flops, counters and multiplexers etc. have been further discussed for clearify the further chapters.

REVIEW QUESTIONS

1.1 Which was the first microprocessor that was launched ? Also, give the name for the first 8 bit microprocessor?

1.2 Describe the traditional diagram of a computer? What are the functions of the CPU ?

1.3 What is the difference between a microprocessor and the CPU ? How are they similars?

1.4 What effect does the introduction of a microprocessor have on the traditional block diagram of a computer.

1.5 Explain the organization of a microprocessor based microcomputer? Explain the functions of each of the corresponding subunits?

1.6 What is a system bus? What are the different buses that comprise the system bus? Explain in brief the basic components of a microprocessor ?

1.7 What are peripherals? Name a few input devices and a few output devices?

1.8 What is RAM and ROM? What are the main differences between the two? What are the different types of RAM and ROM devices?

1.9 Describe the 74245 Octal Transceiver IC ? How does it operate and why is it considered to be such an important IC ?

1.10 What are the basic elements that are used to design

(*i*) Registers

(*ii*) RAM

(*iii*) ROM

(*iv*) Multiplexers and Demultiplexers

●●●

Chapter-2

THE 8085 MICROPROCESSOR

INTRODUCTION

A microprocessor is a semiconductor chip that serves as the central processing unit (CPU) of a computer. Thus we can say that microprocessor is the brain of a computer. A microprocessor consists of arithmetic and logic unit (ALU) and the control unit. The 8085 includes on its chips most of the logic circuitry for performing computing task and for communicating with peripherals. However, its eight bus lines are time shared by the low-order address and data i.e. multiplexed. The microprocessors are usually characterized by speed, word length, architecture and instruction sets. In this chapter, we will discussed the 8085 architecture, instruction set, addressing modes and assembly programming in detail.

The 8085 is an 8-bit general purpose microprocessor that is capable of addressing 64K of memory. It has forty pins and can operate with +5V single power supply and with a 3 MHz single-phase clock. The main characteristic features of the 8085 microprocessor are explained in following section.

2.1. FEATURES OF THE 8085 MICROPROCESSOR

The following are the main features of 8085 microprocessor:

1. It has 8-bit data bus and 16-bit address bus.
2. It is 8-bit microprocessor i.e., it can process, accept or provide 8-bit data simultaneously.
3. It has 16 address lines, thus can access (2^{16}) 64 K bytes of memory.
4. It has 8-bit I/O addresses to access (2^{8}) 256110 ports.
5. It operates with +5V single power supply connected at V_{cc} power supply and ground is connected to Vss'
6. The maximum clock frequency of 8085 microprocessor is 5 MHz. Normally it operates at 3 MHz clock frequency.
7. It has following addressing modes: Register, immediate, direct, indirect and implied.
8. In 8085 microprocessor, the data bus ($D_0 - D_7$) is multiplexed with lower 8-bit address bus ($A_0 - A_7$), thus it requires an external hardware (latch) to separate data lines from address lines.
9. It has five hardware interrupts: TRAP, RST 7.5, RST 6.5, RST 5.5 and INTR.
10. It provides control signals ($IO/\overline{M}$, $\overline{RD}$, $\overline{WR}$) to control the bus cycles, thus external bus controller is not required.

11. It has 8-bit accumulator, flag register, six general purpose registers (B, C, D, E, H, L), and two special purpose 16-bit registers (SP and PC).
12. To allow serial communication it has serial I/0 lines i.e., SOD and SID.
13. It can perform following arithmetic and logical operations:
    8-bit binary addition, 16-bit binary addition with and without carry, 2 digit BCD addition, 8-bit binary subtraction with and without borrow, logical AND, OR, Ex-OR, complement, shift operation, etc.
14. It supports a mechanism that enhances its interrupt handling capability.
15. The 8085 microprocessor can be used as three chip microprocessor using IC 8155 and IC 8355.

2.2. THE 8085 BUS STRUCTURE

(UPTU-Semester Exam 2005-2006)

A group of wires that are bundled together to serve a purpose are called buses. The 8085 microprocessor has three sets of communication buses named; the address bus, the data bus, and the control bus. The group of these buses is called system bus. Figure 2.1 illustrates the bus structure of 8085 microprocessor.

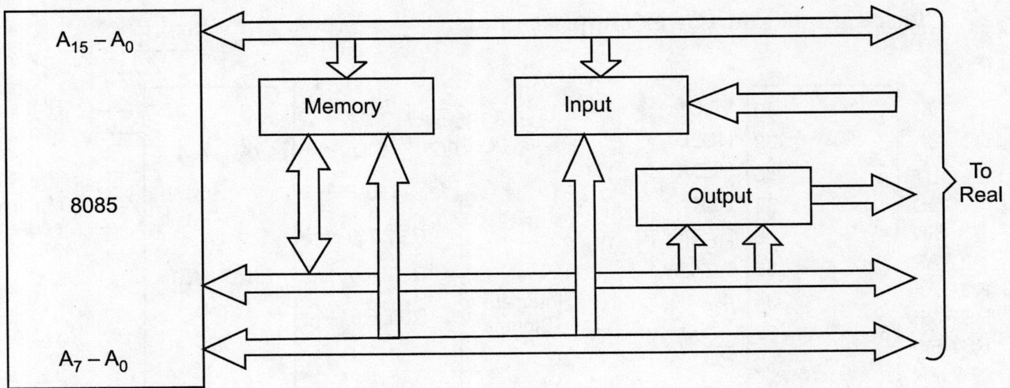

Figure 2.1. The 8085 Bus structure

2.2.1. Address Bus

The buses that are used by microprocessor for sending out the address of a memory location are called the address bus. The 8085 has 16-bit address bus ($A_0 - A_{15}$) to access 16-bit address and capable of addressing 2^{16} = 65536 i.e., 64 K memory locations. The address buses are unidirectional i.e., data can flow in one direction from the microprocessor to the peripheral devices. These can also be used for sending port address when microprocessor reads data from or write data to a port.

2.2.2. Data Bus

The 8085 microprocessor has 8 parallel lines data bus. The data bus can transfer data from CPU to memory and vice versa, thus data buses are bidirectional bus. The number of data lines used in data bus is equal to the size of data word being read or written. The data buses are also used to write data to or read it from the memory or I/O ports.

2.2.3. Control Bus

Control buses are that lines on which timing signal and control signal are travel. The microprocessor sends signals on the control bus to enable the outputs of addressed memory devices or I/O port devices. Memory read, memory write, I/O read, and I/O write are some of the control signals. These signals provide the information about the type of devices with which the microprocessor has to communicate.

2.3. LOGIC PIN DIAGRAM OF 8085 MICROPROCESSOR

(UPTU-Semester Exam 2005-2006)

The 8085 is an 8-bit general purpose microprocessors, it consists of 40 pins and operates with single power supply. Logic pin diagram of 8085 is shown in Figure 2.2. It consists of following functional groups:

1. Address bus and data bus
2. Status signals and control signals
3. DMA request signals
4. Reset signals
5. Interrupt signals
6. Serial input/output signals
7. Clock signals and Power supply signals.

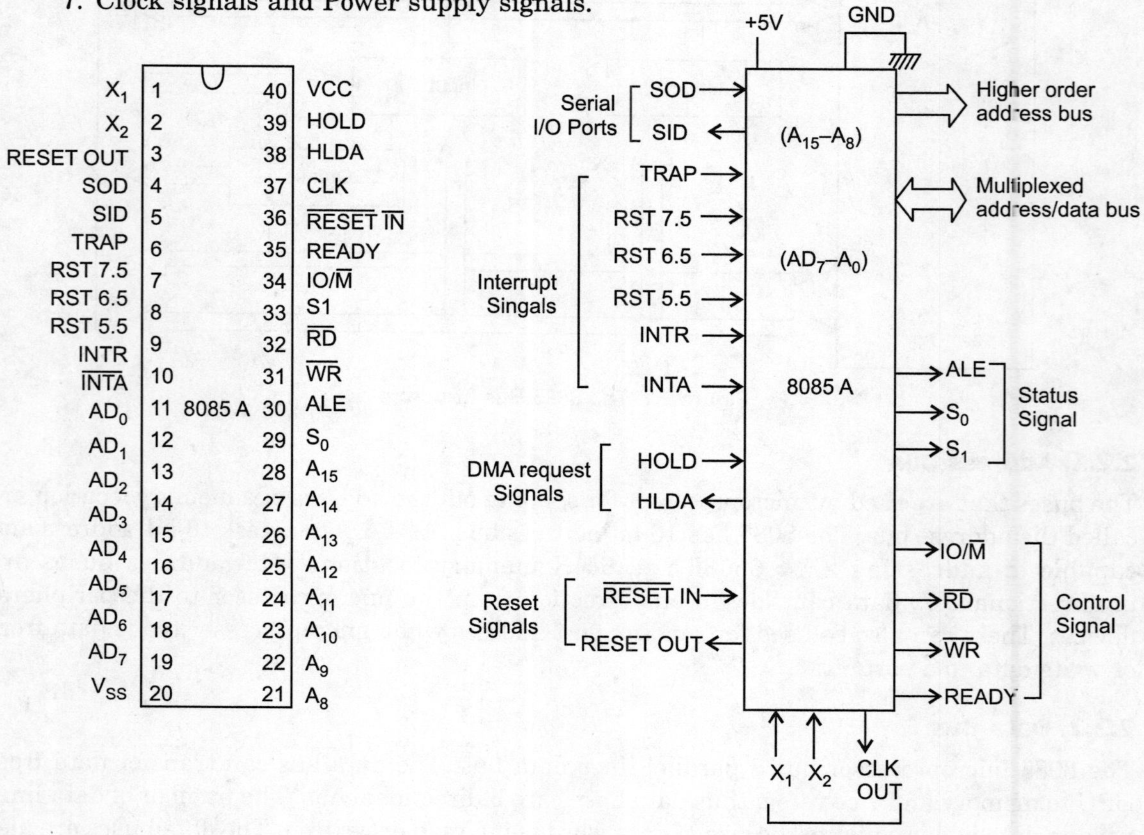

Figure 2.2. Logic pin diagram of 8085 microprocessor

2.3.1. Address Bus and Data Bus

1. Address bus ($A_8 - A_{15}$) : These eight signal lines ($A_8 - A_{15}$) are unidirectional output, tristate and used as higher order 8-bits of 16 address bits. Thus, these 8 lines are used only to select a memory or an I/O device.
2. Multiplexed Address/Data bus ($A_0 - A_7$) : These signal lines ($A_0 - A_7$) are tristate, bidirectional and serve a dual purpose. These signal lines are used as address as well as data bus. Thus, the lower order 8 bits of 16 bit address bus of the 8085 are multiplexed with data bus. During operation cycle, in earlier part these are used as lower addressbus and in later part used as data bus.

2.3.2. Control and Status Signals

These signals consist of two control signals ($\overline{RD}$ and $\overline{WR}$), three status signa is (10/ $\overline{M}$ S_1 – and S_0) to determine the nature of operation along with a special signal (ALE) to tell the beginning of the operation.

1. Address Latch Enable (ALE) : This is an output signal that gives information about the contents of ($A_0 - A_7$). It is a positive going pulse generated during the first cycle of a machine cycle. When it is high it shows that contents of ($A_0 - A_7$) are address and data when it is low. ALE signal is used to separate ($AD_0 - AD_7$) (i.e., demultiplex) to ($A_0 - A_7$) and ($D_0 - D_7$)· This operation is performed by connecting a latch to ($AD_0 - AD_7$) lines and this latch is controlled by ALE signaL
2. Status Signals (S_1 and S_0) : These status signals are used to give information of operation performed by the microprocessor. These signals have four different conditions during machine cycle as shown in Table 2.1.

Table 2.1. 8085 Machine Cycle Status

	Operation	S_0	S_1
1	Opcode fetch (Instruction read from memory)	1	1
2	Read (Data read from memory)	0	1
3	Write	1	0
4	Halt	0	0

3. Input Output/Memory (IO/$\overline{M}$) : This status signal is used to differentiate between I/O and memory operation. When = 1 (high), it indicates an I/O operation and when = 0 (low), it shows memory operation.
4. Read ($\overline{RD}$) : This is an active low read signal which indicates that the selected I/O or memory device is to be read and data are available on the data bus.
5. Write ($\overline{WR}$) : This is an active low write signal which indicates that the data on the data bus are to be written into selected memory or I/O location.
6. Ready: This is an active high input control signal that is used by microprocessor to detect whether a peripheral is ready for the data transfer or not. If not the processor waits till the signal goes high. The main function of this signal is to synchronize the 8085 microprocessor with slower peripherals. Table 2.2 shows the status of all the machine cycle (operations) performed by 8085 when (S_1 and S_0) is combined with 1O/$\overline{M}$.

Table 2.2. 8085 Machine cycle status and control signals

Control signals	Status			Machine Cycle
	1O $\overline{M}$	**S_1**	**S_0**	
$\overline{RD}$ = 0	0	1	1	Opcode fetch
$\overline{RD}$ = 0	0	1	0	Memory Read
	0	0	1	Memory Write
$\overline{WR}$ = 0	1	1	0	I/O Read
$\overline{RD}$ = 0	1	0	1	I/O Write
$\overline{WR}$ = 0	1	1	1	Interrupt Acknowledge
	Z	0	0	Halt
$\overline{INTA}$ = 0	Z	X	X	Hold
$\overline{RD}$, $\overline{WR}$ = Z and $\overline{INTA}$ = 1	Z	X	X	Reset
Note: Z = Tri-state (high Impedance) and X = Unspecified				

2.3.3. DMA Request Signals

1. **HOLD**: Hold is an active high input signal that is used by another controller to request microprocessor to use address, data and control signals.
2. **HLDA :** The 8085 microprocessor has one signal called HLDA (Hold Acknowledge) inresponse to HOLD, to acknowledge the requesting device. When HLDA is active it shows that microprocessor has received HOLD request and will hand over the buses in next cycle. The other controller will use buses and remove HOLD signal upon the completion of work and due to this microprocessor will also make HLDA low. The 8085 microprocessor takes control of buses half clock cycle after HLDA goes low.

2.3.4. Reset Signals

1. **$\overline{\text{RESET IN}}$:** This is an active low input reset signal. When $\overline{\text{RESET IN}}$ = 0, it clears program counter, i.e., 0000 and makes address, data and control lines tri-stated. The status of internal registers and flags is unpredictable after reset. The CPU is held in the reset condition as long as $\overline{\text{RESET IN}}$ is applied. The microprocessor starts executing instructions from 0000 H onward after reset.
2. **RESET OUT :** This is an active high output signal which indicates that the microprocessor is reset. This signal can be used to reset other devices.

2.3.5. Interrupt Signals

The 8085 has five interrupt signals that can be used to interrupt a program execution as follows:

1. **TRAP :** This is an active high, non-maskable level and edge triggered interrupt and has highest priority. The microprocessor performs internal restart automatically at address 0024 H when TRAP line is active.
2. **Restart Interrupts (RST 7.5, RST 6.5 and RST 5.5) :** These are active high and triggered, maskable interrupt that transfer the program control to specific memory locations. They have higher priorities than the INTRinterrupt with following priority RST 7.5, RST 6.5, RST 5.5. When RST 7.5, RST 6.5, or RST 5.5 is active microprocessor, performs internal restart automatically at address 003C H, 0024 H, 002CH, respectively.
3. **INTR :** INTR is an active, high level triggered general purpose interrupt. It has lowest priority.
4. $\overline{\textbf{INTA}}$ **:** $\overline{\text{INTA}}$ is used by the microprocessor to acknowledge an interrupt.

2.3.6. Serial I/O Signals

1. **SID (Serial Input Data) :** This is an active high, serial input port pin to accept serial 1-bit data under software control. The SID pin data is loaded in bit D_7 of accumulator when a RIM instruction is executed.
2. **SOD (Serial Output Data) :** This is an active high, serial output port pin to transfer serial I-bit data under software control. The SOD pin is set or reset depending on D_7 and D_6 bits of accumulator when a SIM instruction is executed.

2.3.7. Clock Signals and Power Supply

1. $\mathbf{X_1\ X_2}$ **:** A crystal (or RC, or LC network) is connected at these two pins to derive the internal clock generator circuit. The frequency is divided by 2 and used as operating frequency. Usually the 6.014 MHz crystal is connected to these pins, this is divided by 2. Thus the operating frequency of 8085 microprocessor is 3.07 MHz.
2. **CLK OUT :** This is an output signal, can be used as the system clock for other devices.
3. $\mathbf{V_{CC}}$ **:** + 5 V power supply.
4. $\mathbf{V_{SS}}$ **:** Ground reference.

2.4. THE 8085 MICROPROCESSSOR ARCHITECTURE (BLOCK DIAGRAM)

(UPTU Semester Exam 2005-2006)

The internal architecture (block diagram) of the 8085 microprocessor is shown in Fig. 2.3. This block diagram for sake of simplicity, can be divided into following groups:

1. Arithmetic and logic unit
2. Registers
3. Address buffer
4. Instruction decoder and machine cycle encoder
5. Address/data buffer
6. Timing and control circuitry
7. Serial I/O control
8. Interrupt control
9. Increment/decrementer Address Latch.

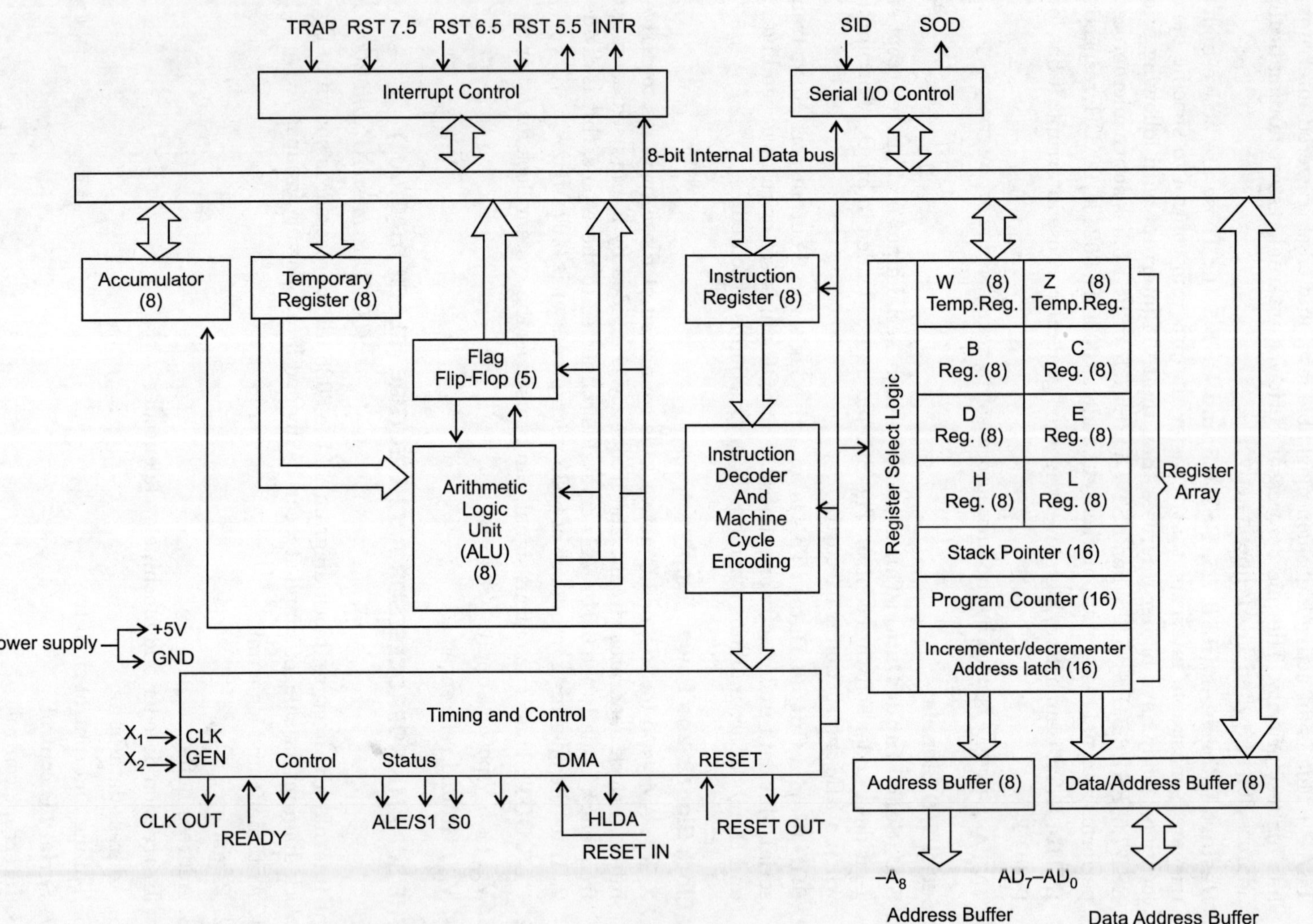

Fig. 2.3. The 8085A microprocessor architecture (functional block diagram)

2.4.1. The Arithmetic and Logic Unit (ALU)

The arithmetic and logic unit performs the computing function i.e. all logic and arithmetic operations are performed at ALU. Arithmetic and logic unit consists of the accumulator, the temporary register, the arithmetic and logic circuits, and five flags. ALU performs the arithmetic such as addition, subtraction and logic operation like AND, OR, EX-OR, etc. The ALU is not available to the user. Its word length depends upon the width of an internal data bus that usually is 8-bit. The ALU is always controlled by timing and control circuits. ALU accepts operands from accumulator and temporary register and stores the result of arithmetic and logic- operations in accumulator or temporary resister.

1. **Accumulator :** Accumulator is one of general purpose registers of the 8085 microprocessor and called as A register. The accumulator is a multi purpose register because it can be use for data storage, arithmetic, logical, and input/output operations. Accumulator is used to store the results of operation performed by arithmetic and logic unit and can also be used as a general purpose register to store 8-bit data.
2. **Temporary Registers :** Since it provides operands to the ALU, it is also called as 8-bit operand register. Although temporary data register is not available to the user but the ALU can store intermediate results in it. It is also used for operations such as register to register data transfer, etc. However, the ALU get input from accumulator but the temporary registers also serve as one input to ALU. During an arithmetic or logic operation these registers are used by the control unit.
3. **Status or Flag Register :** The flag register of the 8085 microprocessor includes five flags and it is connected to ALU. Flags are made of flip-flops and show some condition produced by the execution of an instruction. During an operation, performed by ALU, the result is transfer on internal data bus and status of result will be stored in flip-flops. Fig. 2.4 illustrates the different flags and their position in flag register.

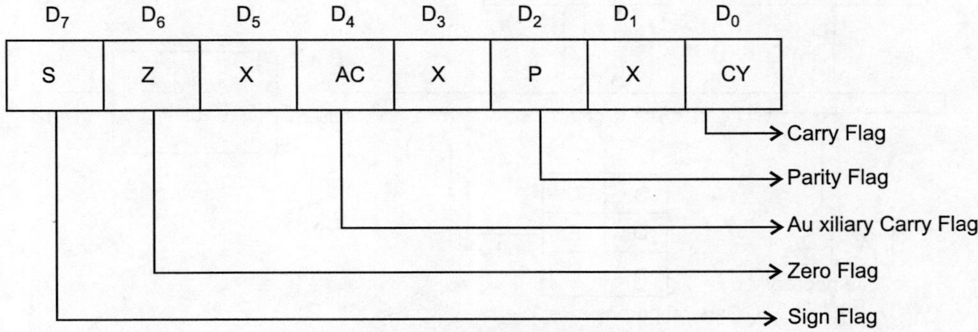

Figure 2.4. 8085 Flag register

- **The Carry Flag (Cy) :** The carry is set (1 or High) when arithmetic operations result in carry or borrow is used in subtraction, otherwise reset (Low or 0).
- **The Parity Flag (P) :** If the result of arithmetic or logic operation has an even number of l's, the parity flag is set (High or 1). If it has odd number of l's, the parity flag is reset (Low or 0).
- **The Auxiliary Flag (AC) :** The auxiliary flag is used only internally by microprocessor for BCD (binary coded decimal) operation and is not available for the user to change

the sequence of a program with a jump instruction. This flag is set (1 or High), when a carry is generated, in an arithmetic operation, by digit D_3 and passed on to digit D_4 otherwise flag is reset (Low or 0).

- **The Zero Flag :** The zero flag is modified by results in the accumulator as well as in the other register. This flag is set (1 or High); when the result of ALU operation is 0, otherwise reset (Low or 0).
- **The Sign Flag :** Since in 8085 microprocessor negative number is represented by the 2's complement notation, the sign flag is used to indicate the sign of the result i.e., 1 for negative sign and 0 for positive sign. Thus the flag is set (l or High), when high bit of result is 1 i.e., the result is negative number, otherwise the flag is reset (Low or 0).

2.4.2. Register Section

The architecture of the microprocessor depends upon the number and type of the registers used in microprocessor. Register section of 8085 includes parallel in parallel out registers and also called as scratch pad memory because it can store data and address of memory. The registers include a set of binary storage cell/flip-flops with reading and writing facilities. The registers are used for temporary storage of instruction and data/address. Thus, the number of bits in a register is equal to data or address or instruction size depending on the application. The register section consists of 8-bit resisters: B, C, D, E, H, L, and two 16-bit registers program counter (PC) and stack pointer (SP).The Fig.2.5 shows the register section of the 8085 microprocessor and classified as :

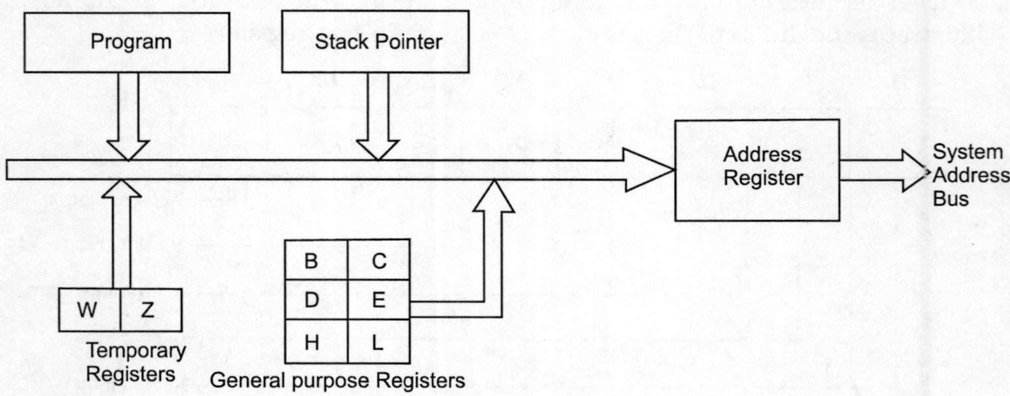

Figure 2.5. Register section of 8085

1. **General Purpose Registers :** The 8085 microprocessor consists of six 8-bit (each) general purpose registers named as B, C, D, E, H, and L. These registers individually can be used to store 8-bit and 16-bit data can be stored with register-s pair. The valid register pairs available in the 8085 microprocessor are BC, DE, and HL. These registers are available to the user and can also be programmed by the user. The programmer can use these registers to hold data, result of arithmetic and logical operations and address of data memory. The

HL register pair can be used as data pointer and memory pointer. It holds the address of a 16-bit address of a memory location when used as memory pointer.

2. **Special Purpose Registers :** The special purpose registers are used for special use. These are instruction register, program counter, and stack counter.

- Instruction Register: The register that holds the opcode of instruction from memory which has to be decoded and executed is called instruction register. It is not accessible to the user. The CPU stores the opcode in instruction register and further sent to the instruction decoder-to select one of the 256 alternatives.
- **Program Counter (PC) :** Program counter is a 16-bit register, at a given time it stores the address ofthe next instruction to be fetch i.e., it holds the memory address of the next instruction to be executed. During instruction fetch operation, the processor places contents of program counter on the address bus and fetch first byte of the instruction from that memory location. Then the processor increments program counter to point the next byte of the instruction. The program counter is set to the address of the first 'instruction to be fetch and executed when reset is activated. The reset address of the 8085 microprocessor is 0000H. In case of jump and call instructions, the address followed by the jump and call instructions is placed in the program counter. If the condition is fulfilling then the processor fetch the next instruction from the new address specified by JUMP or CALL instruction otherwise processor continues with the next instruction after JUMP or CALL instruction.
- **Stack Pointer (SP) :** The stack pointer is a 16-bit register to define the stack starting address and usually it points at top of the stack. Thus it is used to keep track of data stored on stack and the stack is a reserved area of the memory where register pair information can be stored or taken back under software control. The stack point is decremented after each stack write operation and incremented after each stack read operation. A transfer type instruction can load the stack pointer with initial value that is highest address of an assigned stack in memory. Thus stack should be always initialized in data memory.

2.4.3. Address Buffer

This is an 8-bit unidirectional buffer and used to derive the higher order address bus ($A_8 - A_{15}$). It is used to tri-state the higher order address buses under certain conditions such as reset, hold, halt and when address lines are not in use.

2.4.4. Instruction Decoder and Machine Cycle Encoder

This section is used to accept opcode from instruction register, decode it and provide the decoded information to control logic. The instructions are recognized in this block means what operation has to perform, who is going to perform it, how many operand byte are in instruction, etc. The status of machine cycle is resolute by machine cycle encoder means it gives the information about which machine cycle is currently being executed.

2.4.5. Address/Data Buffer

This is an 8-bit bidirectional buffer and used to derive the lower order address bus ($A_0 - A_7$) and data buses. This buffer is used to tri-state to address/data lines under certain conditions such as reset, hold, and halt.

2.4.6. Timing and Control Circuitry

This unit synchronizes all the microprocessor operations with the clock and generates the control signal necessary for communication between the microprocessor and peripherals. Thus, this is the controlling section for all internal and external circuits in the microprocessor system and it operates with a reference clock signal. It accepts information from instruction decoder and generates microsteps to perform it. The control section of a microprocessor is illustrated in Fig.2.6 in which a quartz crystal (LC or RC circuit) is used to determine the clock frequency. The contents of the instruction register are in binary form and the decoding network called matrices converts these cont nts in meaningful form. The control matrix provides the internal signals for controlling operation and data between registers. It also generates the essential timing signals for microprocessor operation.

2.4.7. Serial I/O Control

Usually the data transferred on the lines is parallel data, but for some applications, it is advantageous to use serial data transfer. In the 8085 microprocessor, this is achieved by SID and SOD signals. The data on these lines is accepted or transferred by serial I/0 control unit. The 8085 has two special instructions RIM and SIM to perform serial data transfer.

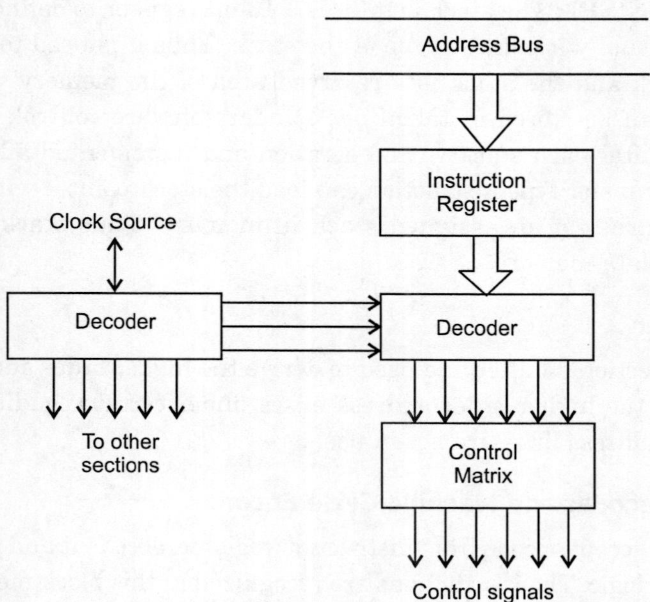

Figure 2.6. Control section of 8085 microprocessor

2.4.8. Interrupt Control

This unit accepts different interrupt request input like TRAP, RST 7.5, RST 6.5, RST 5.5, and INTR. This section informs the control logic to take action in response to each interrupt upon valid receipt.

2.4.9. Incremental/Decrementer Address Latch

This 16-bit register performs the increment or decrement in the content of program counter and stack pointer registers.

2.5. THE 8085 MACHINE CYCLES AND BUS TIMING DIAGRAMS

Different 74 instruction types are executed in the 8085 microprocessor and each instruction has two parts: operation code, called as opcode, and operand. The opcode is a command such as add, subtract, etc. and the operand is an object to be operated on, like a byte or the contents of a register. The 8085 has to perform various operations, to execute an instruction, like Memory Read/Write-and I/O Read/Write. All instructions can be divided into a few basic machine cycles and these machine cycles are divided into precise system clock periods. The time required by the microprocessor to complete the operation of accessing memory or I/O device is called as a Machine cycle. Operation like opcode fetch, memory read, memory write, I/O read, and I/O write, are performed in a machine cycle. The microprocessor external communication functions can be divided into three categories:

- Memory Read and Write
- I/O Read and Write
- Request Acknowledge.

These functions are further divided into various operations (machine cycle), shown in Table 2.2. Each instruction has one or more of these machine cycles, and each machine cvcle is divided into T-states. Microprocessor performs an operation in a specific time period i.e. specific clock cycle. Each clock cycle is called as T-state. The number of T-state required to perform an operation is also called machine cycle.

2.5.1. Opcode Fetch Machine Cycle

Opcode fetch is the first operation in any instruction. Before the microprocessor cah begin to execute the instruction, the microprocessor needs to get (fetch) this machine code from the memory register where it is stored. The fetch operation takes the following steps:

- Send the address of the next instruction to the memory.
- Acquire the instruction code from the memory.
- Update the program Counter.

To fetch the byte, the microprocessor performs the following steps:

Step 1 : In T_1 state the program counter places the 16-bit memory address on the address bus. Figure 2.7 illustrates that at T_1 the high-order memory address is placed on the address lines ($A_{15} - A_8$); low-order memory address is placed on the bus ($A_7 - A_0$), and the ALE signal goes high. During the same T_1 state, microprocessor also sends status signals IO/$\overline{M}$, S_1 and S_0, IO/$\overline{M}$ specifies whether it is a memory or I/O operation. S_1 specifies whether it is read/write operation, S_1 and S_0 together show read, write opcode fetch, machine cycles, or whether it is in HALT state. From the table 2,2, it may be noted that for opcode fetch machine cycle status signals are I0/$\overline{M}$ = 0, S_1 = 1 and S_0 = 1.

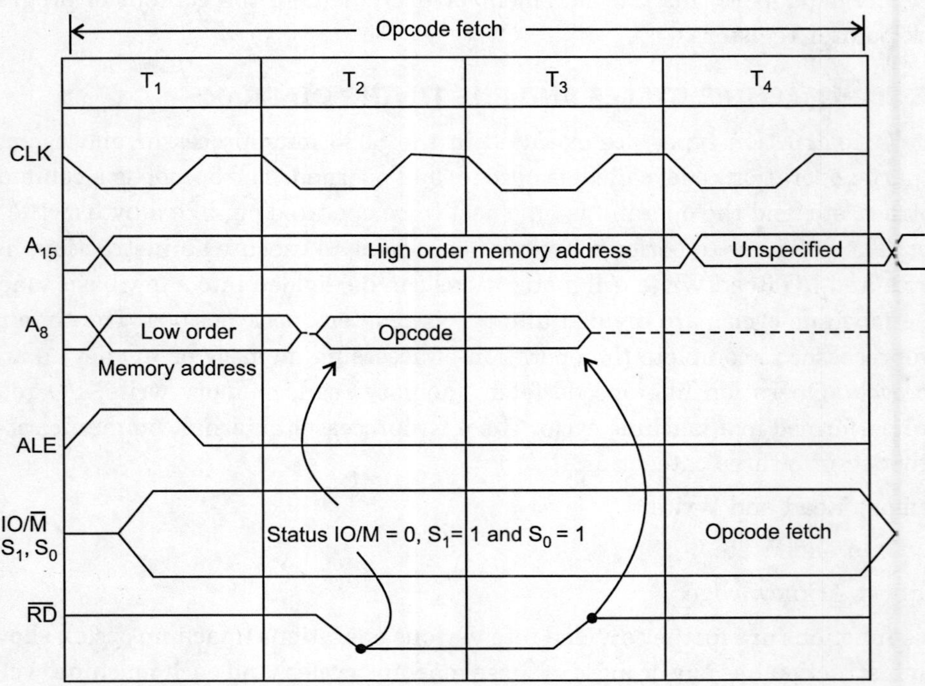

Figure 2.7. Opcode fetch machine cycle

Step 2 : During the state T_2, the control signal $\overline{RD}$ is sent out to enable the address memory location. In T_2 state, low order address disappears from the (AD_0 – AD_7) lines and memory device places the contents of address memory location on the data bus (AD_0 – AD_7).

Step 3 : During T_3 state, the 8085 loads the data from data bus into instruction register and makes RD high which disables the memory device.

Step 4: In T_4 state, the microprocessor decodes the opcode and on the basis of the instruction received encodes it into various machine cycles and also for completing the task specified by the instruction. During this state because $\overline{RD}$ is high, the address and data bus remains at HIGH IMPEDANCE.

2.5.2. Memory Read Machine Cycle

Memory read instruction has two bytes: the first is the opcode and 'the second is the data byte. The microprocessor needs to read these bytes first from memory and thus it requires at least two machine cycles for this operation. The first machine cycle is opcode fetch and the second machine cycle is Memory read as shown in Figure 2.8. Therefore, this instruction requires seven T-states for execution of these two machine cycles; four T-states for opcode fetch and remaining three for READ operation. In the memory read machine cycle, the microprocessor places the address on address lines from the stack pointer, or general purpose register pair or program counter. During the read process, microprocessor reads data from the addressed memory location. The following steps illustrate the execution of memory: read machine cycle.

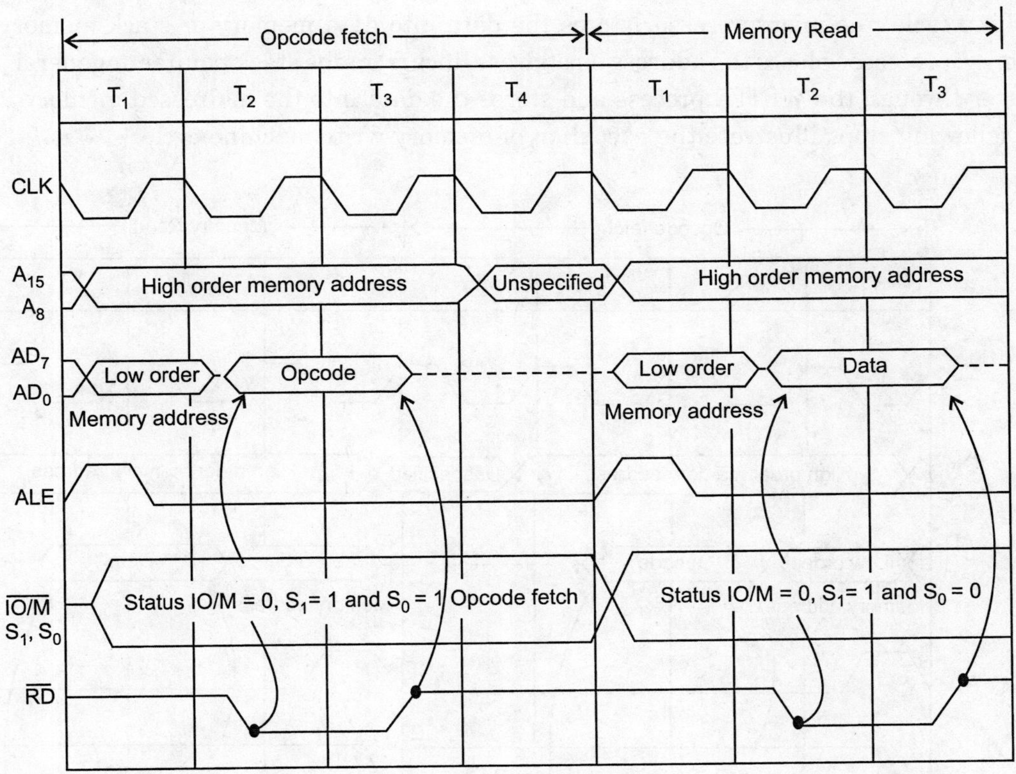

Figure 2.8. Memory read machine cycle

Step 1 : In the T_1 state, the 8085 places the address on the address lines from stack pointer, or general purpose register pair or program counter and activates the ALE signal to latch lower order byte of address. It may be noted that during the memory read the status signals are $1O/\overline{M} = 0$, $S_1 = 1$ and $S_0 = O$.

Step 2 : During this T_2 state, the microprocessor sends $\overline{RD}$ signal to enable the address memory location and then memory device places the contents of memory location on the data bus ($AD_0 - AD_7$).

Step 3 : During the T_3 state, microprocessor loads the data in specific register from the data bus and makes $\overline{RD}$ too high, which disables the memory device. The microprocessor reads and stores the data byte in the accumulator during T_3.

2.5.3. Memory Write Cycle

Memory write instruction has two bytes: the first is the opcode and the second is the data byte. Therefore, this instruction requires seven T-states for execution of these two machine cycles, four T-states for opcode fetch and remaining three for WRITE operation. In the memory write machine cycle, the microprocessor stores the data into data memory or stack memory. Thus, the microprocessor places the address on address lines from the stack pointer or general purpose register through the WRITE process and stores the data into the addressed memory location. The following steps illustrate the execution of memory write machine cycle.

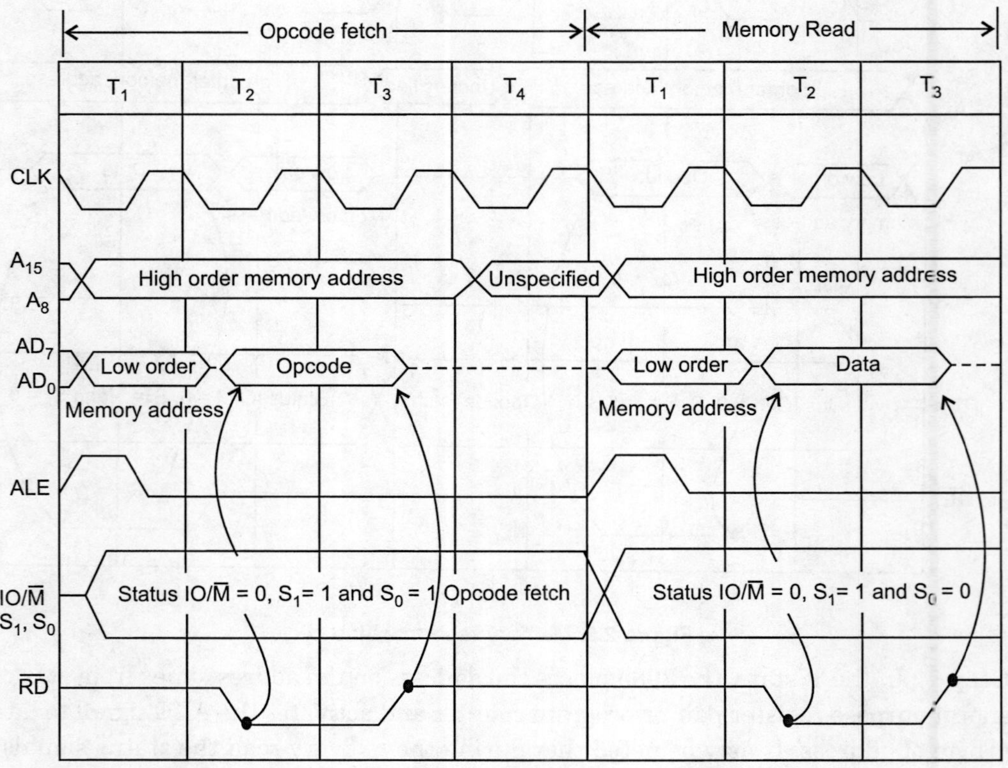

Figure 2.9. Memory Write machine cycle

Step 1: In the T_1 state, the 8085 places the address on the address lines from stack pointer or general purpose register and activates the ALE signal to latch lower order byte of address. It may be noted that during the memory read the status signals are IO/ $\overline{M} = 0$, $S_1 = 0$ and $S_0 = 1$.

Step 2 : During the T_2 state, the microprocessor place the data on data bus and sends $\overline{WR}$ signal to low for writing data into specified memory location.

Step 3 : During the T_3 state, microprocessor activates $\overline{WR}$ signal high which disable th memory device and terminates the write operation. The Fig. 2.9 illustrates the memory writ operation.

2.5.4. I/O Read and I/O Write Cycle

I/O read and I/O write cycles are similar to the previously discussed procedures for memo read and write cycles, respectively. The difference here is the IO/$\overline{M}$ signal is high for I/O rea and I/O write cycle. From the Table 2.2, IO/$\overline{M}$ signal shows that it is an I/O operation. Fig. 2.1 (a) and (b) illustrate the timing diagram for I/O read and I/O write cycles, respectively.

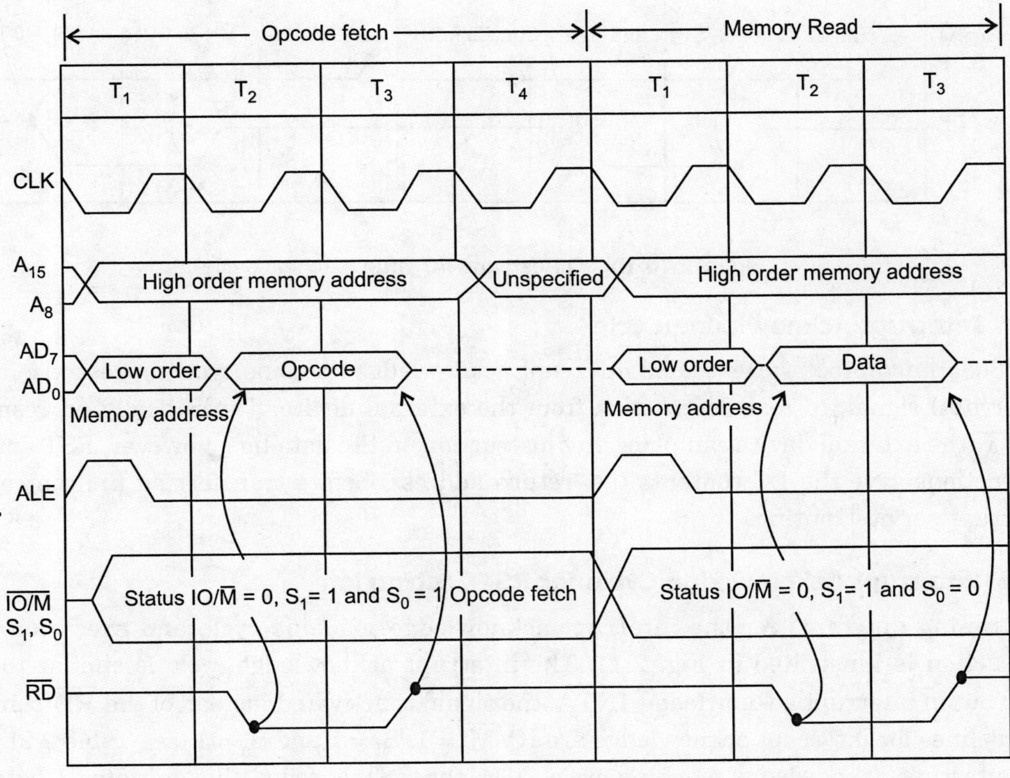

Figure 2.10. (a) I/O Read and I/O write machine cycle

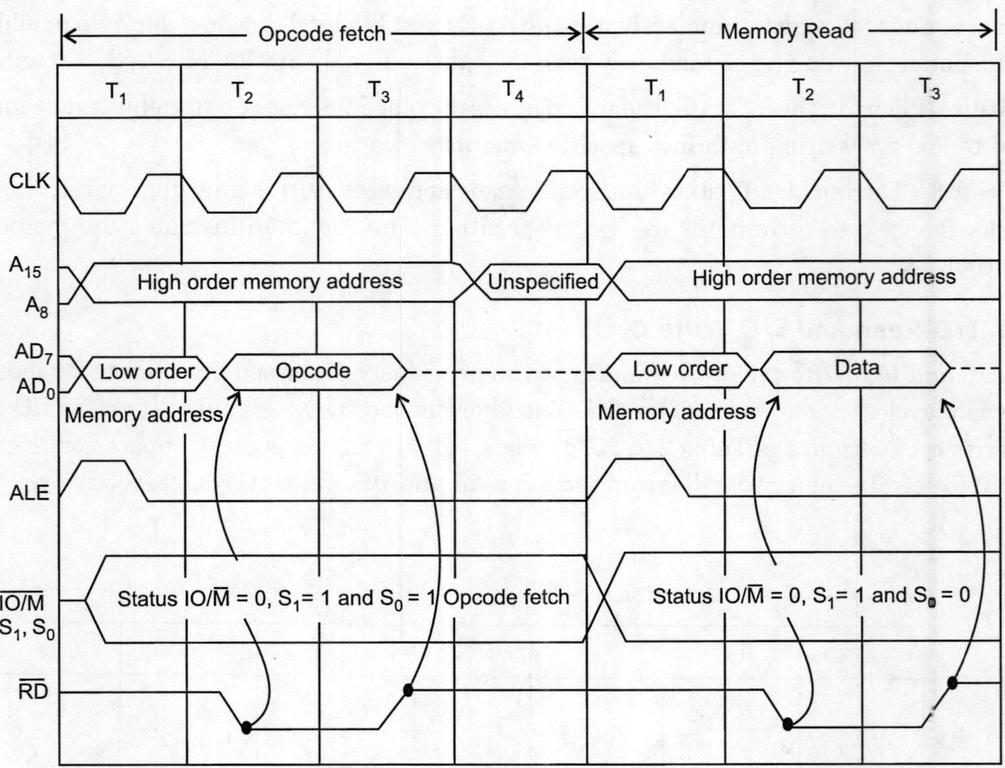

Figure 2.10. (b) I/O Read and I/O write machine cycle

2.5.5 Interrupt Acknowledge Cycle

The 8085 microprocessor executes interrupt acknowledge machine cycle, in response to INTR (interrupt) signal, to read instruction from the external device. Theoretically, in response to $\overline{INTA}$,the external device can place any instruction on the data bus, however, RST and CALL instructions save the PC contents (i.e. return address) before transferring the control to the interrupt service routine.

2.5.6. Interrupt Acknowledge Cycle for RST Instruction

The timing diagrams for the interrupt acknowledge machine cycle and execution of RST instruction is illustrated in Fig. 2:11. The interrupt acknowledge cycle is similar to opcode fetch but in interrupt acknowledge $\overline{INTA}$ the signal is activated instead of the $\overline{RD}$ signal. The status lines for interrupt acknowledge are $IO/\overline{M} = 1$, $S_1 = 1$ and $S_0 = 1$ (see Table 2.2). During the interrupt acknowledge machine cycle (M1), the RST is coded that initiates 1 byte CALL instruction to the specific vector location provided for INTR. The machine cycle (M_2) and (M_3) are memory write cycle to store the contents of program counter on the stack and new cycle instruction, then, begin.

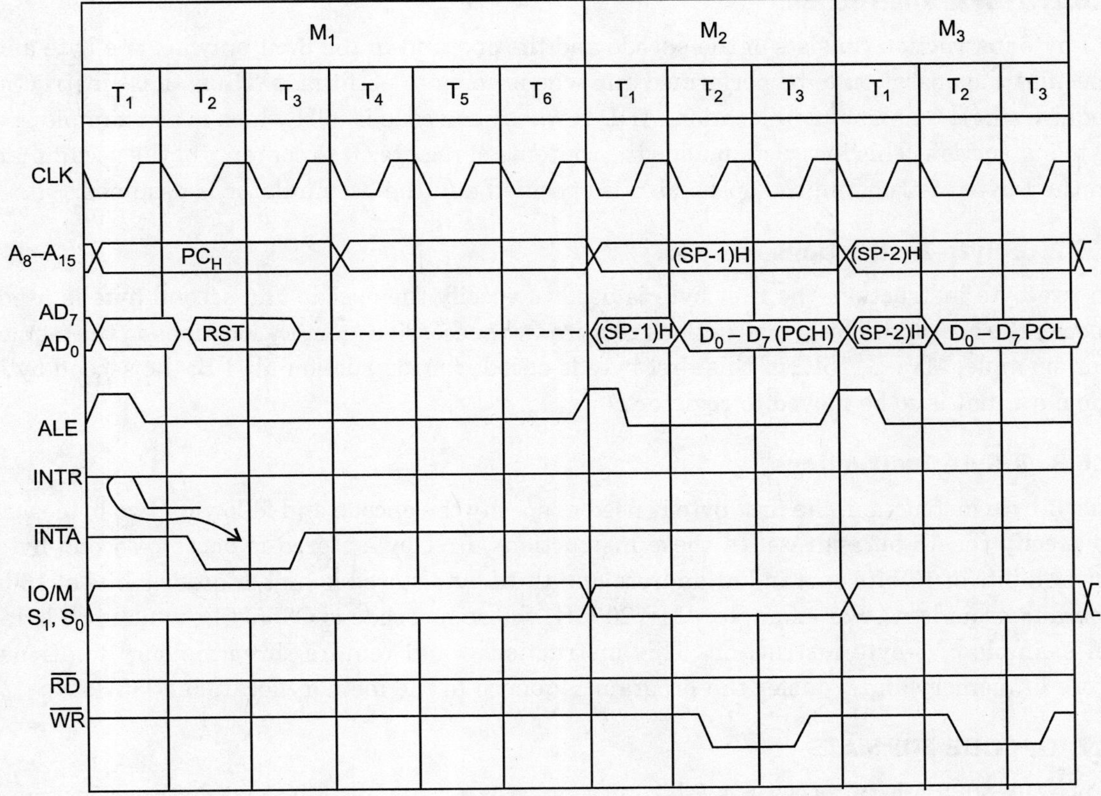

Figure 2.11. Timing diagram of reset (RST) instruction

2.6. INSTRUCTION FORMAT

An instruction is a command to the microprocessor to execute a given task on specific data. The entire group of instructions, called the instruction set, decides what functions the microprocessor can perform. Each instruction has two parts: operation code, called as opcode, and operand. The opcode is a command such as add, subtract, etc. and the operand is an object to be operated on, like a byte or the contents of a register. The operand can be specified in a number of ways. It consists of:

(*i*) 8 bit/16-bit internal general purpose registers.

(*ii*) A memory location.

(*iii*) 8-bit port addresss/16-bit memory address.

(*iv*) Implicit operand: the operand is not specified, instead it is assumed in register.

The instructions are ofthree types depending on the number of bytes or bit patterns required to specify an operation and the data:

(*i*) One byte instructions.

(*ii*) Two byte instructions.

(*iii*) Three byte instructions.

2.6.1. 1-Byte Instructions

A l-byte instruction consists of the opcode and the operand in the 8-bit only i.e. one byte and it specifies the operation to be performed and who is going to perform it. Thus, these instructions require single (one) memory location. ADD B whose hex code is 80H which is an example of one byte instruction. This instruction adds the contents of register B to contents of the accumulator. In the l-byte instruction, the operand B is specified and the accumulator is assumed.

2.6.2. 2-Byte Instructions

In two-byte instruction, the first byte is used to specify the opcode and second byte is used to specify the operand. These instructions require two successive memory locations in the memory. For example, MVI B, 15H, in this first byte is opcode for instruction MVI B; the second byte is the data that is to be moved to register B.

2.6.3. 3-Byte Instructions

In a 3-byte instruction, the first byte is used to specify the opcode and following two byte is used to specify the 16-bit address. In these instructions, first byte stored is opcode, second byte is lower 8-bits of 16-bits operand or address and third byte stored is higher order 8-bits of 16-bits operand or address. For example, JMP 2055H, whose hex code is C35520H, which is 3-byte, is an example of 3-byte instruction. This instruction would require three memory locations to store in memory and transfer the program sequence to the memory location 2055H.

2.7. OPCODE FORMATS

Since the 8085 microprocessor is 8-bit microprocessor, thus has 8-bit·opcode. Each instruction consists a unique opcode. The opcode includes the operation information, type of operation to be performed, and register to be used. Thus the opcode is fixed for each instruction. In the design of the 8085 microprocessor chip, all operations, registers, and status flags are identified with specific code as given in Table 2.3.

Table 2.3 (a) Specific codes of 8085 microprocessor

Code	Registers	Code	Register Pairs
000	B	00	BC
001	C	01	DE
010	D	11	AFORSP
011	E		
100	H		
101	L		
111	A		
110	Reserved for Memory related operation		

Table 2.3 (b)

Information operation

I_0	I_0	I_0	Operation
0	0	0	Read/set interrupt mask
0	0	1	Immediate operation
0	1	0	Load/store
0	1	1	Increment/Decrement
1	0	0	Increment single register
1	0	1	Decrement singleregister
1	1	0	Immediate operation on single register.
1	1	1	Register shifting/Miscellaneous.

Table 2.3 (c)

Arithmetic logic unit operation

AL	AL	AL	Operation
0	0	0	ADD
0	0	1	ADD with carry
0	1	0	SUB
0	1	1	SUB with borrow
1	0	0	Logical AND
1	0	1	X-OR
1	1	0	Logical OR
1	1	1	Compare

Table 2.3 (d)

Branch condition

CB	CB	CB	Operation
0	0	0	JNZ
0	0	1	JZ
0	1	0	JNC
0	1	1	JC
1	0	0	JPO
1	0	1	JPE
1	1	0	JP
1	1	1	JM

Table 2.3 (e)

Branch operation

BO	BO	BO	Operation
0	0	0	Conditional return
0	0	1	Simple return/miscellaneous
0	1	0	Conditional jump
0	1	1	Unconditional jump / miscellaneous
1	0	0	Conditional CALL
1	0	1	Simple CALL/miscellaneous
1	1	0	Special A/L operations
1	1	1	Special unconditional jump

The Fig. 2.12 illustrates, some opcode formats.

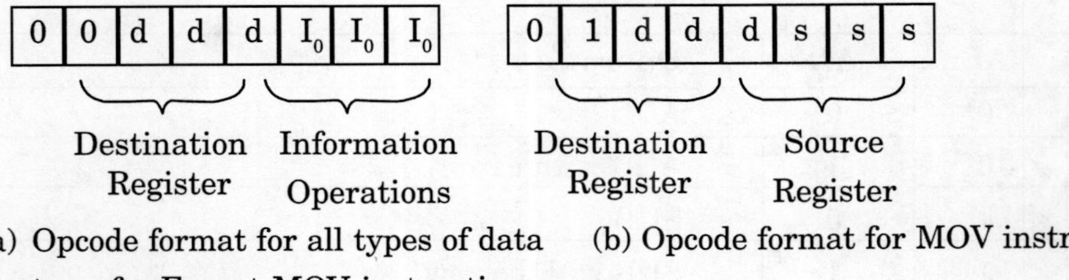

(a) Opcode format for all types of data transfer Expect MOV instruction

(b) Opcode format for MOV instruction

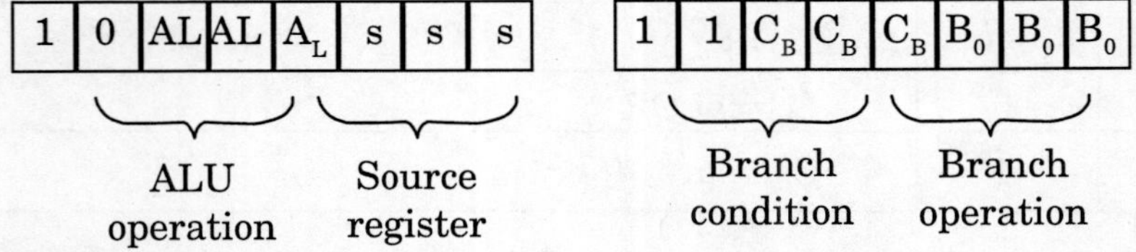

(c) Opcode format for arithmetic and logical group

(d) Opcode format for branch, stack and machine control group.

Figure 2.12. Opcode formats

2.8. ADDRESSING MODES

The various formats for specifying operands are called the addressing modes. The 8085 microprocessor instruction set has the following addressing modes.

- **Immediate addressing :** in the immediate addressing mode, the data (S/16·bit) is specified in the instruction itself and these instructions are either 2·byte or 3·byte long. This type of mode can be understood by the following example. For example, while eating in a hostel mesh, we may make the request to pass the butter. So our request specifies the exact item and the instructions which contain the letter "I" are the examples for immediate addressing modes such as MVI A, Data, LXI H, data, etc.

- **Register Addressing :** In register addressing mode, the source, and destination operands, or both to be contained in general purpose registers of the 8085 microprocessor. Since in this type of addressing mode, it is not necessary to access the memory location for operands, thus has faster execution. This is analogous if we request to pass bowl. Examples are MOV D, B, ADD B, PCHL, etc.
- **Direct Addressing :** In the direct addressing modes, 16-bit address of the operand is specifying within the instruction itself. The instructions in the direct addressing mode are usually 3·byte instructions, in which first byte is opcode, second and third bytes are 16 bit address. It is just like if we request to give the item number 12 on the menu. The examples are LDA location, STA location, etc.
- **Indirect Addressing :** In indirect addressing mode, the memory address, where two operands are located, is specified by the contents of a register pair. This is just like if we order to serve what Mohan has. Examples are, LDAX B,MOVM, A, etc.
- **Implied addressing :** In implied addressing mode, opcode specifies the address of the operands. It is also called as inherent addressing mode. Usually, the implied addressing mode instruction is one byte instruction and data is supposed to be present generally in accumulator. Examples are CMA,RAL, etc.

2.9. THE 8085 MICROPROCESSOR INSTRUCTION SET

An instruction is a binary pattern designed inside a microprocessor to perform a specific function.

The entire group of instructions, called the instruction set, decides what functions the microprocessor can perform. The 8085 microprocessor instruction set, on the basis of the nature of function performed, can be categories in the following groups:

- Data transfer group
- Arithmetic group
- Logic group
- Branch group
- Stack, Input/Output and machine control groups.

2.9.1. Data Transfer Group

One of the primary functions of the 8085 microprocessor is copying data from a register (I/O or memory) called source, to another register (I/O or memory) called destination. The various types of data transfer are possible such as between registers, specific data to register or memory, memory location and register, I/O device and the accumulator, a register pair and stack, etc. The term data transfer is used for copying function. The contents of source are retained without any modification during the transfer or copy operation. The table 2.4 illustrates some data transfer operation.

Table 2.4. Data transfer operation instructions

Instruction	Opcode	Operand	Bytes	M-cycle	T-state	Addressing mode	Descriptions
MVI R, data	MVI	R, data	2	2	7	Immediate	This instruction directly loads a specific register with 8-bit data given within the instruction.
MVIM, data (8) 0	MVI	M, data	2	3	10	Immediate	This instruction directly loads an 8-bit data given within the instruction into memory location specified by the contents of HL register pair.
MOV R_d, R_s	MOV	Rd, Rs	1	1	4	Register	This instruction copies data from the source register (R_s) into destination register (R_d). The Rd and R_s are general purpose registers like A, B, C, D, E, H, and L. The contents of source register remain unchanged after execution of the instruction.
MOVM,R	MOV	M,R	1	2	7	Indirect	This instruction copies data from the source register into memory location specified by the contents of HL register pair.
MOVR,M	MOV	R,M	1	2	7	Indirect	This instruction copies data from the memory location specified by the contents ofHL register pair into destination register.
LXIR_p, data 16-bit	LXI	Rp, data	3	3	10	Immediate	This instruction loads 16-bit data specified within the instruction on the specified register pair or the stack pointer.
STAaddress	STA	Address	3	4	13	Direct	This instruction stores the contents of register A into the memory location whose address is specified within operand. Content of register remains unchanged.

(Contd ...)

LDA address	LDA	Address	3	4	13	Direct	This instruction copies the contents of memory location whose address is specified within operand to the accumulator. Contents of memory location remain unchanged.
SHLD address	SHLD	Address	3	5	15	Direct	This instruction stores the contents of register L into the memory location whose address is specified by 16-bit address within the operand and content of register H are stored in the next memory location by incrementing the operand. The contents of register pair remain unchanged.
LHLD address	LHLD	Address	3	5	16	Direct	This instruction copies the contents of the memory location whose address is specified within the instruction into register Land content of next memory into the register H.
STAXR_p	STAX	RP	1	2	7	Register indirect	This instruction copies the contents of accumulator to memory location whose address is specified by the register pair. RP is either BC or DE pair only.
LDAXR_p	LDAX	RP	1	1	4	Register	This instruction copies the contents of memory location whose address is specified by the register pair to the accumulator. RP is either BC or DE pair only.
XCHG	XCHG		1	1	4	Register	This instruction exchanges the contents of the register H with D and of register L with E.

Example 2.1. Write a program to exchange the contents of memory locations 2155H and 3244.

Solution :

Instruction	Comments
LDA 2155H	; Get the content of memory location 2155H into accumulator.
MOVB, A	; Move the content from register A to register B.
LDA 3244H	; Get the content of memory location 2155H into accumulator.
STA 2155H	; Store the content of accumulator at address 2155H.
MOV A, B	; Move the content from register B to register A.
STA 3244H	; Store the content of accumulator at address 3244H.
HLT	; Stop the execution.

Example 2.2. Write a program to load data A6 into accumulator and then transfer the same data to B register and H register.

Solution:

Instruction	Comments
MVI A, 05 H	; Initialize A register = 05 H
MOV B, A	; Transfer the content from register A to register B.
MOV H, A	; Transfer the content from register A to register H.
HLT	; Stop the execution.

2.9.2. Arithmetic Group

The 8085 microprocessor performs various arithmetic operations like addition, subtraction, increment and decrement etc.

- **Addition :** In the addition group of instructions, the contents of a register or the contents of a memory location or any 8-bit number can be added to the contents of accumulator and the sum is stored in the accumulator. The direct addition of the contents of a two 8-bit number is not possible
- **Subtraction :** In the subtraction group of instructions, the contents of a register or the contents of a memory location or any 8-bit number can be subtracted from the contents of accumulator and the results stored in the accumulator. The subtraction is performed in 2's complement, and the results if negative, are expressed in 2's complement. The direct subtraction of the contents of a two 8-bit number is not possible
- **Increment/Decrement :** In the increment/decrement group of instructions, the contents of a register or the contents of a memory location or any 8-bit number can be incremented or decremented by 1. Similarly, the 16-bit contents of a register pair like Be, DE, and HL, can be increment or decrement by 1. These increment/decrement operations differ from the addition and the subtraction in an important way that these can be performed in anyone of the registers or in a memory location. Table 2.5 illustrates some arithmetic group of instructions.

Table 2.5. Arithmetic operation instructions.

Instruction	Opcode	Operand	Bytes	M-cycle	T-state	Addressing mode	Descriptions
ADDR	ADD	R	1	1	4	Register	This instruction adds the contents of specificregister to the contents of accumulator and the sum is stored in the accumulator. The register R are general purpose registers like A, B, C, D, E, H, andL.
ADDM	ADD	M	1	2	7	Direct	This instruction adds the contents of memory location specified by the contents of HL register pair to the contents of accumulator and the sum is stored in the accumulator.
ADD Data	ADD	Data	?	2	7	Immediate	This instruction adds the 8-bit data specified to the contents of accumulator and the sum is stored in the accumulator.
ADCR	ADC	R	1	1	4	Register	This instruction adds the contents of register R along with carry to the contents of accumulator and the sum is stored in the accumulator.
ADCM	ADC	M	1	2	7	Direct	This instruction adds the contents of memory location specified by the contents of HL register pair to the contents of accumulator and the sum is stored in the accumulator.
ACI data	ACI	Data	2	2	7	Immediate	This instruction adds the 8·bit data specified and contents of carry to the contents of accumulator and the sum is stored in the accumulator.
$DADR_P$	DAD	R_P	1	3	10	Register	This instruction adds the contents of specified register pair to the contents of HL register pair and the sum is stored in HL pair. Register pair RP is 16-bit register pair like BC, DE, HL or stack pointer

(Contd ...)

Subtraction Group

SUBR	SUB	R	1	1	4	Register	This instruction subtracts the contents of specific register R from the contents of accumulator and the result is stored in the accumulator. The register R are general purpose registers like A, B, C, D, E, H, and L.
SUBM	SUB	M	1	2	7	Direct	This instruction subtracts the contents of memory location specified by the contents of HL register pair from the contents of accumulator and the result is stored in the accumulator.
SUI data	SUI	Data	2	2	7	Immediate	This instruction subtract the 8-bit data specified from the contents of accumulator and the result is stored in the accumulator.
SBBP	SBB	R	1	1	4	Register	These instructions subtract the contents of register R and borrow flag from the contents of accumulator and the result is stored in the accumulator.
SBBM	SBB	M	1	2	7	Direct	These instructions subtracts the contents of memory location specified by the contents of HL register pair and borrow flag from the contents of accumulator and the result is stored in the accumulator.
SBI data	SBI	Data	2	2	7	Immediate	This instruction subtracts the 8-bit data specified and contents of borrow from the contents of accumulator and the result is stored in the accumulator.
DAA	DAA		1	3	10	Implied	This instruction is used to make sure that the result of adding two packed BCD format is BCD number. It operates only on A register. If the value of low-order four bits ($D_3 - D_0$) in accumulator is greater than 9 or if auxiliary carry flag is set, the instruction adds 6 to the low order four bits. If the value of higher order bit (D7--D4) in the accumulator is greater than 9 or if auxiliary carry flag is set, the instruction adds 6 to the higher-order four bits.

(Contd...)

Increment and Dect:ement Group

INRR	INR	R	1	1	4	Register	These instructions increment the contents of register R by 1 and the result is stored in the register.
INRM	INR	M	1	3	10	Direct	These instructions increment the contents of memory location specified by the contents ofHL register pair by 1 and the result is stored back at the same register.
$INXR_P$	INX	Rp	1	1	6	Register	This instruction increments the contents of specified register pair by 1 and the result is stored in same register pair. Register pair RP is 16-bit register pair like BC, DE, HL or stack pointer
DCRR	DCR	R	1	1	4	Register	These instructions decrement the contents of register R by 1 and the result is stored in the same register.
CDRM	DCR	M	1	3	10	Direct	These instructions decrement the contents of memory location specified by the contents ofHL register pair by 1 and the result is stored back at the same register.
$DCXR_P$	INX	R_P	1	1	6	Register	This instruction decrements the contents of specified register pair by 1 and the result is stored in same register pair. Register pair Rp is 16-bit register pair like BC, DE, HL or stack pointer.

Example 2.3. Write a program to add two 8-bit numbers.

Solution:

Instruction	**Comments**
LXI H, D111 H	; HL points to the memory location D111 H
MOVA,M	; Transfer the contents oflocation D111 H into the accumulator.
INXH	; Increment HL to point to next memory location i.e. D112H
ADDM	; A = A +M i.e. perform the addition
INX H	; Increment HL to point to next memory location i.e. D113H
MOVM, A	; The result in accumulator is transfer at memory location i.e. D113H
HLT	; Stop the execution.

Example 2.4: Write a program to subtract two 8-bit numbers.

Solution:

Instruction	**Comment**
LXI H, DOOOH	; HL points to the memory location D111 H
MOVA,M	; Transfer the contents oflocation D111 H into the accumulator
INXH	; Increment HL to point to next memory location i.c. D112H
SUBM	; A = A – M i.e. perform the subtraction
INXH	; Increment HL to point to next memory location i.e. D113H
MOVM,A	; The result in accumulator is transfer at memory location i.e. D II3H
HLT	; Stop the execution.

Example 2.5. Write a program to add two I6-bit numbers.

Solution :

Instruction	**Comment**
LXI H, DOOO H	; HL points to the memory location DOOO H
MOVA,M	; Get the contents oflocation DOOO H into the accumulator.
INXH	; Increment HL to point to next memory location i.e. DCOlH
MOVB,M	; Get the contents DOOI H into register B i.e. MSB of first number.
INX H	; Increment HL to point to next memory location i.e. D002H
MOVC,M	; Get the contents D002 H into register C i.e. LSB of 2nd number.
INXH	; Increment HL to point to next memory location i.e. D003H
MOVD,M	; Get the contents D003 H into register C i.e. MSB of 2nd number.
SUBC	; Compute LSB subtraction A = A – C
MOVL,A	; Store the result in register L.
MOVA,B	; Copy the contents of MSB of first number to the accumulator.
SUBD	; Compute MSB subtraction A = A –D
MOVH,A	; Store the result in register H.
SHLD 2504H	; Store the result at memory locations 2504 Hand 2505 H
HLT	; Stop the execution.

2.9.3. Branch Group

The branch instructions are most powerful instructions because either unconditionally or under certain conditions they allow the microprocessor to change the sequence of the program. The 8085 microprocessor is a sequential machine, i.e. it executes machine codes from memory location to the next. The branch instructions instruct the microprocessor to go to a different memory location, and the microprocessor continues executing machine codes from that new location. The address of this new memory location is either specified explicitly or supplied by microprocessor or by extra hardware. These instructions are the key to the flexibility and versatility of a computer. The branch instructions are classified in three categories such as Jump instruction, call, return instructions, and Restart instructions as shown in Table 2.6.

2.9.4. Logic Operation

The 8085 microprocessor is basically a programmable logic chip and can perform all the logic functions of the hard-wired logic through its instruction set. All logic operations are performed in relation to the contents of the accumulator and result is stored in accumulator. The instructions of all logic operations are illustrated in Table 2.7.

2.10. ASSEMBLY LANGUAGE PROGRAMMING

The 8085 microprocessor is capable of executing a few operations at very high speed. The instruction set discussed in previous section shows that microprocessor can transfer data into and out of the CPU, perform logical and arithmetic operation, and can make decisions based on the results. However, the microprocessor cannot perform simple calculation or data transfer by itself, it must be told in clear terms, in complete details, what to do and in what order. The computer hardware consists of the physical components connection and logic circuits that participate in decoding and execution of operation codes. A program is a sequence of instructions written to explain a computer to perform a specific function. The instructions are selected from the instruction sets of the microprocessor.

2.10.1. Programming Steps

A program has following steps to write an assembly language programming:

Step I : Define the problem to be solved

The problem for which you have to write a program must be clearly specified such as what are the inputs, what is the operation to be executed, and where have store the results otherwise you can't start writing a program.

Step II: Solution plan

To achieve a specified result you must have a solution plan i.e. how you are taking inputs, which method you are opting and what are steps involved in this method, and how will you output the results. Thus you must have a proper solution program ..

Step III : Flow chart

The steps involved in the problem analysis and the sequence can be represented in block diagram, called flowchart. In other words, the flowchart is a pictorial representation of various actions and computations that are taken to perform a specific task. Flowchart is used to assist and clarify the thinking process and communicate the program logic to others. With the program structure clearly defined in blocks, the flowchart can be converted directly into a set of instructions, using any programming language, Figure 2.13 illustrates the commonly used symbols in flowchart.

Table 2.6. Branch group instruction.

Jump instruction								
Instruction	**Condition for jump**	**Opcode**	**Operand**	**Bytes**	**M-cycle**	**T-state**	**Addressing mode**	**Descriptions**
JMP address	Un-conditional	JMP	16-bit Address	3	3	10	Immediate	This instruction loads the PC (Program Counter) with the address given within the instruction and resumes the program execution from this location.
JC Addrs JNCAddrs JP Addrs JMAddrs JPEAddrs JPOAddrs JZAddrs JNZAddrs Addrs= address	CY = 1 Cy = 0 8 = 0 8 = 1 P = 1 P = 0 Z = 1 Z = 0	JC JNC JP JM JPE JPO JZ JNZ	16 bit 16-bit 1G-oit 16-bit 16-bit 16-bit 16-bit 16-bit Address	3 3 3 3 3 3 3 3	2 (if condition is not true) and 3 if condition is true	7	Immediate	Jump on carry Jump on carry Jump on positive Jump on minus Jump on parity even Jump on parity odd Jump on zero Jump on not zero In conditional JUMP instruction, the jump is made only when condition is true or satisfied. It will just check and further execute the next instruction after it if condition is false or not satisfied.
PCHL	PCHL		None	1	1	6	Register	This instruction transfers the contents of Hand L registers to program counter. The content of Hand L registers transfer to high-order and low order, respectively. This instruction is equivalent to a 1 byte unconditional JUMP instruction;HL register pair provides the address of jump. This means that the program sequence is transferred to address specified by HL register pair.

(Contd...)

CALL Instruction

Instruction	Condition for call	Opcode	Operand	Bytes	M-cycle	T-state	Addressing	Descriptions mode
CALL address	Un-conditional	CALL	Address	3	5	18	Immediate	This instruction is used to transfer the program control to a subprogram or sub-routine, i.e. whenever this instruction is executed, the program control is transferred to the address specified in the instruction.
CZ	Z=1	CZ	Address	3	2 if condition is	9	Immediate	CALL on zero
CNZ	Z=O	CNZ	Address	3	not true) and			CALL if not zero
CP	8=0	CP	Address	3	5 if condition is			CALL if positive
CPO	P=O	CPO	Address	3	true			CALL if parity odd
CPE	P=1	CPE	Address	3				CALL if parity even
CM	8=1	CM	Address	3			18	CALL on minus
CC	Cy= 1	CC	Address	3				CALL if carry
CNC	CY=O	CNC	Address	3				CALL if no carry
								The conditional CALL instruction calls the subroutine at a given address if a specified condition is true. Before CALL it stores the address of the instruction next to the call on the stack and decrements stack pointer by two.

RET Instructions

Instruction	Condition for call	Opcode	Operand	Bytes	M-cycle	T-state	Addressing mode	Descriptions
RET	Un-conditional	RET	None	1	3	10	Register indirect	This instruction transfers the program control from the subroutine to calling (main) program. The return address is popped/ taken from stack (where the call instruction has stored its PC contents i.e. return address) this address is loaded in PC and the program execution begins at address taken from stack.
RZ	Z=1	RZ	Address	1	(if condition is	6	Immediate	Return on zero
RNZ	Z=O	RNZ	Address	1	not true) and 3			Return if not zero
RP	S = 0	RP	Address	1	if condition is	12		Return if positive
RPO	P = 0	RPO	Address	1	true			Return if parity odd
RPE	P = 1	RPE	Address	1				Return if parity even
RM	8 = –1	RM	Address	1				Return on minus
RC	CY = 1	RC	Address	1				Return if carry
RNC	CY = O	RNC	Address	1				Return if no carry These instructions return the control to the main program if specified condition is satisfied.

RET Instructions

Instruction	Opcode	Operand	Bytes	M-cycle	T-state mode	Addressing	Descriptions
ANAR	ANA	R	1	1	4	Register	This instruction logically ANDed the content of accumulator with the contents of the operand specified in the register and the result is stored in accumulator.
ANAM,	ANA	M	1	2	7	Register	This instruction logically ANDed the content of accumulator with the contents of the memory location specified by the HL register pair and the result is stored in accumulator.
ANI, Data	ANI	Data	2	2	7	Immediate	This instruction logically ANDed the content of accumulator with the 8·bit data specified in the instruction and the result is stored in accumulator.
XORR	XOR	R	1	1	4	Register	This instruction logically XORs the content of accumulator with the contents of the operand specified in the register and the result is stored in accumulator.
XORM	XOR	M	1	2	7	Indirect	This instruction logically XORs the content of accumulator with the contents of the memory location specified by the HL register pair and the result is stored in accumulator.
XRI data	XRI	Data	2	2	7	Immediate	This instruction logically XORs the content of accumulator with the 8·bit data specified in the instruction and the result is store in accumulator.
ORAR	ORA	R	1	1	4	Register	This instruction logically ORs the content of accumulator with the contents of the operand specified in the register and the result is stored in accumulator.
ORAM	ORA	M	1	2	7	Indirect	This instruction logically ORs the content of accumulator with the contents of the memory location specified by the HL register pair and the result is stored in accumulator.

ORI data	ORI	Data	2	2	7	Immediate	This instruction logically ORs the content of accumulator with the 8-bit data specified in the instruction and the result is stored in accumulator.
CMPR	CMP	R	1	1	4	Register	This instruction compares the content of specified register with. the contents of the accumulator using subtraction and sets the condition flags as a result of subtraction.
CMPM	CMP	M	1	2	7	Indirect	This instruction compares the content of accumulator with the contents of the memory location specified by theHL register pair and sets the condition flags as a result of subtraction. The comparison is done using subtraction.
CPI data	CPI	Data	2	2	7	Immediate	This instruction compa-res the content of accumulator with the 8-bit data specified in the instruction and sets the condition flags as a result of subtraction. The comparison is executed using subtraction.
STC	STC	None	1	1	4	Implied	This instruction sets the carry flags.
CMC	CMC	None	1	1	4	Implied	This instruction inverts the value of the carry flag.
CMA	CMA	None	1	1	4	Implied	This instruction inverts the value of contents of the accumulator. The result of this instruction is complement and complement is performing using a NOT operation with each bit.
RLC	RLC	None	1	1	4	Implied	This instruction rotates all the bits of the accumulator by one bit position to the left. The bit moves out of the MSB is circled back into the LSB. The data bit moves out of MSB is also copied into carry flag.
RAR	RLC	None	1	1	4	Implied	This instruction rotates all the bits of the accumulator by one bit position to the right. The bit moves out of the LSB is rotated around the MSB. The data bit moves out of LSB is also copied into carry flag.
RAL	RAL	None	1	1	4	Implied	This instruction rotates all the bits of the accumulator by one bit posrtir-n to the left along with carry. Bit B_0 is placed in the carry flag and carry is placed in LSB B_7.
RAR	RAR	None	1	1	4	Implied	This instruction rotates all the bits of theaccumulator by ono bit position to the right along with carry. Bit B_0 is placed in the carry flag and carry is placed in bit B_7.

Stack Control Group

SPHL	SPHL	None	1	1	4	Register	This instruction transfers the contents of Hand L registers into the stack pointer register. The contents of Hand L registers transfer to high-order and low order bytes of stack pointer, respectively.
PUSH R_p	PUSH	R_p	1	3	12	Register	This instruction copies the contents of the specified register pair into the stack pointer in the following sequence: the stack pointer is decremented by 1 and the content of higher order register of register pair are copied into that position. The stack pointer is again decremented by 1 and the content of lower order register of register pair are copied into that position.
PUSHPSW	PUSH	PSW	1	3	12	Indirect	This instruction stores the contents of the Program Status Word and accumulator into the stack pointer in the following sequence: the stack pointer is decremented by 1 and the content of accumulator is copied into the memory location specified by stack pointer. The stack pointer is again decremented by 1 and the flag register is copied into the memory location specified by stack pointer.
POP R_p	POP	R_p	1	3	12	Register	This instruction copies the contents of the memory location specified by the stack pointer into low order register. The stack pointer is incremented by 1 and the contents of that memory location are copied to higher order register. The stack pointer is again incremented by 1. The process of POP is exactly opposite to PUSH. Thus the contents stored by PUSH are taken back using POP instruction.
POPPSW	POP	PSW	1	3	12	Indirect	This instruction copies the contents of the memory location specified by the stack pointer into flag register. The stack pointer is incremented by 1 and the contents of that memory location are copied to accumulator. The stack pointer is again incremented by 1.
XTHL	XTHL	None	1	5	16	Register	This instruction exchanges the contents of memory location specified by stack pointer with contents of L register and the contents of the next memory location with the contents of H register. The contents of stack pointer remain unaltered.

Stack, Input/Output and machine control groups

Instruction	Opcode	Operand	Bytes	M-cycle	T-state	Addressing mode	Descriptions
EI	EI	None	1	1	4	None	This instruction sets the interrupt enable flip- flop to enable all maskable interrupts. The interrupt enable flip-flop is reset to re-enable the interrupts after an interrupt is acknowledged.
DI	DI	None	1	1	4	None	This instruction resets the interrupt enable flip- flop to disable all maskable interrupts.
SIM	SIM	None	1	1	4	None	This instruction enables the desired interrupt and also sends out serial data through SOD pin. For this instruction command byte should be loaded in accumulator.
RIM	RM	None	1	1	4	None	This instruction copies the status ofthe interrupt into the accumulator and also reads serial data through SID pin.
NOP	NOP	None	1			None	No operation is performed by the microprocessor upon the receipt of this instruction.
HALT	HALT	None	1			None	This instruction halts the microprocessor.
NOP	NOP	None	1			None	No operation is perforD1.ed by the :microprocessor upon the receipt of this instruction.
HALT	HALT	None	1			None	This instruction halts the microprocessor.
R8TN	R8T	N					This instruction transfer::. the program control to the specific memory location as given below. This instruction is like a fixed address CALL instruction. These addresses are also known as vector addresses and these are calculated by multiply 8 to the RST number. Just like Call instruction RST instruction saves the current program counter contents on the stack before transferring the program control to the instruction following the vector address.

							Instruction code / Vector address RSTO — O × 8 = 0000H RSTI — 1 × 8 = 0008H RST2 — 2 × 8 = 0010H RST3 — 3 × 8 = 0018H RST4 — 4 × 8 = 0020H RST5 — 5 × 8 = 0028H RST6 — 6 × 8 = 0030H RST7 — 7 × 8 = 0038H
IN address	IN	8-bit port address	2	3	10	Direct	This instruction transfers the data at the port address whose address is specified in the instruction into the accumulator.
PUT address	OUT	8-bit port address	2	3	10	Direct	This instruction sends the contents of the accumulator to the output port whose address is specified in the instruction.

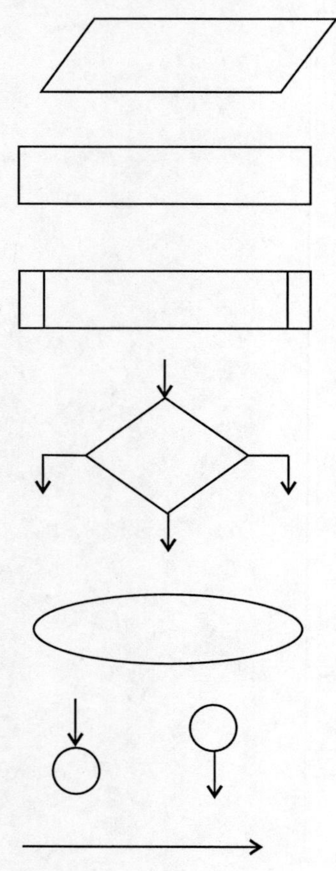

Parallelogram-is used for input/output units.

Rectangle-is used to indicate process operation.

Double Sided Rectangle-is used to indicate predefined processes i.e., subroutines operation.

Diamond Unit-is used to indicate decision unit.

Oval-is used to indicate start and stop unit.

Circle with Arrow-is used to represent continuation.

Arrow-is used to indicate the direction of flow.

Figure 2.13. Flowchart symbols

Step IV: Program

Once a flowchart has been prepared, and then put on the instructions in place of flowchart blocks.

Step V: Debugging

After writing the program the next step is debugging the program. Debugging is basically the process of checking the program. Find the codes for instruction feed in 8085 system and execute the program. Then check the obtained result. If it is correct then OK otherwise you have to debug your program to find error.

2.11. HOW TO WRITE, ASSEMBLE, AND EXECUTE A SIMPLE PROGRAM

For example write a simple program to store 8-bit data in registers.

Problem Statement

Write a program to load data 06 H into accumulator then transfer the same data to register B and H.

Explanation:

- We have immediate data 06 H. To load this data into the accumulator we will use the MVI A, 06 H instruction.
- To transfer the data to B register we will use the instruction MOV B, A.
- To transfer the data to H register we will use the instruction MOV H, A.

Problem analysis

Step I : Load the accumulator with immediate data 06 H.

Step II : Copy the contents of accumulator to the register B.

Step III : Copy the contents of accumulator to the register H.

Flowchart : Flowchart is drawn below representing the above steps

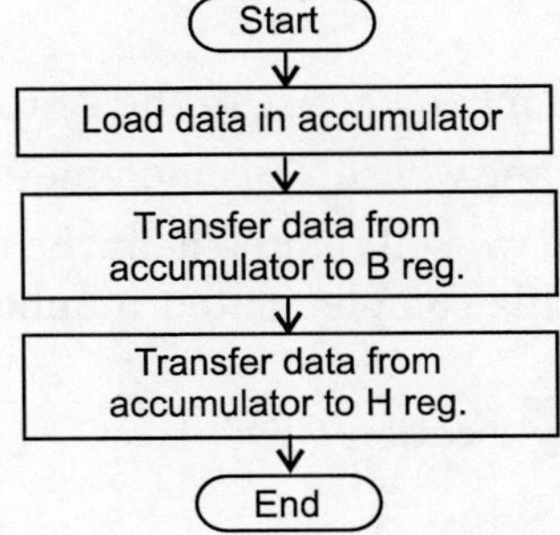

Assembly Language Program: To write an assembly language program, we need to translate the blocks shown in flowchart above into 8085 operation and then subsequently into mnemonics. To translate these steps into assembly and machine languages, you should review the instruction set. Translation of each block into mnemonics with comments is shown follow :

	Instruction	Comments
Block 1:	MVI A, 06 H	; Load register A with 06 H.
Block 2:	MOV B, A	; Copy the contents of Accumulator into register B.
Block 3:	MOV H, A	; Copy the contents of Accumulator into register H.
Block 4:	HLT	; Stop the execution.

To convert the mnemonics into Hex code, we need to see the code in the 8085 instruction set: this is called either manual or hand assembly.

Mnemonics	Hex code
MVIA, 06 H	3E
	06
MOVB, A	47
MOVH,A	67
HLT	76

Storing the hex code in the memory : The microprocessor trainer kit has a key pad to enter the hex code in the memory. Typical steps to enter the program:

(*i*) Reset the system by pushing the RESET key.

(*ii*) Enter the first memory address where the program should be stored.

(*iii*) Enter each machine code by pressing hex keys.

(*iv*) Repeat step 3 until the last machine code, 76H.

Executing the program: To execute the program, we need to instruct the microprocessor where the program begins i.e. we have to enter starting address of the program. Now we press the Execute key to begin the execution. As soon as the Execute function key is pressed, the microprocessor loads the starting memory of the program in the program counter and the program control is transferred from the Monitor program to our program.

Example 2.7. Find the 2.'s complement of a number.

Solution: Explanation:

- 2's complement of number means add 1 to the l's complement of that number.
- We will first load the number whose 2's complement is to be found in the accumulator.
- Using the CMA instruction we will complement the accumulator. This is l's complement of the number. Add 1 to this complemented number to get the 2's complement of the number.
- Store the result at memory location ClOO H.

Assembly Language Program:

Instruction	**Comments**
LDA C200H	; Get the number.
CMA	; Complement the number.
ADI,OlH	; Add 01 in the number.
STA, ClOOH	; Store the result.
HLT	; Stop the execution.

Example 2.8. Write a program to find l's complement of the number.

Solution: Explanation:

- One's complement of number means to invert each bit of that number. So our task is to complement each bit of number.
- We will first load the number whose 1's complement isto be found in the accumulator.
- Using the CMA instruction we will complement the accumulator. This is l's complement of the number.
- Store the result at memory location ClOO H.

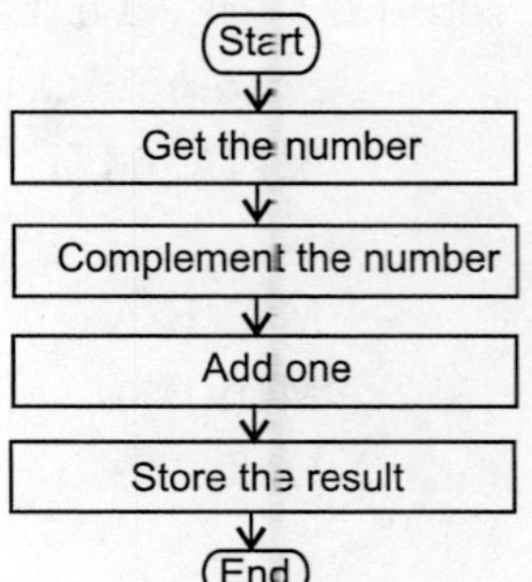

Assembly Language Program:

Instruction	**Comments**
LDA C200H	; Get the number
CMA	; Complement number

```
STA C100H             ; Store the result
HLT                   ; Stop the execution.
```

Example 2.9. Add two 8-bit BCD numbers.

Solution: Explanation:

- We are given two BCD numbers at memory locations DOOO Hand D001 H. Let these numbers be 02 Hand 09 H.
- We have to add the byte at memory location DOOO H (02 H) with the byte at memory location D001 H (09 H).
- Initially we will store the first number in the accumulator.
- Using ADD instruction add the two contents. Result of addition will be stored in the A register.
- Using DAA adjust the result to valid BCD number.
- Store the result at memory location D002 H.

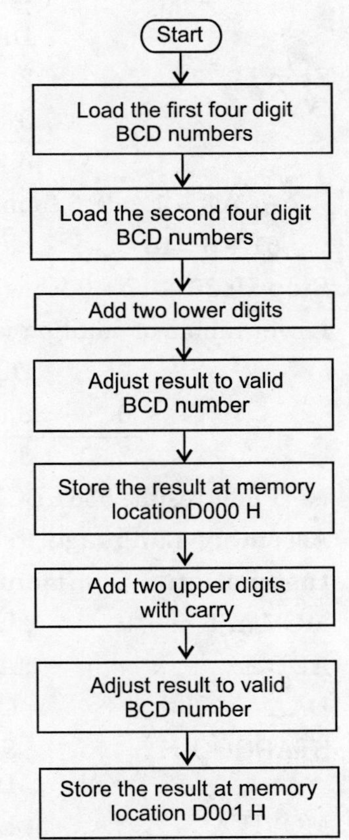

For example : DOOO H = 02 02 H

H

DOOI H = 09 + 09 H

H

DAA OB H (invalid BCD)

+06H

11 H (valid BCD)

Assembly Language Program:

Instruction	**Comment**
LXIH, DOOOH	; HL points to the memory Location DOOO H
MOVA,M	; Get the contents of location DOOO H into the accumulator.
INX H	; Increment HL to point to next memory location i.e. DOO1H
ADDM	; A = A + M i.e,' compute the addition
DAA	; Convert the HEX to valid BeD result.
INXH	; Increment HL to point to next memory location i.e. D002H
MOVM,A	; Store the result at memory location D002H
HLT	; Stop the execution.

Example 2.10. Add two 16-bit BCD numbers.

Solution: Explanation:

- Consider that two words are available in register pairs HL and DE. We have to add these two words.
- Using add instruction, add the contents of the lower two bits i.e. add Land E.
- The result of this addition is stored in the A register.
- DAA instruction is then used to convert the result to valid BCD. Store the result at memory location DOOO H.
- Now, add the contents ofMSB along with carry if generated in LSB addition.
- The result of MSB addition is stored in the A register. Adjust this result tc valid BCD number. The final result isthen stored at the memory location D001 H.

e.g.: HL = 3629,BCD

DE = 4738 BCD'

Step I

```
   2  9
+  3  8
   ----
   6  1   and auxiliary carry = 1
```

As AC = 1, add 6 to make result valid.

61 + 6 = 67

Step II: 36 + 47 + 0 (Carry of LSB) = 7 D.

Lower nibble of addition is greater than 9, so add 6.

```
   7  D
+  0  6
   ----
   8  3
```

Final result: 8367 BCD.

Assembly Language Program:

Instruction	**Comment**
MOV A, L	; Load the LSB of first number in A
ADD E	; A = A +E i.e. compute the LSB addition
DAA	; Convert the HEX to valid BCD result.
STADOOO H	; Store result of LSB addition at location D000 H
MOVH,A	; Load the MSB of first number in A
ADCD	; A = A +D + CY i.e. compute the MSB addition
DAA	; Convert the HEX to valid BCD result.
STADOOI H	; Store result of LSB addition at location D00l H
HLT	; Stop the execution.

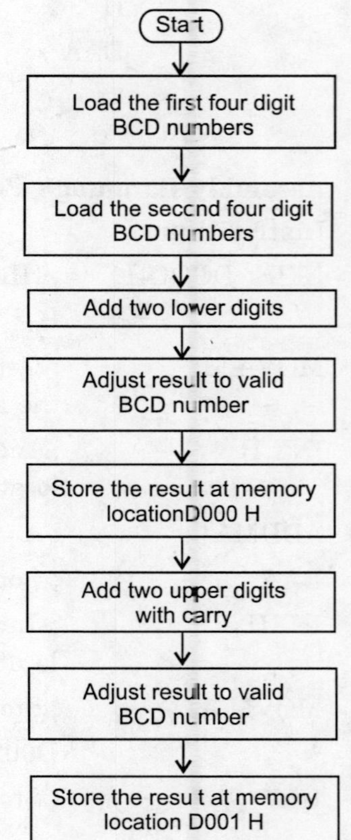

Program 2.11. Calculate the sum of series of numbers.

Solution: Explanation:

- We are given a series of numbers. The length of series is stored at memory location DOOO H. We will initialise register C as counter with the length of the series.
- We will initialise the accumulator with OOH, so that the sum can be stored in the accumulator.
- The series begins at DOOI H. So, we will initialise HL the register pair as memory pointer to point the series.
- Using add instruction; add the contents of the accumulator with the contents of memory location pointed by HL register pair. The result of this addition will be stored in the A register.
- Then we will increment HL to point to next memory location of the series. Decrement the count in register C. Continue this process till the count is zero i.e. all the numbers in the series are added.
- Store the result at memory location E000 H.

Example: Let D000 = 05 H i.e. series is of 5 numbers.

D001 = 08H

D002 = I2H

D003 = 74H

D004 = 34H

D005 = 04H

Result = 08 + 12H + 74 H + 34 H + 04 H = C6H

E000 H = C6H.

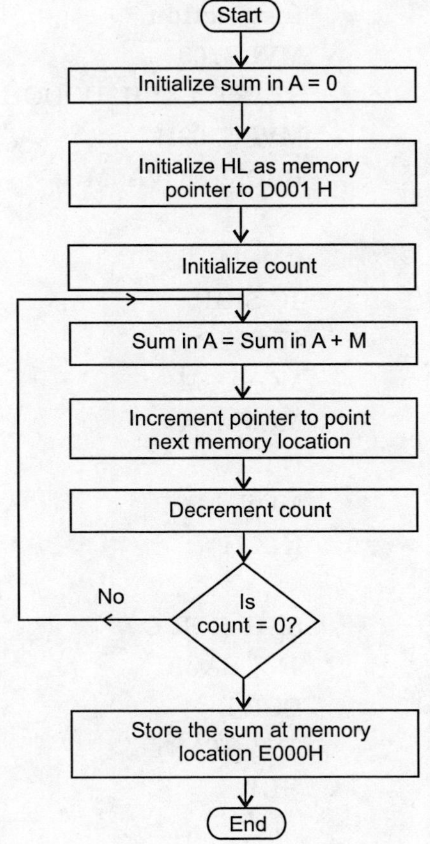

Assembly Language Program:

Instruction	**Comment**
LDA D000	; A = contents of location D000 H
MOV C, A	; Initialize counter
XRAA	; sum = 0
LXI H, D001H	; Initialize HL as memory pointer for the series
L1: ADD M	; SUMinA = SUMinA+M
INX H	; Increment pointer
DCR C	; Decrement counter
JNZ L1	; If counter 0 repeat
STA E000H	; Store the result at memory location E000H
HLT	; Stop the execution.

Example 2.12. Write a program to sort the numbers in ascending order.

Solution: Explanation:

- Consider that a block of N words is present. Now we have to arrange these N words in ascending order, Let N = 4 for example. We will use HL as pointer to point the block of N words.
- Initially in the first iteration we compare first number with the second number. If first number < second number, don't interchange the contents, otherwise if first number> second number swap the contents.
- In the next iteration we go on comparing the first number with third number. If first number < third number, don't interchange the contents. If first number> third number then swapping will be done.
- Since the first two numbers are in ascending order the third number will go to first place, first number in second place and second number will come in third place in the second iteration only if first number> third number.
- In the next iteration first number is compared with fourth number. So comparisons are done till all N numbers are arranged in ascending order. This method requires approximately N comparisons.

Assembly Language Program:

Instruction	**Comment**
MVI B, 09	; Initialize counter 1
START: LXI H, DOOOH	; Initialize memory pointer
MVI C, 09H	; Initialize counter 2
BACK: MOVA, M	; Get the number in accumulator
INX H	; Increment memory pointer
CMPM	; Compare number with next number
JC SKIP	; If less, don't interchange.
JZ SKIP	; If equal, don't interchange.
MOVD, M	; Otherwise swap the contents
MOVM, A	;
DCXH	; Interchange numbers
MOVM,D	;
INX H	; Increment pointer to
	; next memory location
SKIP:DCR C	; Decrement counter 2
JNZ BACK	;1f not zero, repeat
DCR B	; Decrement counter 1
JNZ START	; If not zero, repeat
HLT	; Terminate program execution.

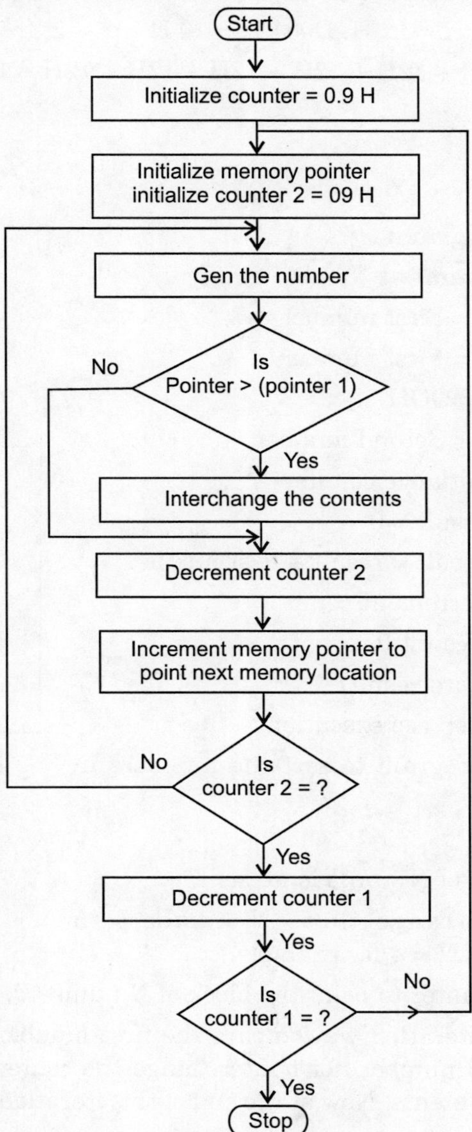

Example 2.13. Multiply two 8-bit numbers.

Solution: Explanation:

- Consider that a byte is present at the memory location DOOO Hand second byte is present at memory location D001 H.
- We have to multiply the bytes present at the abovetwo memory locations. "
- We will multiply the numbers using successive addition method.
- In successive addition method, one number is accepted and other number is taken as a counter.The first number is added with itself, till the counter decrements to zero.
- Result is stored at memory locations E000 H and E001 H.

For example: D000 H = 12 H, D001 H = 10 H

Result = 12H + 12H + 12H + 12H + 12H + 12H + 12H + 12H + 12H + 12H

Result = 0120 H

E000 H =20 H

E001 H = 01 H .

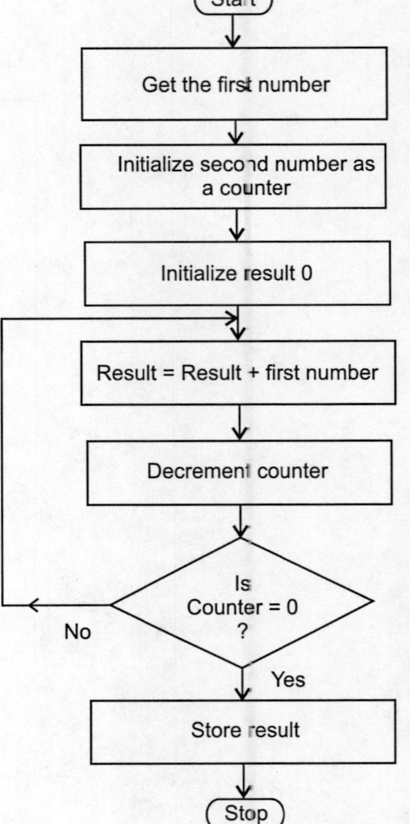

Assembly Language Program:

Instruction	Comment
LDA2500H	; A = First number
MOVE,A	; E = First number
MVID,OO	; D =OOH
LDA 2501H	; A = Second number
MOV C,A	; Initialize counter
LXI H, OOOOH	; Result = 0
BACK:DADD	; Result = result + first number
DCRC	; Decrement count
JNZBACK	; If count 0 repeat
SHLD 2550H	; Store result
HLT	; Stop the execution.

Example 2.14. Write a program to sort the numbers in descending order.

Solution: Explanation:

- Consider that a block of N words is present.
- Now we have to arrange these N numbers in descending order, Let N = 4 for example.
- We will use HL as pointer to point the block of N numbers.
- Initially in the first iteration we compare the first number with the second number. If first number> second number don't interchange the contents. If first number < second number swap their contents. Now at the end of this iteration first two elements are sorted in descending order.
- In the next iteration we will compare the first number along with third.Tf first > third don't interchange contents otherwise swap the contents. At the end of this iteration first three elements are sorted in descending order. Go on comparing till all the elements are arranged in descending order. This method requires approximately N comparisons.

Assembly Language Program:

Instruction	Comment
MVI B, 09	; Initialize counter 1
START: LXI H, DOOOH	; Initialize memory pointer
MVI C, 09H	; Initialize counter 2
BACK: MOV A, M	; Get the number in accumulator

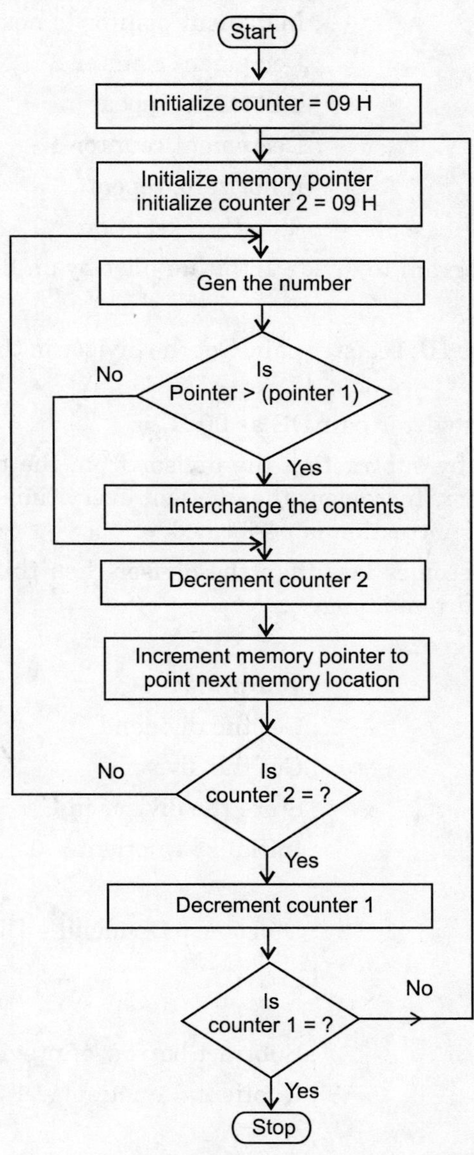


```
INXH          ; Increment memory pointer
CMPM          ; Compare number with next number
JNC SKIP      ; If more, don't interchange
JZ SKIP       ; If equal, don't interchange
MOVD,M        ; Otherwise swap the contents
MOVM,A        ;
DCXH          ; Interchange numbers
MOVM,D        ;
```

```
INX H              ; Increment pointer to next memory location
SKIP:DCR C         ; Decrement counter 2
JNZ BACK           ; if not zero, repeat
DCR B              ; Decrement counter 1
JNZ START          ; If not zero, repeat
HLT                ; Stop the execution.
```

Example 2.15. Write a program to divide 16-bit number by an 8-bit number.

Solution: Explanation:

- Get the dividend in the HL register pair. Get the divisor in the accumulator and store it in register C.
- Initialize quotient in register pair DE as 00 H.
- Perform the division by subtracting the divisor from the dividend, till the dividend is greater than the divisor. Increment the quotient every time the dividend is greater than the divisor when the subtraction is performed.
- When the dividend becomes less than the divisor then this dividend is the remainder. Store the quotient and remainder.

Assembly Language Program:

Instruction	**Comment**
LHLD DOOOH	; Get the dividend
LDA D002H	; Get the divisor
MOV C, A	; Store the divisor in C
LXI D,OOOOH	; Initialize Quotient = 0
BACK: MOV A, L	;
SUB C	; Division = Dividend – Divisor
MOVL,A	;
JNC SKIP	;
DCR H	; Subtract borrow of previous subtraction
SKIP:·INXD	; Quotient = quotient + 1
MOVA,H	;
CPI 00	; Check if dividend < divisor
JNZ BACK	;
MOVA,L	; If no, repeat
CMPC	;
JNC BACK	;
SHLD E002H	; Store the remainder
XCHG	;
SHLD EOOOH	; Store the quotient
HLT	; Stop the execution.

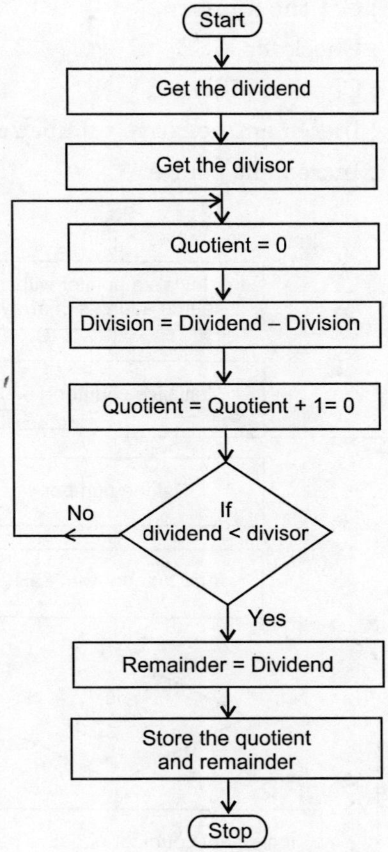

Example 2.16. Write a program to find the number of negative numbers in the array:

Solution : Explanation:

- We have an array of N numbers. For e.g. we initialize the count with 5.
- Also we initialize a pointer to point the elements in the array.
- We will check for the MSB. If the MSB of number is 1, the number is negative.
- Increment count for counting the negative numbers.
- Increment the pointer to point to the next element. Check if MSB is 1. Repeat process till all the numbers are scanned.
- Store the count of negative numbers.

Assembly Language Program:

```
Instruction          Comment
LDADOOOH             ;
MOVC,A               ; Initialize count
MVI B, 00            ; Initialize the Negative number count = 0
LXI H, D001H         ; Initialize pointer to array
```

```
BACK: MOV A, M     ; Get the number
ANI80H             ; Check for MSB
JZ SKIP            ; Check if MSB = 1
INR B              ; Increment negative number count
SKIP: INX H        ; Increment pointer
```

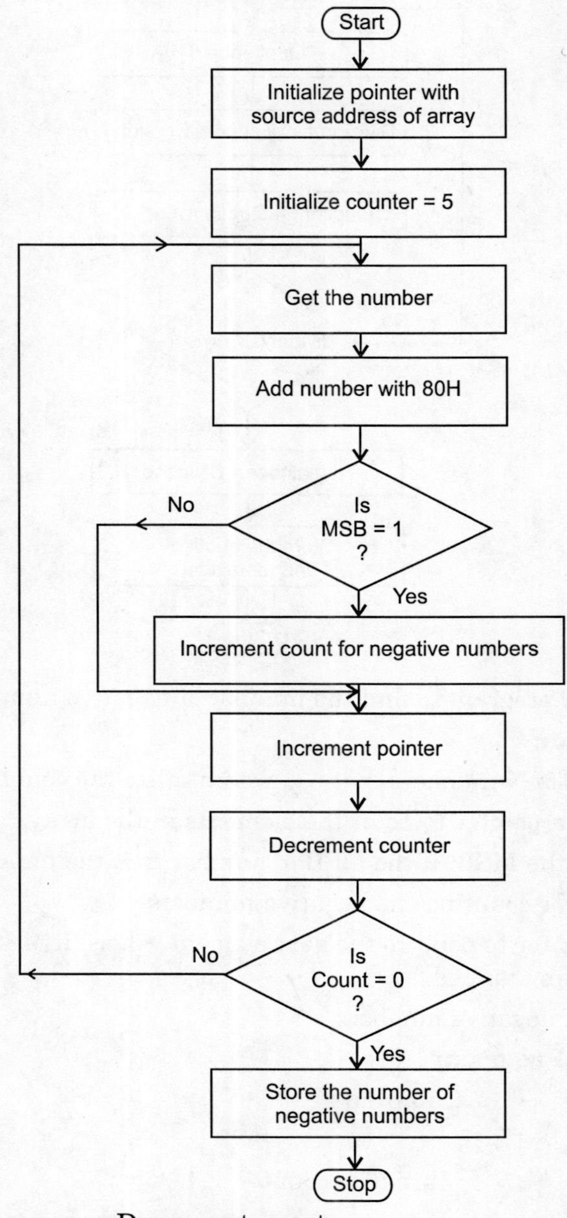


```
DCRC               ; Decrement count
JNZ BACK           ; If count 0, repeat
MOVA,B             ;
```

```
STA EOOOH          ; Store the result
HLT                ; Stop the execution.
```

Example 2.17. Write a program to find maximum number in the array.

Solution: Explanation:

- For example: we have an array of 10 numbers. So we initialize the counter with 10. Also we initialize a pointer to point these numbers.
- Compare first number with initial maximum number i.e. zero. If number> maximum number, save number otherwise, increment pointer to compare next number. Decrement counter, compare till all the numbers are compared. Store the maximum number in memory location E000 H.

Assembly Language Program:

Instruction	**Comment**
LDADOOOH	;
MOVC,A	; Initialize counter
XRAA	; Maximum = Minimum possible value = 0
LXI H, DOOIH	; Initialize pointer
BACK: CMPM	; Is number> maximum
JNC SKIP	;
MOVA,M	; If number> maximum
	; then interchange.
SKIP: INX H	;
DCR C	;
JNZ BACK	;
STA E000H	; Store maximum number
HLT	; Stop the execution

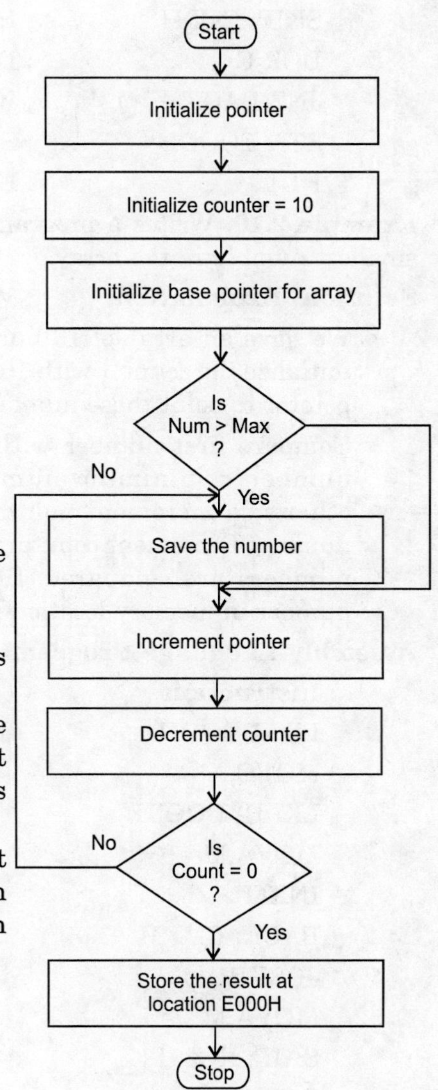

Example 2.18. Write a program to calculate the usm of the series of even numbers.

Solution : Explanation :

- Initialize the usm to zero. Load the count of numbers in register C. Intialize pointer to start of series.
- To check whether the number is even we will ADN the number with 01 H. If the result of ANDing is zero it indicates that the number is even. We will add this even number to the initialised sum.
- Decrement count. Increment source pointer to next location. Continue the process till sum of all even numbers is found. Store the sum at memory location E000 H.

Assembly Language Program:

Instruction	**Comment**
LDA DOOOH	;
MOV C, A	; Initialize counter

```
MVI B, OOH          ; Sum = 0
LXI H, DOOlH        ; Initialize pointer
BACK: MOV A, M      ; Get the number
ANI, 01H            ; Mask Bitl to Bit7
JNZ SKIP            ; Don't add if number
                      is ODD
MOV A, B            ; Get the sum
ADD M               ; SUM = SUM + data
MOV B, A            ; Store result in B
                      register
SKIP: INX H         ; Increment pointer
DCR C               ; Decrement counter
JNZ BACK            ; If counter 0, repeat
STA EOOOH           ; Store sum
HLT                 ; End.
```

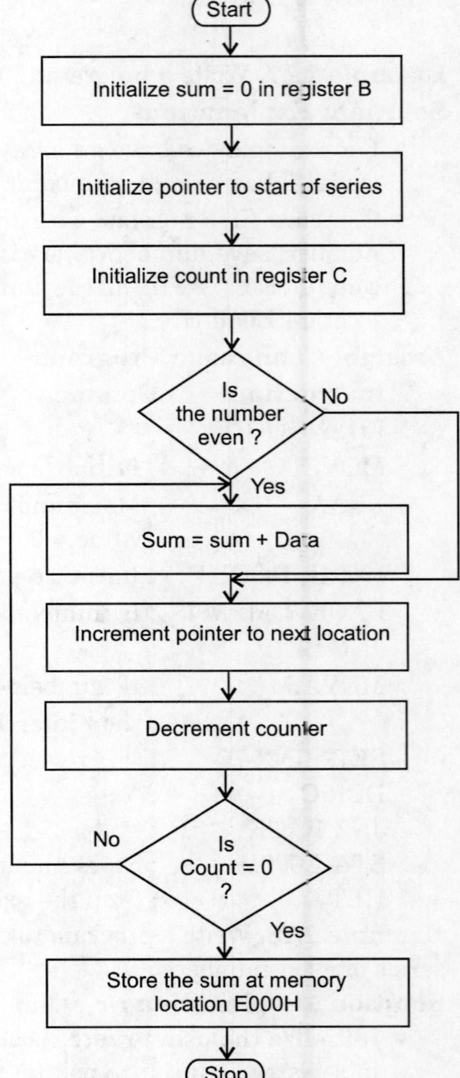

Example 2.19. Writes a program to find minimum/ smallest number in the array.

Solution: Explanation:

- We have an array of 10 numbers for e.g. so we initialize the counter with 10. Also we initialize a pointer to point these numbers.
- Compare first number with initial number. If number < minimum number, save number otherwise increment pointer to compare next number. Decrement counter, compare till all the numbers are compared. Store the maximum number in memory location EOOO H.

Assembly Language Program:

Instruction	**Comment**
LDADOOOH	
MOVC,A	; Initialize counter
LXI H, DOOlH	; Initialize pointer
MOVA,M	;
INXM	;
BACK: CMPM	; is number < minimum
JCSKIP	;
MOVA,M	; If number < minimum then interchange
SKIP: INX H	;
DCRC	;

```
JNZBACK          ;
STA EOOOH        ; Store minimum number
HLT              ; Stop the execution.
```

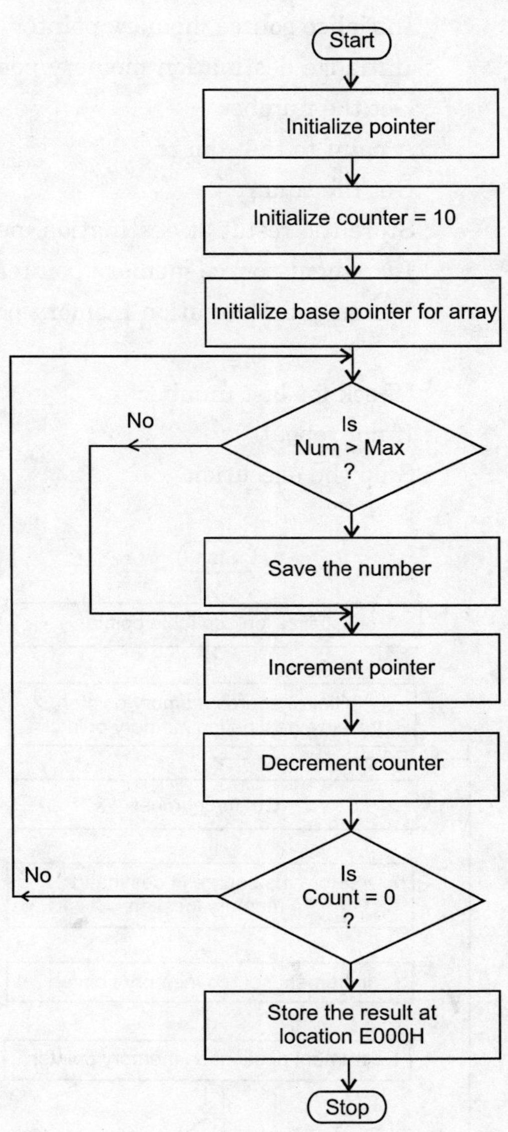

Example 2.20. Write the program to find the square of numbers.

Solution: Explanation:

- To find the square of a number we multiply the number with itself.
- Store the result.

Assembly Language Program:

Instruction	Comment
LXI H, DIOOH	; Initialize lookup; table pointer
LXI D, DOOOH	; Initialize source memory pointer
LXI B, D200H	; Initialize destination memory pointer
BACK: LDAX D	; Get the number
MOVL,A	; A point to the square
MOVA,M	; Get the square
STAXB	; Store the result at destination memory location
INXD	; Increment source memory pointer
INXB	; Increment destination memory pointer
MOVA,C	;
CPI05H	; Check for last number
JNZBACK	; If not repeat
HLT	; Stop the execution.

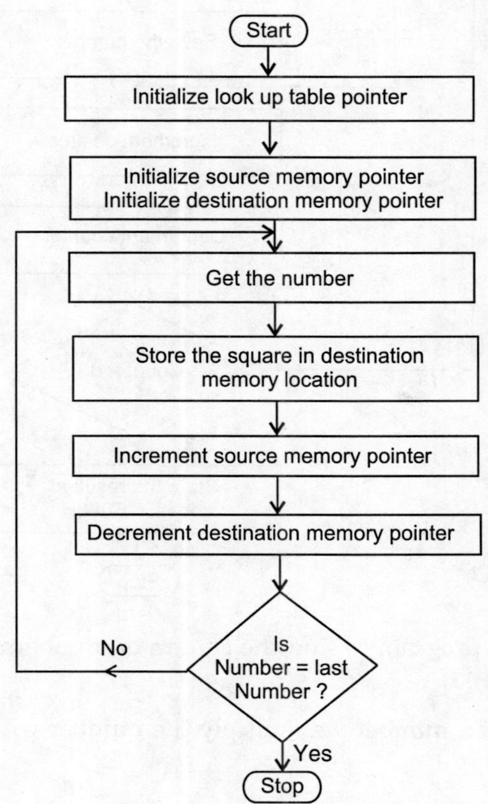

2.12. SUBROUTINES

A subroutine is a group of instructions written separately from the main program to perform a function that occurs repeatedly in the main program. For example, if a time delay is required between three successive events; three delays can be written in the main program. This repetition of the same delay instruction can be avoided by using the subroutine program. Thus, the delay instructions are written separately from the main program and can be called by the main program when needed.

To implement the subroutine program, the 8085 microprocessor uses two instructions: CALL (call a subroutine), and RET (return to main program from the subroutine). The CALL instruction is used in the main program while RET used at the end of the subroutine to return in main program.

Example 2.21. Write a program to control a traffic light signal controller.

Solution: Explanation: Here we have to write a program to provide the given on/off to three traffic lights (Green, Yellow, and Red) and two pedestrian signs (Walk and don't walk). The signal lights and signs are turned on/off by the data bits of an output port as shown below:

	Lights	**Data bits**	**On time**
1.	Green	Do	15 seconds
2.	Yellow	D2	5 seconds
3.	Red	D4	20 seconds
4.	Walk	D6	15 seconds
5.	Don't Walk	D7	25 seconds

The traffic and pedestrian flow are in the same direction; the pedestrian should cross the road when the green light is on. The on/off times for the traffic signals and pedestrian signs are as follows:

Time sequence In seconds	**Don't Walk**	**Walk**		**Red**		**Yellow**		**Green**	**Hex Code**
0	D7	D6	Ds	D4	D3	D2	Dl	Do	
15	0	1	0	0	0	0	0	1	= 41 H
20	1	0	0	0	0	1	0	0	= 84H
40	1	0	0	1	0	0	0	0	= 90H

The above table shows that the green light and walk sign can be turned on by sending data byte 41H to the output port. The 15 second delay can be generated by using a l-second subroutine and a counter with count of 15_{10}, Similarly, other lights can be turned on by sending the appropriate bytes and delay as shown in flowchart.

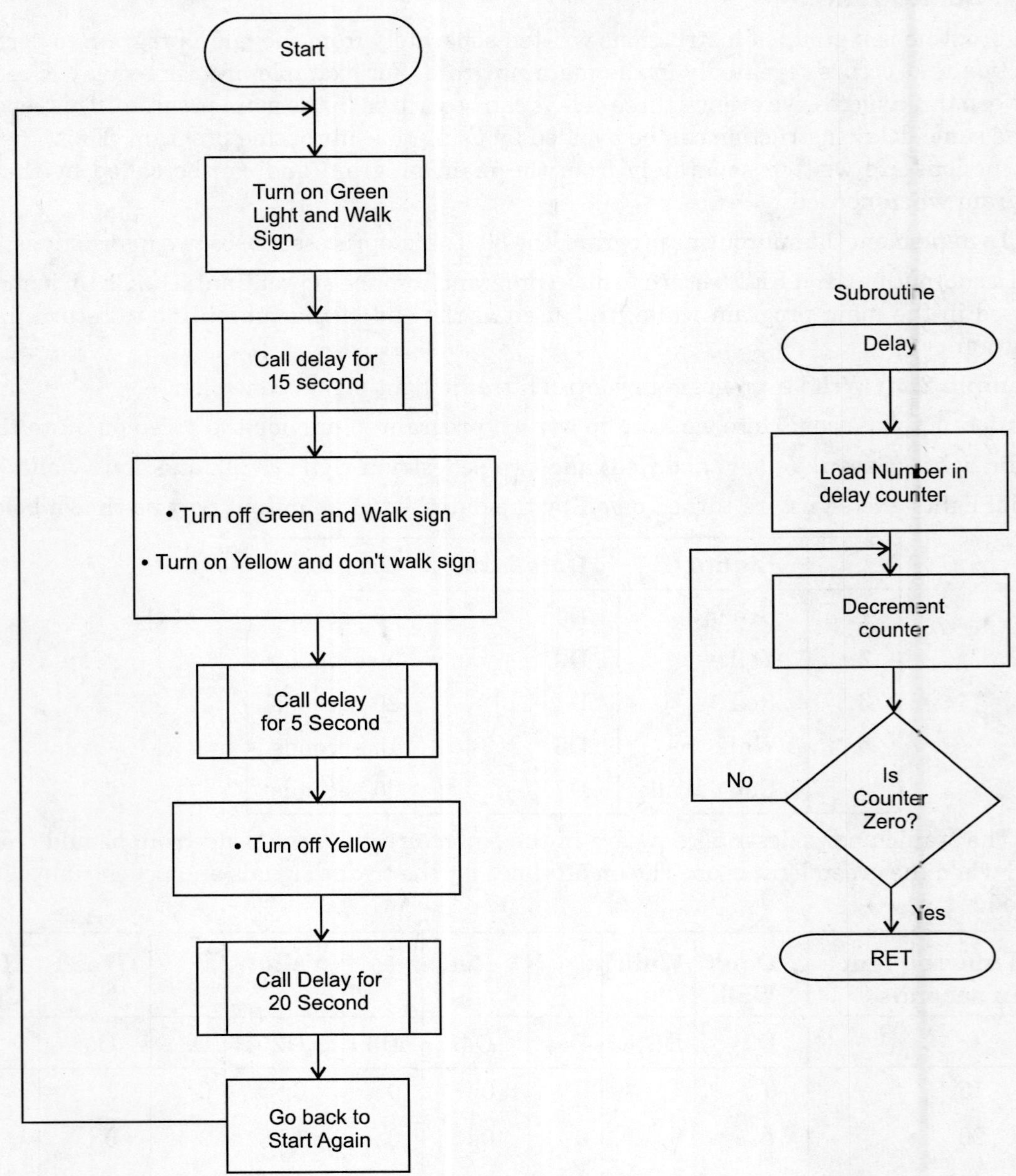

Assembly Language Program:

Instruction	**Comment**
LXI SP, XX99	; Initialize stack pointer at location XX99H
START: MVI A, 41H	; Load accumulator with the bit pattern for green light and walk sign
OUT PORT#	; Turn on green light and Walk sign
MVI B, OFH; Use	B as a counter to count 15 seconds.

```
CALL DELAY      ; Call delay subroutine located at XX50H
MVI A, 84H      ; Load accumulator with the bit for yellow light and don't walk sign
OUT PORT#       ; Turn on Yellow light and don't walk sign
MVI B, 05H      ; Use B as a counter to count 5 seconds
CALL DELAY      ; Call delay subroutine located at XX50H
MVI A, 90H      ; Load accumulator with the bit for Red light and don't walk sign
OUT PORT#       ; Turn on Red light, don't walk sign, and keep yellow light turned off
MVI B, 14H      ; Set up the counter for 20-seconds delay
CALL DELAY      ; Call delay subroutine located at XX50H
JUMP START      ; Go back to location START to repeat the sequence.
```

DELAY: This is a 1-second delay subroutine that provides delay according to the parameter specified in register B.

```
Instruction             Comment
DELAY: PUSH D           ; Save contents of DE and accumulator
PUSHPSW
SECOND: LXI D, COUNT    ; Load register pair DE with a count for 1-second delay
LOOP: DCXD              ; Decrement register pair DE
MOVA,D                  ;
ORAE                    ; OR (D) and (E) to set zero flag
JNZ LOOP                ; Jump to loop if delay count is not equal to 0
DCRB                    ; End of 1-second delay; decrement the counter
JNCSECOND               ; Is this the end of time needed? If not, go back to repeat
                        ; 1-second delay
POP PSW                 ; Retrieve Contents of saved registers
POPD                    ;
RET                     ; Return to main program
```

2.13. PARAMETER PASSING

Usually it is a common requirement of the calling program to pass some data as input to the subroutine. The procedure in which various informations have to pass between a calling program and a subroutine is called parameter passing. Therefore, it is very important to document a subroutine clearly and carefully. The documentation should have at least the following:

(*i*) Function of the subroutine

(*ii*) Input/Output parameters

(*iii*) Register used or modified

(*iv*) List of the other subroutines called by this subroutine.

2.13.1. Function of the Subroutine

It is important that user should clearly and precisely understand the function of subroutine without going through the instructions.

2.13.2. Input/Output Parameters

The parameters passed to a subroutine are known as inputs and parameters returned to calling programs are known as outputs. When many parameters must be used, R/W memory locations are used to store the parameters and HL registers are used to specify parameter locations. Similarly the stack is also used to store and pass parameters.

2.13.3. Registers used or Modified

Registers used in a subroutine also may be used by calling program. Therefore, it is necessary to save the register contents of the calling program on the stack at the beginning of the subroutine and to retrieve the contents before returning from the subroutine.

2.13.4. List of Subroutines failed

If a subroutine is calling other/subroutines, the user should be provided with the list. The user can check what parameters need to be passed to various subroutines and what registers are modified in the process.

2.14. COUNTER

A counter can be designed simply by loading an appropriate number into one of the registers and using the INR (Increment by one) or the DRC (Decrement by one) instructions. To update the count, a loop has to be established and each count is checked to determine whether it has reached the final number; if not the loop is repeated. Figure 2.14 illustrates these steps in the flow chart.

2.15. TIME DELAY

The procedure used to design a specific delay is similar to that used to set up a counter. A register is loaded with number, depending on the time delay required, and then the register is decremented until it reaches zero by setting up a loop with a conditional jump instruction. The loop causes the delay, depending upon the clock period of the system.

2.15.1. Time Delay using one Register

Figure 2.15 illustrates the flowchart of a time delay loop. A count is loaded in a register and the loop is executed until the count reaches zero. The time delay is most widely used in digital clock, process control, serial communication etc. An 8085 microprocessor with 2 MHz clock frequency will execute the instruction MVI in 3.5μs as follows:

$$\text{Clock frequency of the system } f = 2\text{MHz}$$

$$\text{Clock period T} = 1/f = 112 \times 10^{-6} = 0.5\ \mu\text{s}$$

$$\text{Time to execute MVI} = 7\ \text{T-state} \times 0.5 = 3.5\ \mu\text{s}$$

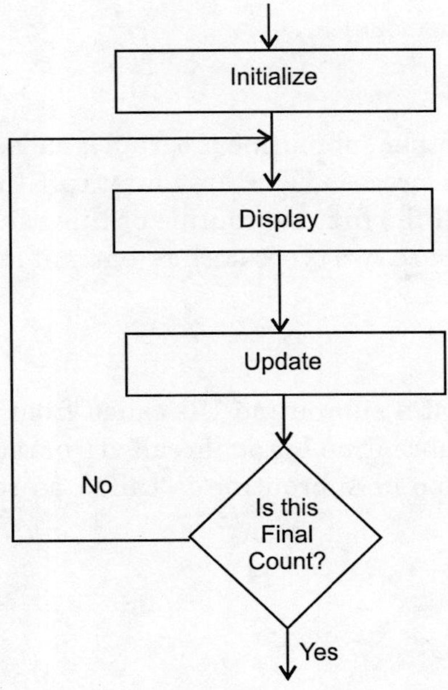

Figure. 2.14. Flowchart of a counter.

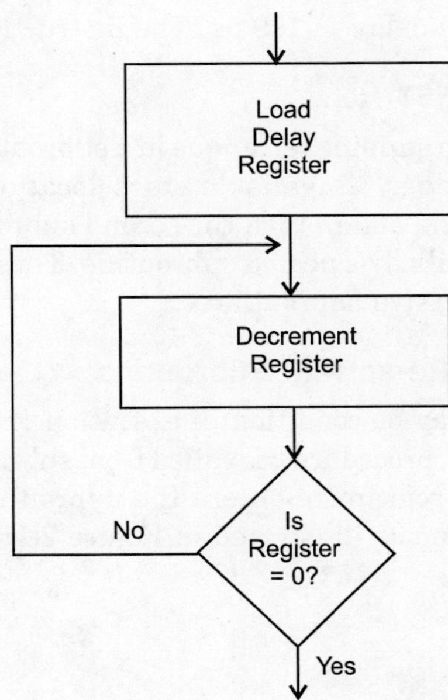

Figure 2.15. Flowchart of a time delay loop.

2.15.2. Time Delay using a Register Pair

The time delay can be considerably increased by setting a loop and using a register pair with 16-bit number. The 16-bit number is decremented by using the instruction DCX. The following subroutine can be used for time delay.

Label	Instruction	Comments	T-states
	LXI D, Count	; Load 16-bit count	10
LOOP	DCXD	; Decrement count by one	6
	MOVA,D		4
	ORAE	; Logically OR D and E	4
	JNZLOOP	; if result ≠ 0, continue	10/7

The time delay in the loop can be calculated as follows: the loop has four instructions DCX, MOV, ORA, and JNZ and takes 24 clock periods for execution. The loop for example, repeated 2384 times, which is equal to 9092 in decimal.

If time period of the system is 0.5 us, the delay in the loop is

$$T_L = (0.54 \times 2.4 \times 9092)$$
$$\approx 109 \text{ ms (without adjusting for the last cycle)}$$

T_D = 109ms + To

Total delay ≈ 109 μs (The instruction LXI add only 5μs)

2.16. NESTING

The programming technique in a subroutine calling another subroutine program is called nesting. The number of available stack locations limits this process. The first may call the second subroutine and in turn the second subroutine may call the first subroutine or third subroutine. This is called as nested subroutine. These subroutines are two types such re-entrant subroutine and recursive subroutine.

2.16.1. Re-entrant Subroutine

There may be condition in execution of a program that a subroutine 1 is called from the main program, procedure 2 is called from subroutine 1 and subroutine 1 is again called from subroutine 2. The program re-enters the subroutine 1. This type of subroutine is called as re-entrant subroutine as illustrated in Figure 2.16.

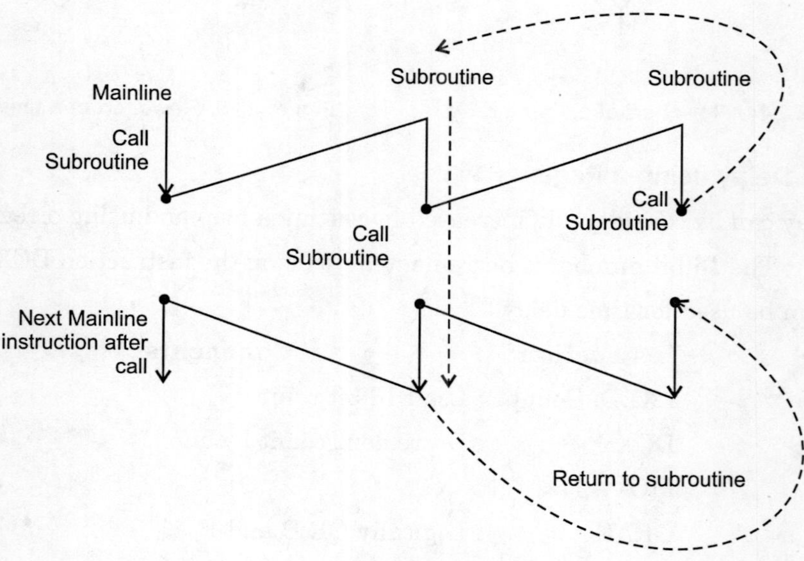

Figure 2.16. Program execution for re-entrant subroutine

2.16.2. Recursive Subroutine

A procedure that calls itself is called a recursive subroutine. These are used when we work with complex data structures called trees and take a long time to execute and occupy stack space. We normally define the recursive depth (N) for this type of subroutines. Figure 2.17 illustrates the flow diagram for N = 5.

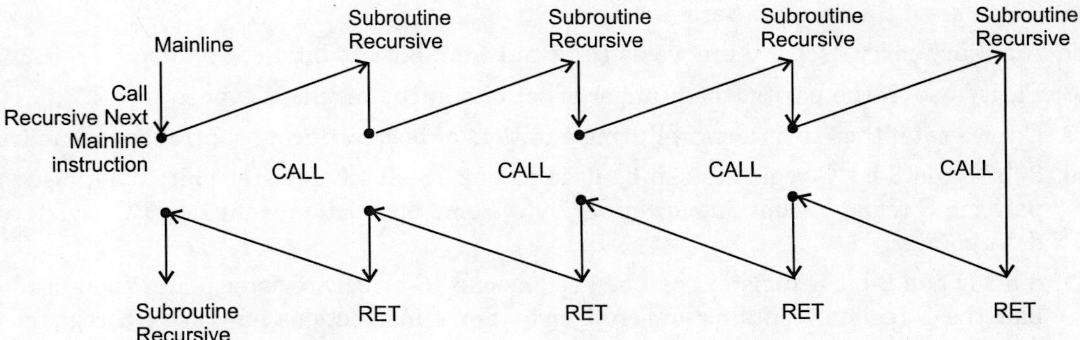

Figure 2.17. Flow diagram for recursive subroutine with N = 5

2.17. KEY POINTS TO REMEMBER

1. Intel 8085 had single 5 Volt power supply.
2. Clock oscillator and system controller were integrated on the chip.
3. The CPU included serial I/O port.
4. The processor has 5 interrupts. They are presented below in the order of their priority (from lowest to highest):
5. INTR is maskable 8080A compatible interrupt. When the interrupt occurs the processor fetches from the bus one instruction, usually one of these instructions. One of the 8 RST instructions (RSTO – RST7).
6. The processor saves current program counter into stack and branches to memory location N * 8 (where N is a 3-bit number from 0 to 7 supplied with the RST instruction).
7. CALL instruction (3 byte instruction). The processor calls the subroutine, address of which is specified in the second and third bytes of the instruction.
8. RST5.5 is a maskable interrupt. When this interrupt is received the processor saves the contents of the PC register into stack and branches to 2Ch (hexadecimal) address.
9. RST6.5 is a maskable interrupt. When this interrupt is received the processor saves the contents of the PC register into stack and branches to 34h (hexadecimal) address.
10. RST7.5 is a maskable interrupt. When this interrupt is received the processor saves the contents of the PC register info stack and branches to 3Ch (hexadecimal) address.
11. Trap is a non-maskable interrupt. When this interrupt is received the processor saves the contents of the PC register into stack and branches to 24h (hexadecimal) address.
12. All maskable interrupts can be enabled or disabled using EI and DI instructions. RST 5.5, RST6.5 and RST7.5 interrupts can be enabled or disabled individually using SIM instruction.
13. Accumulator or A register is an 8-bit register used for arithmetic, logic, I/O and load store operations.
14. Flag is an 8-bit register containing 5 1-bit flags:

15. Sign - set if the most significant bit of the result i:s set.
16. Zero - set if the result is zero.
17. Auxiliary carry - set if there was a carry out from bit 3 to bit 4 of the result
18. Parity - set if the parity (the number of set bits in the result) is even.
19. Carry - set if there was a carry during addition, or borrow during subtraction/comparison.
20. 8-bit Band 8-bit C registers can be used as one 16-bit BC register pair. When used as a pair the C register Contains low-order byte. Some instructions may use 'BC register as a data pointer.
21. 8-hit D and 8-bit E registers can be used as one 16-bit DE register pair. When used as a pair the E register contains low-order byte. Some instructions may use DE register as a data pointer.
22. 8-bit Hand 8-bit L registers can be used as one 16-bit HL register pair. When used as a pair the L register contains low-order byte. HL register usually contains a data pointer used to reference memory addresses.
23. Stack pointer is a 16 bit register. This register is always increment/decrement by 2.
24. Program counter is a 16-bit register.
25. Addressing modes
    - Register - references the data in a register or in a register pair.
    - Register indirect - instruction specifies register pair containing address, where the data is located.
    - Direct.
    - Immediate - 8 or 16-bit data
26. 8085 instruction set consists of the following instructions:
    - Data moving instructidns.
    - Arithmetic - add, subtract, increment and decrement.
    - Logic - AND, OR, XOR and rotate.
    - Control transfer - conditional, unconditional, call subroutine, return from subroutine and restarts.
    - Input/Output instructions.
    - Other - setting/clearing flag bits, enabling/disabling interrupts, stack operations, etc.

REVIEW QUESTIONS

2.1. What are the different components of the 8085 microprocessor?

2.2. Draw the pin diagram of 8085 and specify the function and direction of information flow of address bus, data bus, and control bus.

2.3. With help of a neat diagram explain the architecture of 8085 microprocessor.

2.4. List the features of 8085 microprocessor.

2.5. Why are the AD_0–AD_7 lines multiplexed?

2.6. Explain the different conditional flags in 8085.

2.7. Explain addressing modes of 8085 microprocessors.

2.8. Explain the operation of each instruction of 8085.

(*i*) DCR M

(*ii*) STA address

(*iii*) LXI H, address

(*iv*) NOP

Also mention their instruction group and addressing mode.

2.9. Explain the instructions SIM and RIM.

2.10. Explain the classification of instruction set of 8085 microprocessor with suitable example.

2.11. Explain the subroutine.

2.12. Write the note on flow chart.

2.13. Write the program to sort the 10 given numbers from memory location 2500H in ascending order.

2.14. Write a program to count the number of 1's in a byte stored in a register.

2.15. Write a program to find the square root of the numbers.

●●●

Chapter-3

ARCHITECTURE AND ADDRESSING MODES OF THE 16BIT 8086 MICROPROCESSOR

INTRODUCTION

The intel 8086 microprocesser was the most advanced microprocesser when it was released. This 16 bit microprocesser was implemented in advanced N-channel, depletion load, silicon gate (HMOS) technology and is available in three clock rates: 5, 8 and 10 MHz. The designers of this processor made numerous advances in processor technology such as pipelining and caching. The 8086 is designed for operation in single processor and multiprocessor configurations. Multiprocessor configuration systems use more than one 8086 microprocessor to achieve higher performance levels. The 8086 microprocessor can directly address 1 M byte of memory.

The 8086 microprocesser was, as we have earlier said, a 16 bit processer. By this term what we mean is that firstly it has a 16 bit wide data bus which, effectively means that it can operate on 16 bit of data from memory and ports. Secondly, the ALU and registers were designed to work on 16 bits of data.

Since the 8086 microprocessor has a 20 bit wide address bus, it can effectively thus address a maximum of 220 or 1048576 memory locations where each ofthese locations references to a byte wide memory register. Therefore, we can say that the 8086 can directly access 1 Mbyte of memory.

One thing about the 8086 operation that we have to always remember is that when the first byte of a 16 bit data is at a even memory address, the 8086 can read (or write) 16 bits (2 bytes) in one cycle. However, in the other case, that is the first byte of a 16 bit data is at an odd memory location, the 8086 CPU will read (or write) the first byte in one bus cycle followed by the second byte in the next bus cycle. This effectively cummulates to the fact that must be always remembered:

1. A 16 bit data will be read in one cycle if located at an even memory address
2. A 16 bit data will be read in two cycles if located at a odd memory location.

With the 8086 processor, Intel also released the 8088 microprocessor. The 8088 is a _ scaled down version of the 8086 that is internally the 8088 is a 16 bit microprocessor having the same ALU and register file as the 8086 microprocessor. It also has the same 20 bit' address bus but the difference between the 8086 and 8088 lies in the width of the external data bus. The 8088 has a 8 bit data bus thereby enabling it to read or write only one byte of data from memory or I/O ports.

3.1. THE 8086 INTERNAL ARCHITECTURE

A microprocessor's internal features and their uses need to be discussed and it is advisable to be through in these before in depth programming and cycle timing can be discussed. Here we shall briefly but thoroughly discuss the 8086 Architecture.

Since the 8086 followed the 8085, designers pressed for the need for processor speed up. By speedup, we means to effectively increase the execution rate of a microprocessor (or inshort its speed*). General terminology for increase in speed (or execution rate) of a microprocessor is called speedup. The intel 8086 used advanced processing techniques to speed up execution. For this, the. 8086 was internally divided into two functional units:

1. THE EXECUTION UNIT (EU)
2. THE BUS INTERFACE UNIT (BIU).

Both, these units are designed to work simultaneously. It is therfore, obvious that the processing time required for an instruction will be reduced in comparison to a processor having only a single unit working at a time. The reason for this is in the division of work between the two, where, the BIU is designed to function as the external interface of the processer. Effectively thus the BIU generates addresses and sends them on the address bus and then reads (or writes) data from memory (and ports) through the data bus. Therefore the BIU is the interface for the external buses to the EU. In short the BIU is the interface of the EU to the outside world.

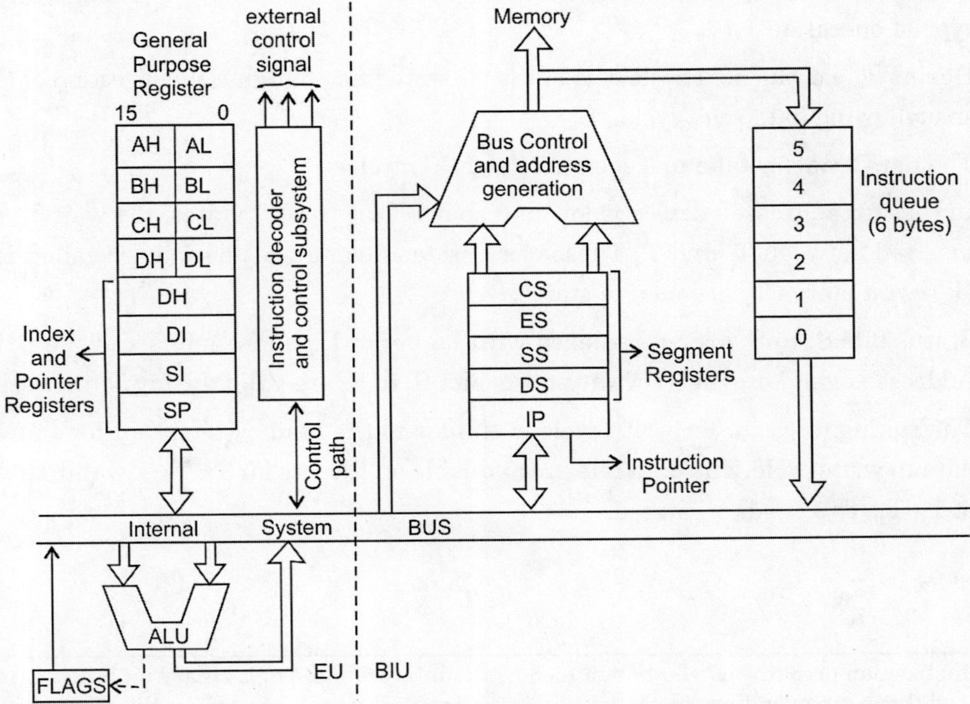

Figure 3.1. The 8086 Internal Architecture

The Execution unit of the 8086 handles all controlling functions such as telling the BIU

* The microprocessor execution rate and speed shall be used interchangely throughout this text except where specified.

where to fetch instructions (or data) from, where to write data to, and does the instruction decoding* as well as does all of the instruction execution.

The division of the 8086 architecture into these two units can be easiliy seen in figure 3.1. Now we shall consider these two units in detail.

3.1.1 General Bus Opreation

In the 8086 microprocessor, the address and data buses are combined for the maximum utilization of the processor pins. These buses are commonly known as a time multiplexed address and data pins. Thus, it available in a 40 pin standard DIP package. The bus can be de-multiplexed using a few latches and transceivers wherever required. Basically, all the processor bus cycle consist of at least 4 clock cycles. There are referred to as T_1, T_2, T_3, and T_4. The general bus operation cycle shown fig 3.2 can be explained as follows:

1. The address is transmitted by the processor during T_1 cycle. It is present on the bus only for one cycle.
2. The negative edge of this ALE pulse is used to separate the address and the data or status information. In maximum mode, the status lines So, S_1 and S_2 are used to indicate the type of operation.
3. During T_2, i.e. the next cycle, the bus is tri-stated for changing the direction of the bus for the following data read cycle.
4. The data transfer take place during T_3 and Trcycles
5. In case an addressed device is low and shows "Not Ready" states, the wait T_w cycles are inserted between T_3 and T_4 These clock states during wait period are called idle status (T_1), wait status T_w or inactive states.
6. Status bits S_3 to S_7 are multiplexed with higher order address bits and the BHE signal. Address is valid during TI while status bits S_3 to S_7 are valid during T_2 through T_4.
7. The timing diagram for write cycle is similar to the read cycle except for the validity of data in write cycle. The data bits are available on the bus for T_2. T_3, T wand T_4. At the end of T_1, T_4, The bus is tri stated.

* Instruction decoding can be understood as the interpretation of a binary bit pattern (instruction) and then corresponding generation of required control signals to perform the necessary operations by the microprocessor.

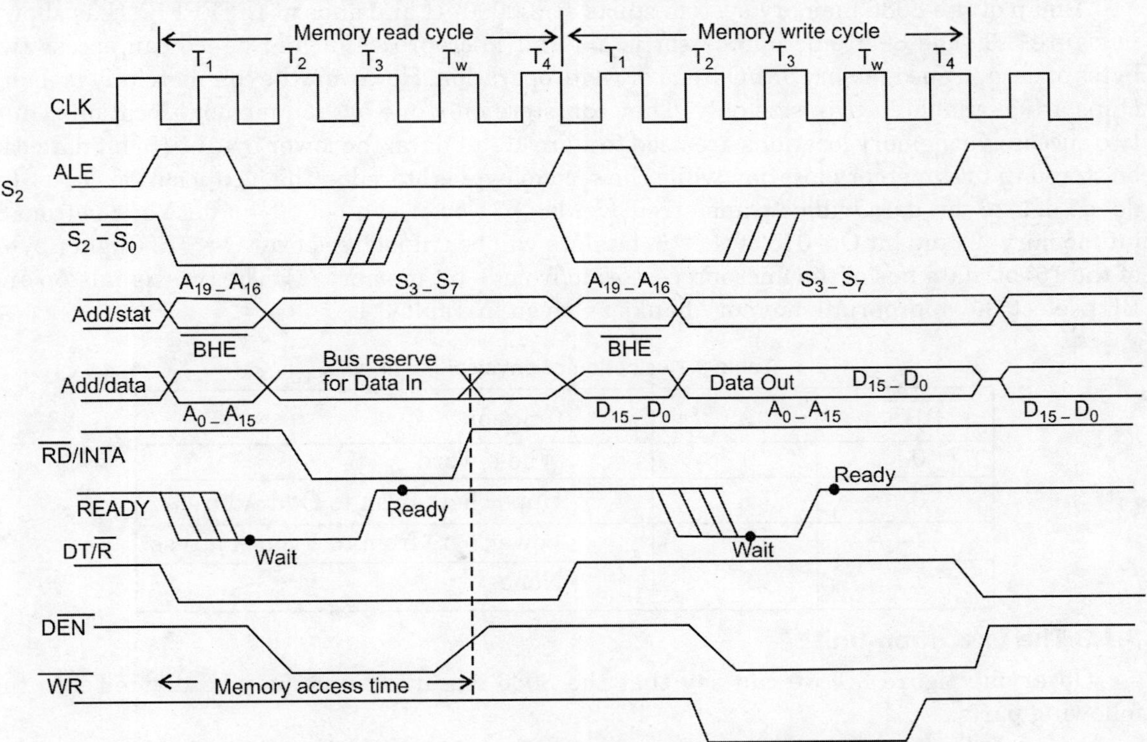

Figure 3.2 Timing diagram of general bus operation of 8086 microprocessor

3.1.2 PHYSICAL MEMORY ORGANISATION

In on 8086 bused system, the 1M bytes memory is physically organized as an odd bank and an even bank, each of 512 k bytes, addressed in parallel by the Processor. Byte data with an even address is transferred on D_7–D_0, while the byte data with an odd address is transferred on b_{15}–b_8 bw bs bus lines. The processor provides two enable signals, BHE and Ao for selection of either even or odd or both the banks. If microprocessor fetches a word from memory then there are following different possibilities:

1. Both the bytes may be data operands.
2. Both the bytes may be containing op code bits.
3. One of the bytes may be op code while the other may be data.

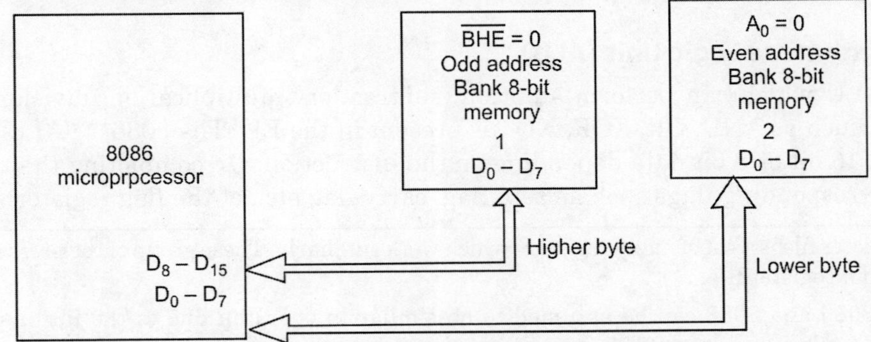

Figure 3.3 Physical memory organization of 8086 microprocessor.

A map of the 8086 memory system starts at (00000) H and ends at (FFFFFH) H as shown in figure 3.3. The 8086 microprocessor is a 16-bit microprocessor and hence can access two bytes of data in one memory or I/O Read or write operation. However, the commercially memory chips are available in byte size only. They can store only one byte in memory location. Thus, two successive memory locations are used to store 16-bit data; the lower byte of 16- bit data can be stored in first memory location while the second byte is stored in the next location. Thus, bit do- d7 of a 16- bit data will be transferred over bo- b7 (lower byte) of 16- bit data bus to/from 8- bit memory (2) and bit Os- 015 of the 16- bit data will be transferred over Os- 015 (higher byte) of the 16- bit data bus of the microprocessor, to/from 8- bit memory (1). The two signals Ao and BHEselect the appropriate memory banks as given in Table 3.1.

Table 3.1 selection of physical memory

$\overline{BHE}$	A_0	function
0	0	Whole word
0	1	Upper byte from to Odd Address
1	0	Lower byte from to Even Address
1	1	None

3.1.3 The Execution Unit

Observing figure 3.1 we can say that the 8086 execution unit may be divided into the following parts:

(a) Instruction Decoder and Control Subsystem.

(b) Arithmetic Logic Unit (ALU).

(c) FLAG Register.

(d) General Purpose Register (GPR) File**.

(e) ex and Pointer Registers.

3.1.1.1. Instruction Decoder and Control Subsystem

The EU contains the instruction decoder which translates the received instruction from the BIU and performs the necessary operations. The instruction decoder and the control system perform the proper sequencing and release the corresponding synchronization signals for the proper operation of the microprocessor to achieve the task specified by the instruction being decoded.

3.1.1.2. Arithmetic Logic Unit (ALU)

A 16 bit ALU which can perform addition, subtraction, multiplication, division and logical operations such as AND, OR, XOR, NOT is present in the EU. The 8086*** ALU can be made to work on 16 bit or 8 bit data depending on the instruction. On completing the task the ALU sets the corresponding flags such as zero flag, carry flag etc., of the flag register.

** A register file is a set of many registers which work similiarly. These group of registers are collectively called a register file.

*** The 8086 and 8088 can be assumed to be similar in operaton due to the internal architecture

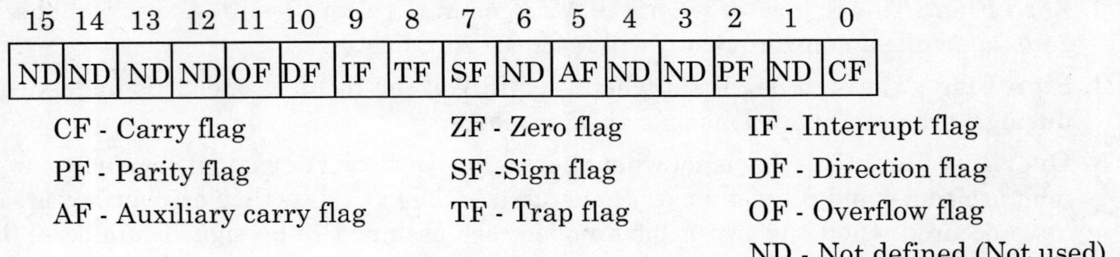

Figure 3.2. 8086/8088 Flag register

3.1.1.3. FLAG Register

The 8086/8088 flag register is represented in figure 3.2. The 8086 has a 16 bit flag register which contain 9 active flags. These 9 active flags can be divided into two types.

(a) **Conditional flags :** Conditional flags are set or reset on the basis of the result generated from the ALU.

(b) **Control flags :** The 8086 contains three control flags which are used for controlling the microprocessor. These flags are set or reset with the help of instructions that are put in a program and are used to control program flow.

Specifically, we shall now first discuss the respective flags.

1. **Carry Flag :** This conditional flag is set when a carry is generated in the ALU as shown by the example given below.

```
e.g.,   ADD   F F F F h,      A A A A h
Sol.          F F F F
              AAAA
             ----------
              (l)AAA 9
               ↑
               |
              carry
```

thus carry flag = 1 [Carry flag is set to shown the carry generated]

In the above operation we find that after two 16 bit numbers are added their sum is 17 bits wide. This excess bit which cannot be contained in a 16 bit register or memory location is used to set the carry flag thereby .can be used by a programmer to check the state of a result.

2. **Auxillary Carry Flag :** This flag is set when a carry is generated after the 4th bit when data being operated is 8 bit wide. When the data being operated is 16 bits wide the carry if generated after the 8 bit will set the auxillary carry flag (ACF).

```
e.g.   carry        1 1 1 1      1 1 1      carry after 4 bits (shown by the arrow)
       Byte 1       1 1 1        1 1 1 1    sets AF = 1.   I
     + Byte 2     + 0 0 0 0      0 0 0 1    However, if this carry is a
                  -----------------------
                    1 0 0 0      0 0 0 0    then AF = O
```

The auxillary carry flag is used during arithmetic operations ion BCD numbers.

3. **Zero Flag :** This flag is set when the ALU generates a zero result. A zero would set the zero flag while a non-zero result will reset the zero flag.
4. **Sign Flag :** The sign flag is set when the MSB' of the result is one. This flag is used as during arithmetic operations on signed numbers.
5. **Overflow Flag :** This flag is set when there is an overflow. For example when two number n bit are added and there is a carry generated in the (n - 1) bits then an overflow is said to have occured when the two n bit numbers are assumed to be signed numbers. (Refer section 1.3.2).

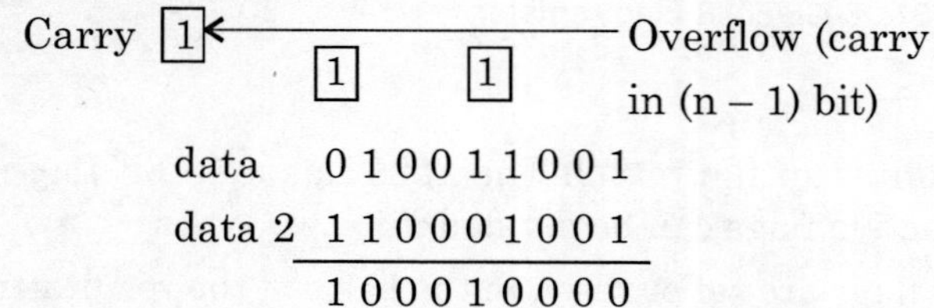

This flag is used when two signed numbers (where the MSB bit is used as a sign bit) are added.

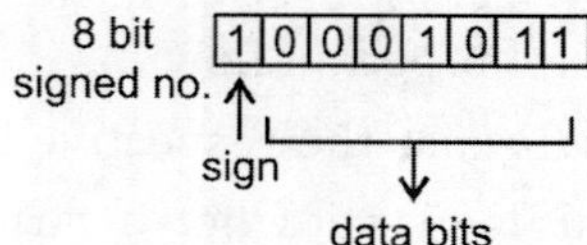

Here seven (7) bits are used as the data bits, thus a carry from these 7 bits if generated would also be added to the MSB (the sign bit) it would result in wrong results. This carry is used to set the overflow flag so that a programmer can check the overflow flag and thereby be informed that the result is invalid (wrong). A programmer can then initiate necessary steps to correct the results.

6. **Parity* Flag :** Parity flag is set when the ALU output has even parity and reset when the ALU output has odd parity.
7. **Trap Flag:** Used for single stepping through a program**.
8. **Interrupt Flag :** When the interrupt flag is set a program can be interrupted for other more important operations. When it is reset the program execution cannot be interrupted. (Refer to Section on Interrupts)
9. **Direction Flag :** This is used with string operation***.

Of the above flags, flags 1 through 6 are conditional flags while flags 7 to 9 are control flags.

---

* Parity signifies the number of bits in a corresponding state. Suppose if 6 bits in a byte are 1 we assume even parity. While if an odd number of bits are 1 we assume odd parity.

** This flag, is discussed in detail in the chapter discussing Interrupts.

*** Refer to Chapter on Data Movement Instructions.

3.1.1.4. General Purpose Registers

The EU has four 16 bit general purpose register called the AX, BX, CX and DX Registers. The AX register is also called the Accumulator and is one ofthe most important registers as you shall read on later. When the 8086 ALU is programmed to work on 8 bit data then these registers can be divided as in figure 3.3.

These 8 registers can be individually be used to store 8 bit data. BX For 8 bit operations AX, register is the accumulator. The Accumulator is a very important register as it has many features that others do not have. The accumulator will be studied in detail as we discussed for the 8086 instructions. There are several advantages of using internal registers for data storage:

1. Since the CPU data transfer speed is generally higher than memory thus data transfer is faster when during registers to ALU than memory to ALU.
2. Secondly, since data is already in the EU (i.e., in these registers) this data can be used faster than data from memory by the ALU.

3.1.1.5. Pointer and Index Registers

In addition to the four 16 bit general purpose registers (AX, BX, CX and DX) the execution unit contains three other registers collectively called the Index and pointer registers.

The index registers are the 16 bit SI (Source index) and 16 bit DI (Destination index) CX registers. The third is the 16 bit BP (Base pointer) registers. All the three can be used for temporary storage of data. All these three registers can only be used to store 16 bit data unlike the General purpose registers can only be used to store 16 bit data as well as 8 bit : data registers. Though these registers can be programmed for data storage but their actual design is used to store the offset address of a data byte (or word) in the data segment. The use of these registers as an offset address storage register is shown in chapter 4.

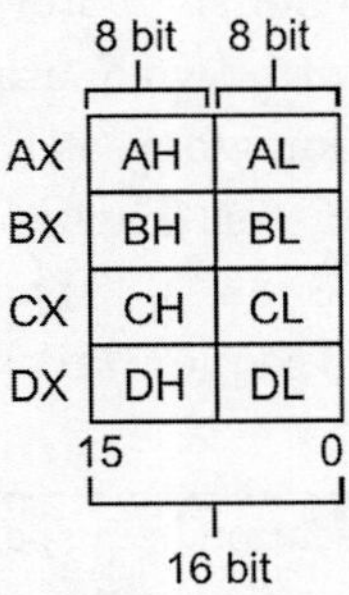

Figure 3.3. The 8086 Registers

For example instead of storing the offset address (16 bit) in memory for a data byte/word located in data segment we can store this address in the SI or DI registers and use this register as the offset address. Effectively then the address will be calculated by adding the contents of SI (or DI) to the segment base so as to produce the effective address of the data byte (or word) in the data segment.

3.1.2. Bus Interface Unit (BIU)

The bus interface unit can be divided into the following units

1. Instruction Queue.

2. Segment Registers.
3. Instruction Pointer.
4. Bus control and address generator.

These functional parts can be very clearly seen in figure 3.1.

3.1.2.1. The Queue

The 8086 processer used a method called Pipelining* for speed up. This is very clearly demonstrated in figure 3.4. The instruction queue of the 8086 is a six byte first in first out (FIFO) register and can be understood as a cache memory.

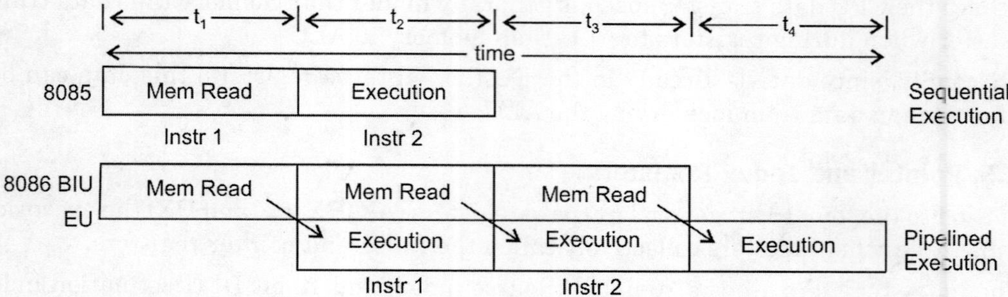

Figure 3.4. Pipelininq of 8086

When the EU is executing or decoding an instruction which does not require the use of external buses or the segment registers, the BIU can be thought of as being idle. Since the 8086 BIU is independently functional the BIU fetches the next instruction bytes for the EU and stores them in the queue. Now when the EU becomes free it can directly access (read) the instruction from the queue (which in obviously faster when compared to the 8085 processor). It is thus observed that the 8086 can perform two operations at the same time, that is a memory read (of instruction bytes) and a execution whereas the 8085 can only do one of the above. Thus it is obvious that the 8086 was much faster. This is demonstrated in the figure 3.4. Observe that during time intervals t_2 and t_3 the 8086/8088 is doing two operations, one by the BIU and one by the EU. This is faster compared to sequential execution where one operation is performed after the other.

Note: It is assumed that the Instruction execution being performed by the EU doesnot require the system buses

3.1.2.2. Segment Registers

Another new concept that was introduced in the 8086 was memory segmentation. Memory segmentation allowed the 8086 micorprocessor to be able to access only four 64 K byte segments within its 1 M byte memory range at any given time. These four segments are defined by the data values stored in the four segment. registers, which are named as follows :

1. CS – code segment Register
2. SS – stack segment Register

* Pipelining can be correlated to a car manufacturing assembly line where each worker works on a different aspect of the car be manufactured.

3. DS – Data segment Register
4. ES – Extra Data segment Register.

The use of these registers is discussed in more detail in the topic memory segmentation. By default the 8086 assumes that the code (or instructions) will be fetched from the memory segment addressed by the CS register value. Similarly, the 8086 by default uses the DS register to address the memory segment that-shall hold data bytes to be used by the program and the SS register to address the memory segment that shall be used as stack memory.

Therefore, the default segment registers usage is shown below with the registers that shall contain the offset address

Segment register	Memory segment	Deault offset register
CS	Code	IP
DS	Data	All except IP, BP, SP
SS	Stack	SP, BP
ES	Extra	All except IP,SP,BP

3.1.2.3. Instruction Pointer (IP)

The instruction pointer stores the offset of the physical address (due to memeory segmentation) in the segment of memory as defined by the segment registers. This, in short, means that the upper 16 bits of the 20 bit address is stored in the segment register while the offset in the memory segment is stored in the IP dr the pomter and index registers. This is dischssed in more detail in the section 3.3.2.

3.1.2.4. Bus Control and Address Generation

The bus control and address generation logic subunit is one which handles the external bus as well as calculates the effective address from where the instructions has to be fetched. The address generation circuit is basically an adder where the addend is the segment register and the audend the IP (or index registers). The addend is always appended with four zeros (or multiplied by 10H) before being put in the adder. The output of the adder is the effective memory address.

Now consider the following data

l000H : Addend (Segment Register)

2000H : Audend (IP or Index Register)

Thus, Addend X10H + Audend = Memory address

or

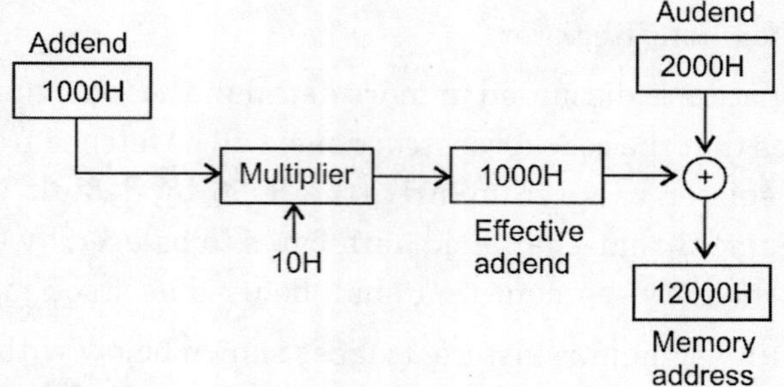

Figure 3.5. Memory Address Generation

We shall study the segmented memory addressing of 8086 where all this will be discussed.

3.2. SEQUENTIAL MEMORY ACCESS

Designers of computer systems generally use two types of memory organization. These are:

1. Linear Addressing or organization.
2. Segmented Addressing or organization.

3.2.1. Linear Addressing

The earliest of memory organizations was .the linear addressed memory organizaticn. In linear addressing the system memory to available to the processer as one linear array. This was used in the earlier 8085 processor and is illustrated in figure 3.6. In the 8085 the instruction pointer (IP) was 16 bit wide and was effectively the address of the memory location as the 8085 had a 16 bit address bus.

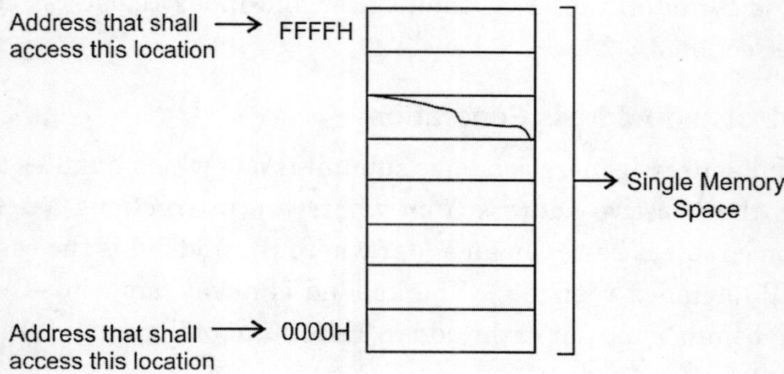

Figure 3.6. Linear Addressing

3.2.2. Segmented Addressing

Segmented memory addressing on the other hand divides the memory into many segments. For example, the 1 MB memory is divided into smaller segments of 64 Kb. Each of these segments can be considered as a linear memory space as discussed above. Now each of these segments is addressed by a segment register. However since the segment register is 16 bit wide and the memory needs 20 bits for an address the 8086 appends four bits segment register to obtain the segment address. Therefore, to address the segment 10000H by, say the SS register, the SS must contain 10000H (as the 4 appended zeros would automatically added by 8086/8088).

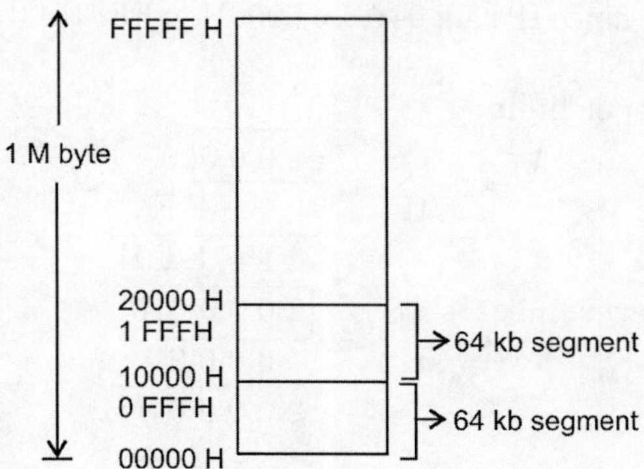

Figure 3.7. Segmented Memory Space

Thus,

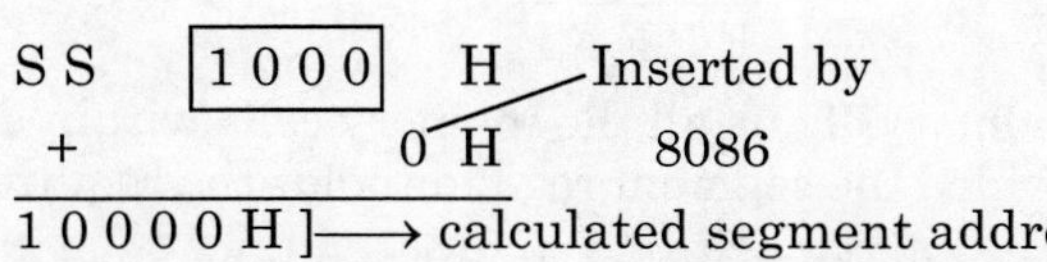

Thus if a segment registers has any arbitary value XXXX H the 8086 assumes it points to XXXX 0 H (or a 20 bit address where XXXX is the segment registers value). This is known as the segment base address. Effectively this means that the segment registers is used to reference to the zeroth address of the segment. Now when another register is added to the (segment) base address a valid 20 bit address is produced where these added registers can address the whole 64 Kb segment.

SS	1 0 0 0	0	H ←	base address
SP	A B B	C	H ←	offset address
1	A B B	C		Physical address

The register that is added is said to contain the offset address. This is very clearly understood as shown in figure 3.8.

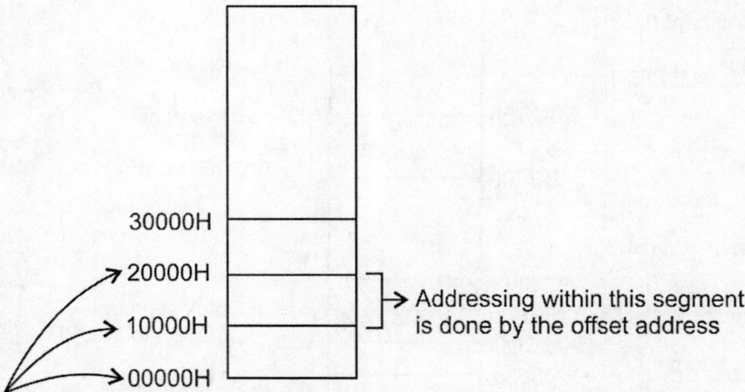

Figure 3.8. Segment addressing by different base addresses

Ques. Suppose an 8086 code segment CS registers contain 2000 H and the IP is added to this what is the range at the memory that can be addressed?

Ans. The upper and lower limits are calculated by assuming the upper and lower values that IP can hold. Therefore since IP ranges from 0000H to FFFFH (because it is a 16 bit register), the range of address is

Upper limit :

CS	2000 0
IP	FFFF
	2FFFFH

Lower limit :

CS	20000 0
IP	0000
	200000H

Thus range is

Upper limit	2FFFF
–Lower limit	20000
	FFFFH

This is equivalent to 64 Kb of memory. Thus all the memory cells within a segment can be addressed by the offset address (provided the segment register holds the required base address).

Thus it can be effectively combined to the fact that at any time the 8086 divides it memory space into 4 chunks of 64 Kb. Where each of these chunks can be addressed by the four segment registers. Thus at a time there can be only 4 active segments (as there are four segment registers) A typical segmented memory space of the 8086 is illustrated below in figure 3.9. By changing these segment registers values during the processor operation we can utilize different spaces in the 1 MB memory.

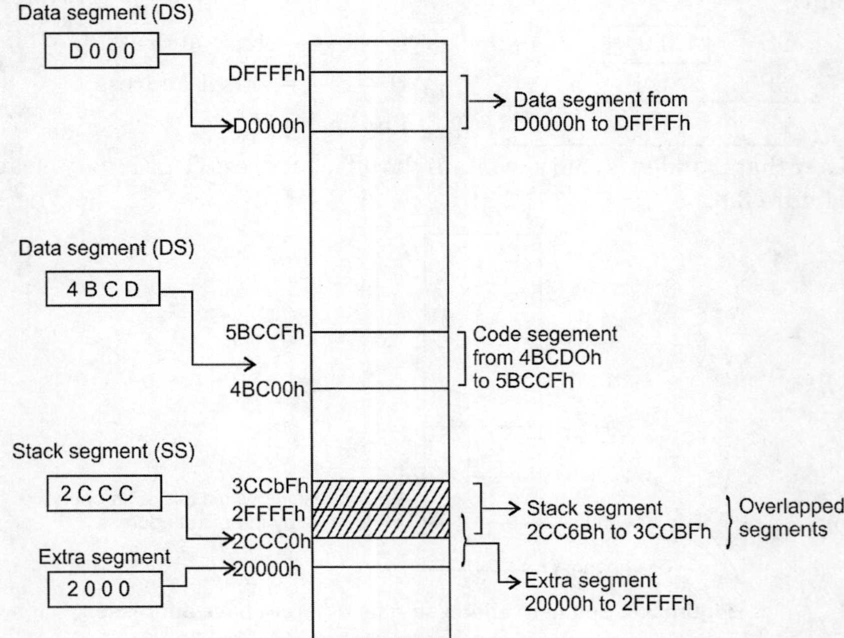

Figure 3.9. A typical segmented memory space for 8086

A Point to note is that the four segments need not point to different segments but can be same when

ES = CS = DS = SS

or can be partially overlapped as in figure 3.9 where the stack segment and extra segment are overlapped from the address 2C C C 0 H to 2 F F F F H. As earlier discussed the addresses within a segment can range from 0 0 0 0 H to F F F F H. The address within the segment is called offset or logical address.

eg.,

CS	2 0 2 0 0	base address
IP	F 2 D 2	offset address
	2 F 4 d 2	physical address

Now the physical address is the address that is actually sent on the address bus (that is it is 20 bits wide). The offset or logical address is the offset from the location zero of a segment. This segment is the location pointed by the base address or the segment register.

Now consider the following example:

For extra segment

C 2 2 2 0 H	base address ES
2 3 2 3 H	Offset BP
C 4 5 4 3 H	physical address

for Stack segment

B F 3 D O H	base address SS
5 1 7 3 H	Offset SP
C 4 5 4 3 H	physical address

Now here we observed that two different logical address point to the same memory address. This is due to the overlapping of two segments and should be avoided as here the program data and stack data values can be caused to be overwritten on each other resulting in wrong results.

3.2.3. Advantage of Memory Segmentation

Though you may at this find memory segmentation to be slightly complex but as you advance through this last section you shall find out about its advantages. However, here we shall discuss the advantages only superficially. The first advantage that memory segmentation has is that only 16 bit registers are required both to the store segment base address as well as offset address. This makes the internal circuitry easier to build as it removes the requirement for 20 bits register in case the linear addressing method is used.

The second advantage is relocatability. For example in a normal computer system time sharing process is used. Time sharing involves allocating a fixed amount of time for programs of different users. However, memory segmentation is a boon for time shared systems as all user program can be loaded into memory. The 8086 can then be made to work on a time sharing basis on each of these program by just reloading the segment address of each of these programs in the segment registers. Thus segmentation allows the programmer to separately use each of the program loaded in memory by just reloading the 8086 segment registers.

3.3. PROGRAMMING LANGUAGES

Now for a computer or a processor to run (or execute) programs, we have to store data and instructions in memory, which the processor fetches, decodes and then executes. There are many different ways in which a programmer may write programs for a computer.

3.3.1. Machine Language

When a programmer writes programs in the instruction set of the processor, he is said to have written a program in machine language. The instruction set of a processor is a total operations that a microprocessor can perform. Each of these operations has its own unique binary code. A processor has hundreds of instructions such as add, multiply, subtract, divide, compare, move to etc. Now each of these instruction has it's own unique binary code. The 8086 instruction can have a minimum size of 8 bits to a maximum size of 32 bits. You can imagine the difficulty, a programmer will face in memorizing hundreds of such binary number and then writing a program for the microprocessor. This type of programming is called machine language or low level programming as the programs are written in a form that the microprocessor directly understands.

	Memory Address	Contents	Contents (Hex)	Operation
e.g.	0 0 1 0 0 H	1 0 1 1 0 0 0 0	B 0 H	M O V A, 1 2 H
	0 0 1 0 1 H	0 0 0 1 0 0 1 0	1 2H	
	0 0 1 0 2 H	0 0 0 0 0 1 0 0	0 4 H	AD D A X, 0 8 H
	0 0 1 0 3 H	0 0 0 0 1 0 0 0	0 8 H	

When the above program win be executed by the microprocessor we would have said to have written a machine language program as we stored the instructions and data in memory in the form that the microprocessor understands.

3.3.2. ASSEMBLY LANGUAGE

Assembly language' is another technique to write programs for a microprocessor. Here we use two, three or four letter mnemonics to represent instructions. Instead of writing "00000100" (which we know is the binary equivalent for the instruction ADD), we simply replace it by ADD mnemonic. The mnemonics are called operation codes or op codes as they specify the operation to be performed.

After writing the op codes we also want to write the memory address, port address or name of register on which the operation (instruction) is to be performed. For this we write the operand field. For example

```
MOV   A,B
```

MOV is the op code as this specifies the operation is going to be data transfer.

A, B Specifies the op code field or that the data transfer will take place between A and B.

After writing an assembly language we want to convert it to machine language so that it can be understood and operated upon by the microprocessor. There are two ways of doing so. One way to decode it by using the data sheets. The second easier way is to use an assembler. An assembler basically converts these assembly language programs to machine code which can then be loaded into the processor memory for execution. Common Assemblers that are used with the 8086 microprocessor are Microsoft Assembler (MASM) and Borland Turbo Assembler (TASM).

3.3.3. High Level Languages

Another way of programming is to use high level languages. The instructions here are more English like words such as print etc. An example of high level languages are C and C ++. Each of the program statements (such as printf in C) represent many machine code instructions (sometimes even thousands). A compiler is then used for translation to machine code. A compiler converts these statements into their respective machine codes which can be thus loaded to a computers memory and run. It is obvious that programming in high level language is easier but there are certain applications where programs must be run very quickly (such as control systems in factories) where the programs for the processor are written in assembly or even machine language.

3.4. DATA ADDRESSING MODES

The 8086 microprocessor accesses data or instructions in memory by first sending the address of the memory location*. This is followed by control signals which specify to the memory whether the data is to be read from or written to memory. This is followed by the 8086 to put data on its data bus if the operation is a memory write or the memory device to put data on the 8086 data bus if the operation is a memory read operation. The way the 8086 generates an address for the memory location that is to be accessed any point of time is called the addressing mode.

We shall use the 8086 MOV instruction which performs a data copy operation to illustrate the 8086 addressing modes.

MOV destination, source

where the source specifies the source register (or memory location) and destination specifies a destination register (or memory location).

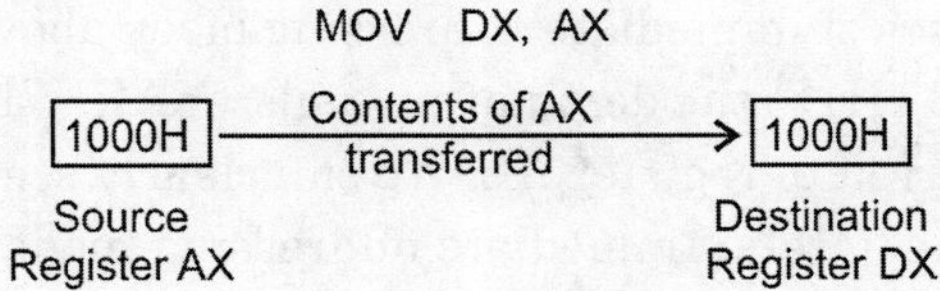

Figure. 3.10. Pictorial view of MOV OX, AX instruction

It should be clearly understood that MOV instruction copies data from the source and put this data into destination. However, it does not alter the contents of the source. (Refer figure 3.10)

There are many ways of data addressing that the 8086 or 8088 microprocessor follow and we shall discuss each in detaiL

3.4.1. REGISTER ADDRESSING MODE

This is the simplest addressing mode that the 8086 uses. In this type of addressing scheme the source and destination for data are both registers.

MOV AX, BX

* The memory location can also be a register.

This type of addressing can be used for 16 bit or 8 bit of data. Therefore the data length that can be operated on 16 bits or 8 bits wide. A point to note here is that we cannot mix two different data lengths, that is, we cannot have a destination register which is 16 bits wide and a source: register whose width is 8 bits or vice versa. The lengths of the source and destination registers have to be same.

Instruction	Reg. Size	Operation
MOV AH, AL	8 bit	Copies AL to AH
MOV BH, BL	8 bits	Copies AL to AH
MOV AX, BX	16 bits	Copies BX to AX
MOV SP, BP	16 bits	Copies BP to SP
MOV DI, BP	16 bits	Copies OP to 01

Figure 3.11. Example of using register addressing mode instructions

When. we execute MOV AX, BX the contents of BX register are copied to AX but the contents of BX remain unchanged" Register addressing mode allows the use of any two of the AX, BX, CX. DX. SP, BP, SI, DI registers for 16 bit data transfer and any two of AH, AL, BH, BL. CH, CL and DH, DL for 8 bit data transfer.

3.4.2. Immediate Addressing Mode

In the immediate addressing mode the data that is to be executed on, follows the instruction stored in memory immediately.

e.g. MOV AX,-2432 H

MOV AL, 12H

We have shown two cases of immediate addressing mode above. In the first case we want to transfer 16 bits of data {2432 H) to the destination register AX. While in the second instruction8 bits of data is to be transferred to register AL. We can clearly see that the data bytes (or word) follow the instruction immediately. Immediate addressing mode can be used for moving data immediately to memory locations or registers.

3.4.3. Direct Data Addressing

Most instructions of the 8086/8088 can use the direct data addressing mode. The direct data addressing mode simply specifies the location of the data which is to be operated on by the instruction. There are two forms of the direct data addressing mode.

1. Direct Addressing
2. Displacement Addressing

The address of the data is calculated by adding the displacement to default segment register which is to be used as the segment address for that addressing mode.

3.4.3.1. Direct Addressing

An instruction executing in the direct addressing mode operates between a memory location in the data segment and the AX (16 bit) or AL (8 bit) register. Consider the following MOV instruction which uses this type of addressing mode. These instructions can be two or three bytes long.

MOV AL, [1234 H]*

The MOV instruction will move the contents of the address formed by DS :[1234 H] to the register AL. The four forms of the MOV instruction using direct addressing are listed below in figure 3.12.

Assembly	Opcode	Size	Operation
MOV, AL, [1234H]	AO	8 bits	Copies the contents of DS: [1234H] to AL
MaV [1234H], AL	A2	8bits	Copies the contents of AL to memory location OS: [1234H]
MaV [2345H], AX	A3	16 bits	Copies the contents of AX to memory location specified by OS: [2345 H]
MaV AX, [2345H]	A1	16 bits	Copies the contents of DS: [2345 H] to AX

Figure 3.12. Direct addressingmodes

3.4.3.2. Displacement Addressing mode

Since, the direct data addressing mode is a very commonly used addressing mode for the instructions used in a program, intel designers also included the displacement addressing mode in 8086/8088 microprocessor. Displacement addressing mode can use any of the 8086 registers (excluding segment registers) unlike the direct addressing mode where only the AX or AL register can be used. In this type of addressing the instruction formed is four bytes long. Consider an example.

MOV CL, [1234 H]*

MOV DX, [4567 H]

The figure 3.13 lists a few of displacement addressing instructions for MOV.

Assembly	Size	Operation
MOV CH, [1000 H]	16 bits	Copies contents of location OS: [100H] to CH register
MOV BL, [1000H]	8 bits	Copies contents of OS: [1000H] location to BL register
MOV [1000-u SP	16bits	Copies contents of SP to memory location OS: [1 aaaH]
MOV [1234 H], ES	16 bits	Copies contents of ES to memory location OS : [1234 H]

Figure 3.13. Displacement addressing instruction for MaV

3.4.4. Register Indirect Type Addressing Mode

When the offset address of the data to be addresed is located in the BP, BX, DI, SI registers, then the addressing mode is called Register indirect addressing mode. An example of this instruction is MOV AX, [SI]**. Assuming DS to hold 2000 Hand SI to hold 1234 H.

On execution of the above instruction (which uses register indirect addressing mode) the effective address is calculated as DS : [SI] = [2000 H] : [1234 H] = 21234 H. After this address is calculated the contents of memory location 21234H is transferred to AL and the contents of memory location 21235H is transferred to AH. Thus MOV AX, [81] copies the contents of the address specified by DS : SI to the register AX. When an instruction uses BX, DI or SI register to hold the offset address of the memory location, the microprocessor assumes that the segment register to be uses's DS (data segment register). If an instruction uses the BP register to old the offset address of the memory location to be accessed, the processor assumes the segment address is contained in register SS (Stack segment). Figure 3.14(b) shows some of the different Register indirect MOV instructions.

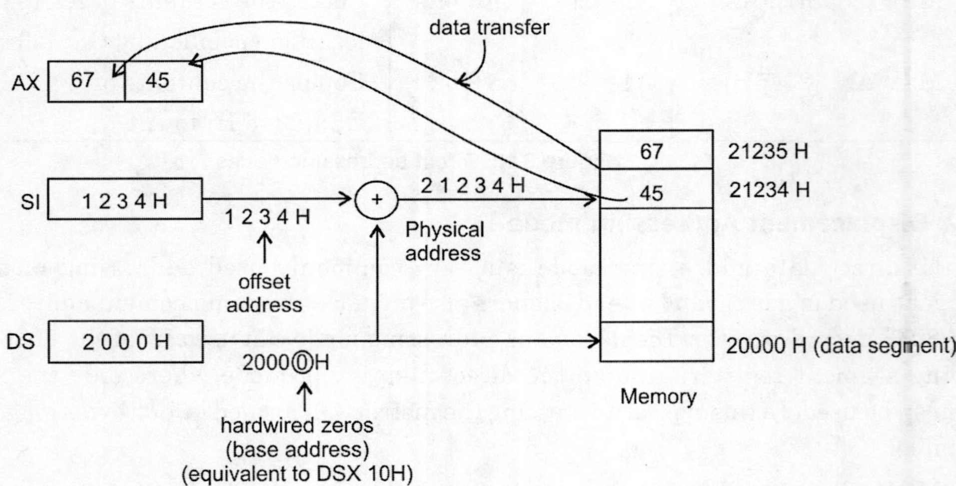

Figure 3.14 (a). Operation of MOV AX, [SI] where DS = 2000H and 81 = 1234 H.

Assembly	Size	Operation
MOV CX, [BX]	16 bits	Copies contents of DS : [BX] to CX.
MOV [BP], AL	8 bits	Copies contents of AL to SS : [BP] address
MOV [DI], DL	8 bits	Copies contents of DL to DS: [DI] address
rMOV CL, [SI]	8 bit	Copies contents of CL to DS : [SI] address

Figure 3.14(b). Register indirect instructions of MOV

3.4.5. Base Plus Index Register Addressing

This addressing mode is an enhanced Register Indirect addressing mode. As in register indirect addressing, the base plus Index register addressing indirectly addresses data by the use of registers to store the offset address. In this type of addressing mode one of the base registers (BX or BP) is combined with one index register (SI or DI) to form the memory offset address. The base register holds the beginning location of the memory array (in a segment) while the index register holds the relative position of an element in the array. If the BX register is used as the base register the DS is default memory segment while if BP register is used as the base register then the default segment register will be the SS register.

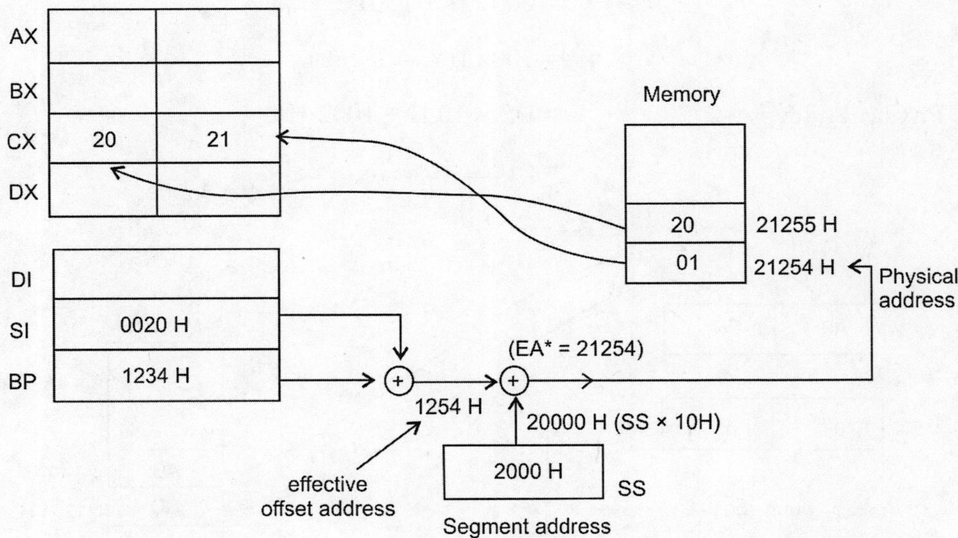

Figure 3.15. Example of base plus index regular addressing showing operations for MOV ex, [BP + Si] instruction. The default segment used is SS as BP hold the offset.

SS= 2000H, SI = 0020H, BP = 1234H

Assembly	Size	Operation
MOV CX, [BP + SI]	16 bits	transfers contents of SS : [BP + SI] address to ex
MOV AL, [BX + DI]	8 bits	transfer contents of DS : [BX + DI] address to AL
MOV [BP + SI], DX	16 bits	transfer contents of DX register to SS: [BP + SI] address

Figure 3.16. Examples of base plus index register addressing mode

3.4.6. Register Relative Data Addressing Mode

Register relative data addressing mode is similar to base plus index register addressing mode. The difference here lies that a relative data is added to the contents of one register(BX,BP, SI,

Dl) to obtain the effective offset address. The relative data that is added can be a 8 bit data or a 16 bit data. This data is also stored in the memory and also known as displacement data. For example consider the instruction.

Displacement data

MOV AX, [BX + 1001H]

Assuming BX has 0001 H and DS has 2000 H.

Effective offset address = BX + 1001 H

= 0001 H + 1001H

= 1002 H

Physical address = DS × 10 H + 1002 H

= 21002 H.

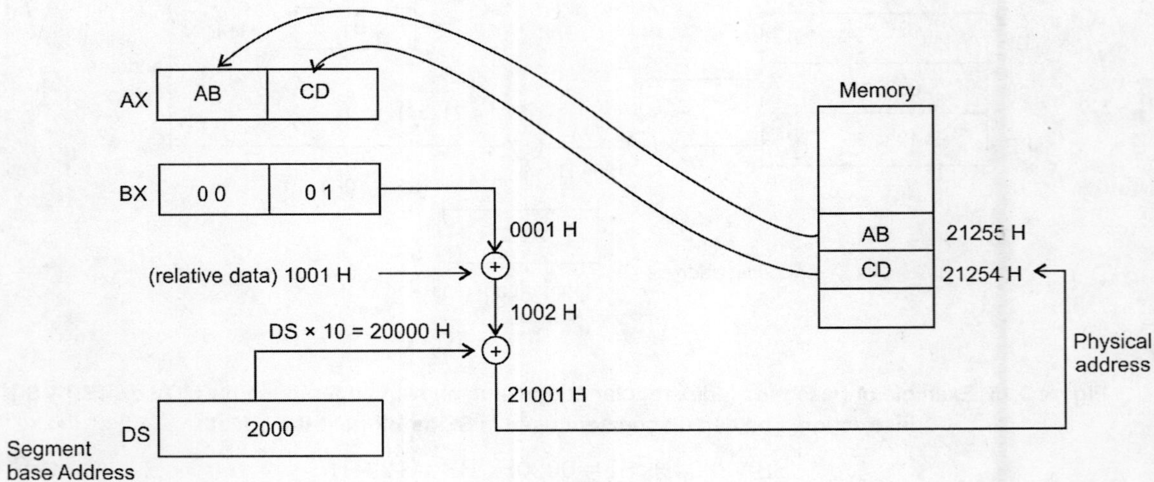

Figure 3.17. Operations of MOV AX, [BX + 1001 H) where BX = 0001 Hand DS = 2000 H

Thus, the AX register is loaded with the contents of memory location specified by DS : [BX + 1000 H];

3.4.7. Base Plus Index Relative Data Addressing

The base plus index relative data addressing mode is similar to base plus index and register relative data addressing mode. It combines features of both these instructions. A base register [BX, BP] is added to a index register [SI, DI] as well as a data displacement (8 bit or 16 bit data). There are two types of base plus index relative data addressing instruction. The first

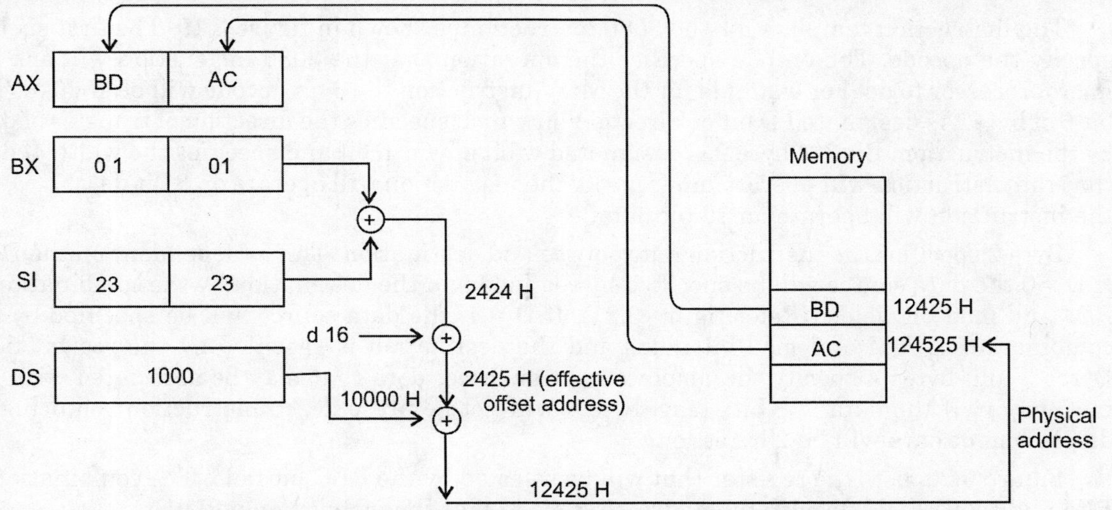

Figure 3.18. Operation of base plus index relative data addressing mode for

MOV AX, [BX + SI + d16] where BX = 0101 H, SI = 2323 H, d16 = 0001 Hand DS = 1000 H

type involves the use of a base and index register and a displacement data which is 8 bits wide to form the address. The second involves the base and index register and a displacement data which is 16 bits long. Given below are examples of both these types of instructions.

MOV AX, [BX + SI + d 8]

MOV AX, [BX + SI + d 16]

In the above two general examples the first corresponds to an address of DS: [BX + SI + d8] where d8 is an 8 bit displacement while the second corresponds to an address ofDS: [BX + SI + d 16] where d 16 is an 16 bit relative data. The operation of this instruction is shown in figure 3.18.

3.5. WRITING MACHINE CODE INSTRUCTIONS FOR 8086/8088

An 8086/8088 instruction is broken down as follows. This is the basic format for an 8086 instruction.

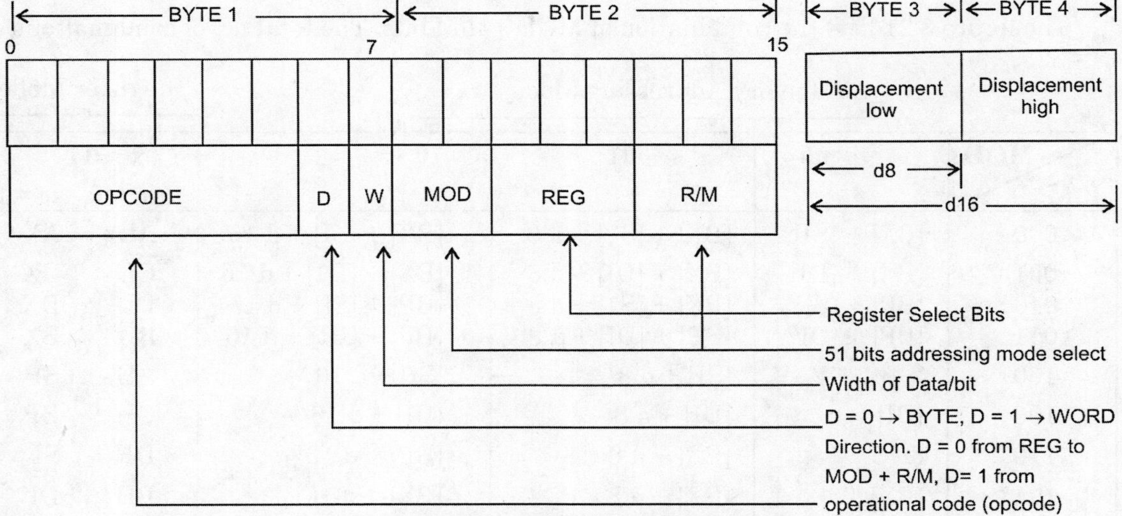

Figure 3.19. 8086 Instruction Template

The instruction template for the 8086 instruction is shown in figure 3.19. The first six bits specify the opcode. The opcode specifies the operation that the 8086 instruction will ask the microprocessor to do. For example for the MOV instruction the 6 bit opcode will be 100010. The bit 6 of byte 1 is designated D bit or direction bit which specifies the direction of transfer of data by the instruction. Bit 7 of byte 2 is designated w bit of width bit and specifies the width of data that the instruction will operate on. If $w = 0$, the instruction will operate on 8 bit data. If $w = 1$, the instruction will operate on 16 bit data.

Byte 2 specifies the instruction data source and destination. This is dependent on the D bit If D = 0 the data source will be specified by reg field and the destination by the combination of R/M and mod bit fields. (Refer figure 3.21). If D = 1, the data source will be specified by the combination of R/M and mod bit fields and the destination is specified by the reg bit field. Bytes 3 and bytes 4 specify the displacement or direct data that may be associated with the instruction. If the data is 8 bits long, byte 4 will not be present. An instruction requiring no displacement data will be 2 bytes long.

Figure 3.20. lists the register that will be selected by the REG bit field. The combination of R/M and MOD Field decides the addressing mode that the instruction will use.

CODE	**REGISTER**	
REG	***w* = 1**	***w* = 0**
000	AX	AL
001	CX	CL
010	DX	DL
011	BX	BL
100	SP	AH
101	BP	CH
110	S1	DH
1 1 1	D1	BH

Figure 3.20. Register selected by reg bit field

The figure 3.21 lists the combination of MOD + R/M bits. The total no. of combinations are $2^5 = 32$.

MOD / **R/M**	Memory Addressing Mode **00**	**01**	**10**	Register Mode **11** *w* = 0	*w* = 1
000	[BX] + [SI]	[BX] + [S1] + d 8	[BX] + [SI] + d 16	AL	AX
001	[BX] + [DI]	[BX] + [DI] + d 8	[BX] + [D1] + d 16	CL	CX
010	[BP] + [SI]	[BX] + [S1] + d 8	[BP] + [SI] + d 16	DL	DX
011	[BP] + [DI]	[BP] + [DI] + d 8	[BP] + [DI] + d 16	BL	BX
100	[SI]	[SI] + d 8	[SI] + d 16	AH	SP
100	[DI]	[DI] + d 8	[DI] + d 16	CH	BP
110	d 16	[BP] + d 8	[BP) + d 16	DH	SI
111	[BX]	[BX] + d 8	[BX] + d 16	BH	DI

Figure 3.21. Mod and R/M bit field combinations

Figure 3.21 shows Addressing mode selected by an instruction for different combintaion of MOD and R/M fields. The default segment register used is DS except in the case when BP register is used to form the offset address. Whenever BP register forms part of offset address the SS segment register is used as the default segment register. The default segment register can be changed by use of segment override prefix (section 3.6).

It can be clearly observed from figure 3.20 that when

(*a*) MOD = 1 1 → Then the addressing mode used is Register addressing where one register is specified by REG and the second by R/M and MOD bits.

(*b*) MOD = 0 0
R/M = 110 → Then the addressing mode used is direct data addressing with *d* 16 being added to Data segment (DS) to produce the effective address d 16 here is the 16 bit direct address.

(*c*) MOD = 0 0
R/M = 100 or 101 → Then the addressing mode used by the 8086 will be register indirect mode.

(*d*) MOD = 0 0
R/M = 000 to 011 → Then the addressing mode will be base plus index register addressing mode.

(*e*) MOD = 01
R/M = 000 to 011
or MOD = 10
R/M = 000 to 011 → Then the addressing mode will be base plus index relative data addressing mode.

(*f*) MOD = 01
R/M = 100 to 111
or MOD = 10
R/M = 100 to 111 → Then the addressing mode will be register relative data addressing mode.

For the memory mode operation specified by the combination of MOD and R/M the default segment base register is the DS (data segment) except in the cases where BP is used to form the offset address. Then the default segment will be the stack segment SS register. If we want to use any other segment register for the base address we can use the segment override prefix to bypass the default segment address. Then the. addressing mode will be base plus index relative data addressing mode.

Then the addressing mode will be register relative data addressing mode.

3.6. SEGMENT OVERRIDE PREFIX INSTRUCTION

For cases where the programmer wants to use a segment register other than the default segment register defined for the instruction, he can do so by using the segment overide prefix instruction. The segment overide prefix instruction format is given below

001	REG	10

Segment override prefix (8 bits)

To change a default segment to one desired can be done by loading the segment overide prefix before the instruction whose segment is to be changed. Then the default segment of the instruction will be changed to that specified by REG. The segment register that will then be used by the instruction is listed in figure 3.22.

Segment Register	CODE (For REG)
CS	01
OS	11
ES	00
SS	10

Figure 3.22. Segment register codes for use by segment override prefix

Suppose we want to change the segment base address from the default DS register to the ES register; we would have to precede the instruction with a segment override prefix of 00100110.

3.7. EXAMPLES OF CODED INSTRUCTION

Let us code a few of the 8086 instructions using the format discussed above.

(a) MOV AX, BX

Opcode for mov = 1 0 0 0 1 0

width = 1 [As AX and BX are 16 bit register]

Now we can see there are two ways to coding the instruction.

First is if D = 0, then BX must be specified by REG and AX by MOD + R/M combination then transfer will take place from REG to MOD + R/M

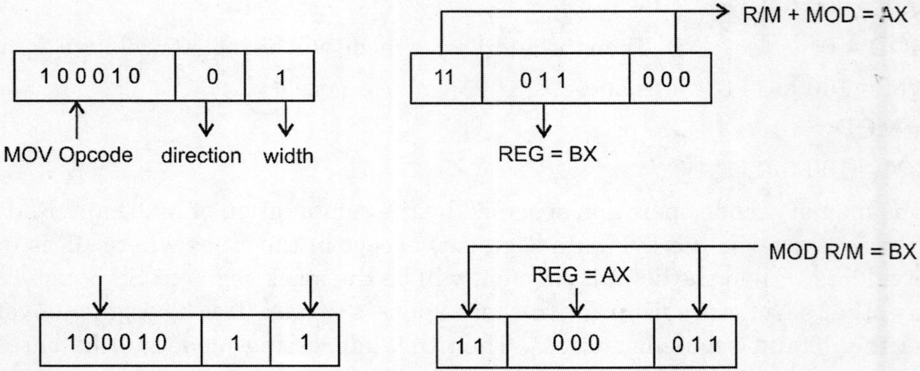

Thus we have formed two instructions code for instruction MOV AX, BX. It must be clearly understood that one instruction can be computed in more than one bit patterns for the 8086. This instruction (is as earlier discussed) is an instruction using register to register addressing mode.

(b) MOV AL, [DI]

Now let us code a register indirect addressed instruction. Note that AL is a 8 bit register so data transfer is 8 bit will, thus w = 0

Now opcode = 1 0 0 0 1 0

$w = 0$

D = 1

REG = 0 0 0]→ for AL when w = 0

MOD = 00

RIM = 01

Thus the machine code for MOV AL, [DI] is

1 0 0 0 1 0	1	0

0	0	0	0	0	1	0	1

(c) MOV AX, d 16

This is clearly a direct addressing mode based instruction where d 16 is direct data. Now

opcode = 1 0 0 0 1 0

w = 1

D = 1

REG = 000

MOD = 00

RIM = 110

thus in memory we well store

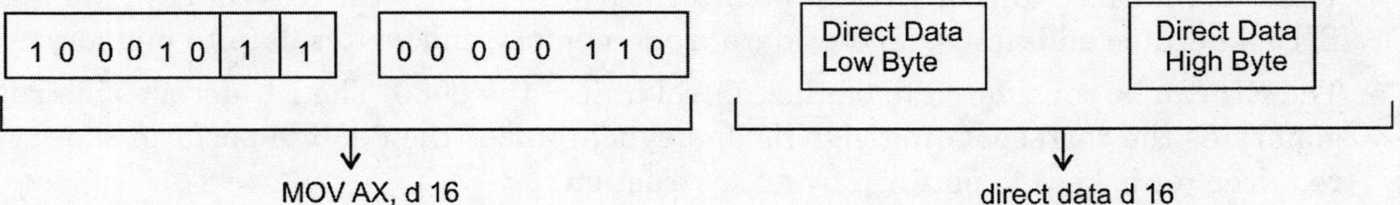

Now we are very clear with 8086 instruction therefore, we shall be easily able to code instruction for the 8086/8088 microprocessor. However suppose 'we have to perform the instruction where we donot want to use the default segment.

(d) MOV BP, [BX + SI]

We know the default segment for BX + SI is DS suppose we want to use the CS segment register as the segment base address. Thus we will have to use the segment override prefix as the default segment register used for MOV BP, [BX+ SI] is the SS register. Now to choose the CS register to hold the segment address we would have to chose the value of reg bits in the segment override prefix as 01. (refer figure 3.22).

Now for MOV BP, [BX + SI]

opcode = 100010

w = 1

D = 1

RIM = 000

MOD = 00

REG = 101

thus instruction in of bit pattern

1 0 0 0 1 0	1	1

0 0	1 0 1	0 0 0

However when we load this instruction in memory we will do as follows

00001H	2EH →	Segment Override so as to change DS to CS
00002H	8BH ⎤→	MOVBP, [BX+SI]
00003H	28H ⎦	Hexadecimal representation of above instruction

On execution first 2EH is executed which is the segment override prefix. This converts the default segment for next instruction to CS. After 8 Band 28 H are executed [Hex for above instruction] the memory location CS : [BX + SI] are transferred. to BP.

3.8. KEY POINTS TO REMEMBER

1. The Intel 8086 microprocessor has a 20 bit address bus, a 16 bit data bus while the 8088 has a 20 bit address bus and a 8bit data bus.
2. This makes the 8086 a 16-bit processor while the 8088 is an 8 bit microprocessor. However internally the 8086 and 8088 are similar.
3. The 8086 (or 8088) can be considered to be made of two functional units which can operate simultaneously. These two units are
   - Execution unit
   - Bus Inter face unit
4. The Bus interface unit or BIU is the interface of the EU to the external system bus. The BIU generates addresses, fetches data from memory and writes data to memory.
5. The EU can be considered to be the control unit of the 8086. The EU decodes instructions generates the corresponding signals and synchronizes the 8086 to perform the task it is required (or asked) to perform by an instruction.
6. The EU can be broken down into
   - Instruction decoder and control subsystem
   - ALU
   - Flag Register
   - General Purpose Register
   - Index and Pointer Registers
7. The BIU can be divided into the following parts
   - Instruction decoder
   - Segment Registers
   - Instruction Pointer
   - Bus control and address generator unit.
8. The 8086 contains nine active flags which are of two types
   - Conditional flags: examples ZF, CF, SF, OF, ACF, PF
   - Control flags: examples TF, DF, IF
9. The ALU of the EU is the unit where all arithmetic and logic operations are performed.
10. Instruction Decoder and control subsystem decodes instructions and generates the required control and synchronization signals.

11. There are four 16 bit general purpose register in the 8086 namely AX, BX, CX, DX. The AX register is also called the accumulator. In addition the 8086 also contains three other registers namely SI, DI (Index Registers) and BP (Base Pointer register)register.
12. The 8086/8088 uses pipelining as shown in figure 2.4 for increasing its execution rate.
13. The Queue is a 6 byte FIF0 register which can be considered as a cache memory. The BIU fetches upto 6 byte of instructions from memory and stores it in the queue for the EU to execute.
14. The segment registers are used to hold the segment base address of the 8086 memory segment.
15. The 8086 uses a memory addressing scheme called segmented addressing. Segmented addressing divides the available memory into segments.
16. The 8086 memory space can be divided into four-segments each of 64 Kb.
17. The 20 bit 8086 address is produced as follows: Physical Address = segment base address × 10H + offset address
18. Programs for microprocessor such as 8086/8088 are written in three different types of languages which are
    - Machine Language
    - Assembly Language
    - High Level Language
19. Machine languages use the instruction set of the microprocessor. These programs can be directly run (executed) by the processor. Programming in machine language is very tedious.
20. Assembly language use mnemonics (such as MOV, ADD) to represent the machine instruction set. The mnemonics are chosen so as to explain the operation of the instruction. Programming in assembly language is simples than machine language.
21. The simplest way to write program is by use of high level language such as C, C++. Here the commands are more english like such as printf. Each of these commands is translated into a number of machine codes by the compiler.
22. Data is accessed from memory by the microprocessor. To retrieve data from memory an address is sent. The different ways an address is formed by the microprocessor are called the addressing modes of the microprocessor.
23. There are eight different types of addressing modes supported by the 8086/8088 microprocessor. These are

    (a) register addressing mode .

    (b) immediate addressing mode

    (c) direct data addressing mode

    (d) displacement addressing mode

    (e) regiseer indirect addressing mode

    (f) base plus index addressing mode

    (g) register relative addressing mode

    (h) base plus index relative data addressing mode

24. The 8086/8088 instruction has the general format given below

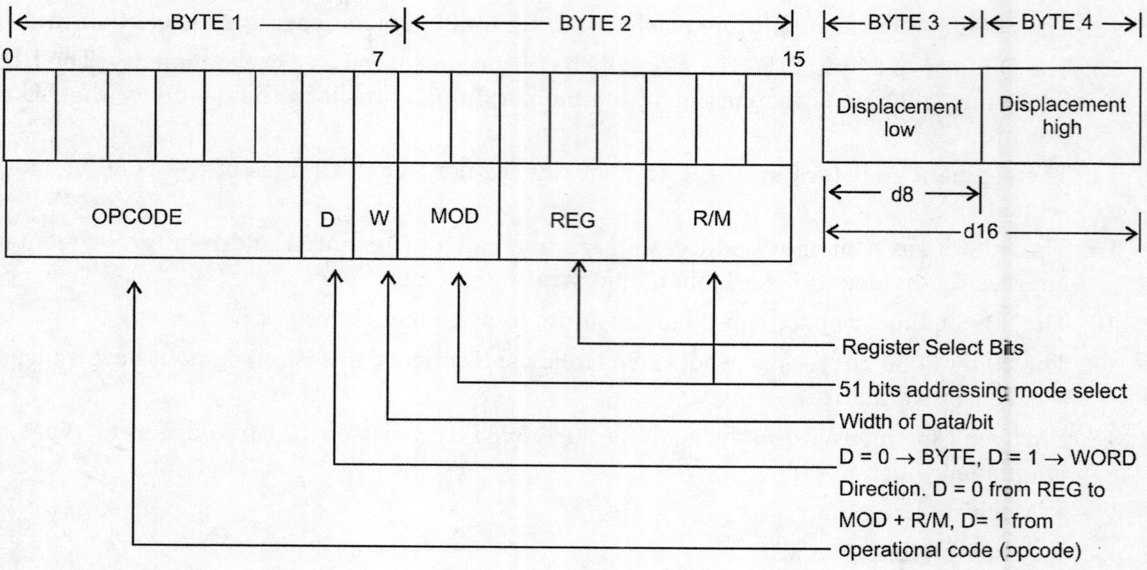

25. There are 32 different ways an instruction can reference data as shown in figure 3.21.
26. For memory addressing mode of figure 3.21, the default segment address is contained in the data segment (DS) register except for the cases when BP register forms part of the offset 'address. When the BP register is used to hold the offset address then the default segment register is the SS (stack segment)
27. To use a segment base address except the default cases (as discussed in point 9), the 8086/8088 instruction set contains the segment override prefix. The segment override prefix must precede the instruction whose default segment base register is to be overrided. The format for segment override prefix is

REVIEW QUESTIONS

3.1. What determines whether the microprocessor is an 8 bit, 16 bit or 32 bit microprocessor?

3.2. What is segmented memory addressing? What are its advantages?

3.3. What is the size of the 8086 address bus? What is the effective memory space it can address using this address bus?

3.4. At any time what is the maximum number of memory that an 8086 processor can use. (Hint:Number of segments = 4, Hence total space of one segment multiplied by number of segments).

3.5. What is pipelining? How is 8086 architecture designed to incorporate pipelining?

3.6. What is the use of the BIU and EU? Describe each's functions?

3.7. Describe the 8086 instruction queue? What are its uses and advantages?

3.8. Why are the microprocessor general purpose registers preferred for use of temporary data storage over the 8086 memory?

3.9. If the first instruction of a program is to be fetched from 12340H what should the value of CS register be? The sixth instruction to be fetched is at memory location 1234FH. What should the value of IP register be?

(Hint: Assume IP to be OOOOH for first part).

3.10. Calculate the physical addresses represented by

(a) 1234H: 000 2H

(b) 2670H: 2222 H

(c) F2F2H: 1234 H

3.11. What are the uses of flags? What are the different flags present in the 8086 microprocessor? Write the use of each?

3.12. Describe the 8086 internal architecture in detail?

3.13. List the differences and similarities between the 8086 and 8088 microprocessor?

3.14. Explain the difference between linear addressing and segmented memory addressing?

3.15. What are the main differences between the general purpose registers, index and pointer registers?

3.16. Comprehensively, explain the different subunits of the Execution unit and Bus interface unit?

3.17. The stack segment register SS contain 2424H while the SP register contains 2121H. What is the physical address of the start of stack.

3.18. A data byte is to be fetched from the memory location 42C20H. What are the different combinations for segment base address and offset address that can be used to access this memory location.

3.19. What addressing mode is used the following instruction

(a) MOV AX, BX

(b) MOV AX, [DI]

(c) MOV [SI + BX], AX

(d) MOV.EX, [SI + DI -+ dS]

(e) MOV CX, 2342H

3.20. List the operation performed by MOV AX, [DI]. Also graphically show the process. Assume contents of DI = 4232H and of DS = 2000H. Assume any other data required.

3.21. Code the following instruction using Sec. 3.3. and Appendix A for

1. ADD AX, BX

2. MOV CX, [DI]

3. MOV DX, [SI + BP]

4. MOV CX, 2432H

5. MOV BX, [1234H]

3.22. It is desired to move a data from the ES (extra segment) to a register using MOV AX, [BP] where BP = 2343H, SS = 2000H and ES = 1000H. It is found that the data that is entered into AX is incorrect. Explain what is · the solution to remove this error.

3.23. What is the use of the direction bit and ,width bit of the 8086/8088 instruction template.

3.24. Suppose DS = 0200H, BX = 0300H, DI = 400H, SI = 2000H. Determine the memory address accessed by each of the following instructions.

(a) MOV AL, [1234H]

(b) MOV AX, [BX]

(c) MOV AX, [BX + DI]

(d) MOV CS, [BX + SI]

3.25. What is wrong with a MOV [BX], [DI] instruction?

3.26. What is wrong with a MOV AX, DL instruction?

3.27. List the 16 bit registers as well as the different 8 bit registers that are used for register addressing.

3.2S. Suppose DS = 1000H, SS = 2000H, BP = 1000H and Dl= 0001H. Determine the memory address accessed by each of the following instructions

(a) MOV AL, [BP + DI]

(b) MOV CX, [DI]

(c) MOV DX, [BP]

Also form the instruction codes for each of these instructions.

3.29. Suppose DS = 2000H, SS = 2100H, BP = 0200H and SI = 1500H. Determine the memory accessed by the following instructions

3.30. Explain the different addressing modes of the 8086/808S microprocessor.

3.31. The mod bits = 01 and RIM bits = 110 in a particular instruction. Explain the addressing mode that will be used. Also explain the default segment register that will be used.

3.32. What is the use of the segment override prefix. Suppose in the above instruction it is required to use the segment register as CS to hold the segment base address. What would be the steps to do so.

3.33. Graphically explain the execution of the following instructions

1. MOV AX, [SI]

2. MOV BX, [DI + dS]

3. MOV BX, [BP + DI + dS]

4. MOV DX, [1234H]

3.34. What do you understand by machine language.

3.35. Explain the differences between assembly and high level languages.

3.36. What is understood by addressing modes of a microprocessor?

3.37. Differentiate between direct and displacement addressing modes?

3.38. Calculate the memory address the following instructions will access. Also explain the addressing modes that are used by each instruction

1. MOV CX, [1234H]
2. MOV AX, [2222R]
3. MOV DX, [BP]
4. MOV DX, [BP + DI]
5. MOV DX, [BP + DI + 01R]
6. MOV AX, [BX + SI + 200R]
7. MOV AX, 1234H

Assume Bp = 0100H, Dl = 0200H, SI = 2000H, BX = 1200H, SS = O100H, DS = 2000H

3.39. Calculate the instruction code for the following instruction (use appendix A)

1. ADD BX, CX
2. ADD CX,[BP + 1OOR]
3. SUB [BX + SI + 20H] , AX
4. SUB [BP], AX
5. AND [BX], CX
6. AND AX, [BP + SI + 02H]

●●●

Chapter-4

THE 8086/8088 MICROPROCESSOR CPU MODULE DESIGN

INTRODUCTION

In the preceding chapters we have read about the 8086/8088 microprocessors internal architecture. However to design an 8086/8088 microprocessor based system we have to have. Comprehensive knowledge about the 8086 and 8088 microprocessor IC module. In addition the 8086 microprocessor requires a few dedicated IC's to function as a complete computer. In this chapter we shall introduce to you the 8086 and 8088 microprocessor IC pin outs, their pin assignments and functions. We shall also study the 8284 clock generator IC. Appendix B contain comprehensives manufacturer data sheets enabling students who want to word on project based on the 8086 microprocessor to be able to properly design such systems. Here we shall also attempt to provide you knowledge of the 8086 differed system, a minimum and maximum made configuration and its operation. However a reference to an 8086 microprocessor will be valid to the 8088 microprocessor due to internal similarities which exist.

4.1. THE 8086/8088 CPU MODULE PIN OUT

On the first glance the 8086 or the 8088 microprocessor may appear to be a very complex IC. Figure. 4. 1 (a) shows the pin out of an 8086 microprocessor while fig. 4.1.(b) shows the pin out of the 8088 microprocessor. You may at the first glance observe no difference between the two microprocessor IC packages.

Both microprocessor are packaged in a 40 pin dual in line package (DIP). As discussed earlier the 8086 microprocessor has a 16 bit data bus while the 8088 has a 8 bit microprocessor, though both are very similar to each other.

4.1.1. DC CHARACTERISTICS

Both the 8086 and 8088 microprocessors require a supply voltage of + 5V with a tolerance ± 10%. The 8086-1 and 8088 - 1 versions can be operated at a supply voltage of 5V with a tolerance of ± 5%. These microprocessor can be operated in temperature ranging from O°C to 70°C. Since this range is insufficient for certain applications there are extended temperature versions of both these microprocessor which can be operated between temperature range from –40°C to 140°C.

Both these microprocessors draw a maximum supply current of 340 mA. The 8086 – 1 version draws a maximum supply current of 360 mA. These microprocessor have an average power dissipation value of about 2.5W.

4.1.3 ARCHITECTURE OF THE 8088 MICROPROCESSOR

The register set of 8088 is exactly similar as that of 8086. The architecture of 8088 is also similar to 8086 except for two changes (a) 8088 has 4- byte Instruction queue and (b) 8088 has 8- bit data bus. The function of each block is the same as in 8086. The architecture of 8088 is shown in figure 4.1.

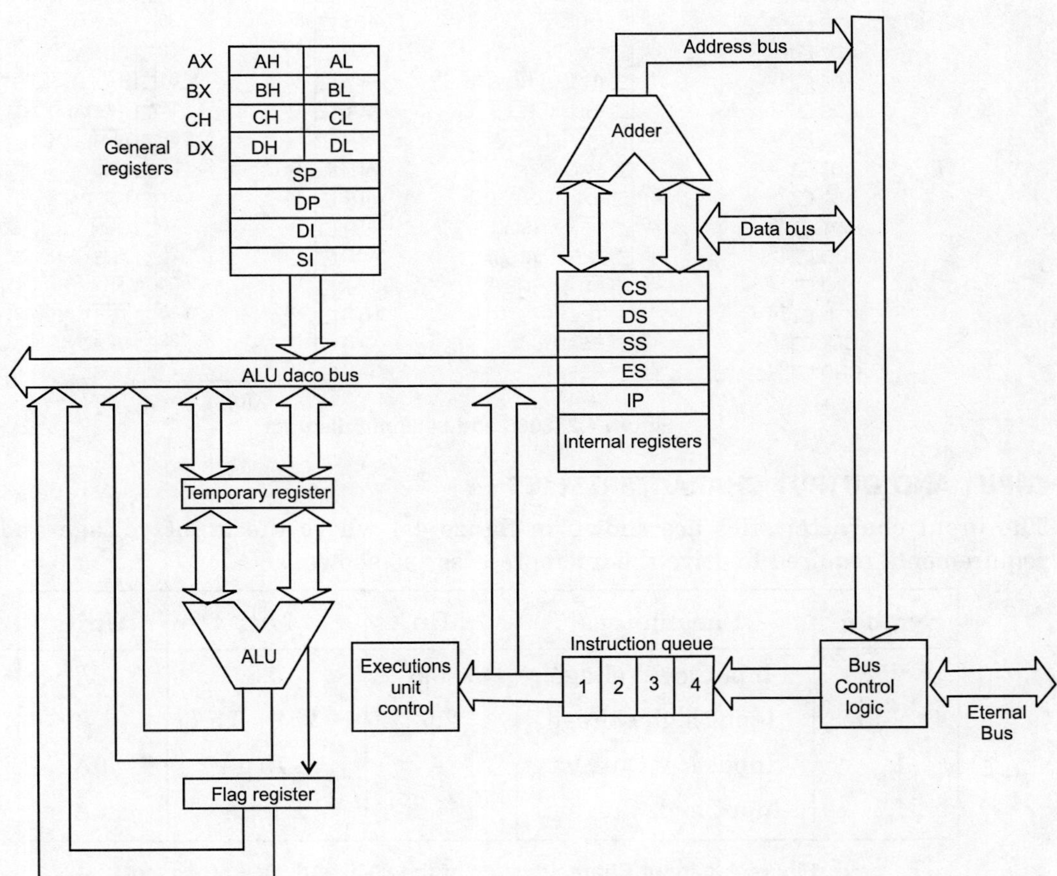

Figure 4.1: Architecture of 8088 microprocessor.

Since the addressing capability of 8088 microprocessor is 1 M byte, therefore, it needs 20 address bit, i.e. 20 addressing lines. The segmented memory scheme is used while handling this 20- bit address and the complete physical address forming procedure is the similar to the 8080 microprocessor. The memory organization and addressing methods of 8088 is also similar to 8086 microprocessor, while it physically interfacing memory. The complete memory is homogeneously addressed as a bank of 1 M byte memory location using the segmented memory scheme. This change in hardware is completely transparent to software, as a result of the modified data bus; the 8088 can access only a byte at a time. This fact reduces the speed of operation of 8088 as compared to 8086 but the 8088 can internally process the 16- bit data.

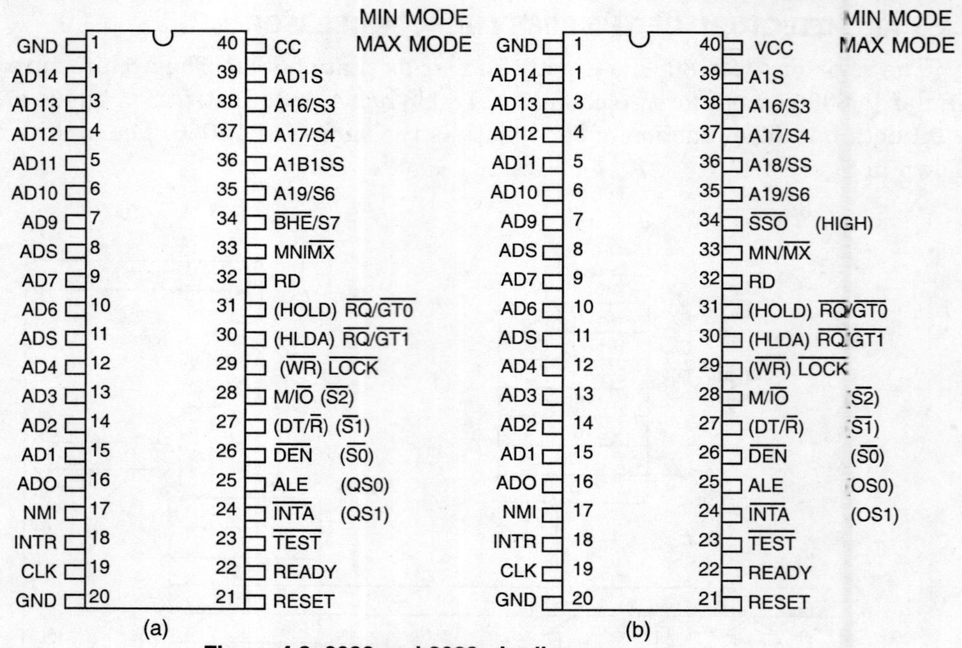

Figure 4.2. 8086 and 8088 pin diagram

INPUT AND OUTPUT CHARACTERISTICS

The input characteristics are shown in figure 4.4 where the input voltage and current requirements required to drive the microprocessor as shown.

Symbol	Parameters	Min	Max	Units
V_{IL}	Input low Voltage	– 0.5	0.8	V
V_{IL}	Input high voltage	2.0	Vcc + 0.5	V
I_{IL}	Input low Current	–	± 10 μA	μA
I_{IH}	Input high Current	–	± 10μA	μA

Figure 4.3. Input Characteristics of the 8086 and 8088 processor

Output characteristics are shown in figure 4.4. However the 8086 and 8088 microprocessor have low noise immunity due to which a 8086 buffered system is preferred when drawing large number ofloads. Buffered 8086 system is discussed in detail later.

Symbol	Pararneters	Min	Max	Units
V_{OL}	Output low voltage	–	0.45	V
V_{IH}	Output high voltage	2.4	–	V
I_{IL}	Output low Current	–	2.5	mA
I_{IH}	Output high Current	–	400	μA

Figure 4.4. Output characteristics for the 8086 and 8088 microprocessors.

Here we shall briefly explain the 8086 and 8088 microprocessor IC pins, their functions and names as well as a brief description of their purpose. Appendix B lists the pin definition as given by Intel Corporation.

4.1.3. PINS SIGNAL ASSIGNMENTS (8086)

Here we shall briefly explain the 8086 and 8088 microprocessor IC pins, their functions and names as well as a brief description of their purpose Appendix B lists the pin defintions as given by intel corporation.

Pin Assignment	Pin No.	Description
AD_{15}–AD_0	39,2–16	This is the multiplexed data and address bus. These bits contain 16 bits of data or address depending on the ALE logic value. When ALE = 0, these pins contain 16 bits of data where AD_0 is LSB bit and AD_{15} is MSB bit of data. When ALE = 1, these pin contain the right most 16 bits of the memory address. For I/O operations AD_7–AD_0 contains the port address when ALE = 1.
A_{19}/S6–A_{16}/S3	35–38	This is the upper order address bus bits multiplexed with status bits. Initially, these pins contain the upper order address bits (when ALE = 1). During the next clock after ALE goes, low these pins contains the status bits as shown in figure 4.4 which shows the truth table for S_6–S_3 status bits.
$\overline{RD}$	32	This is the 8086 read control signal. The read signal is active when it goes low. When RD goes low the microprocessor receives data on AD_{15}–AD_0 from memory or I/O devices.
RESET	21	When the RESET pin is held high for four or more clock periods, it resets the microprocessor. This clears all registers and loads the CS register with FFFFH and the IP register with 0000H .. This also resets the IF flag to zero.
CLK	19	This Pin is used to provide the clock signal for processor synchronization. The clock signal must have a duty cycle of 33% as specified by Intel for proper operation.
V_{CC}	40	This Pin is used to supply the +5.0 V input supply voltage.
GND	1,20	These are the 8086 ground pins. For reliable operation both these pins must be grounded.

S3	S4	Decoded Function
0	0	Extra Segment being accessed
0	1	Stack Segment being accessed
1	0	Code segment being accessed
1	1	Data segment being accessed

S_6 = l(Always)

S_5^* = IF (Interrupt flag status bit)

Figure 4.5. Function of S_3, S_4, S_5, S_6 Status bits

Pin Assisgnment	Pin No	Description
MN / $\overline{MX}$	33	The voltage level at this Pin determines whether the 8086 or 8088 microprocessor will operate in minimum or maximum mode. For minimum mode operation of the microprocessor this Pin should be connected to + VCC and for maximum mode operation this pin should be connected to GND voltage levels.
NMI	17	This is Non-maskable Interrupt Pin. This is a positive edge triggered input and cannot be disabled. This is Interrupt Request Pin used for interrupting the 8086.
INTR	18	When the INTR Pin is held high the 8086 checks the interrupt flag bit in the 8086 flag register (IF). If IF = 1 then the current instruction execution is completed and then the $\overline{INTA}$ pin becomes active. The processor is then said to have been interrupted and program execution is relocated to memory address specified by the interrupt request.
READY	22	This is an input that is used to control 8086 operation. This Pin is used to synchronize slower peripheral devices to the 8086 microprocessor. When READY = 0, the 8086 microproccssor enters a wait state where it remains idle. (performing no operation)., when READY = 1, the microprocessor functions normally.
$\overline{TEST}$	23	The $\overline{TEST}$ Pin is used in conjunction with the WAIT intruction. On excution of a wait instruction, the 8086 microprocessor checks the TEST pin status. If $\overline{TEST}$ = 0 the wait instruction executes as a NOP (No operation) instruction. If $\overline{TEST}$ = 1, the wait instruction will cause the microprocessor to becomes idle until $\overline{TEST}$ goes low and becomes zero.
$\overline{BHE}$ / S7	34	The bus high enable Pin is an active low and used to enable the higher order 8 bits of data bus during a memory operation. The 87 status bit is always 1.**

* S_5 status bit will show the state of the IF flag bit. Thus, IF = 0 when S_5 = 0, when IF = 1 then S_5 = 1.

Minimum Mode Pins Signal Assignment

We have earlier referred to a term minimum mode and maximum mode operation of the 8086 microprocessor. The 8086 or 8088 is mode to operate in minimum** mode when the MN/$\overline{MX}$ pin is connected directly to + 5V while the maximum*** mode operation occurs when the MN/ $\overline{MX}$ pin is connected directly to ground.

Pin Assignment	Pin No.	Description
M/$\overline{IO}$	28	This output Pin is used to select memory or I/O devices. When the address in the 8086/8088 refers to an I/O device then this Pin is driven low while if it is accessing a memory location then this Pin is in high (1) state.
$\overline{WR}$	29	This is the write signal output Pin. When this Pin goes low it indicates that the data bus contains valid data that is to be stored in memory or outputed to a I/O device.
$\overline{INTR}$	24	Interrupt Acknowledge Signal. When the INTR goes high and the microprocessor this Pin goes low. This Pin stays low till the interrupt routine is executed and is high during normal operation of the processor (i.e., processor is not interrupted).
ALE	25	Address Latch Enable Pin goes high when the 8086/8088 multiplexed data bus contains valid address bits i.e., when ALE is 1, the multiplexed address data bus contains address bits. When ALE = 0, the multiplexed address data bus is used as a data bus i.e., the multiplexed address data bus will behave as a data bus and the data that is presently present on MAXIMU this bus is valid data bits.
DT/$\overline{R}$	27	This is labelled Data Transmit/Receive signal. When the 8086 microprocessor is transmitting data on its bus then DT/$\overline{R}$ = 1 and when microprocessor is receiving data on its buses then DT/$\overline{R}$ = O. This is used as an enable signal for data buffers that may be used in an 8086/8088 based system.
$\overline{DEN}$	26	Data bus enable signal goes low when the microprocesser address data bus is to be used as a data bus. It is can be used to enable buffers for buffering the data bus.
HOLD	31	When the external peripherals request a DMA (Direct Memory Access), they do so by placing a high signal on this Pin. When is placed on the microprocesser stops execution and relenquishes control of the address, data and control bus by placing them in a high impedance state. It also tristates

		AD15 – ADO, A19/S6 – A16/S3, $\overline{RD}$, $IO/\overline{M}$, $M/\overline{IO}$, $\overline{WR}$ Pins. After doing so HLDA is set to l. If Hold = 0, the microprocesses executes nor mally.
HLDA	30	Hold Aknowledge bit is placed 1 when the processor enters the DMA state. All Pin that have tri stated remain so till the HOLD Pin becomes O.

$\overline{S}_2$	$\overline{S}_1$	$\overline{S}_0$	Control Function*
0	0	0	Interrupt Acknowledge
0	0	0	I/O Read
0	1	0	I/O write
0	1	1	Halt
1	0	0	Opcode fetch
1	0	1	Memory Read
1	1	1	No operation

Figure 4.6. Status Bit function

QSl	QSO	Decoded Function
0	0	Queue is an idle state
0	1	First byte of opcode has entered queue
1	0	Queue empty
1	1	Subsequent byte of opcode

Figure 4.7. Instruction Queue Status Bit Function

Maximum Mode Pin Assignment

When the $MN/\overline{MX}$ Pin is grounded the microprocessor (8086/8088) enters maximum mode operation. This mode is used for multiprocessor systems or systems containing the 8087 numeric coprocessor when interfaced to 8086 or 8088 microprocessor.

Pin Assignment	Pin No.	Description
$\overline{LOCK}$	29	This pin is placed low when the LOCK instruction is used. It is interfaced to active high chip enable Pins of peripherals so as to lock them out of the system (therefore disabling them)
QS1, QS0	24, 25	These are the queue status bits. Figure 4.6 shows the truth / table and are used by the 8087 coprocessor.

* The control functions are arbitarily decoded by the 8288 bus controller which performs the corresponding task hence with

$\overline{RQ}$/GT1 and $\overline{RQ}$/GT0	30, 31	When DMA transfer mode is required during maximum mode operation then these request/grant Pin are used. They are bidirectional pins and are used for requesting a DMA. When the processor grants a DMA it is also conveyed through these bits.
$\overline{S2}$, $\overline{S1}$ and $\overline{S0}$	26 – 28	The status bits are used by the bus controller for the current bus cycle decoding. Their truth table is shown in figure 4.6.

IO / $\overline{M}$	DT / $\overline{R}$	$\overline{SSR}$	Function
0	0	0	Interrupt acknowledge
0	0	0	Memory read
0	1	0	Memory write
1	0	0	Opcode fetch
1	0	1	I/O Read
1	1	1	I_0 write
1	1	1	Passive

Figure 4.8. 8088 decoded status bits for 8088 bus synchronization

Pin Signal Assignments (8088)

As discussed earlier there are only few differences between the 8086 and 8088 microprocessor. The major difference ,being that the 8088 has an 8 bit external data bus. This in effect qualifies if as a 8 bit CPU. Though in the true sense the 8088 is not really a 8 bit CPU because internally it can handle 16 bit of data just like the 8086. It should be clearly understood that internally the 8088 works exactly as an 8086 microprocessor. Here we discuss only the Pins that are different from 8086 such as the address and data Pins, $\overline{SSO}$ Pin and 101$\overline{M}$ Pin.

Pin assignemnt	Pin No	Description
AD7 – AD0	9 –16	The 8088 multiplexed address data bus. It is 8 bits wide and carries the lowest 8 bits of the address of memory or I/0 port address when ALE = 1, when ALE = 0 this bus functions as a data bus and can carry 8 bits of data.
A15 – A8	39, 2 – 8	These carry the next 8 bits of address and the remain contant throughout. the bus bycle.
10/M	28	The memory/I0 port selection Pin. If this Pin is high it selects an I/0 device and low it is

		used to select a memory chip. This Pin is directly connected to chip enable of a memory chip or port device.
$\overline{SSO}$	34	The status line output pin out is a status bit which is sent by 8088 microprocesser. The $IO/\overline{M}$, $DT/\overline{R}$ and $\overline{SSO}$ together carry the function of the current bus cycle the truth table shown in figure 4.7.

4.2. CLOCK GENERATION

Before we attempt to design an 8086 based computer, we will obviously have to provide basic input signals such as clock and reset synchronization. The intel designers with the 8086 released the 8284 clock and Reset synchronization IC. The advantages of using the 8284 clock generator shall become clear as you go along. It is an important IC and reduces the number of devices required for generation of a stable clock signal. It also provides RESET and READY synchronization signals which are required for propel' operation of the 80888 and 8086 microprocessors. Figure 4.8 shows the pin out of the 8284 clock generator IC. The 8284 clock generator is available in an 18-pin integrated circuit and is for use specifically with the 8086 and 8088 microprocessor. We shall now list each pin and its function.

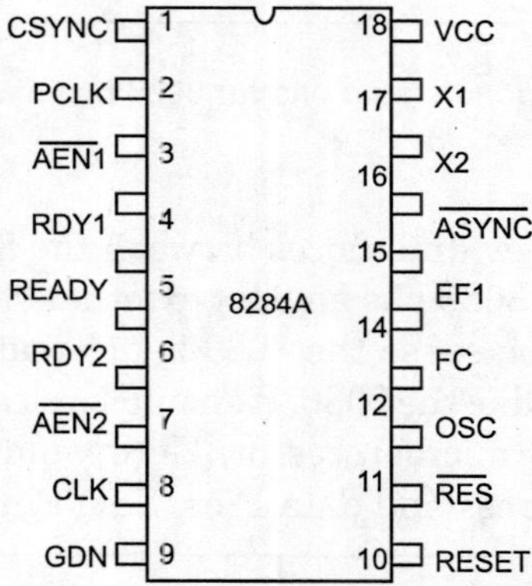

Figure 4.9. The pin-out of the 8284A clock generator

4.2.1. The 8284a Clock Generator Pin Functions

CSYNC : This is the clock synchronization input. When an external frequency input is used for clock generation this Pin must be held high. If this Pin is grounded, the internal crystal oscillator is used for timing.

PCLK : The PCLK output is basically used for peripheral clock synchronization. Since peripherals generally are much slower than the microprocessor, the 8284 supplies the PCLK signal at one sixth the frequency of the EFI or crystal clock frequency. This signal is half that of the CLK output that is provided to the microprocessor and has a 50 percent duty cycle.

$\overline{AEN1}$ and $\overline{AEN2}$: These are the address enable Pins. These Pins are used in conjunction with the RDYI and RDY2 input pins to cause wait states in the microprocessor. The wait states also require the READY Pin of the 8086 microprocessor to be attached to the 8284.

RDY1 and RDY2 : The bus ready input pins which cause wait states in an 8086/8088 system, when synchronized with $\overline{AEN}_1$ and $\overline{AEN}_2$ as well as READY output pin of 8086.

CLK : The clock output Pin which is directly connected to the 8086/8088 CLK Pin for clock timing. The clock output is one third the crystal or EFI input frequency that is a 9 MHz EFI input will give an output signal of 3 MHz at CLK Pin. The output signal has a 33 percent duty cycle as required by the 8086/8088 processor.

GND : This connects to the system ground for the clock generator IC grounding.

RESET : This RESET output pin connects to the 8086 RESET and provide a synchronized RESET output signal which resets the 8086/8088 microprocessor.

$\overline{RES}$ The reset input is an active low input to the 8284 A. It is generally connected to an RC coupling network which resets the 8086 microprocessor when the complete system is switched on. When the RES input pin goes low it effectively will send a synchronized reset signal to the 8086 microprocessor and hence reset it.

OSC : The oscillator output Pin is a TTL level signal that is at the same frequency as the EFI input or crystal frequency and is used to provide EFI input to other 8284 IC's in a multiprocessor based system.

F/C : This is a Pin for selection of whether to use an external frequency circuit or the crystal frequency of the 8284 clock generator to produce the clock output. If this Pin is low the internal cystal oscillator network provides the signal that is used for timing and when F/C is pulled high the EFI input is used.

EFI : External frequency input Pin. When F/C is pulled high the signal that is supplied to this Pin is used for the 8284 clock output as well as internal timing of the 8284 IC.

$\overline{ASYNC}$: This Pin also called ready synchronization decision input and selects the number of synchronization levels for RDY1 and RDY2. If this Pin is high one level of synchronization is chosen for RDY1 and RDY2 signals.

READY : The READY output directly connects to the 8086 microprocessor ready input pin and is used as output signal which allows the 8086 microprocessor to enter wait states.

X_1 and X_2 : The Xtal oscillator input Pins. These are used to connect a quartz crystal of thrice the desired clock frequency output i.e., for a 5 MHz clock, a 15 MHz crystal will be connected. This crystal is used as the resonating clock source for the 8284 clock register and to synchronize all its internal circuitry.

V_{CC} : The power supply pin which connects to +5.0V ±10% power source.

4.2.2. OPERATIONS OF 8284

Internally the 8284 clock generator can be divided into three sections as shown in figure 4.10. These section s are :

1. RESET SECTION
2. CLOCK SECTION
3. READY SYNCHRONIZATION

CLOCK SECTION

When a crystalis attached to terminals Xl and X2 it makes the 8284 internal oscillator oscillate at the same frequency as the crystal. This generated square wave it fed to an AND gate which is controlled by the state of F/C. If F/C is zero this signal to fed to the divide by three counter and ifF/C is one, the EFI signal is fed to the counter. The divide by three counter simply converts the frequency of the square wave to one by third of the input frequency (Xtal or EFI frequency). This signal is buffered arid fed to the CLK output giving us the required clock frequency for driving a 8086 microprocessor system. It can be seen the output ofthe divide by three counter is re-fed to a divide by two counter, whose output will be one sixth the input signal frequency. This is our PCLK signal.

RESET SECTION

This section is extremely simple to comprehend. It consists of a schmitt trigger buffer and a D type flip flop (negative edge trigger). Since the D flip flop is negative edge trigerred and is triggered by the negative transition of the CLK signal it will apply the RESET signal during a 1 to O clock transition of the CLK output. Since the 8086 samples the RESET pin on a 0 to 1 (+ve edge) transition of the clock thus this signal will be meet the requirements of the 8086 processor. The figure 4.11 shows an 8086 interfaced with an 8284 IC with RESET and CLK signal connected. An RC coupling circuit is attached with a time constant 10 ms. This RC coupling circuit is coupled to $\overline{RES}$ Pin. When the system is switched on $\overline{RES}$ signal is low there by ensuring that the 8284 clock generator resets the 8086/8088. After about 10 ms (the time constant of the RC circuit) the $\overline{RES}$ will rise to logic one therefore ensuring that the 8086 RESET Pin goes low thereby enabling normal operation of 8086.

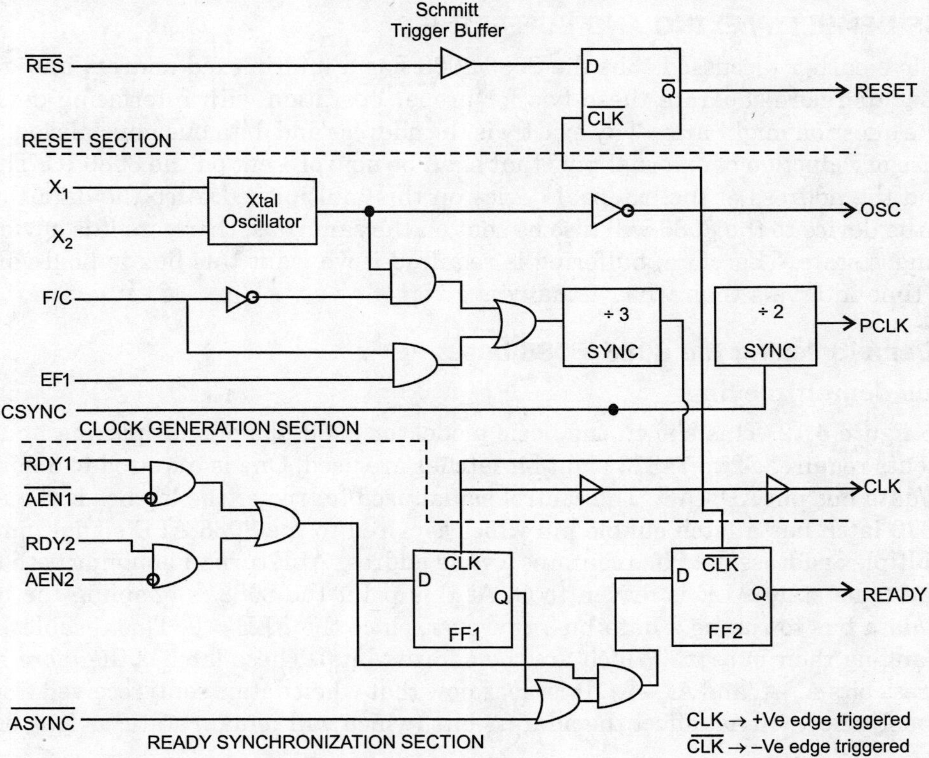

Figure 4.10. 8284 Internal Block diagram

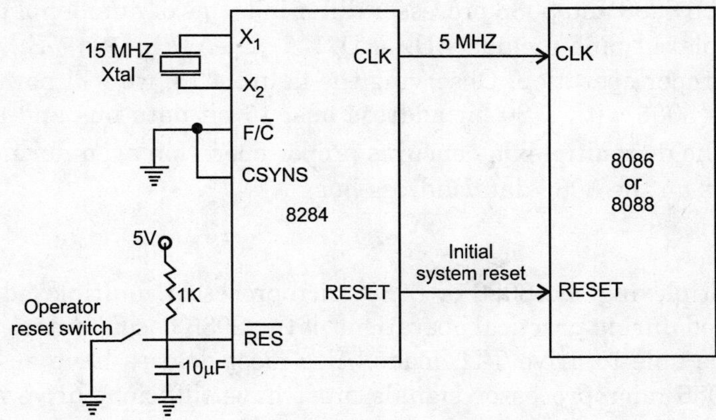

Figure 4.11.8284 and 8080 Interface for clock and reset

Since the RESET Pin of the 8086 require the input to remain at 1 for atleast 50 ms the above circuit ensures this criteria which is met due to a large time constant of the RC network. Since the RC network has a time constant of 10 ms the RESET output of 8284 clock generator will remain high for 10 ms therefore ensuring proper resetting of 8086. An operator reset switch is also provided. On pressing the switch the capacitor shall discharge therefore repeating the above process inturn resetting the 8086/8088 microprocessor.

4.3. BUS DEMULTIPLEXING AND BUFFERING

As we have earlier discussed that the 8086/8088 has a multiplexed address and data bus we first must therefore separate these two for proper operation with interfacing device. At this point, the question might arise why exactly is the address and data bus multiplexed. The reason is because of reduction of external pins that shall be now present on the 8086 IC. The 8086 will first send the address of the external device on this multiplexed. Also the data that must be sent by the device to the 8086 will also be sent on the same bus, therefore it is obvious that the bus changes state. Therefore, buffering is required if we want this bus to hold valid data for greater time intervals than what actually is given.

4.3.1. Demultiplexing the 8086/8088 buses

8088 bus demultiplexing

Refer to figure 4.12. This shown the basic model for 8088 bus demultiplexing and shows the components required, Two 74373 eight bit latches are used. One is required to demultiplux the address/data bus pin AD_0–AD_7 The control signal used to trigger the latches is the ALE signal. The 74373 latch has a latch enable pin which is wired to the 8086 ALE signal pin. When the 8088 multiplex address/data bus contains a valid address ALE turns 1 enabling both the latches. After one tome state ALE is driven to O. At this point the 8088 is enabling the multiplexed address/data bus to run as a data bus. However, since the ALE = 0. This disables the latches from changing their output. (Which are the address bits). These latches therefore have stored the address bits A_7–A_0 and A_{19}–A_{16}. It is clear now that when datais senti received from memory/ IO obviously they will not effect the address bits (which will remain static as long as ALE = 0).

Demultiplexing the 8086

The concept discussed above shall remain same for 8086 except for the addition of another 74373 octal latch. The 8088 and 8086 processor differ in terms of number of pin multiplexed. In the 8086 the multiplexed pins include AD_{15}–AD_0 , A_{19}/S_6–A_{16}/S_3, $\overline{BHE}/S_7$. All these must be demultipluxed for proper operation. Observing the figure 4.13, we will now notice that the 10 devices observe this 8086 with a 20 bit address bus, 16 bit data bus and a 3 bit control bus $(\overline{WR}, \overline{RD}, M/\bar{I}o)$, The de multiplexing ensures proper operation even during times when data glitches and changes on the 8086 data/address bus.

4.3.2. Buffering

However the demultiplexing the 8086 or 8088 microprocessor multiplexed doesnot solve the problems encountered during practical operations of the 8086 based system. We do require the microcomputer to be able to drive TTL loads. Also most devices have an input capacitance therefore all the 8086 microprocessor signals must have.sufficient drive capability to drive external output loads.

A solution to this problem would be to buffer the 8086 or 8088 microprocessor buses so that they may be able to drive high capacitive loads. The dem ultiplexed address bus doesnot require buffering as the octal latch will buffer it (as shown in figure 4.11 and 4.13). However the data bus must be buffered. Secondly the control signal such as $\overline{RD}$, $\overline{we}$ and IO/$\overline{M}$ (or M/$\overline{IO}$) must also be provided with sufficient drive ratings therefore should also be buffered. An 8086 or 8088 microprocessor which has all its buses (address and data bus) as well as its control signals buffered is called a fully buffered 8086 system. The data bus must use a bidirectional buffer such as 74245 (discussed in chapter 1) as it allows data signals to flow in both directions.

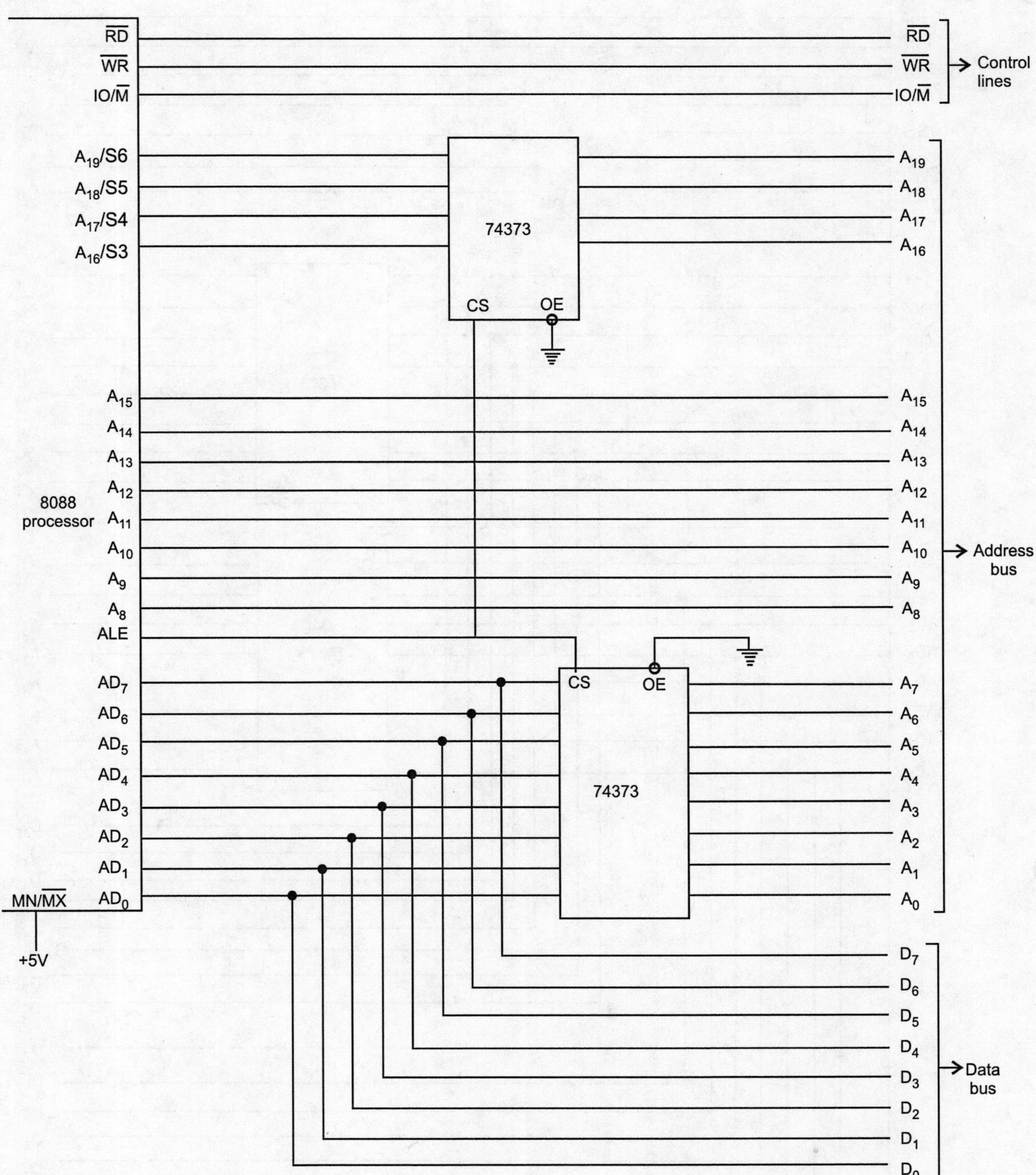

Figure 4.12 8088 bus demulitiplexing using 74378 8 bit latches

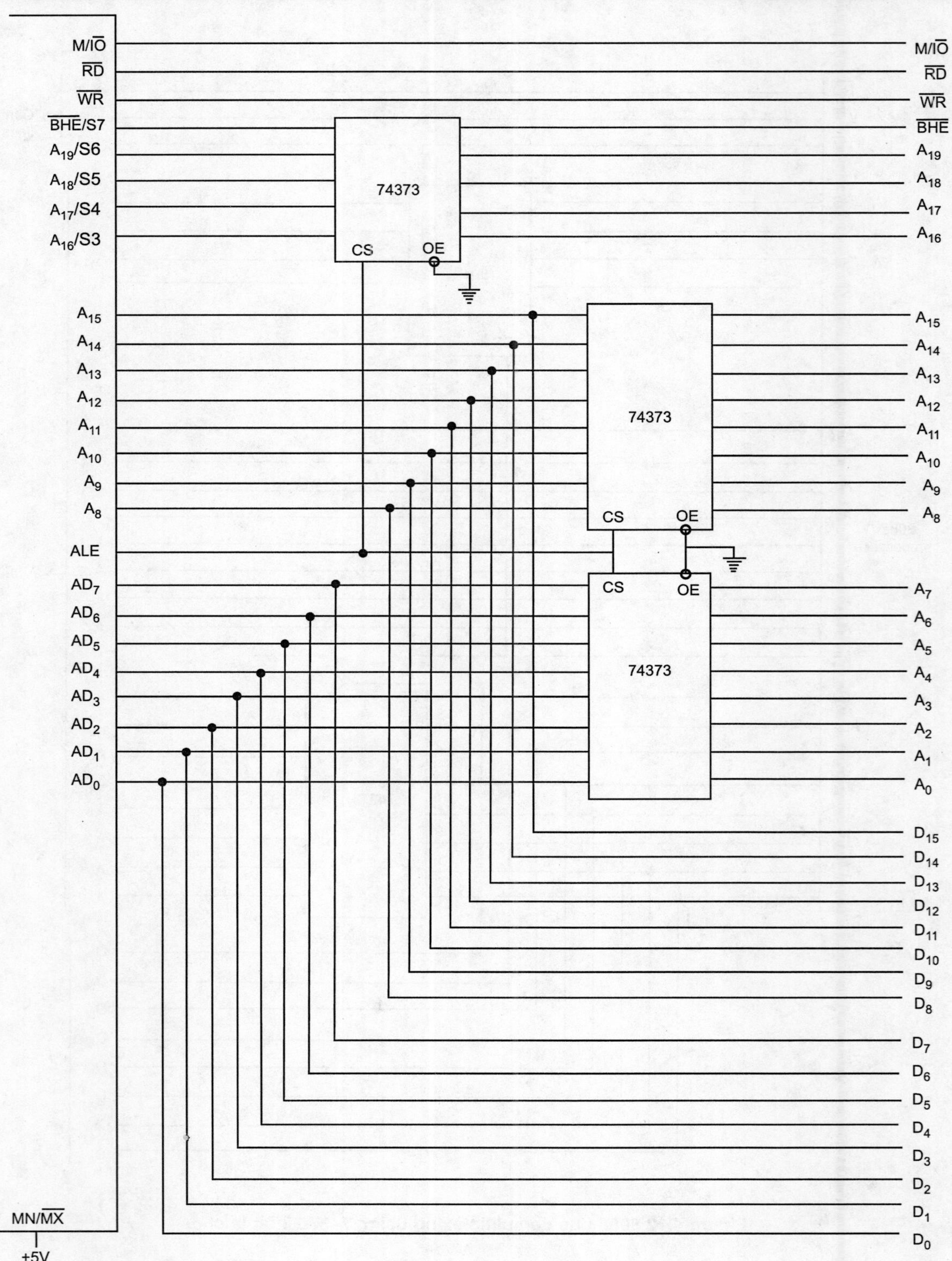

Figure 4.13 8086 bus demulitiplexing

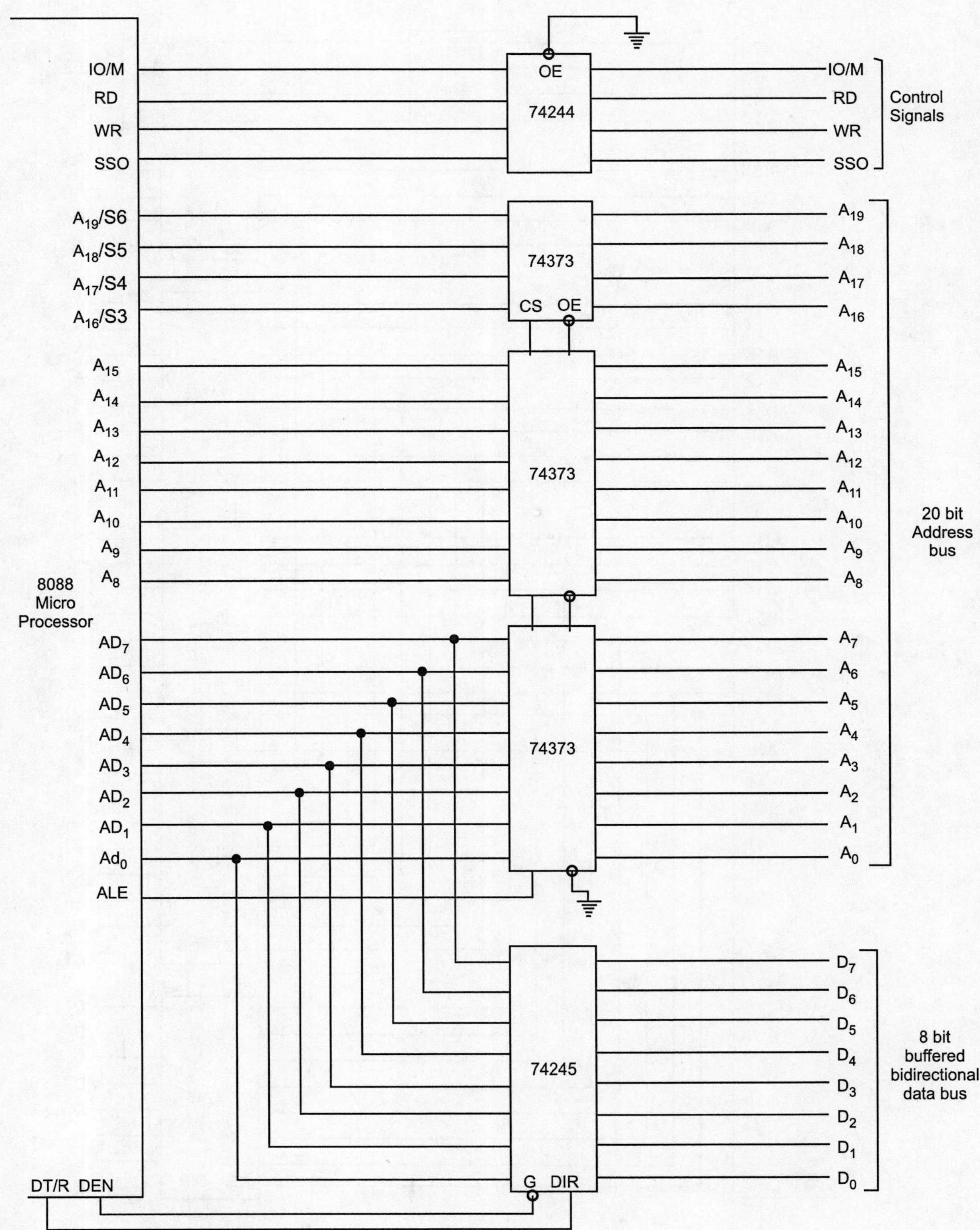

Figure 4.14. Fully buffered 8088 based system

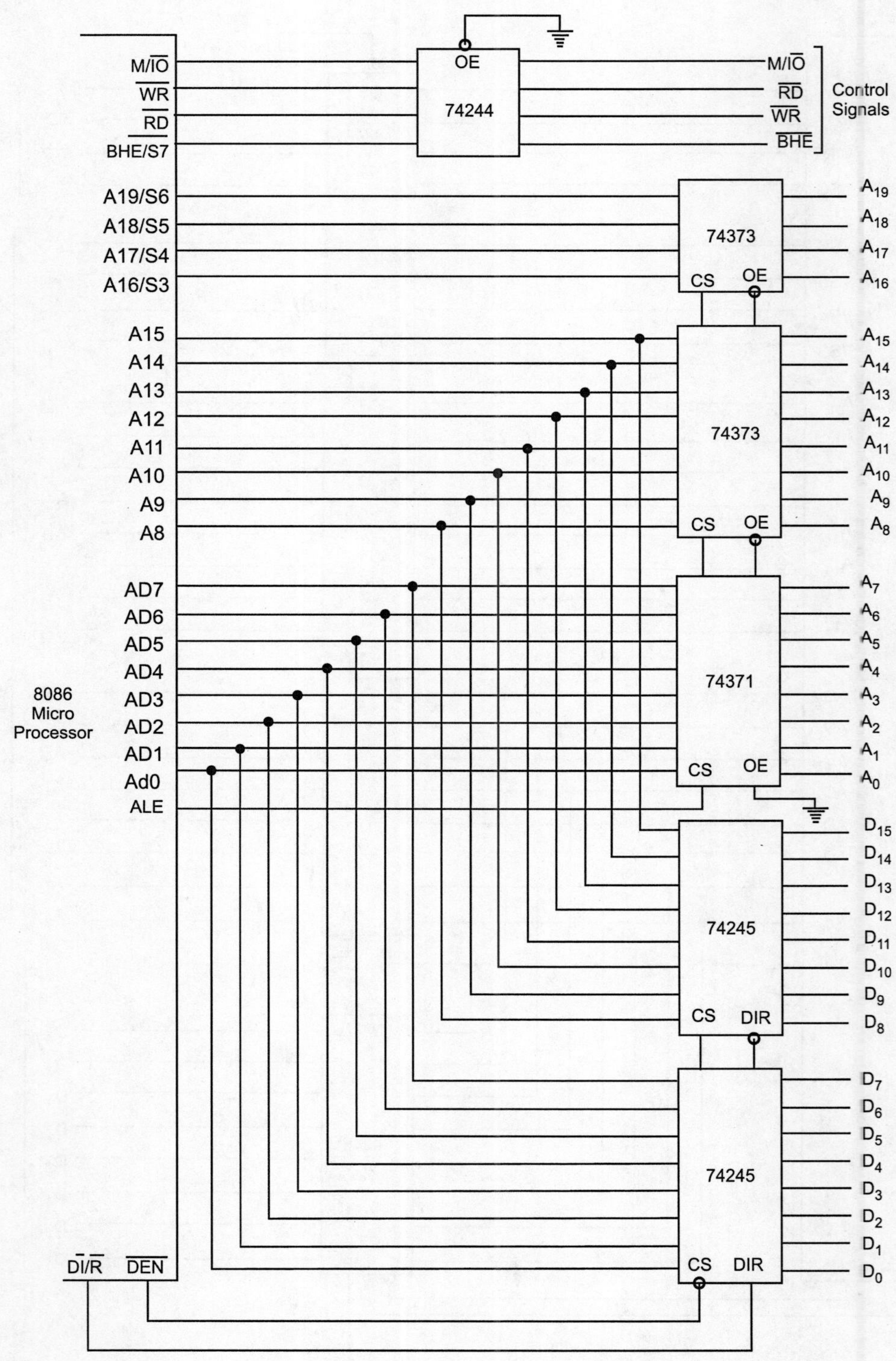

Figure 4.15. The Fully buffered 8088 microprocessor

Fully buffered 8086 and 8088 systems are shown in figure 4.14 and 4.15. The octal transceiver are used for buffering the data bus and has its direction signal pin controlled by the DT/$\overline{R}$. As usual DT/$\overline{R}$: is low when a. read operation is to take place enabling the 74245 latches to direct data toward the 8086 microprocessor. During write cycles the DT/$\overline{R}$. goes high enabling the 74245 latch to transmit data towards devices. The 74245 is also enabled or disabled by the $\overline{DEN}$ pin. The use of the 74244 buffer IC's is also required for control signals because if we use the 8 bit 74373 octal IC that is connected to the upper order address bus then the control signals will not be able to be transmitted to the devices on this latch because it will be disabled after the first cycle by the ALE signal. ($\overline{RD}$, $\overline{WR}$ signals undergo changes much later).

The 74373 attached to the address bits A_8 to A_{15} in figure 4.13 can be replaced by the 74244 IC as these bits do not carry any information except the address bits for the 8088 microprocessor.

4.4. READ AND WRITE CYCLE TIMING

All the hardware that we have discussed till now is obviously sequential logic circuitry. All sequential logic circuitry is synchronized with respect to a synchronization signal which we call the clock signal (CLK). However bus timing is more complicated than simple sequential circuit synchronization as we have to synchronize a large number of signals. Take for example the ALE has to be synchronized correctly so that the system designer may be able to correctly seperate the address and data bits that the microprocessor is inputting. Similarly other signals such as $\overline{RD}$ (or $\overline{WR}$), status bits (S7, S6, S5, S4 and S3, S2, S1), IO/M $\overline{(8088}$ microprocessor) are sent out at predetermined intervals of time with respect to the basic synchronizing signal (i.e., CLK). Due to the above reasons it becomes necessary for a hardware designer to understand bus timing so that he can chose the necessary devices (which have to obviously be compatible to the 8086 or 8088 bus timing) so as to form a system of use. However, we shall here discuss only the minimum mode read and write cycles (The maximum mode timings are listed in appendix B).

4.4.1. Simplified Read and Write Timing

The 8086 and 8088 microprocessor when accessing memory and I/O devices do operations that are performed in time intervals that remain constant. These constant time intervals or periods are called bus cycles. Each bus cycle is divided into exactly four clock periods". Let us assume that we are operating an 8086 microprocessor at 5MHz. Therefore

Microprocessor clock frequency = f = 5MHz

Microprocessor clock period = $\frac{1}{f} = \frac{1}{S_{MHz}}$ = 200ns

(or time state)

1 bus cycle = 4 time states = 4 × 200 ns = 800 ns

Thus a 8086 or 8088 microprocessor will access memory or I/O devices in periods of 800ns. thus the maximum rate with which an 8086 microprocessor can read or write data to and from memory (or I/O devices) is 1.25 million transfers per second (At 5 MHz).

Now let us discuss the read and write bus cycles. We know that a bus cycle is composed of four time state but then what exactly happens during these four time states. Simplified read

* a clock period or time state or T state is one time period ofthe clock signal that is connected to tl.e microprocessor eLK input pin from the 8284A clock generator.

and write cycles are shown in figure 4.15 and 4.16. We shall name the four time states Tl, T2, T3, and T4 and now discuss (with great simplicity) what would be the basic operations per formed by the microprocessor.

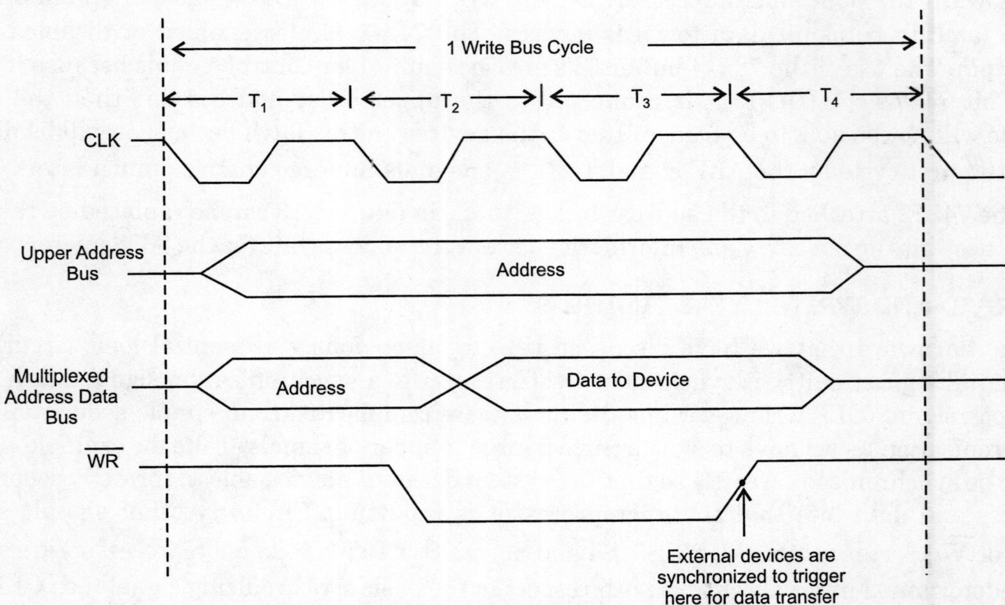

Figure 4.16. Simplified 8086/8088 write cycle

During the first time state T1, the microprocessor sends out the 20 bit address of the memory or I/O locations through the address bus. Control signals such as ALE, DT/$\overline{R}$ and IO/$\overline{M}$ (or M/$\overline{IO}$) are also sent out so that the microprocessor can synchronize the required circuitry for demultiplexing the address data bus as well as provide sufficient delay time for external devices to be enabled before the access signals are sent (such as $\overline{RD}$, $\overline{WR}$ etc.).

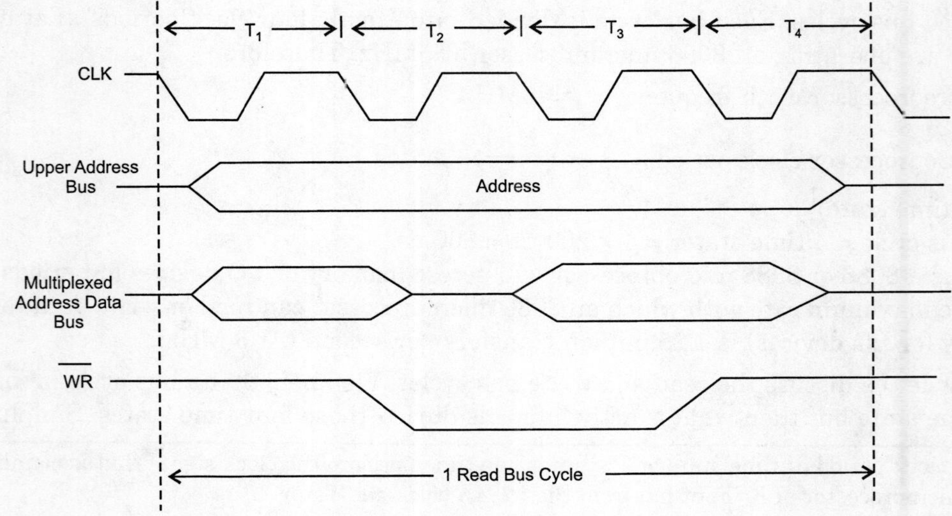

Figure 4.17. Simplified 8086/8088 Read Cycle

In the next time state T2, the processor sends out contr.ol signals such as RD,WR and DEN so that the external devices may be caused to perform the operation requested. During a write cycle the data to be written to memory (or I/0) is also output so that the data bits are set up on the input pins of the devices as the write (WR) signal is listed (Therefore avoid conflicts of setup time). The READY signal is sampled towards the end of T2 and if the READY signal is low the microprocessor enters a wait state (i.e., T3 becomes a. no operation time state or wait state. This is used to synchronize slower devices).

During a read bus cycle the data bus is sampled at the end of T_3 and the data held on the bus is transferred to the microprocessor. The T4 time state is the bus reset time state. During this time state all the buses are deactivated in preparation for the next bus cycle. During this period, the data bus is also in the process being read (sampled) if the current bus cycle is a read cycle. Also the transition on the WR signal (from low to high) is used to activate external devices to be written to (in case of the write cycle).

4.4.2. The 8086 Read Bus Cycle

Now we shall discuss the read bus cycle which is shown in figure 4.18. It might seems slightly complicated but on close observation with figure 4.19 (a) and (b) we will start to get a clearer picture. Fig. 4.19 lists the names of the various time divisions that the read bus cycle has been broken down into.

Here we shall describe to you the procedure for calculating the maximum access time that is allowed for accessing (read operation) a memory device. All memory devices must be able to operate well within this time frame for proper operation of the 8086 based system.

Now let us calculate the maximum time that can be made available for memory access. Now we have earlier read that the microprocessor samples the data bus at the end oftime state T3. However in figure 4.17 if you see you shall find a line going straight down at T3. Observe the data bus you find at this point a time T_{CLDX} which is the Data in hold time (figure 4.19(a). Therefore, this is the hold time required so that the proper sampling of the data bus may occur. Now this point which is at the end of T3 is exactly 3 Time states (i.e., 600ns) from the start of Tl.

Now a memory address appear on the address bus after a time T_{CLAV} after the start the start oftime state Tl. Now the memory should be allowed an access time from the time the address has appeared on the bus to the time It will be sampled.

$\therefore$ Memory access time = 600ns – T_{CLAV}

However the data must be setup before the sampling period by the microprocessor. Again observe figure 4.17. Where you shall find the T_{DVCL} or data setup time which is the setup time required before sampling of data bus. Now the memory obivously has to place its data before this setup time required by the processor. Therefore the maximum memory access time allowed will be

Maxim urn memory access time = 600ns – TCLAV – TDVCL

= 600 – 110 – 30 (From fig. 4.18(a) and (b) at 5MHz clock)

= 460 ns

Now a memory chip that is to be interfaced to an 8086 microprocessor should have an access time much lower than the calculated 460 ns.ln the buffered system, the buffers too will introduce a propagation delay therefore ideally memory chips interfaced to the 8086 should have access times well below 400 ns.

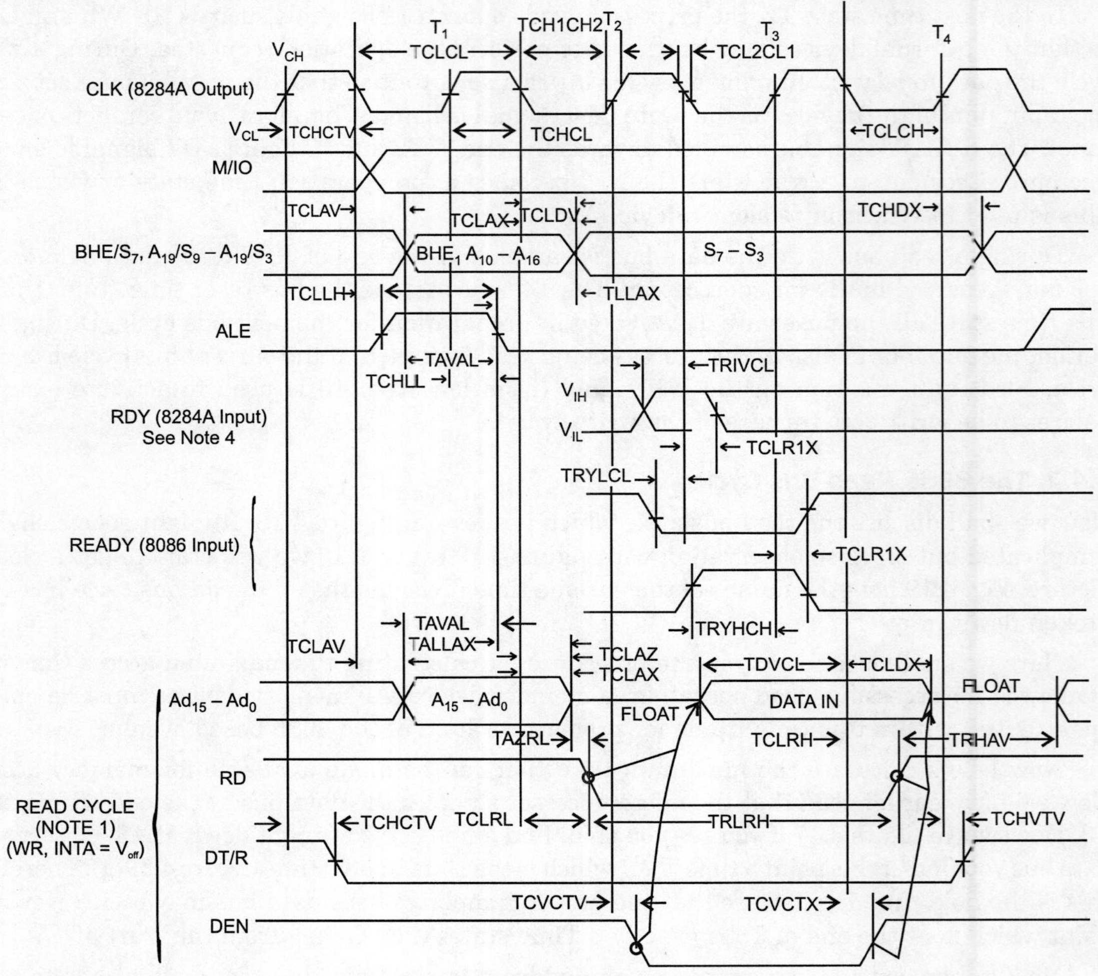

Figure 4.18. The 8086 Read Bus Cycle

4.4.3. The 8086 Write Bus Cycle

Figure 4.20 illustrates the 8086 minimum mode write cycle. There are very few differences between the read and write bus cycles. The $\overline{RD}$ strobe signal has been here replaced by the $\overline{WR}$ strobe while the data bus contains valid data which has to be written to memory. The $\overline{DI}/R$ signal remains at V_{OH} (logic 1) throughour the write bus cycle. The data is writen to memory device at the low to high transition of the WR strobe (during end of T3). Observe figure 4.19 where we observe that the memory write time has to less than the total time between the time of the $\overline{WR}$ strobe transition and the total data hold time after this transition. Thus memory devices must be able to transfer data in the limiting time period T_{WHDX} which is equal to T_{CLCH} – 30 = 118 – 30 = 88ns*. (Figure 4.19(a) and 4.19(b). Therefore memory devices must be able to written to in times less than 88ns when the 8086 microprocessor is operated at 5 MHz clock. Most memory Ie's available have hold times which range from 1 to 10 ns which is withm the limiting value of 88ns.

* On a 8086 operating clock frequency of 5 MHz

A.C. CHARACTERISTICS: (8086: TA = ODC to 70DC, VCC = 5 V ± 10%)
(8086·1: TA = ODC to 70DC, VCC= 5V± 5%)
(8086·2: TA = ODC to 70DC, VCC = 5 V ± 5%)

MINIMUM COMPLEXITY SYSTEM TIMING REQUIREMENTS

		8086		8086-1		8086-2		Units	Test
Symbol	**Parameter**	**Min**	**Max**	**Min**	**Max**	**Min**	**Max**		**Conditions**
TCL	CLK Cycle Period	200	500	100	500	125	500	ns	
TCLCH	CLK Low Time	118		53		68		ns	
TCHCL	CLK High Time	69		39		44		ns	
TCHICH2	CLK Rise Time		10		10		10	ns	From 1.0 Vto 3.5V
TCL2CLI	CLK Fall Time		10		10		10	ns	From 3.5 to 1.0V
TDVCL	Data in Setup Time	30			5		20	ns	
TCLDX	Data in Hold Time	10			10		10	ns	
TRIVCL	RDY Setup Time into	35			35		35	ns	
	8284A(See Notes I, 2)								
TCLRIX	RDY Hold Time into	0			0		0	ns	
	8284 (See Notes I, 2)								
TRYHCH	READY Setup	118			53		68	ns	
	Time into 8086								
TRYLCL	READY Hold								
	Time into 8086	30			20		20	ns	
TRYLCL	READY Inactive to CLK	–8		–10		–8		ns	
	(See Note 3)								
THVCH	READY Setup Time	35			20		20	ns	
TINVCH	INTER, NMI, TEST	30			15		15	ns	
	Setup Time (See Note 2)								
TILIH	Input Rise Time		20		20		20	ns	From
	(Except CLK)								0.8V to 2.0V
TIHIL	Input Fall Time		12		12		12	ns	From
	(Except CLK)								2.0V to 0.8V

Figure 4.19(a) Timing Responses

A.C. CHARACTERISTICS (Continued)
Timing Reponses

Symbol	Parameter	8086		8086-1		8086-2		Units	Test Conditions
		Min	Max	Min	Max	Min	Max		
TCLAV	Address Valid Delay	10	110	10	50	10	60	ns	
TCLAX	Address Hold Time	10		10		10		ns	
TCLAZ	Address Float Delay	TCLAX	80	10	40	TCLAX	50	ns	
TLHLL	ALE Width	TCLCH·20		TCLCH·I0		TCLAX		ns	
TCLLH	ALE Active Delay		80		40		50	ns	
TCHLL	ALE Inactive		85		45		55	ns	
TLLAX	Address Hold Time	TCHCL·20		TCHCL·I0		TCHCL·I0		ns	
TCLDV	Data Valid Delay	10	110	10	50	10	60	ns	
TCHDX	Data Hold Time	10		10		10		ns	
TWHDX	Data Hold Time After WB	TCHCL·I0				TCLCH·25		ns	
TCHCTV	Control Active Delay	10	110	10	50	10	70	ns	
TCHCTV	Control Active Delay 2	10	110	10	45	10	60	ns	* CL = 20 - 100 pF for all 8086 outputs (In addition to 8086 selfload)
TCVCTX	Control Inactive Delay	10	110	10	50	10	70	ns	
TAZRL	Address Float to READ Active	0		0		0		ns	
TCLRL	RD Active Delay	10	165	10	70	10	100	ns	
TCLRH	RD Inactive Delay	10	150	10	60	10	80	ns	
TRHAV	RD Inactive to Next Address Active	TCLCL·45		TCLCL·35		TCLCL·40		ns	
TCLHA	HLDA Valid Delay	10	160	10	60	10	100	ns	
TRLRH	RD Width	2TCLCL·75		2TCLCL·40		2TCLCL·50		ns	
TWLWH	WR Width	2TCLCL·60		2TCLCL·35		2TCLCL·40		ns	
TAVAL	Address Valid to ALE Low	TCLCH·60		2TCLCL·35		TCLCH·40		ns	
TOLOH	Output Rise Time		20		20		20	ns	From 0.8V to 2.0V
TOHOL	Output Fall Time		12		12		12	ns	From 2.0V to 0.8V

NOTES:

1. Signal at 8284 A shown for reference only.
2. Setup requirement for asynchronous signal only to guarantee recognition at next eLK
3. Applies only to T2 state. (8 ns into T3).

Figure 4.19 (b) timing Response of the 8086

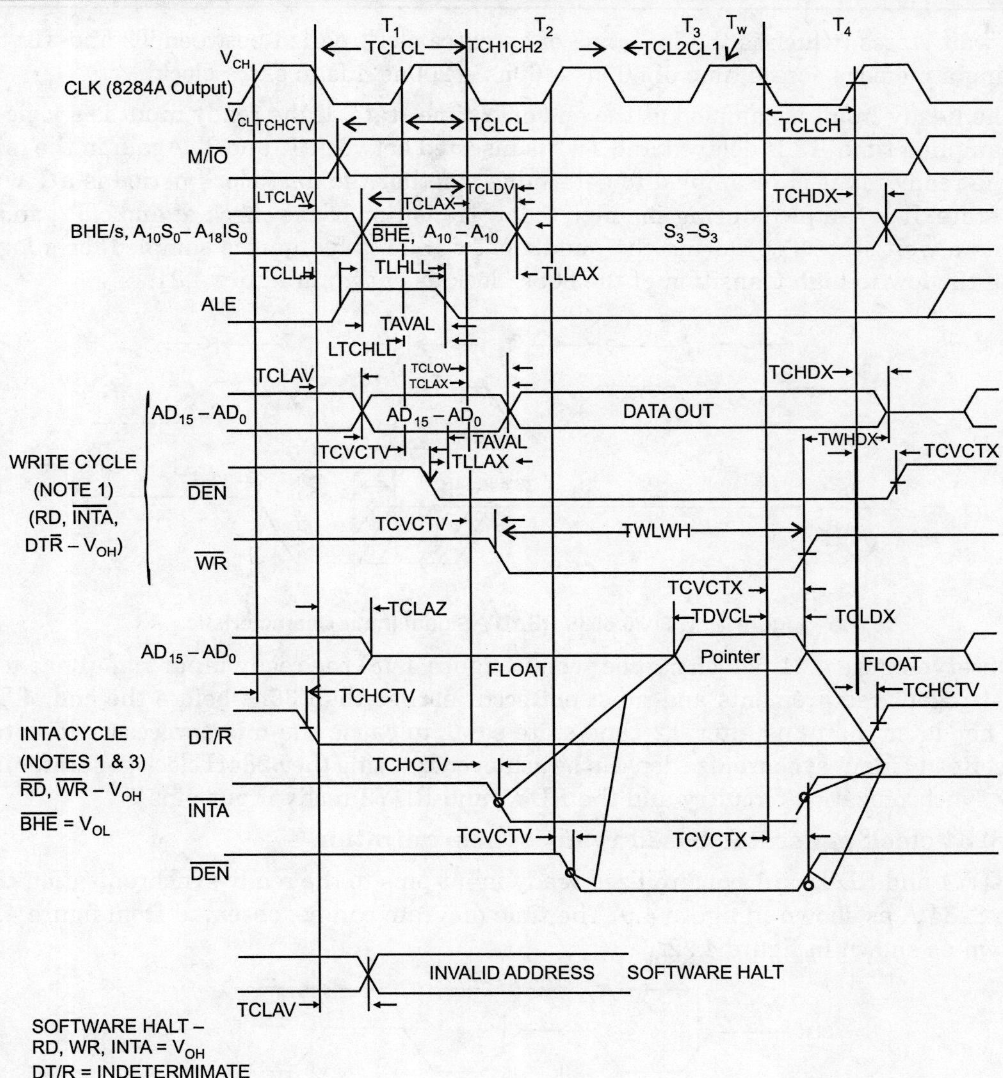

Figure 4.20. The 8086 minimum mode write bus cycle

4.5. READY AND WAIT STATE CYCLE SYNCHRONIZATION

As we have read earlier the microprocessor can be made to enter wait states so as to interface to slower memory and I/0 devices. Suppose we have to design a system based an the 8086 microprocessor and have to interface a memory device where access time is 560ns. Obviously this is not achievable by the normal 8086 read bus cycle (as the time allocated for a maxim um memory access time of 460 ns). However the use of ready input can cause the microprocessor to enter wait states (which is the insertion of an extra clock period between T2 and T3) thereby allowing a memory access time of 600ns (460ns + 200ns due to extra clock).

This is not achievable by the normal 8086 read bus cycle (as the time allocated for a maximum memory access time of 460 ns). However the use of ready input can cause the microprocessor to enter wait states (which is the insertion of an extra clock period between T2 and T3) thereby allowing a memory access time of 600ns (460ns + 200ns due to extra clock).

The Ready Input is sampled at the end ofT2 time state. If the ready input is a logic zero at this sampling time, T3 is delayed and Tw* is inserted between T_2 and T_3 Again in the middle of T w the ready input is resampled to determine whether the next clock period is a T w or T 3' Time state. It is sampled during the high to low transition of the check at and ofT2, and if zero inherits a wait state Tw. During the middle ofTw the ready input is sampled for a logic high during the low to high transition of the 8086 clock as shown in figure 4.21.

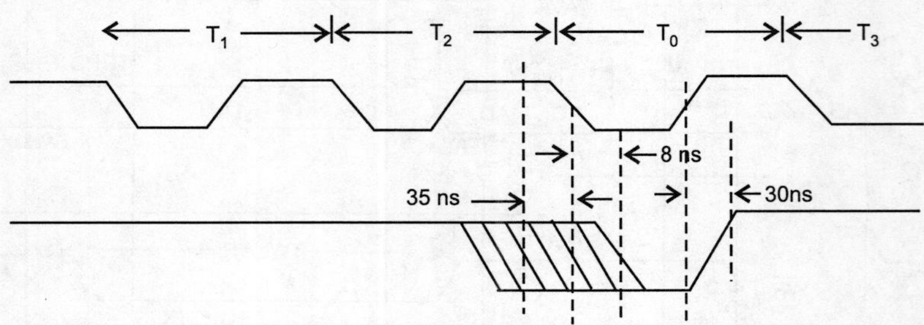

Figure 4.21. 8086/8088 READY Signal input Characteristics

Observe figure 4.21 (which can be seen in figure 4.18) the ready input signal has to match very stringent requirements and must not occur earlier than 35ns before the end of T2 time state nor later than 8ns after T2 time state so as to cause the microprocessor to enter wait states. Ready input synchronization can be achieved by using the 8284H clock generator internal ready synchronization circuitry and the RDY1 and RDY2 ready input pins.

The 8284 clock generator based ready synchronization

The RDY1 and RDY2 are synchronized ready input pins to the ready synchronization circuitry of the 8284 A as shown in figure 4.9. The time diagram can be. observed from figure 4.18 and redrawn as shown in figure 4.22.

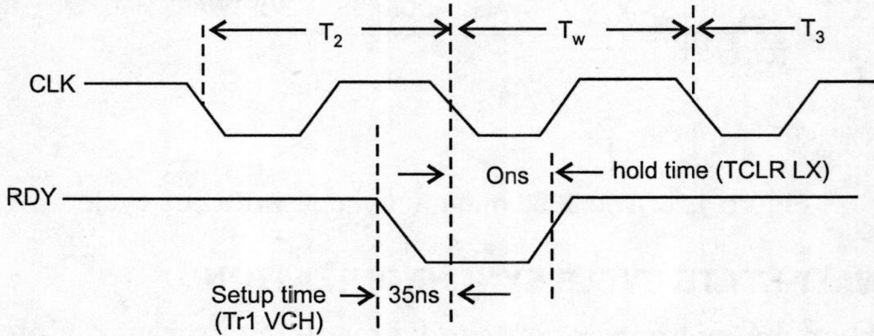

Figure 4.22. Ready input timing for 8284 A (Appendix B: 8284A data sheet)

We show you the internal functional diagram of 8284 A clock generator in figure4.22 where the lower halfis ready synchronization circuitry. We see that the RDY1 and $\overline{AEN}$ 1 are anded and fed to one input of the OR gate while the second input of the OR gate is fed by the RDY2 and $\overline{AEN}$ 1 signal which are fed to the second AND gate. In order to present a logic 1 to either of the D-Flip Flop we require either the RDY 1 and $\overline{AEN}$ 1 signals to be active or the RDY2 and $\overline{AEN}$ 2 signals to be active.

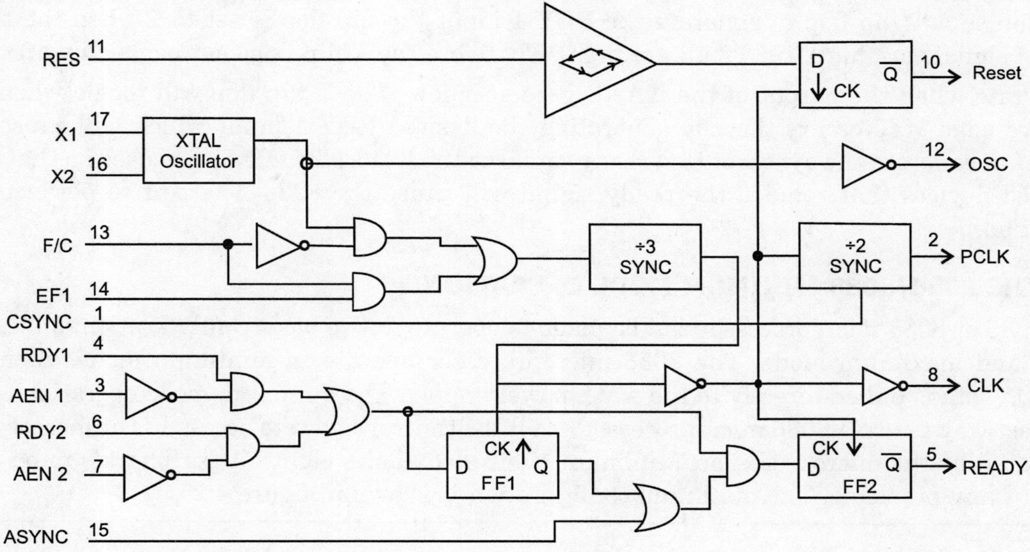

Control Pin	Logical	Logical 0
F/C	External Clock	Crystal Drive
RES	Normal	Reset
RDYl, RDY2	Bus Ready	Bus not Ready
$\overline{\text{AEN1}}$, $\overline{\text{AEN2}}$	Address Disabled	Address Enable
$\overline{\text{ASYNC}}$ Synchbronization	1 Stage Ready Synchronization	2 Stage Ready

Figure. 4.23. 8284A internal functional diagram

Now assuming ASYNC to be a logic one there by enabling single stage ready synchronization. Figure 4.24 shows you a simple. READY synchronization circuit for introducing single wait states.

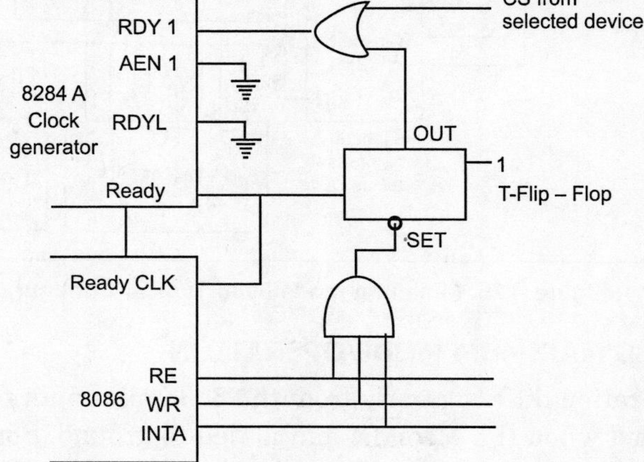

Figure 4.24. A Ready Sychronization Circuit for 8284 A

We use a T flip flop to generate the RDY 1 input the flip flop is set to 1 when the three control signals are high (till T2 all are high). Therefore the T flip flop can toggle only after T2 time state when the output of the NAND gate goes low. The T flip flop will toggle when the positive edge ofT2 occurs thereby generating the desired RDY 1 input which will cause the 8284 A to generate the synchronized ready signal to the 8086. The OR gate is used so that only selected devices that require the ready signal will cause the 8284 A circuit to perform the operation.

4.6. THE 8086/8088 MINIMUM MODE OPERATION

The 8086 or 8088 microprocessor can be made to operate in two basic configu.ration: minimum mode and maximum mode. The 8086 microprocessor operate in minimum mode when the MN/$\overline{MX}$ pin is pulled directly to the + V_{CC} power supply. The minimum mode operation is the simplest way to use a 8086 microprocessor as all the control signal to access I/O or memory are generated by the microprocessor. Minimum Mode also allows eight bit peripherals to be used without any problems. The minimum configuration is shown in figure 4.25.

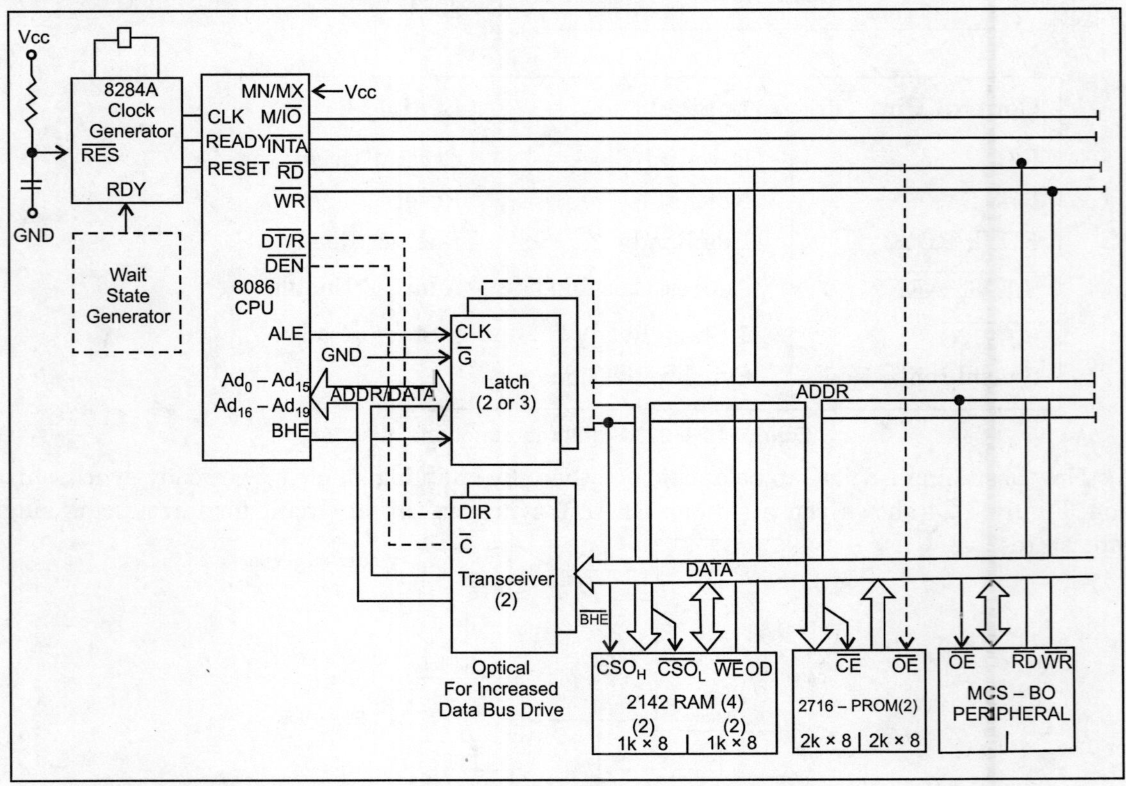

Figure 4.25. Minimum mode 8086 Typical Configuration

4.7. THE 8086/8088 MAXIMUM MODE OPERATION

Maximum mode operation differs greatly from the 8086 minimum mode operation. This mode of operation is entered when the MN/$\overline{MX}$ pin is tied to ground. Some control signals such as $\overline{RD}$, $\overline{WR}$ etc must now be externally generated. The 8288 bus controller is designed for use with the 8086 and generates all external control signals for the 8086 based system as shown in figure 4.26. Maximum mode operation is designed for the 8086 system which requires the 8087 coprocessor to be interfaced.

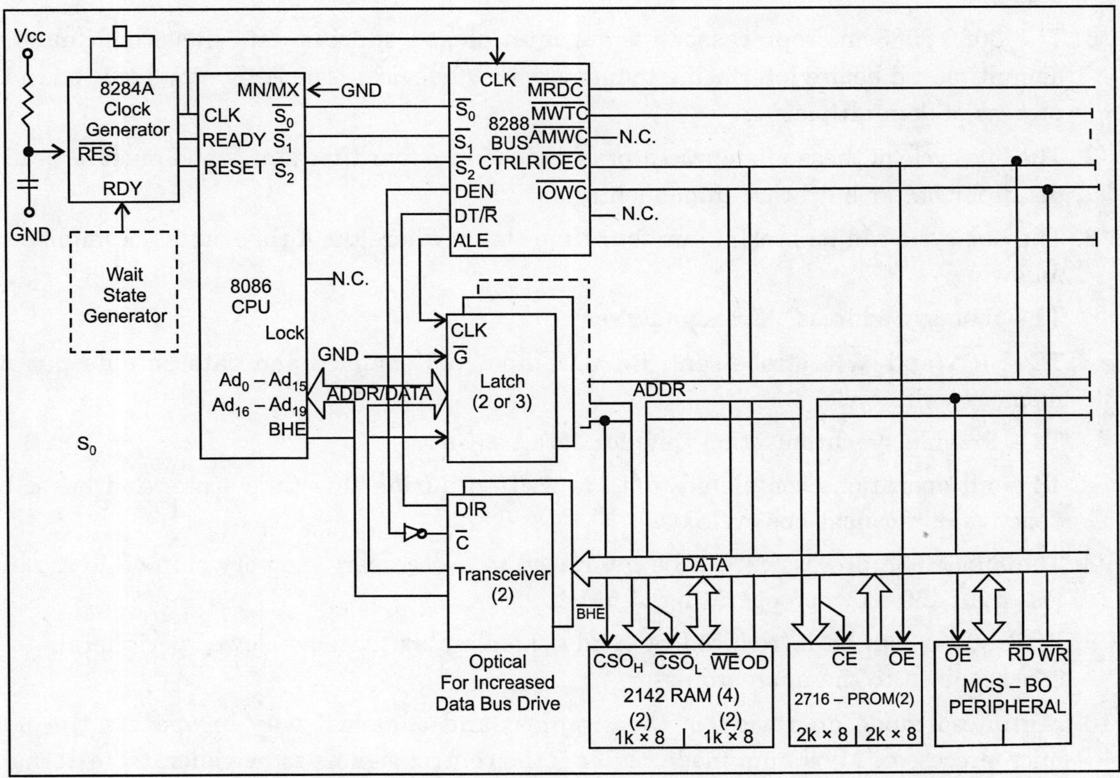

Figure 4.26. Maximum Mode 8086 Typical Configuration

4.8. KEY POINTS TO REMEMBER

1. The major differences between the 8086 and 8088 microprocessor are
   (a) size of data bus
   (b) IO/$\overline{\text{M}}$ pin on 8088 instead of M/$\overline{\text{IO}}$ on 8086
   (c) SSO pin on the 8088 in lieu of BHE/S7 pin on the 8086
2. The 8086 microprocessor requires a single power supply of + 5V with tolerance of ± 10%. The 8086 draws a maximum supply current of 340mA and has a power dissipation rating of 2.5 W.
3. The 8284A clock generator is designed for use with 8086 and 8088 microprocessors. The 8284A is need to provide the system clock, ready and reset synchronization for the 8086.
4. The 8284 A produces a clock with frequency one third of the resonant crystal attached to the Xtal pins i.e. a 15 MHz crystal will give a 5 MHz system clock. The PCLK or peripheral clock signal is half the clock signal generated by the 8284A.

5. The Reset pin is used to reset the 8086/8088 microprocessor. When the microprocessor is reset the CS register gets loaded to FFFFH and the IP to OOOOH. The IF flag is also disabled.
6. The 8086/8088 microprocessor have a multiplexed address data bus which must be demultiplexed before interfacing to memory or I/0 devices. The ALE signal is used in the process of dem ultiplexing.
7. The bus cycle of these microprocessor is divided into four time states and must be read in detail for better hardware implementation.
8. The bus cycle can be broken into four time states where the 4-time state accomplish the following:

   T1 – memory address, ALE signals sent

   T2 – $\overline{RD}$ and $\overline{WR}$ strobe sent, READY input pin sampled and data on data bus if a memory write cycle.

   T3 – Provide synchronisation time for data transfers.

   T4 – all operations completed (data is written during this time state) and buses are deactivated for next bus cycle
8. The 8086/8088 processors can be interfaced to a necessary memory chip of access time less than 460 ns when operating at 5MHz.
9. Wait states can be introduced by ready signal so as to allow slower peripherals to be synchronized to the microprocessor.
10. Minimum mode operation is the simplest and cheapest way to operate the 8086 microprocessor. Maximum mode requires the control signals to be generated externally. This requires the 8288 bus controller to be used.

REVIEW QUESTIONS

4.1. Describe the difference between the 8086 and 8088 microprocessor.

4.2. What would be the invalid components that appear on the address data bus of the 8086 while it is active?

4.3. What purpose do the status bits S_3, S_4, S_1 and S_2 serve?

4.4. Explain the operation of the $\overline{TEST}$ pin with respect to the wait instruction?

4.5. What are conditions required for the CLK signal for reliable 8086 processor operation?

4.6. Describe the requirement for de multiplexing of the address data bus?

4.7. What is the requirement of the buffered systems?

4.8. How the 8088 microprocessor be caused to service.

1. A DMA Operation
2. Interrupt
3. Wait state

4.9. Which three pins are used to convery the information of the present external control signals that must be generated by 8288 during maximum mode operations? List the truth table?

4.10. Explain the operation of

1. $\overline{\text{LOCK}}$ pin
2. $\overline{\text{TEST}}$ pin
3. RESET pin
4. MN/$\overline{\text{MX}}$ pin
5. DT/$\overline{\text{R}}$ pin
6. $\overline{\text{DEN}}$ pin

4.11. What are the conditions that can be decoded from QSO/QS1 pins of the 8088?

4.12. Why is the 8284A clock generator used with the 8086/8088 microprocessor.

4.13. What are the uses and differences between the following 8284 A pins.

1. CLK
2. PCLK
3. EFI
4. F/$\overline{\text{C}}$

4.14. If the crystal attached to the 8284A has a resonant frequency of 30MHz what would be the frequency of signals at following pins

- PCLK
- CLK

4.15. How is the 8086 data address bus demultiplexed ?

4.16. Describe the 8086 bus read and write cycles in detail?

4.17. What is a bus cycle?

4.18. What are the different operations that occur during the four time states of an 8086/8088 bus cycle.

4.19. Calculate the memory access time that can be allowed for the following processor operating at 5MHz.

- 8086
- 8086 – 1

4.20. How is the READY synchronization achieved? What are the timing synchronization required for the ready signal to be able to cause wait states? .

4.21. What would happen if the READY pin is grounded?

4.22. Explain the use .of the 8284 clock generator as a reset signal synchronization element. Also explain how it is used to synchronize the ready signals?

4.23. Discuss the 8284A internal architecture in detail.

4.24. Discuss the difference between minimum mode and maximum mode operation?

4.25. What is the use of the following 8284A pins

- RDY 1 and RDY 2
- PCLK
- OSC
- $\overline{\text{RES}}$
- CSYNC

4.26. Explain the following 8086/8088 pins

- DT/R
- $\overline{\text{DEN}}$
- $\overline{\text{SSO}}$
- $\overline{\text{S2}}$, $\overline{\text{S1}}$, $\overline{\text{SO}}$
- A_{19}/S_7, A_{18}/S_6, A_1/S_5, A_{16}/S_4

●●●

Chapter-5

ASSEMBLY LANGUAGE PROGRAMMING*

INTRODUCTION

In the following chapters you shall read about the 8086/8088 instruction set. You will note that the instructions are one byte, two byte or more longer. You will also observe the difficulty of programming the 8086/8088 using machine language. For easier programming for users there are programs called assemblers which simplify programming of the 8086. Here in this chapter we shall explain assemblers, assembly language programming fundamentals and teach you how to write programs in assembly language.

5.1. ASSEMBLERS

Assemblers are low level languages. Here we use code symbolic instructions, each of these generates the respective machine language instruction. However, programming in higher level languages, such as C, C++ is more easier compared to assembly language but assembly lan- guage have some advantages which are:

(*i*) Provides greater control over direct handling hardware.

(*ii*) Generates smaller more optimized code.

(*iii*) Faster execution time for programs.

The low level languages commonly called assembly language uses an assembler to translate the symbolic instruction into an object file (.OBJ). Then a linker program is used to convert this object file into executable machine language files. (.EXE or .COM files).

Commonly available assemblers are MASM, (Microsoft Assembler) and TASM (Turbo Assembler). There are many version of each of these assemblers. The assemblers contain their own linker program. (MASM linker program is the MLINK or ML.EXE file).

5.2. ASSEMBLY LANGUAGE PROGRAMMING

Now, we shall discuss the syntax of the commonly available assemblers for 8086.

5.2.1. COMMENTS

Comments are introduced in a program to increase its readability and to help understand programs. It also serves as an aid when one's program is being read or used by another. A comment begins with a semicolon (;). Whenever a semi colon is encountered in a program the assembler assumes all characters to the right of it are comments.

Comments may be introduced as an a single line or may follow instructions. The following are examples

1. ; this is a comment

* At this end of the chapter we expect you to read thoroughly section 5.11.

2. ADD AX, CX; Add instruction comment

A comment appears only in the source program listing and generates no machine code. Any number of comments may be introduced in a program.

5.2.2. Reserved Words

Those students who have read C or C++ will clearly understand what RESERVED words are. Reserved words are specific characters or strings of characters which are reserved by the assembler for its specific purposes. These reserved words can only be used under specific conditions else will generate errors in the program execution.

We now list the reserved words by category.

1. Instructions: These are operations whose machine code is executed by the microprocessor (for example here it is the 8086/8088). Refer to chapters 6,7,8 for complete set of the 8086 instructions that are used by the assembler and their syntax.
2. Directives: Directives are special purpose instructions which direct or command the assembler to do a certain operation. Directives include the END, SEGMENT Commands.
3. Operators: Operators such as FAR, SIZE provide settings to the assembler for the program.
4. Pr edefine d Symbols: Predefined symbols are those which are reserved for specific purposes (such as return of data to program) by the assembler. These include a DATA, MODEL etc.

Using a reserved word for the wrong purpose causes the assembler to generate an error message or in short, termination of program. Thus, this results in the need to remove errors. In short to save time a programmer must take care not to use reserved words wrongfully.

5.2.3. Identifiers

An identifier is a name used to reference items in your program. There are two types of identifiers used in assembly language. The first is the name identifier which references the address of a data item, the second the label identifier which references the address of an instruction. An identifier can use the following characters.

- Alphanumeric letters : A through Z and a through z.
- Digits : 0 to 9 [Not as the first character of identifiers]
- Special Digits (?) question mark
  (_) under score
  ($) dollar
  (@) at
  (.) period* [not as first character]

The first character of the identifiers can be any of the following except the digits and period (.). The assembler treats uppercase and lowercase as same. Examples of valid identifers are HELLO, $A24, PAGE 125 etc.

5.2.4. Statements

An assembly language program consists of a set of statements. The statements themselves can be of two types:

---

* However the identifier may not begin with the period special symbol. Therefore, an identifier such as dataseg is illegal.

1. Instructions: which are converted to object code by the assembler. Eg. MOV, ADD
2. Directives: these statements direct an assembler to perform specific action such a define a data item etc.

The general format of a statement is listed below in figure 5.1.

[Identifier] operation [operand (s)] [; comments]

Figure 5.1. General format of a statement

The entries in figure 5.1, which are enclosed in square brackets are optional and are not required sometimes. Below are example of statements:

Type of Statement	Identifier	Operation	Operand	Comment
Directive	COUNT	DB	2	; directive for
				; count as a
				; data item
Instruction		ADD	AX, BX	; Perform ADD
Instruction		MOV	AL, AH	;Move instructions

Figure 5.2. Examples of Statements

The identifier, operation and operand fields may begin in any column. However for better readability we provide spacing and start each field at different columns. Most assembler editor programs such as MASM 6.0 editor program (CODEVIEW) provides Tab spacing of 8 positions offacilitate spacing.

Now let us discuss figure 5.1 in detail..

(*i*) **Identifier :** As discussed earlier, applies to name of a defined item (or directive) or a label identifer which references an instruction.

(*ii*) **Operation :** The operation must consists of assembly language code instructions. For example for data items DB, DW or DQ defines a field or area for data. For an instruction the 8086/8088 instruction set is used. (e.g. ADD, SUB, MOV etc.)

(*iii*) Operand field provides information for the operation field. For data items the operand defines initial values. For instructions the operand field indicates where the operation is to take place. Figure 5.3. lists example for both.

NAME	OPERATION	OPERAND	COMMENT
A_NO	DB	5H	; Define byte with 5H value
	INC	AX	; Increment AX
	ADD	AX,DX	; Add BX to AX
	JMP	LABEL	; Jump to memory
			; referenced by LABEL

Figure 5.3. Operand field Examples

* All entries in this chapter when enclosed by square brackets refer to optional entries.

5.2.5. DIRECTIVES

Assembly languages support a number of statements that allow the programmer to control the assembly program and its listing. Such statements are called directives. Directives work only during assembly of the program and donot generate any machine code. We shall discuss common directives and you will understand directives better through these examples.

5.2.5.1. List Directives

The list directives help to control the listing of an assembled program. These directives have no function except the above these have no subsequent effect on program execution. PAGE directive

The PAGE directive is included at the start of an assembly program. Its function is to tell the assembler the maximum number oflines to be listed on a page and the maximum number of characters per line. Its general format is

PAGE [Lines]. [Characters]

Now suppose we consider the example

Page 70,120

The PAGE directive will tell the assembler that it is to list 70 lines per page and.120 characters per line. Omission of the PAGE directive at the start of an assmbly program causes the assembler to use the default value as PAGE 50, 80. The use of PAGE directive is only to tell the assembler the PAGE size during printing of the program file.

TITLE Directive

The TITLE directive is used cause the title ofthe program to be printed on line 2 of each page of the program listing. The TITLE directive is used once on the start of program. Its syntax is shown below

TITLE text

The text is the text you want printed.

5.2.5.2. Segment Directive

An assembly program executable such as .EXE file consists of one or more segments. These are stack, Data, Code and Extra segments. The directive for defining a segment is given below in figure 5.4.

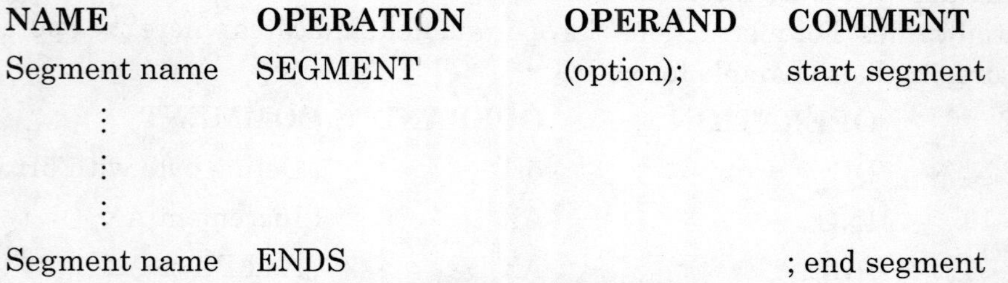

NAME	OPERATION	OPERAND	COMMENT
Segment name	SEGMENT	(option);	start segment
⋮			
⋮			
⋮			
Segment name	ENDS		; end segment

Figure 5.4. Segment Directive Syntax

The SEGMENT directive defines to the assembler the start of a segment with name segment name. The segment name should be unique and follows the rules of the assembler. The ENDS statement follows the name of segment which should be same as the start segment name. The maximum size ofa segment permitted is 64K. (Refer to 8086 memory addressing). A more detail syntax for the SEGMENT directive is given in fig. 5.5.

Segment name SEGMENT alignment combine 'class' 'data'

Figure. 5.5. Detailed syntax of SEGMENT directive

1. Alignment type field: The alignment type field indicative where the segment must begin in memory space. For example, the PARA statement will align the segment at a memory address divisible by 10H (16 decimal). Thus if we consider the example 5.1.

   Segment name SEGMENT PARA

   Ex. 5.1. Alignment of SEGMENT directive

   the segment whose name is segment name will start at a memory address divisble by 10H.

2. Combine type field: The combine type field indicates to the assembler whether to combine the segment with other segments present in the program when they are linked after assembly (i.e. in memory). Combine types are STACK, COMMON, PUBLIC etc. Eg for a commonly encountered STACK SEGMENT in assembly is given below in EX. 5.2.

   Ex. 5.2. Commonly encountered STACK SEGMENT Code

   the other combine type fields are PUBLIC, COMMON, NONE, AT etc.

3. Class type field: The class field enclosed by apostrophes as in figure 5.5. denotes the group type of the segment. Classes are listed below in figure 5.6.

CLASS NAME	SEGMENT LINKED
‘code’	CODE SEGMENT
‘data’	DATA SEGMENT
‘stack’	STACK SEGMENT

Figure 5.6. Microsoft Recommended Classes

5.2.5.3. PROC Directive

The code segment contains instructions (or simply excutable code) for a program. However, for modular programs we may define a code segment with many modules called procedures. These procedures are defined by the PROC directive. A common example for a code segment with a single procedure is listed in figure 5.7.

```
NAME        OPERATION  OPERAND    COMMENT

Code_seg    SEGMENT    PARA    ; code segment starts
               :
Proc_one    PROC       FAR     ; procedure starts
               :
               :


Proc_one    ENDP               ; procedure ends
Code_seg    ENDS               ; code segment ends
```

Figure 5.7. Procedure Directive

In the above example we have defined a procedure with the PROC directive. Here we have used the name of the procedure as proc_one, though the name can be anything but according to rules of the assembler. The operand FAR is related to program control and shall be discussed later. There are many operands type for PROC such as NEAR, SHORT, FAR etc. The ENDP directive indicates the end of a procedure and should contain same name as in the PROC statement. The code segment can contain any number of procedures.

5.2.5.4. ASSUME Directive

An 8086/8088 uses the SS register to address stack memory, the DS register to address data locations and the CS register to address code segment. However the assembler muse be told which is which. The ASSUME directive is used for this purpose. This directive must be included in the start ofthe code segment as follows

OPERATION	**OPERAND**
ASSUME	SS:Stackseg, DS : Dataseg, CS : Codeseg

The assume will till the assembler to use the SS register with the address of the stack segment whose name is stackseg. Similarly, the other operand are associated with the respective registers and segments. The ES register may also be associated with a program segment similarly.

5.2.5.5. END Directive

We have discussed in 5.2.5.2. and 5.2.5.3 the ENDS and ENDP directive. The ENDS directive is used to inform the assembler of the end of a segment while the ENDP directive is used to end a procedure. The END directive is similarly used to end a program. It format is

Operation	**Operand**
END	[proc name]

5.3. THE BASIC ASSEMBLY LANGUAGE PROGRAM

There are two types of executable programs that we shall discuss. The first in the .EXE program which we shall discuss here. The second is the .COM program discussed later in the Chapter. Here we show the basic .EXE program.

```
PAGE 60, 120
TITLE BASICPROG.ASM
;..............................................................................................................
STACKSGONE          SEGMENT PARA STACK 'stack'
...........................
...........................
STACKSGONE ENDS
;..............................................................................................................
DATASG       SEGMENT       PARA 'data'
...........................
...........................
DATASG ENDS
;..............................................................................................................
```

```
14   CODESGMAIN SEGMENT PARA 'code'
15   PROGRA PROC FAR
16   ASSUME SS : STACKSGONE, DS : DATASG, CS : CO DESGMAIN
17   MOV AX, DATASG              ;get address of data segment
18   MOV DS, AS                  ;load DS with data seg address
19   ..........................
20   ..........................
21   MOV AX, HCOOH               ; Request termination
22   INT 21H                     ; exit to DOS command
23   PROGRA ENDP
24   CODESEGMAIN ENDS
25   END PROGRA
```

Figure 5.5. Basic .EXE program

Now let us discuss the program offigure 5.8. in brief detail. We shall examine the program line by line.

Line	Explanation
1	The page directive for listing establishes 60 line and 120 characters per line per page.
2	The title directive identifies the assembly program name as BASICPROG.ASM
3,8,13	Lines 3, 8, 13 are comments to differentiate the program areas (segments) for better readability to the user,
4-7	Defines the stack segment as STACKS GONE
9-12	Defines the data segment as DATASG
14-24	Defines the code segment as CODESGMAIN
15-23	These statements define the only procedure of the code segment as PROGRA
16	The ASSUME directive inform the assembler to associate the STACKSGONE segment with SS register, DATASG segment with data register, CODESGMAIN with the CS register. By associating segments with segment registers the assembler can determine the offset address for items in the stack segment, data items in data segment, instructions in code segment etc.
17-18	These two instructions initialize the DS register to address of data segment. There are two MOV instructions because data cannot be directly copied into segment register. Thus the first MOV instruction loads AX register with the address of data segment. The second MOV instruction copies this address from AX register to DS register.
21-22	The two instructions request an end to the program execution and a return to the DOS.
25	END directive informs the assembler of the end of the program.

Now a point to keep in mind here is that the program is written in symbolic language. For execution of this program we first have to use a assembler to translate it into object code. A linker program then translates the object code into machine executable code (.EXE file). Also when the DOS operating system loads this .EXE file into memory it initializes the CS : IP and SS : SP registers. The DS and ES registers are never initialized and have to be loaded with respective segment addresses by the program itself. This is clearly done in figure 5.8 by the two MOVinstructions of line 17-18. Similarly, the ES register can also be loaded.

5.4. THE DOS EXIT SOFTWARE INTERRUPT

Software interrupts are handled by the operating systems such as DOS, UNIX etc. However, we shall throughout use operating systems and DOS as one entity. Mostly the 8086/8088 or to be more precise the complete X86 family (i.e. 80186, 80286, 20386, 20486, Pentium, Pentium II, III and IV) of processors are designed to optimize the DOS operating system. Another reason for this is that since DOS is the most common and easiest of all operating systems to understand thus we shall use the term operating system in context of DOS. The DOS operating system has a number of software interrupts which are used for a variety of purposes. Our main interest here is the INT21H a commonly used DOS. interrupt. This interrupt is used for many functions such as keyboard input, screen handling, disk I/O and printer output. However. our interest is the DOS interrupt INT 21H function 4CH which is recognized by DOS as a request to terminate program execution.

```
MOV   AX,     4COOH
INT 21H
```

Observe the above two lines of code (as in figure 5.8). The first line loads 4COOH into AX register where 4CH is loaded into AH and OOH into AL register. The next line INT21H requests DOS for a software interrupt 21H. DOS will now check AH register for the interrupt function. Since 4CH is loaded into AH which is the INT21H function for program termination, these two lines will effectively terminate program execution. These two lines which are used in fig. 5.8 are used for the above purpose (i.e, program termination) as discussed.

5.5. SIMPLIFIED SEGMENT DIRECTIVES

A shortcut method of defining segment is provided for in MASM and TASM assemblers. To use the shortcut method in an assembly program we must first initialize the memory model that the program will use. The general format for memory model initialization is

MODEL memory_model_type

There are many memory models whose requirements are listed below in figure 5.9.

MODEL	Number of CODE Segments	Number of Data Segments
TINY*	*	*
SMALL	1	1
MEDIUM	More than 1	1
COMPACT	1	More than 1
LARGE	More than 1	More than 1

Figure 5.9. Memory Models Types

* The tiny model is generally used exclusively for .COM files, The tiny model is only used for programs that can fit in 64K of memory.

The TINY model has its code, data and stack segments in one segment. The TINY model is only used for .COM files. The .MODEL directive generated the required ASSUME directive statement automatically.

Now the short cut directives used to establish the respective segments are

- STACK [size]
- DATA
- CODE [name]

Each ofthese directives causes the assembler to automatically generate the required segment statement and its corresponding ENDS statement. However now the instruction is used to initialize the address of data segment in the DS will have the format as given below:

```
MOV   AX, @data
MOV   DS, AX
```

The stack segment size generated is 1024 bytes or 1 kbyte which can be overrided by giving the stack segment size. Now we shall show an assembly program using the short cut (simplified) segment directives.

```
          PAGE  60, 120
          TITLE PROGBASICSIMPL!FY SEG.ASM
          ;...........................................................................
        • MODEL SMALL
        • STACK32                     ; Define stack segment of size 32 bytes
        • DATA                        ; Define data segment
                 A      DW 20
                 B      DW 40
                 C      DW?
          ;...........................................................................
        • CODE
BEGIN     PROC        FAR
          MOV         AX, @data       ; load address of Data
          MOV         DS, AX          ; segment in DS
          MOV         AX, A           ; load 0020 to AX
          MOV         BX, B           ; load 0040 to BX
          ADD         AX, BX          ; ADD AX and BX
          MOV         C, AX           ; Store sum in C
          MOV         AH, 4CH         ; load function 4CH in AH
          INT         21H             ; exit to dos
BEGIN     ENDP                        ; end procedure begin
          END         BEGIN           ; end program
```

Figure 5.10. A assembly program using simplified segment directives

Here we have overrided the default stack size of 1024 bytes to 32 bytes by use of .STACK 32 directive. Also the ASSUME directive in automatically generated so it is not required to be coded in the program.

5.6. THE .STARTUP AND EXIT DIRECTIVES

The latest version fMASM starting from MASM 6.0 permit the use of simplified directives for program initialization and termination. Here the .STARTUP directive generated instructions to initialize segment registers while .EXIT directive generated the INT 21H function 4CH for program termination.

5.7. DATA DEFINITION DIRECTIVES

The purpose of data segments is to define data areas and input/output area in memory. The MASM and TASM assemblers permit the definition of items of various lengths according to a set of directives that define data. A data item may contain undefined or constant values. The general format for defining data is

[name]	DIRECTIVE	Value

Figure 5.11. Data definition format

The figure 5.11 shows the general data definition format which you can see contains three fields. Now let us discuss each of these fields in brief.

Name : An assembly program references the address of a data item by use of its name. The name field of an item is optional as indicated by square braces.

Directive : The directive is the assembly command to define a data item. The few directives for definining a data item are DB (define byte directive), DW (define word directive), DD (define quad word directive) and DT (define ten bytes directive). Each of these defines a data item of a specific size as indicated by their full forms.

Value : The value field defines the initial value of the data item. The general format for defining an unintialized data items is given in figure 5.12.

```
SIDE_A     DB    ?     ; uninitialized item of
                       ; size one byte
```

Figure 5.12. Defining an uninitialized data item

The directive can be used to define a constant value item as shown in figure 5.13.

```
SIDE_B     DW    25    ; Constant of size two bytes
                       ; Containing value 0019H
```

Figure 5.13. Defining a constant data item

Multiple data items may be declared as

SIDE_C DB 11, 12, 14, 15 ....

The assembler defines these constant in adjacent bytes of memory. A reference to SIDE_C references to the first byte. A reference to SIDE_C + 1 is to second constant 12. Now for example the instruction.

MOV BL, SIDE_C + 2

loads the value 15 into the BL registers.

Duplication of data items is permitted by use of DUP directive. For example

```
DB    10    DUP (10)    ; 10 bytes contain
                        ; OA OA OA OA OA OA OA OA OA OA H
```

Figure 5.14. Duplication of data items

The above example duplicates 10 (OAR) decimal ten times in adjacent memory bytes (due to the directive DB).

5.7.1. Using Data Directive to Load Character Strings

Character are used to enter valid ASCII data containing alphanumeric characters. The DB directive is the only directive format available to an assembly programmer to enter string. The assembler translates the character strings and stores them in ASCII format. The general format to store strings is given below:

```
DB    "Microprocessor intel    ; text entered into memory
                               ; translated in ASCII
DB    "My number is 27         ; Alphanumeric text
```

5.7.2. Numeric Constants Directives

Numeric constants are required for arithmetic values and memory addresses. The constant to be stored is followed by an optional Radix specifier. Most assemblers convert the constant value into hexadecimal and then stores the generated bytes in a reverse order. The numeric formats supported by most assemblers are

Decimal: Decimal format is the default format; It can also be used by specifying the radix specifier as D. An example is listed below

```
A    DB    40       ; load decimal value 40
                    ; into memory location
                    ; referenced by A
B    DB    100      ; load decimal in B
C    ON    4002D    ; load decimal in C
```

Thus, A will be stored with the hex equivalent of 40 decimal. Thus A will contain 25H. Similarly, B will contain OAH. Now C is defined as a double byte and will contain hexequivalent of 4002 Decimal stored in reverse i.e. C contains 5202H. (Observe that the hex equivalent of 4002 decimal is 0252H which is reversed byte wise i.e. 02 is the last byte and 52 the first byte). Hexadecimal: Hex format permits the use of hex numeric format to input data value. However hex numbers must be followd by the H radix speficier

Example:

```
AB    DW    2524H
```

Binary : The binary format permits use of entry of data in the binary number system. We must, use the binary radix specified by B.

Real : We can enter the value in the real format. The number is stored in the floating point format. Refer to chapter 9 on coprocessor. The radix specifier used is R.

5.7.3. Data Directives

Now let us discuss the data directives in detail.

Description Directives	Conventional Directives	MASM 6.0 of Data Items	Size in Bytes
Define Byte	DB	BYTE	1
Define Word	DW	WORD	2
Define Double Word	DD	DWORD	4
Define Far Word	DF	FWORD	6
Define quadword	DQ	OWORD	8
Define tenbytes	DT	TBYTE	10

Figure 5.15. Data Directives

The figure 5.15 shows the data directive that are used by the MASM and MASM 6.0 assemblers.

1. **Define Byte :** DB or Byte → The data item define will consume one byte of memory space. Thus effectively the largest number stored will be within the limits +127 to –127 decimal.
2. **Define Word :** DW or Word → DW or define word defines a data item which will use one word or two bytes of memory. The limits of the maximum number stored is between +32767 and -32768.
3. **Define Double Word :** DD or DWORD → The DD directive defines a data item that is two word or four bytes in length. The limits of the number are +214.7483647 to -2147483648. The assembler converts the DD numeric constant into hex. Also while storing the bytes are reversed.

   Consequently a decimal number 1 2 3 4 5 6 7 8 whose hexadecimal equivalent is 00BC614EH is stored as 4E61BC00H. Graphically

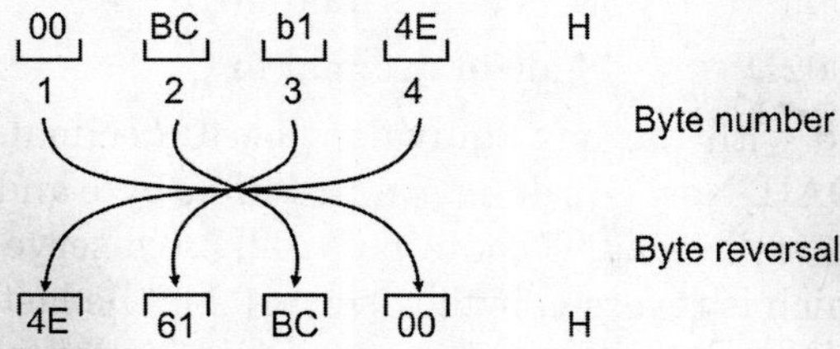

Figure 5.16. Byte Reversal

4. **Define Farword :** DF or FWORD → The define Far word directive defines a data items which is 6 bytes long. The normal use is for the coprocessor instructions.
5. **Define Quadword :** DQ or QWord → The define quadword directive define data items which are 8 bytes or four words long. This is normally used for data input to the 8087 coprocessor.
6. **Define Tenbytes :** DT or TBYTE → This directive defines data items that are 10 bytes

long. This is used for storing data for use by the coprocessor. This is also used for storing data in extended precision format and for the packed BCD numbers.

5.8. THE EQU DIRECTIVE

The EQU directive is used for defining a value which can then be used through the program. Consider the example

```
NO          EQU   10                ; define NO = 10
FIELDMUL    DB    NO    DUP (?);        __1
```

Now the instruction FIELDMUL will substitute NO as equal to ten. This is due to the statement NO EQU 10 which defines NO = 10. Now whenever NO variable is used throughout the program it is replaced by its value by the assembler. Thus the assembler understands statement (1) as (where No replaced by equivalent value)

```
FIELDMUL    DB    10    DUP (?)
```

which defines ten consecutive byte of memory as initialized variables.

5.9. THE .COM PROGRAM

Now let us discuss an important DOS executable file, the .COM file. Till now we have discussed only the .EXE file. However the .COM file is a very important program file.

5.9.1. Differences between .Com and .Exe Programs

An .EXE program can be of any size and may contain any number of segments. However the .COM program is restricted to one segment and a maximum of 64 K bytes of memory. This 64 K byte of space includes the PSP (Program segment prefix) that DOS inserts before loading the .COM or .EXE file. The PSP is a 256 byte block which is automatically loaded by DOS for specific functions. However, DOS does not insert the 512 byte Header block in the .COM file. This header block is inserted in .EXE files and thus makes them larger than .COM files. Another difference is that .COM does not have any relocatable address information.

The .COM program generates a stack automatically. Thus when writing a .COM assembly program the programmer can omit the stack. However, in the case when the program size is large and the stack cannot be accomodated in the 64 Kb program boundary then the assembler will put the stack in higher memory. The .COM program also doesnot have a data segment as the data required is defined in the code segment. However care must be taken to introduce the directive ORG ioon following the .CODE or code SEGMENT statement 80 as to tell the assembler to assemble a program at an offset of 100H*. This is done to by pass the 256 byte PSP that DOS introduces before the .COM file. Also when the COM program is executed the DOS operating system automatically initializes all segment registers with the address of the PSP. Thus you are not required to load the CS and DS registers with address of the memory segments.

5.9.2. Example of a .Com Program

```
            PAGE        60, 120
            TITLE       COM Program
```

* Remember that lOOH is equal to (256)10' This is done to bypass the 256 byte PEP block which will

```
CODESGMAIN    SEGMENT               PARA 'CODE'
              ASSUME                CS : CODESGMAIN, OS: CODESGMAIN
                                    SS : CODESGMAIN, ES : CODESGMAIN
              ORG        100H       ; start execution at end of PSP
BEGIN:        JMP        START      ; bypass program data
;.................................................................
    A         DW         200        ;datadefined
    B         DW         421
    C         DW         ?
;.................................................................
START         PROC       NEAR
              MOV        AX,A       ; MOVE A to AX
              ADD        AX, B      ; add A and B
              MOV        C,AX       ; store sum in C
              MOV        AH,4C      ; Request function 4C of interrupt 21 H
              INT        21H        ; exit
START         ENDP
CODESGMAIN    ENDS
              END        BEGIN
```

Figure 5.17. Basic .COM program

Now let us briefly describe the .COM program offigure 5.17. Also observe the difference between the .EXE program offigure 5.8.

- No defined data or stack segment
- ASSUME statement tells the assembler to load all segment registers to the address of code segment. Also the ORG directive causes execution to begin 256 bytes (100H) from the code segment offset. This is done to bypass the 256 byte of the PSP that is loaded by DOS before the .COM code segment .
- The ORG 100H directive defines the start offset address for program execution. This offset is loaded into IP register .
- A JMP (jump) statement bypasses the data area which is coded in the code segment. Another way to code data items is to write them after instructions so as to remove this JMP statement.

5.9.3. A .Com Program using Simplified Segment Directives

```
          PAGE     60, 120                  ; set page size
          TITLE    .COM simplified model
```

```
          • MODEL SMALL
          • CODE                                   ; code segment
          ORG 100H                                 ; start at end of PSP
BEGIN:    JMP   PROGSTART                          ; bypass data items
;.............................................................................
   A      DW    420    ; define data
   B      DW    222
   C      DW    ?
;.............................................................................
PROGSTART PROC  NEAR
          MOV   AX, A          ; move A to AX
          ADD   AX, B          ; add A and B
          MOV   C, AX          ; store sum in C
          MOV   AH; 4CH        ; Request function
          INT   21H            ; to exit to DOS
PROGSTART ENDP
          END   BEGIN
```

Figure 5.18. Simplified .COM program

5.10. ASSEMBLING, LINKING AND EXECUTION OF AN ASSEMBLY LANGUAGE PROGRAM

Till now we have discussed basic assembly language programming skills that you will require to write assembly program in MASM or TASM assembler. Now however we shall discuss now you assemble, link and execute these assembly programs. However, as we read more about the 8086/8088 instruction set we will discuss more detailed and complex programs.

In this chapter we will explain the procedure for typing an assembly language program on an IBM PC compatible having a DOS based operating system and a MASM or TASM assembler. The symbolic instructions that you code in assembly language is called a source program. Then an assembler program translates or assembles this source program into Machine code (also called object code). Finally, a linker program is used to form the machine addressing for the object program enabling it to run on any computer. This linked program generated by the linker program is called an executable.

5.10.1. Source Program

A source program is typed in using symbolic language. The program is keyed in by using a text editor such as notepad in Windows® or the DOS text editor. After typing in the program we have to save the file on disk. However the source program file instead of being saved as a .txt (text format) file is saved with the .asm (assembly source code format) extension. This is the sogree-program: Figure 5.19 and 5.20 show two source program that have been typed using the notepad-text editor of WINDOWS® and saved with the .asm extension. You can observe the file name as anuj.com Although spacing is not compulsory but it aids readability. However this

.asm file is just a text file i.e, it is not executable by the 8086/8088 processor. The next step involved the assembling of this .asm file into an intermediate .OBJ (Object) file.

5.10.2. Assembling A Ource (.a M) Program

The next task as we have discussed in the last section is to assemble the source file (.asm) to an object (.obj) file. The Microsoft assembler program is MASM.EXE (upto version MASM 5.x) whereas it is the TASM.EXE for Borland turbo assembler. After MASM version 6.0 The microsoft assembler uses ML.exe to assemble . aslll files

Now the general format to assemble a source program is

1. MASM/option source (.asm), [out(.obj)], [list (.lst)], [Cref(.crf)] for MASM ver 5.X and below
2. TASM/option source(.asm), [outt.objj], [list(.lst)], [Cref.(crf)] for turbo assembler
3. MLioptions source (.asm), [out(.obj)], [listt.lstj], [Cref (.crf)), for MASM ver 6.0 and above

[***Note:*** These commands are to be typed at the dos prompt in the folder containing the MASM assembler and linker program]

Where

- **Options :** Provided to set certain setting levels for the assembler.
- **Source :** This is a reference to the source program file name. Thus if suppose akg.asm were to be assembled we would type akg or akg.asm in place of source.
- **Out :** Provides for the path and the name of the object file to be generated.
- **List :** Provides for the path and name of .lst (list) file to be generated. The .LST file contains the source code as well as the object code that is generated. Figure 5.21 and 5.22 show the corresponding .LST files generated for the source code offigure 5.10 and 5.20 when assembled. These .LST files can be viewed under text editors such as notepad or dos text editor.
- **Cref :** This is provided for giving the path and name of the crossed reference file that the assembler generates. The extension for the cross reference files are .CRF for MASM and .XRF for T ASM.

Example: Suppose the MASM working directory in C:\MASM. A file C:\MASM\anuj.asm is to assemble and to generate a object file whose name is akg.obj, a list file whose name is akg.lst and cross reference file whose name is akg.crf. Show the required procedure-to assemble anuj.asm. (Assuming initially you are at C:\)

Answer:

Step 1: Change working directory to C:\MASM

C:\CD MASM.J

Step 2 : Run MASM.EXE and enter the parameters for assembling anuj.asm

C:\MASM\MASM anuj.asm, akg.obj, akg.lst, akg.crf ↵

PAGE 60, 120

```
TITL.E directives usage for segment definition
; PROGRAM adding two numbers
;..............................................................................................................................
        A       DW      120             ; define the two numbers
        B       DW      232
        C       DW      ?               ; unintialized data items
DATASG ENDS
CODESGMAIN SEGMENT PARA 'code'
PROGRAM PROC FAR                ; procedure starts
ASSUME DS : DATASG, CS : CODESGMAIN, SS : CODESGMAIN ; load seg registers
        MOV     AX, DATASG              ; move data segment address to ax
        MOV     DS, AX                  ; transfer address to ds
        MOV     AX, A                   ; store a in ax reg
        ADD     Ax, B                   ; add a and b
        MOV     C, AX                   ; store sum in memory
        MOV     AX, 4COOH               ; request function 4CH of INT21 H of DOS
        INT 21H
PROGRAM         ENDP                    ; procedure ends
CODESGMAIN      ENDS
END PROGRAM                             ; code_.segment ends
```

Figure 5.19. Source program type using an editor such as notepad

```
anuj.asm
{Space}
PAGE 60,120
TITLE ADDITION OF TWO NUMBERS USING SIMPLIFIED MASM SEGMENT
DIRECTIVES
● model small
● stack 64H
● code
        org 100h
begin: jump progstart ,
;..........................................................
        A       DW      120
        B       DW      230
        C       DW      ?
;..........................................................
progstart proc near
```

```
mov ax, a
add ax, b
mov c, ax
MOV AH, 4CH
INT 21H
PROGSTART    ENDP              ; procedure ends
END BEGIN
```

Figure 5.20. Source program for a .COM file using a text editor such as notepad

```
        PAGE 60,120
        TITLE directives usage for segment definition
        ;PROGRAM ADDS TWO NUMBERS FROM DATA SEGMENT AND STORES
SUM IN MEMORY
        ;............................................................................
0000          DATASG SEGMENT PARA 'data'
00000078      A        OW      120       ; define the two numbers
000200E8      B        OW      232
0004 ????     C        OW      ?         ; unintialized data items
                       DATASGENDS
0000          CODESGMAIN SEGMENT PARA 'code'
0000          PROGRAM PROC FAR ; procedure starts
ASSUME DS:DATASG, CS:CODESGMAIN, SS:CODESGMAIN ;load seg registers
0000 B8 - - - - R MOV OS,      DATASG          ; move data segment address to ax
0003 8E 08             MOV              OS, AX          ; transfer addres to ds
0005 A 10000 R MOV              AX, A           ; store a in ax reg
0008 03 06 0002 R      ADD              AX, B           ; add a and b
000C A3 0004 RMOV               C, AX           ; request function 4CH of INT21 H of DOS
0012 CD 21             INT21H
0014                   PROGRAM ENDP
0014                   CODESGMAIN ENDP
END                    PROGRAM
Segments and Groups:
        Name   Length  Align   Combine       Class
        COm:SGMAIN ..........  0014    PARA  NONE  'CODE'
        DATASG .............   0006    PARA  NONE  'DATA'
Symbols:
Name   Type     Value  Attr
A .............  LWORD         0000    DATASG
```

```
B ............. LWORD 0002 DATASG
C ............. LWORD 0004 DATASG
Program         FPROC 0000 CODESGMAIN Length = ·0014
@Filename       TEXT anu

24 Source Lines
24 Total Lines
10 Symbols

51072 + 398720 Bytes symbol space free

0 Warning Errors
0 Severe Errors
```

Figure 5.21. Generated .LST file for source program of figure 5.19

```


                PAGE 60,120
                TITLE Addition of Two Numbers using Simplified MASM segment directives


                .model small
        0064    .stack 64H
        0000    .code
        100     ORG 100h
        0100 EB 07 90   begin: jmp progstart
;...............................................................
        01030078        a       dw 120
        0105 00E6       b       dw 230
        0107 ????       c       dw?
;...............................................................
        0109    progstart proc near
        0109    2E:A10103R              mov     ax, a
        0100    2E:03060105R            add     ax, b
        0112    2E:A30107R              mov     c, ax
        0116    B44C                    MOV AH, 4CH
        0118    CD21                    INT 21H
        011 A                           PROGSTART ENDP
        011A                            END BEGIN
```

Segments and Groups :

Name	Length	Align	Combine	Class
DGROUP	0000	WORD	PUBLIC	'DATA'
DATASTACK	0064	PARA	STACK	'STACK'
TEXT	011A	WORD	PUBLIC	CODE

Symbols:

Name	Type	Value	Attr
A	b. WORD	0103	_TEXT
B	LWORD	0105	_TEXT BEGIN
C..............	L WORD	0100	_TEXT
PROGSTART	N PROC	0109	_TEXT Length = 0011
@CODE			
@CODESIZE	TEXT	0	
@DATASIZE	TEXT	0	
@FILENAME	TEXT	0	

24 Source Lines

24 Total Lines

19 Symbols

50460 + 399316 Bytes symbol space free

0 Warning Errors

0 Severe Errors

Figure 5.22. Generated .LST file for figure 5.20

5.10.3. Debugging a Source Program

The assembler provides for errors that may be made during the typing of a source program. If the syntax discussed earlier are not followed, it is obvious the program will violate the rules of the assembler. We now consider some errors that may be generated. Consider figure 5.23 which shows a source program which has errors

```
PAGE  60, 120
TITLE directives usage for segment definition
; PROGRAM SHOWING SOME ERRORS THAT CAN BE GENERATED
;.......................................................................................................
DATASG SEGMENT PARA 'data'
A        DW            120      ; define the two numbers"
```

```
        B       DW          232
        C       DW          ?       ; unintialized data items
DATASGENDS

CODESGMAIN SEGMENT PARA 'code'
PROGRAM PROC FAR            ; procedure starts
        ASSUME OS : DATASG, CS : CODESGMAIN, SS : CODESGMAIN ; load seg registers
        MOV     OS, DATASG  ; transfer datasegment address to ds register
        MOV     AX, A       ; store a in ax reg
        ADD     AX, B       ; add a and b
        MOV     C, AX       ; store sum in memory
        MOV     AX, 4COO    ; request function 4CH of INT21 H of DOS
        INT 21 H
PROGRAM ENDS                ; procedure ends
CODESGMAIN ENDS             ; code segment ends
END PROGRAM
```

Figure 5.23. A Program typed with errors. The program is same as figure 5. 19 except for the syntax errors

On assembling a program which has syntax error, the assembler generates error statements. The error statements can be observed in figure 5.24 which is the LST file which is generated when MASM 5.0 assembler assembles the source program offigure 5.23. (Note: To observe errors, we must open the .LST file which is generated when we assemble the source program. A .LST file is generated irrespective of the fact that the assembler detects errors in the source program. To open the .LST file all we have to do is open it in a text editor such as notepad).

```
1           PAGE 60, 120
2           TITLE directives usage for segment definition
2           ; PROGRAM SHOWING SOME ERRORS THAT CAN BE GENERATED
4           ;.........................................................................................
5 0000      DATASG SEGMENT PARA 'data'
6 00000078      A       OW      120         ; define the two numbers
7 0002 00E8     B       OW      232
8 0004 ????     C       OW      ?           ; unintialized data items
9               DATASGENDS
   error.ASM(10) : error A2105 : Expected: instruction or directive
10 0000         CODESGMAIN SEGMENT PARA 'code'
11 0000         PROGRAM PROC FAR            ; procedure starts
12          ASSSUME DS:DATASG, CS:CODESGMAIN, SS:CODESGMAIN ;load seg
            registers
            error.ASM (14) : error A2105 : expected instruction or directive
13 0000 BB____R     MOV OS, DATASG ; transfer datasegment address to DS register
```

```
    error.ASM(15) : error A2056 : Immediate mode illegal
14 0003 A 1 OOOOR    MOV    AX, A           ; store a in an reg
15 0006 A30004 R     MOV    AX, B           ; add a and b
16 OOOA A30004 R     MOV    C, AX           ; store sum in memory
17 0000 B8 170C      MOV    AC, 4COO ; request function 4CH of INT21 H of DOS
    error.ASM(19) : error A21 07 : Non-digit in number
18 0010CD21          INT21H
19        PROGRAMENDS; procedure ends
    error.ASM(21) : error A2000 : Block nesting error
20 0012 CODESGMAIN ENDP                     ; code segment ends
    error.ASM(22) : error A2000 : Block nesting error
21        END PROGRAM
Open Segments :
        CODESGMAIN.....................
        DATASG...............................
Segments and Groups:
Name                Length      Align       Combine     Class
CODESGMAIN........  0000        PARA        NONE        'CODE'
DATASG..................  0006        PARA        NONE        'DATA'
Symbols :
        Name            Type        Value       Attr
        A..........     LWORD       0000    DATASG
        B..........     L WORD      0002    DATASG
        C..........     LWORD       0004    DATASG
        Program........ FPROC       000     CODESGMAIN        Length = 0012
        @File name...... TEXT error

        23 Source Lines
        23 Total Lines
        10 Symbols

        51048 + 398728 Bytes symbol space free

        0 Warning Errors
        8 Severe Errors
```

Figure 5.24. The assembled .LST file of the error file showing error statements generated by the assembler

Now let us discuss these errors that are generated line by line for figure 5.24.

Line	Explanation
9	The assembler detects a error as it expects a directive to end the data segment line 9 generate an error as there is no space between DATASGENDS. The assembler assumes this is an identifier. This error can be remove by giving a space between DATASG and ENDS. Then the assembler will detect the directive for the end of data segment.
12	Line 12 generates an error as the assembler assumes that an assume statement will follow. Observe that I have made the mistke of typing ASSUME as ASSSUME. Removing one S will remove this error as them assembler detects the ASSUME directive which it is looking for.
13	Line 13 generates an error because direct data loading of the DS register is illegal (not allowed by the 8086). The error can be removed by placing the following code in place ofMOV DS, DATASG.

```
MOV AX, DATASG
MOV DS, AX
```

Figure 5.25. Code to remove error of line 13

Line	Explanation
	This error is generated because the 8086 only allows register to register data transfer for DS register.
17	This error is generated because the assembler expects the data to be in hexadecimal. We can remove this error by specifying a radix operator H after 4COO.
19,20	Errors ofline 19 and 20 are generated because the ENDP and END directives are at the wrong place. The program does not give correct syntax for ending the procedure PROGRAM and code segment CODESG. The error can be removed by replacing ENDS directive with ENDP ofline 19 and the directive ENDP with ENDS of line 20.

5.10.4. Linking an Obj file

After coupling or assembling an error free source program our next step involves linking the 013J file that the assembler generates. The .OBJ file is a file that only contain machine code the linker program performs the two following steps .

- Combines more than one separate assembled modules(.OBJ files) into one executable program. This may, be two or more different source programs performing different functions but required by one another for useful output.
- To generate a .EXE file and provide it with special instructions so as to enable it to be subsequently loaded by the operating system.

The linker program for MASM is LINK while the borland TASM it is TLINI'C From MASM vel' 6.0 onwards the linker program is the ML.EXE. The general format for generation of an .exe program by using the LINKER program is

LINK name [.obj], out [.exe]

TLINK name [.obj], out [.exe]

where the name is the name of the .obj file which is to be linked into an .exe program. The out is the name of the exe file that is to be generated.

C:\MASS\LINIK ANUJ.OBJ AKG.EXE.J

The above given command will thus link the object file named ANUJ.OBJ into the executeble program named AKG.EXE.

After having assembled and linked an assembly source program (which is free from errors) we have with us an executable program which performs operations according to instructions present in the assembly source program. The executable program thus generated can be executed anywhere and at any time by the DOS operating system.

5.11. KEY POINTS TO REMEMBER

1. We have introduced Assembly Lanugage Programming by use of assemblers to aid in your programming the 8086. Though we have introduced the coding of the 8086 instruction earlier but the process is excruciatingly difficult.
2. For the microprocessor Lab. that you may have, it would be easier for you to write thenecessary program in assembly language. This should be followed by assembling the necessary program by use of assemblers such as MASM/TASM.
3. The assembled .LST file contains a listing of the machine code which can be fed into your 8086 kit. This is an easier process for obtaining machine code for a program. The loaded machine code can now be executed on the kit.
4. Assemblers are low level programming tools to aid the 8086/8088 processor programming. (Note: Assemblers are also available for other microprocessors in the market).
5. Assembler comments are used to increase a programs readability. A comment is always preceded by a semicolon (;).
6. Reserved word are specific characters or string of characters reserved for the assembler for its specific purposes. Wrong usage of reserved words can result in errors.
7. Identifiers are names used to reference items in programs. Identifiers can b o. "ded to reference address of a data item (name identifier) or reference the address of an instruction (label identifier).
8. All assembly language programs consist of a set of statements. Statements are of two types: Instructions and Directives. Instruction Statements are those which are converted to machine code for the 8086 processor to execute. Directives inform the assembler to perform a specific action such as loading a memory address with a data item.

9. Directives are used to perform all assembler control functions such as how an assembler should print a p-egram (LIST) to defining the different segments in memory (SEGMENT directive) Directives are also used to define procedures (PROC), start and end a program (END directive) etc.
10. The key differences between the .COM and .EXE files are
    - size (.COM file is limited to 64 Kb)
    - number of segments
11. The assembly language program is typed in word editors such as notepad in text form. On Saving this file we must save it using the .aSl11 file extension to define it as a assembly language file.
12. An assembly language file (.asm) is assembled by the MASM or TASM executable files. This results in three different output files the .LST, .OBJ and .CRF files.
13. The .LST file can be opened by text editors such as notepad. This file is used for debugging purposes as all errors are displayed with corresponding messages implying what could be p.rrnr
14. The .OBJ file is the object code file. This file is used as a source for creating a dos executable (.EXE) file. Common linker programs are TLINK and MLINK. Linker programs perform the conversion of .OBJ files to .EXE (or .COM) executable files.

REVIEW QUESTIONS

5.1. What is assembly language and what are its advantages?

5.2. What are the different types of assemblers available in the market? What are their respective linker programs.

5.3. What are the differences between the .EXE and .COM files?

5.4. List the two instructions required to initialize the DS register in an assembly language program. Assume the name of the data segment is DATA_SEG.

5.5. What are the uses and reasons for comments in an assembly language program.

5.6. What happens when reserved word are used incorrectly?

5.7. What are the different types of statements in an assembly language program? What are their uses? Also explain the general format for defining a segment.

5.8. Explain the two types of identifiers used in an assembly language program? What are their differences?

5.9. What is the use of the SEGMENT directive? Explain the difference between SEGMENT and PROC directives?

5.10. What could be the possible error if line 16 of figure 5.8. would be detected on program execution? Explain?

5.11. Why are simplified segment directives used? What are the different types of memory Models permitted for assembly language programs?

5.12. What are the automatically generated statements that are assembler produces when we use simplified segment directives?

5.13. Why are the data definition directives used? List all different data definition directives, their uses and different radix specifiers that may be used to store values by these directives.

5.14. Why and what are the uses the ORG lOOH statement which is included in the .COM program? What does the ORG directive do ?

5.15. List the different steps required to

1. assemble a program

2. link a program

5.16. Write a simple assembly program to add two memory locations where each memory location is byte wide. Reference each location by the name data 1 and data 2. Store the result in a memory location referenced by result.

●●●

Chapter-6

DATA MOVEMENT INSTRUCTIONS

INTRODUCTION

Before we attempt to program the 8086/8088 microprocessor it is compulsory to understand the diferent types of 8086/8088 Instructions. All the different instructions that the 8086/8088 microprocessor can executes are collectively called the 8086/8088 instruction set. The set of all instructions that a microprocessor can execute collectively comprise the microprocessor's instruction set. In the preceding chapter we dicussed the various addressing modes as well as the instruction format. Now we shall discuss the various instructions of the 8086/8088 microprocessor

The data movement instructions* form a very important class of instructions because without these type of instructions we would be unable to do anything. The various type of data movement instructions are: MOV, PUSH, POP, XCHG, XLAT, IN, OUT, LEA, LDS, LES, LSS, LANF and string data instructions. (MOVS, LODS, STOS, INS and GUTS). The instructions are used by a programmer to input data from memory to registers or from register to register or from register to memory. Data movement is very necessary as the data has to moved within the processor for execution or the result has to be moved from the processor to memory for result storage. Throughout chapter 6,7, and 8 we shall use the MASM assembler to shows you how these instructions work and how assemblers are used for programming. You can do the same by using the fundamentals of chapter 7 to form a assembly language program and then assemble it by the Assembler. The .LST file ofthe corresponding assembly file shall contain the machine code which we have illustrated throughout.

6.1. MOVE (MOV) INSTRUCTION

The move instruction is the most versatile of all the 8086/8088 data movement instructions. The Move instruction can be used for transferring data between a register and memory location or from a memory location to a register** or between two register. The move instruction can also be used to transfer between a segment register and a general purpose registers (or index or pointer register). A move*** instruction can vary from 2 bytes to 4 bytes in length depending on the addressing mode used. The move instruction can use all the different addressing modes as shown in figure 3.11. (However, all 8 bit displacement are sign extended into 16 bits before being added).

---

* data movement as the name suggest are instructions designed to move data from one place to another (i.e. memory to registers, registers to register or register to memory).

** A reference to a register throughout this text refers to the registers as shown in figure 3.10 except when specified otherwise.

*** Move here shall be a reference to the MOV instruction.

Let us use the basic instruction format and show you the operation of the MOV instruction. The basic form of the MOV instruction is listed below:

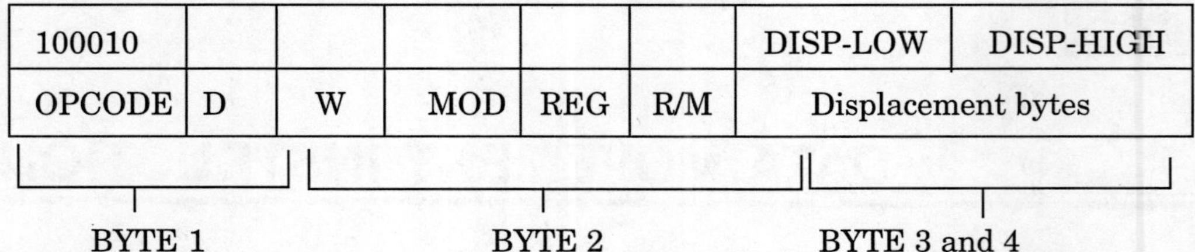

100010						DISP-LOW	DISP-HIGH
OPCODE	D	W	MOD	REG	R/M	Displacement bytes	

BYTE 1 BYTE 2 BYTE 3 and 4

The Opcode is 100010 and the direction bit (0 bit) will decide the direction of data transfer If D = 0, the dati will be moved from the register specified by REG to a register or memory location specified by the MOD and R/M bits. (figure 3.11). Similarly, if D = 1 the process will be reversed.

The MOV instruction also supports data transfer operation wherein the data type transfered is 8 bit wide. This is controlled by the w bit where w = 0, then the MOV instruction will move 8 bit data while if w = 1 the data moved will be 16 bit wide data.

There are many different forms of MOV* instruction as listed below (Refer to Appendix A for machine codes).

Instructions	Operation Performed
MOV REG,MEM	MEM $\longleftrightarrow$ REG
MOV MEM,REG	REG $\longleftrightarrow$ MEM
MOV SEGREG, REG	REG $\longleftrightarrow$ SEGREG
MOV REG, SEGREG	SEGREG $\longleftrightarrow$ REG

The MOV instruction is a very simple instruction where the data is moved (copied) from the specified source location to the specified destination location. However, the MOV instruction cannot be used for data transfer where both the source and destination locations as both memory locations or between segment registers. That is MOV instruction cannot be used for memory to memory data transfer or segment register to segment register** data transfer. For example of the different coded move instruction refer to section 3.5.

6.2. PUSH/POP INSTRUCTION

The PUSH and POP instructions are important store and retrieve instructions that are designed to work on the stack memory. The stack memory is a LIFO (Last in First Out) type of memory that is initialized in the 8086/8088 system (1 Mb) Memory. There are four forms of the PUSH/POP instructions which are:

1. PUSH/POP Register.
2. PUSH/POP Memory Location
3. PUSH/POP Segment Register
4. PUSH/POP Flag Register

* during assembly language programming MOV instruction specifies as

MOV destination, source

where data movement is from source to destination. The source or destination canbe either a register or memory location.

** Segment registers at no time can be used as a source/destination when the destination/source is a memory location i.e. segment register to memory location transfer is not permitted.

6.2.1. Push Instruction

This is a 16 bit data transfer instruction where two bytes of data are transferred on to the stack. The various types of PUSH instructiuons are:

1. PUSH reg.*
2. PUSH memory. *
3. PUSH seg reg.*
4. PUSH flag.*

The transfer of data is done in the following way. Since, the stack memory location is by default addressed by SS:SP, let us consider an example to simplify the PUSH instruction execution operation. Let SS= 0200 Hand SP= 0722 H. Thus the stack at this instant can be assumed to be addressed SSx 10H + SP = 02722 H. Which may be considered to be the stack top due to the stack memory LIFO property.

Now whenever a data is pushed onto the stack the first data (Most significant) byte is (pushed) moved onto the stack segment memory location addressed by SS: SP-l and the second data byte on to the memory location addressed by SS:SP-2. After storage of data the SP register is decremented by 2. Let us consider the execution of two consective PUSH instructions loaded in Memory as shown in the example 6.1.

Example 6.1. Show the operation of the program code given below.

PUSH AX

PUSH BX

Answer: Assuming the 8086 processor to be in the following state at the start of execution of the first PUSH instruction.

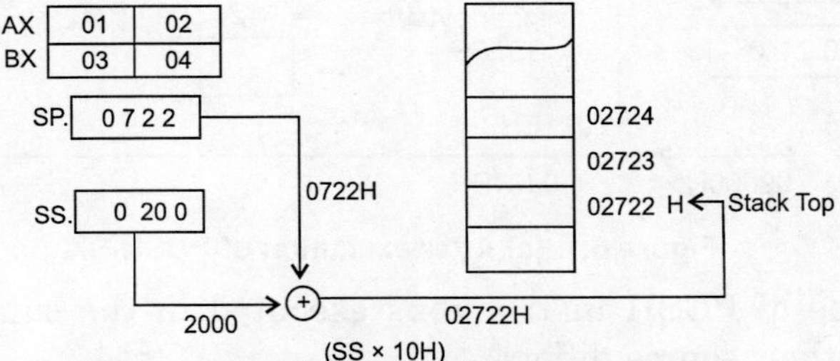

Figure 6.1. (a) Initial condition of processor

Observing the above operation we see that the stack segment address is 0200 Hand SP = 0722 H at the start of the first instruction PUSH AX. On encouneering the first PUSH AX instruction the AH (high or MOST significant byte) is transferred to SS : SP–l and AL register (lower or LEAST significant byte) is transfeered to SS : SP–2. Thus OlH (AH) transferred to 02721H and 02H (AL) to 02720H. After this operation SP is decremented by 2 and thus stores 0720H. Similiarly when the next instruction PUSH BX is executed SS = 0200H and SP = 0720H. Then BH (03H) is transferred to SS: SP–l (0271FH) location and BL (04 H) to ss

* Throughout this text the short form used would by default correspond as (unless specified otherwise)
reg → register (AX, BX, CD, DX, sr, Dr, BP, SP if 16 bit)
Seg reg → segment register (DS, ES, SS or CS)
Flags → Flag register
mem → memory location

: SP-2 (0271 E H) memory location After data transfer the SP register is again decremented by 2 and thus stores 071E H at the end of the instruction execution.

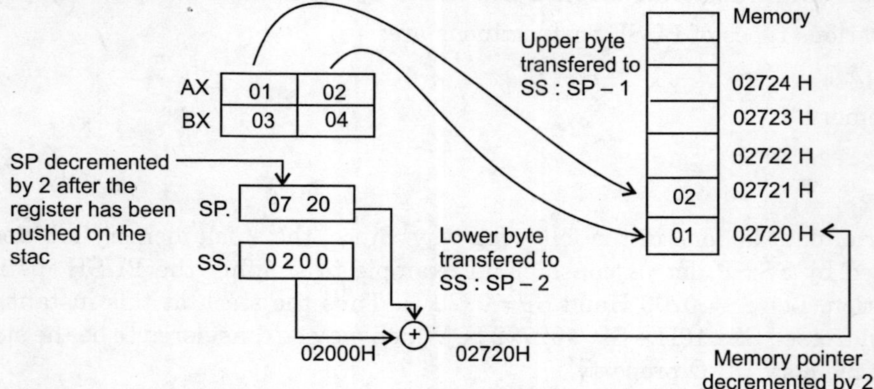

Figure 6.1. (b) After execution of PUSH AX

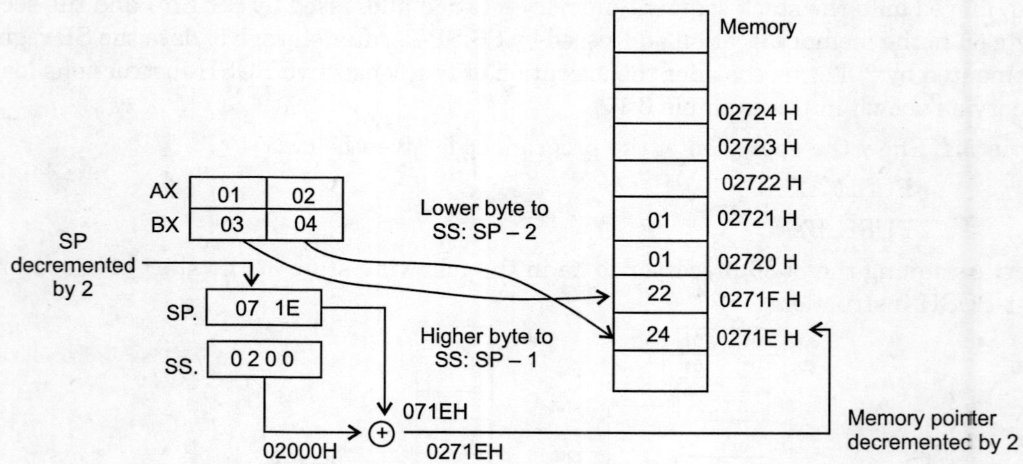

Figure 6.1. (c) After execution of PUSH BX

The different type of PUSH instructions executed in the same manner by the 8086 microprocessor. Only their source differs.

PUSH REG → Pushes the specified register onto the stack

PUSH MEM → Pushes the specified memory location on to the stack

PUSH Seg Reg → Pushes the desired segreg (segment register) on to the stack

PUSH FLAGS → Pushes the flag register onto the stack.

Instruction Format For Push

1. PUSH REG. – | 01010rrr |

   The register specified by rrr is pushed onto the stack. Refer to figure 4.10 for desired register.

2. PUSH MEM – | 11 1 1 1 1 1 1 1 | ×× | 110 mmm | disp low | disp high |

The memory location specified by the XX and mmm (refer to figure 3.10 and figure 3.10 in the chapter 3 addressing modes) is pushed on the stack. If the rim and mod addressing bits require a displacement then it is specified by the disp bits.

3. PUSH Seg Reg – 0 0 0 S S 1 1 0

   The segment register specified by ss is pushed onto the stack (Refer to figure 3.12 for the specified segment register codes)

4. PUSH Flag – 1 0 0 1 1 1 0 0

   The flag register is pushed onto the stack.

6.2.2 Pop Instruction

The POP instruction performs the exact opposite of the PUSH instruction. Here the data is transferred back from the stack onto the specified target register, (or segment register or memory location or to the flag register). The POP Instruction too operates on 16 bits of data. When the POP instruction executes it transfers the SS : SP memory location (byte) to the lower byte and SS : SP +1 memory location to the higher byte of target register I segment register / memory location or flag register. After data transfer the SP register is ineremented by 2.

Example 6.2: Show the operation of program code given below:

```
POP BX
POP AX
```

Answer: Assuming the processor to be in the state as shown in figure 6.2(a) before the first POP instruction. However, observe the AX and BX hold some initial values which will be overwritten during execution.

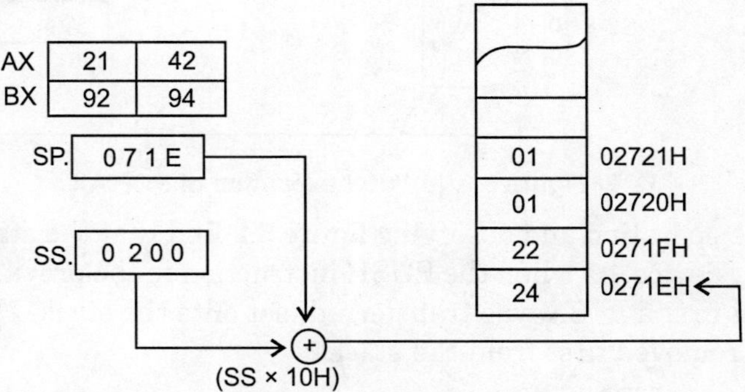

Figure 6.2. (a) Initial* condition of processor

Observing the above fig 6.2(a) we find that the initial stack memory pointer points to SS= 0200H and SP = 07lEH i.e. 0271E H. Now when POP BX instruction is executed BL (lower byte) is loaded with SS:SP memory location data ,:e., 24H and BH is loaded by SS:SP + 1(0271FH) memory location contents /.e., 22H. After this SP is incremented by 2 and therefore stores 0720 H. Now on the execution of POP AX we find the stack memory address points to SS : SP [0200H: 0720H] or 02720H memory address. Then, AL is loaded by the data in 02720H memory address and AH with the data of SS : SP + l or 02721H memory location. Therefore, finally

* Throughout the text the initial state is not static but can be said to be different depending on the previous operations ofthe processor. Here for clarity we assume arbitrary values to pictorially explain the operation (execution) of a specific instruction.

AX stores 0101 H. After this we find that the SP register is incremented by 2 and now stores 0722 H.

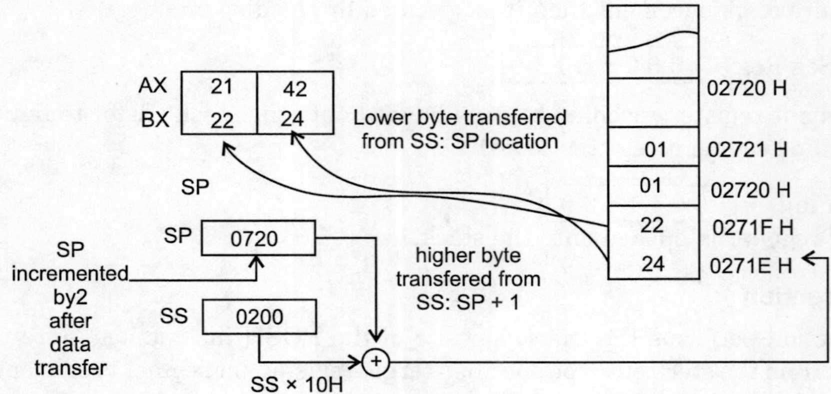

Figure 6.2. (b) After POP BX

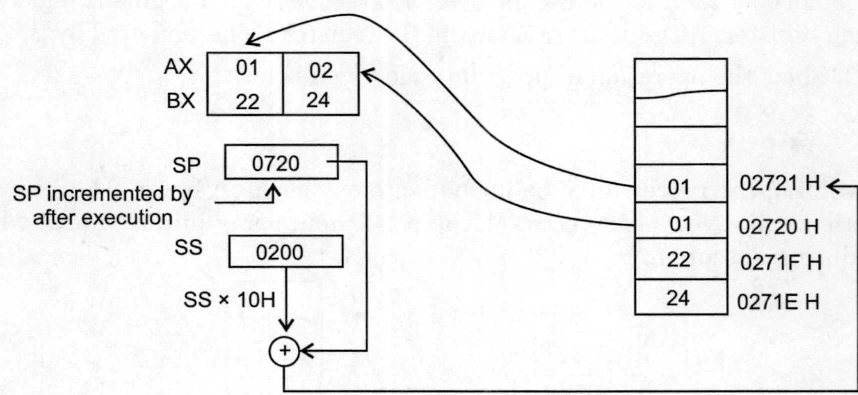

Figure 6.2. (c) After execution of POP AX

From the above operation and observing figure 6.2 find that the stack memory location acts as a LIFO* memory device. As when the PUSH instruction in the previous section were executed AX was transffered first and BX was transferred last onto the stack. However, we observe that BX data has to be removed first from the stack.

The different type of POP instruction are:

1. POP REG → desired register is loaded with data from stack
2. POP MEM → desired memory location loaded with data from stack
3. POP Seg Reg → desired segment register loaded with data from stack
4. POP FLAGS → Flag register is loaded with data from stack.

To refer to instruction machine code for POP instruction refer to Appendix A.

* The stack memory behaves as a LAST IN FIRST OUT (LIFO) memory device. This is a cumulative effect of the PUSH and POP instruction. When the PUSH instruction pushed AX and BX in the previous example and if popped back the BX register should be loaded fn st else the wrong value would be reloaded in BX.

6.3. LOAD EFFECTIVE ADDRESS INSTRUCTION

The load effective address instruction appears in three forms in the 8086/8088 instruction set namely the LEA, LDS and LES. The load effective address instruction is used to load only 16 bit register (or the DS or ES segment registers) with an address. The loaded address is the contents of a memory/register location as specified by the instructions addressing mode selected as shown in figure 4.11.

6.3.1. LEA

The LEA instruction loads a 16 bit register with the offset address of the data specified and is dependent on the addressing mode ofthe instruction. However, comparison of LEA AX, [DI] with MOV AX, [DI] will show you the difference between the two data movement instructions. MOV AX, [DI] will load the AX register with contents of memory addressed by DS : DI*. However, LEA AX, [DI] will cause the AX register to be loaded by the contents of DI register (which is an offset address). Let us consider a few examples.

1. LEA AX, [DI]

The LEA AX, [DI] can be decoded as:

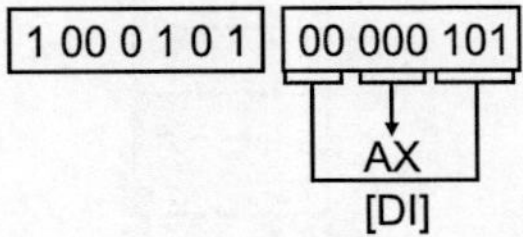

(Refer to figure 3.10 and 3.11. and Appendix A for machine code bit patterns and fields for forming an instruction). This will make the 8086/8088 processor to transfer the offset data that is the contents held in the DI register i.e., [DI] to the AX register. Thus the contents of DI wiil be transferred to AX.

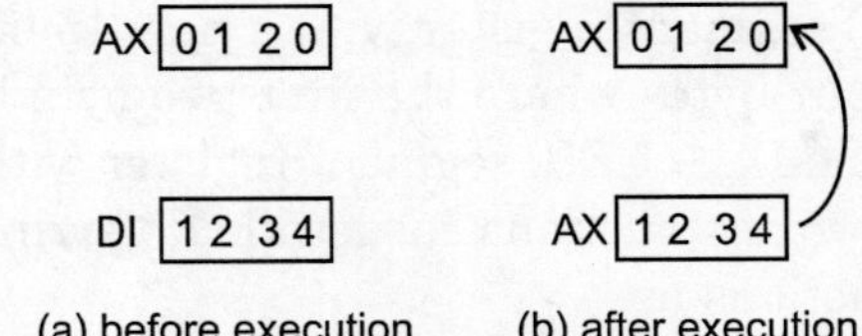

(a) before execution (b) after execution

Figure 6.3. LEA AX, (01) execution

2. LEA BX, 1234 H

This is a direct offset address load instruction where a 16 bit offset will be loaded from memory. The instruction would appear in memory as :

Address	Data	
1 0 0 0 0 H	8D	→ Instruction Load Bx, displacement
1 0 0 0 1 H	1E	
1 0 0 0 2 H	34	→ Displacement 1 2 3 4
1 0 0 0 3 H	12	

Refering to Appendix A we find that the LEA BX, d16 instruction where d16 is a 16 bit direct displacement is specified as.

* Assuming no segment override prefix is used, i.e. default segment register (from figure 3.11) is used i.e. DS segment register.

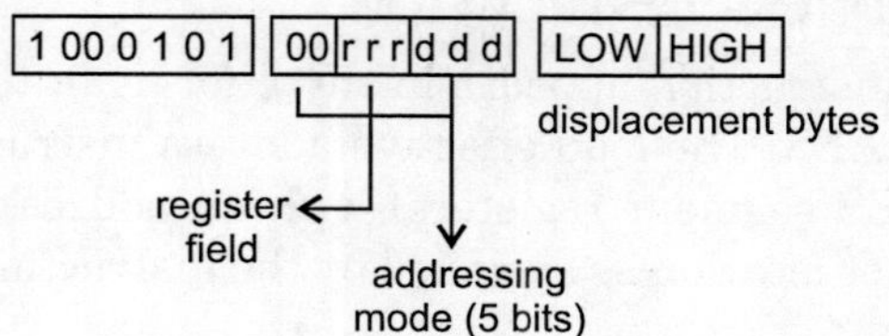

Thus for BX register to be used the register field must be 110 (Refer to figure 3.10) and for direct addressing mode the ddd field should be 110 resulting in the bit pattern as specified by 8D IE H. This will cause the displacement to be moved into the BX register.

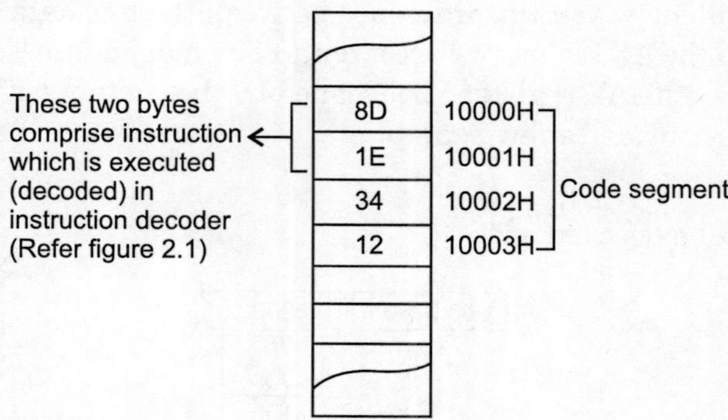

Figure 6.4. LEA BX, 1234H

6.3.2. LDS and LES Instructions

The LDS and LES instructions load any 16 bit register with an offset address stored in memory as well as load the DS (or ES) segment register with a new 16 bit segment base address. Thus the LDS (or LES) transfers four bytes where the first two byte load a specified 16 bit register while the last two byte load the DS (or ES) segment register with a new segment base address value. Let us consider the two instructions in detail. Referring to Appendix A we find the machine code for these instructions as

LDS	reg, mem	1 1 0 0 0 1 0 1 0 0 r r r m m m
LES	reg, mem	1 1 0 0 0 1 0 0 0 0 r r r m m m

where the memory address is reffered for the four bytes that load the specified registers (by rrr field) and the segment register (ES or DS) by the decoding of mmm field (refer figure 3.11 for the memory location value).

Now let us try to understand the functioning by taking an example. Note since mod = 00 we find that only the first column offigure 3.11 can be used for memory addressing.

Example 6.3 : where DS = 0102 H.

DI = 2212 H.

AX = 0101 H.

Answer: We find that the reg field (rrr) for AX are 000. and the (mmm) rim field for Register Indirect addressing is 101 for using the [DI] register. Thus LDS AX, [DI] can be coded as.

Now let us graphical explain the working

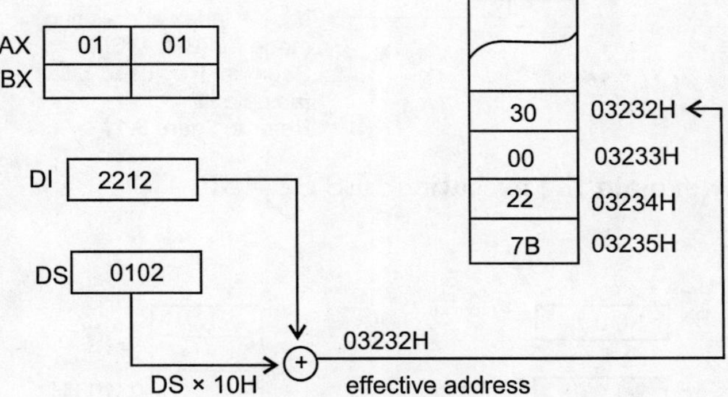

Figure 6.5. (a) before execution

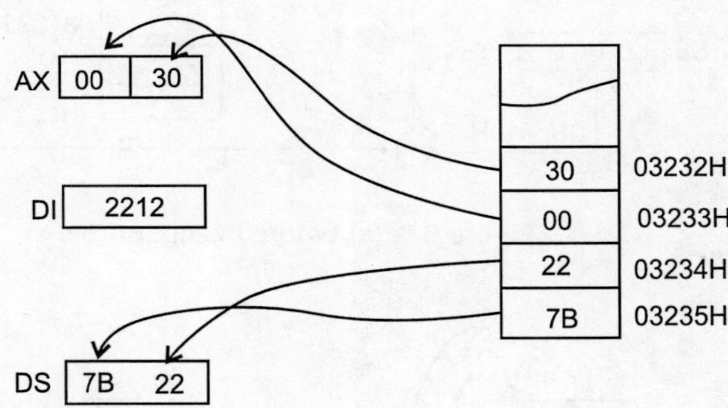

Figure 6.5. (b) after execution

Thus we observe the LDS reg, mem instruction is used to load the specified register (by rrr field) with the contents of the memory location specified (by DS: [mem]) where the lower byte of the specified register is loaded by contents of DS : [mem] and that of higher byte of register with DS: [mem + 1]. Also the DS register is loaded with the contents ofDS: [mem +2] and DS: [mem + 3] memory locations. Now let us consider another instruction of form LES reg, memo

Example 6.4: LES BX, [BP] + [SI] where SS = 0100 H

DS = 2000H

ES = 1002 H

BX = 1001H

BP = 2100H

SI = 0002H

Answer: We find that the reg field (rrr) for BX is 011 and for base plus index addressing from figure 3.E is 010. Thus LES BX, [BP] + [SI] can be coded as

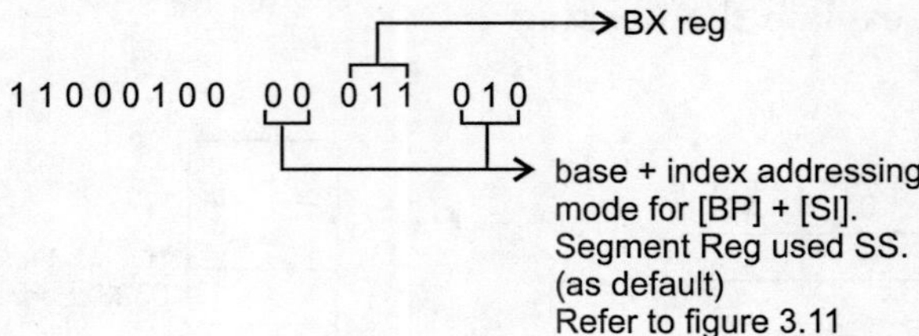

Now let us clearly explain the execution LES BX, [BP] +[SI]

BX I 1 0 0 1 I

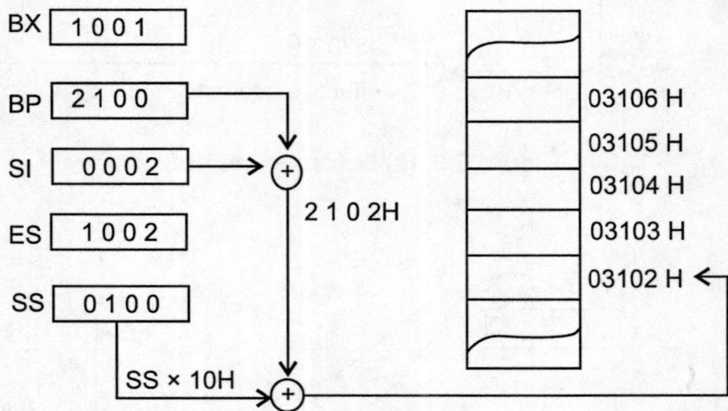

Figure 6.6. (a) before execution

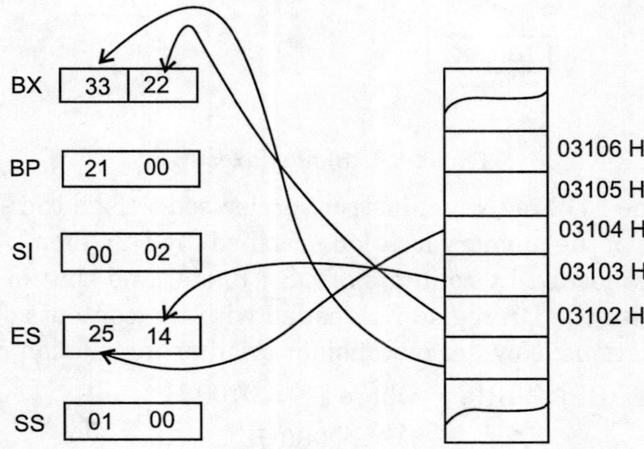

Figure 6.6. (b) after execution

Thus again it is observed by figure 6.6 that the LES is quite similar to LDS instruction.

The above instruction loads BX register with contents specified by SS : [BP + SI]. where the contents of SS : [BP + SI] memory location is transferred to BL and those of SS : [BP + SI + 1] memory location are transferred to BH register. Also the ES segment register is loaded with the contents of SS : [BP + SI + 2] and SS : [BP + SI + 3] memory addresses in that order

Similarly other addressing modes can be used with the LDS and LES instruction by changing the m m m fields. For direct addressing mode we have to make m m m field as 110 and so on (refer to figure 3.11).

6.4. STRING DATA TRANSFERS

Another important class of data transfer instructions is the string data transfer instructions. These include three instructions namely LODS, STOS and MOVS instructions. Each of these instructions allow data transfers that can be either a single byte or a word (double byte or 16 bits) The string instruction are dependent on the direction flag* and use either of the Index registers (SI or DI) for addressing. Thus it is important to understand both the use of the direction flag as well as use of Index registers.

6.4.1. Direction flag

During string operations the direction (DF) flag of the flag register is used to select the auto increment or auto decrement operation for the SI or DI registers. Since string instructions only use the SI or DI registers for addressing incrementing or decrementing the index registers would only cause the effective address to be shifted up or down respectively (as the segment register remains constant). If D = 1 then the SI or DI register is decremented by 1 (for string operations that transfer one data byte) or by 2 (for string operations that transfer one data word). Similarly if D = 0 then the SI or DI register is incremented by either 1 or 2. To enable either of the two modes (auto increment or auto decrement) for string operations we can use the following instruction.]

CLD → Clear Direction flag instruction.

S1'D → Set direction flag instruction.

The CLD instruction clears the Direction flag i.e., DF = 0 or enables the auto increment mode for string operations. Similiarly STD sets the Direction flag i.e., DF = 1 or enables the auto decrement mode for string instructions.

6.4.2. DI and SI register addressing modes for string operation

When the 8086/8088 executes string instructions it uses only the DI or SI register for the offset address of memory. For example the LODS instruction uses the SI register contents for the offset address while the STOS uses the DI register contents as the offset address of memory location. The default segment address base for instructions using DI as offset address is the ES (Extra segment) Register while for instructions using SI as offset address is the DS (Data Segment) Register. Note that the segment register assignment when the SI register is used to hold the offset address can be changed with the segment override prefix. However for the DI register, the segment base register is always the ES register and cannot be changed by the segment override prefix.

---

* direction flag is contained in the flag registers (section 2.2.1.3) at bit position 10.

Offset Address Register	Default Segment base Register	Override Segment Register base
DI	ES	None
SI	DS	ES,SS,CS

Table 6.1. Default Segment base Register

Now let us explain the string data movement instructions in detail.

6.4.3. LODS instruction

LODS instruction simply loads a string data byte (or word) into the AL (or AX) register with the data is being stored at the address referenced by DS : SI memory location. That is the offset address of data byte is stored in the SI register and thus by default the DS register will be used as segment base address. After data transfer to AL(or AX) the contents of SI will increment by 1 (or 2) if DF = 0 or the contents of SI will decrement by 1 (or 2) if DF = 1. The following table 6.2 lists the different form of LODS intruction and the figure 6.7 shows the operation of LODS instruction for auto decrement and increment modes.

Assembly Lang	Name ofInstruction	Operation
LODSB	Load a byte	AL = DS: [SI]
		SI = SI + 1 if DF = 0
		SI = SI - 1 if DF = I
LODSW	Lead a word	AX = DS: [SI]
		SI = SI + 2 if DF = 0
		SI = SI – 2 ifDF = I

Table 6.2. Form of LODS instruction DF = Direction flag

Example 6.5 : Show the operation of LODSB instruction when DF = 0, SI = 0002 H, DS = 2000H.

Answer: Since direction flag DF = 0 it is obvious that the mode selected is auto increment mode. Therefore, the SI register will be incremented after the execution of the LODSB instruction.

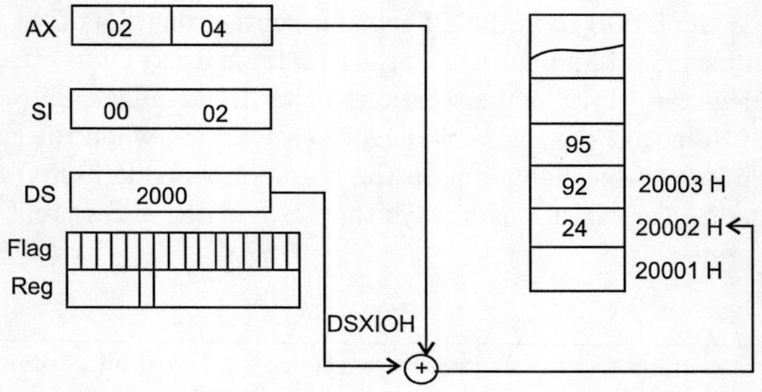

(a) Initial State of Processor

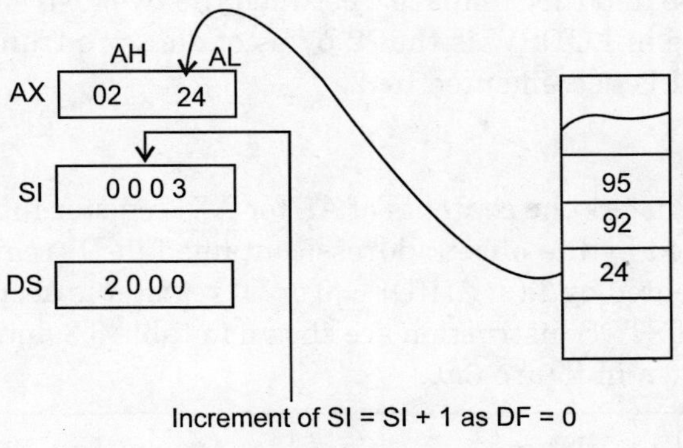

(b) After execution

Figure 6.7. Execution of LOOSB

It is clear from figure 6.7 that the AL register is stored with data at memory location. DS: [SI] and following after execution of the LODSB instruction the SI Register is incremended by 1 as DF = 0 [Auto increment mode].

Example 6.6: Show the operation ofLODSW instruction when DF = 1.

Answer: The only change observed is that DF = 1 [Auto decrement mode]. Also LODSW will transfer a double byte to AX register.

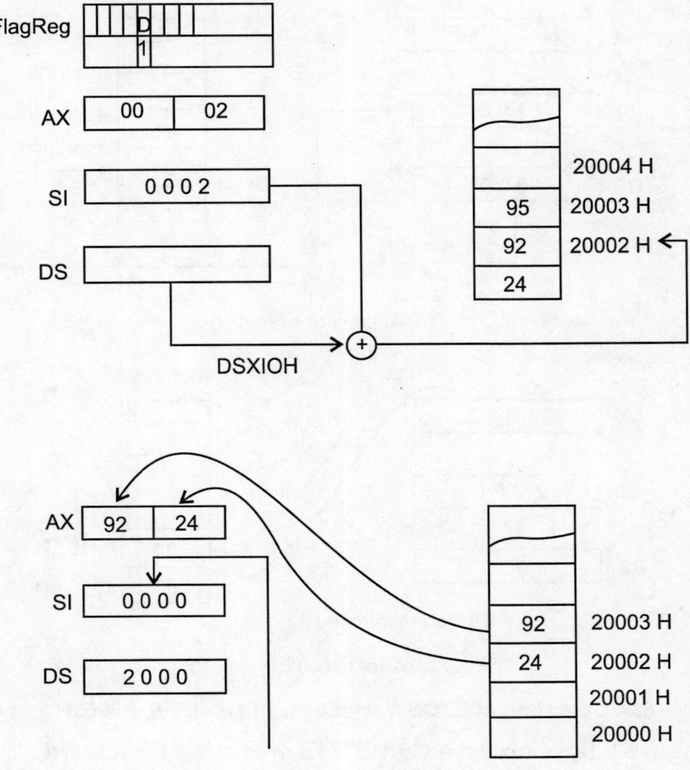

Figure 6.8. Operation of LODSW Instruction

From the figure 6.8 we find the similiarity between the two instructions that is LODSB and LODSW. The only change in LODSW is that 2 bytes of data are transferred. Also since D = 1 (auto decrement mode) SI is decremented by 2.

6.4.4. STOS Instruction

The STOS Instruction transfers the contents of AL (or AX) register into the extra segment (ES) memory location addressed by the offset address contained in DI register. After data transfer the DI register in incremented by 1 (or 2) ifDF = 0 or DI register is decremented by 1 (or 2) if DF = 1. The different forms of STOS instruction are shown in table 6.3 and the operation of STOSW (store string word) is shown in figure 6.9.

Instruction	Name	Operation
STOSB	Store a byte	ES: [DI] = AL DI = DI + l If D = O DI = DI– l If D = ^
STOSW	Store a word	ES : [DI] = AX DI = DI + 2 I fD = O DI = DI – 2 If D = 1

Table 6.3. Types of STOS Instruction

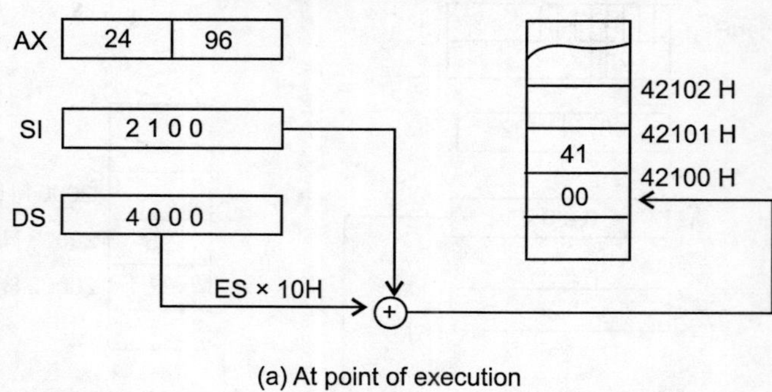

(a) At point of execution

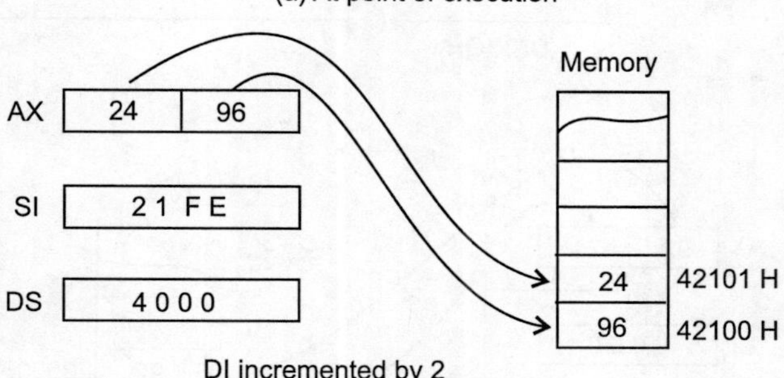

(b) After execution

Figure 6.9. Operation of STOSW with 01 = 21 OOH, ES = 4000H and OF = 0

Clearly from figure 6.9 we observe that STOS operation for a word string data is the exact reverse of LDOS instruction. The AX register contents stored at ES : [DI] memory location wherein AL is transferred to ES: [DI] and AH contents to ES : {DI + 1]. After data transfer the

DI register is incremented by 2 as DF = O. The STOS B instruction operates similiarly except this instruction only transfers a data byte followed by increment of DI register by 1.

6.4.5. MOVS Instruction

A very powerful instruction for data transfer from one location to another is the MOVS instruction. This is the only memory to memory data transfer instruction supported by the 8086 and 8088 microprocessor. This instruction can be used for byte or word transfer from DS : [SI] to ES: [DI]. The function is shown in figure 6.10.

Here it is observed that the data will be transferred from DS : [SI] to ES : [DI]. Thus the data from 20002H will be transferred to 40004HH and 20003H will be transferred to 40005H. (Since it is a word instruction). After data transfer DI and SI both will be incremented by 2 if DF = 0 or decremented by 2 if DF = 1. Now thus the final state of the machine (i.e. the microprocessor) will be as in figure 6.10(b).

As in LODS and STOS, MOVS also can operate on byte or word data. Thus MOVSB transfers a byte from DS : [SI] to ES : [DI] locations and increments or decrements by 1 both SI and DI depending on state of DF flag. Similiarly MOVSW transfers a word of data from DS : [SI] to ES: [DI] memory location followed by increment or decrement of SI and DI register by 2.

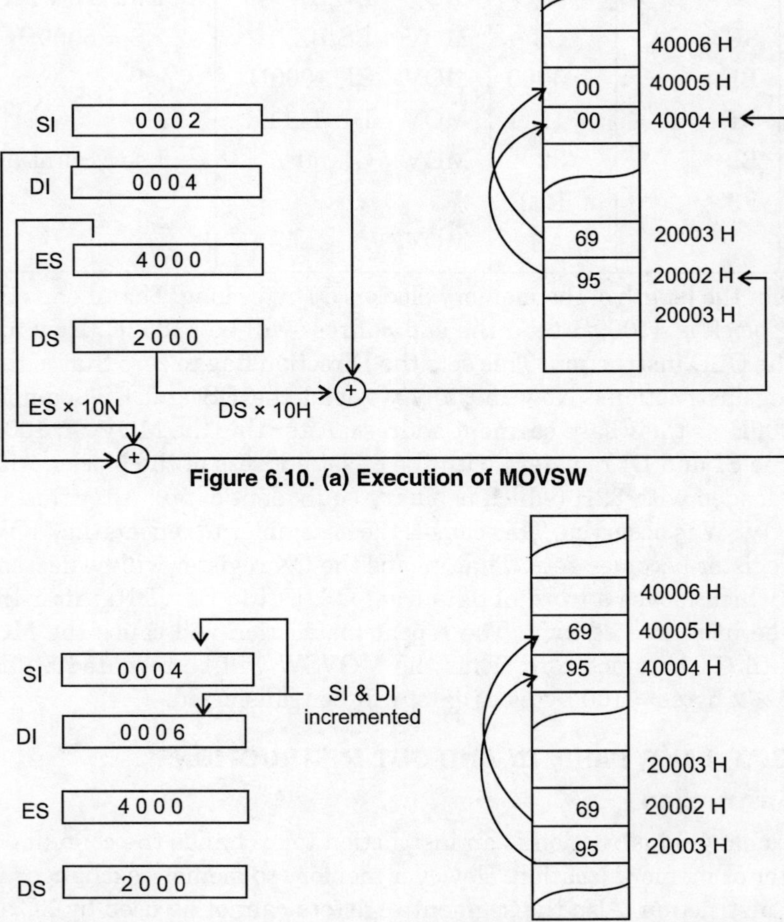

Figure 6.10. (a) Execution of MOVSW

Figure 6.10. (b) Final state of processor after MOVSW (OF assumed to be zero)

6.4.6. Rep Instruction

The Repeat prefix is a prefix that can be added to all string instructions except LODS. REP instruction causes cx to be decremented by 1 after execution of the string instruction. If ex becomes zero then the next instruction in memory is executed, i.e. the loop in broken. However if CX $\neq$ 0 then the string instruction is repeated continuously and the ex register is decremented after each string instruction execution until cx = 0. Thus if we load ex with any arbitary value K, then a REP instruction will cause the string instruction repeat K times.

Example 6.7 : To transfer the data stored in a block of memory of size 50 bytes long starting 41000H to 41032H to memory location 51111H to 51143H.

Memory Address	Machine	Code	Assembly		Comments
0100	FC	CWD			; select
					; increment mode
0101	B8	4000	MOV	AX,4000H	; segment 4000H
0104	8B	5000	MOV	BX, 5000H	; segment 5000H
01e8	8E	D8	MOV	DS,AX	; load DS with 4000H
0110	8E	C3	MOV	ES,BX	; ES = 5000H
0112	BE	0000	MOV	SI, 1000H	
0115	BF	1111	MOY	DI. 1111H	
0118	B9	32	MOV	CX,50	; Load count as 50 Dee
0120	F3	REP			
0121		AS	MOVSW		

Observe that the length of the memory block is 50 bytes long. That means that if the starting address of the block is 41000H then the end address will be 41032H. Now in the program we observe first the CLD instruction. This sets the Direction flag to zero that is the auto increment mode for string instructions. Now initially we load the DS and ES segment registers with 4000H and 5000H as their base segment address. After this the MOV SI, 1000H and MOV DI, 1111H loads the SI and DI registers with the offset address of the blocks. After this the count CX register is loaded with 32H (which is binary equivalent of 50). After this a REP instruction followed by MOVSW is observed. This causes the assembler to repeat, the MOVSW instructions until the CX register becomes zero. That means the CX register will be decremented by 1 after the MOVSW (which moves a word of data from DS: [SI] to ES: [DI]). After this the SI and DI registers will be incremented by 2. The repeat instruction will cause the MOVSW to execute continually until CX becomes zero. Thus the MOVSW will be repeated 50 times till CX = O. That mean 50 × 2 bytes = 100 bytes of data' will be transferred.

6.5. XCHG, XLAT, LAHF, SAHF, IN AND OUT INSTRUCTIONS

6.5.1. Xchg Instruction

The XCHG (exchange) instruction is an instruction to exchange the contents of a register with another register or memory location. However memory to memory exchanges are not permitted by the XCHG instruction. Also the segment registers cannot be used by XCHG instruction.

Assembly	Function
XCHGAL, CL	Exchanges contents of AL with CL
XCHGBP, DI	Exchanges contents of BP with DI
XCHG DX, [SI]	Exchanges contents of DX with memory location DS : [SI]

Table 6.5. XCHG Instruction forms

6.5.2. SAHF and LAHF

SAHF or store AH register into flag register is an instruction used to store the data in the AH register into the lower eight bits of the flag register.

LAHF or load AH register with flag register is used to load the AH register with the data contained in lower eight bits of the flag register.

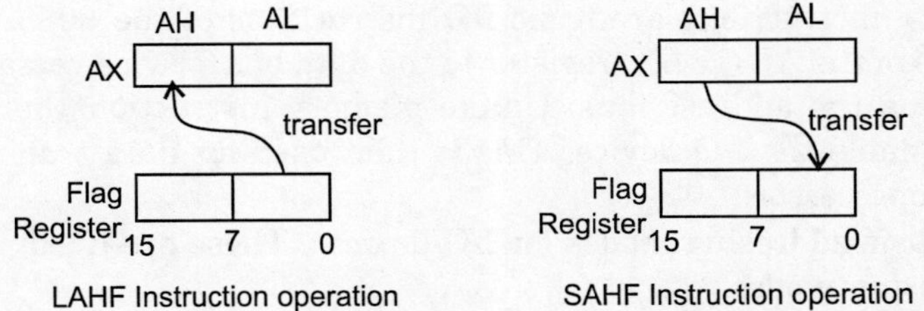

Figure 6.11. SAHF and LAHF instructions

This instructioin can be used to load certain flag condition such as setting the Trap Flag so as to enter single step execution mode.

6.5.3. XLAT Instruction

XLAT instruction or translate instruction is used for code conversion. When an XLAT instruction executes, it first adds AL register to BX register to form a memory offset in the Data segment. Then the contents of the memory location are transferred to AL register. A simple use is a gray code converter. Suppose the memory locations contain the following :

```
00000  00      ;Gray 0
00001  01      ;Gray 1
00002  03      ;Gray 2
00003  02      ;Gray 3
00004  06      ;Gray 4
00005  07      ;Gray 5
00006  04      ;Gray 6
00007  05      ;Gray 7
00008  0C      ;Gray 8
00009  0D      ;Gray 9
00018  BO      05          MOV       AX,0005H
0001A  BB      0000        MOV       BX,0000 4H
```

00010 D7 XLAT

0001E A2 000A MOV 000A,AL

Note that from OOOOOH to 00009 H we have loaded the gray equivalent from 0 to 9. Now we move 05H to AL and OOOOH to BX. Assume DS = OOOOR. Then when XLAT is executed, AL will be added to BX to produce the offset address which will be

Offset address = 05 + 0000

= 0005H

Effective Address = DS × IOH + 0005H

= 00005 H

Thus XLAT will copy contents of memory location DS : [BXl + [ALl to AL register which will be 07H. New 07 H is gray equivalent of 05H (binary). Thus by changing values of AL we can use this program to find the gray code during a bigger program.

6.5.4. IN And OUT Instructions

The IN and OUT inctructions are basic I/O instructions of the 8086. These instructions transfer the contents of the AL (or AX) register to the data bus (or vice versa) as well output the address of the port on the address, bus. Unlike memory instructions here the M/IO* pin is made low so as to enable the I/O device. This is then used for data transfer between an I/O device and the microprocessor.

There are two basic addresing modes for I/O devices. These are fixed port addressing and variable port addressing modes.

The fixed port address scheme directly includes the port address after the instruction. In this an 8 bit port number is included with an instruction for example.

IN AL, 20H

Where 20H is the port address. The 8086 MPU the converts this 8 bit port address into a 16 bit address by zero extending it into a 16 bit address. Thus this would appear as address 0020H on the $A_0 - A_{15}$ pins of the 8086.

The variable port address scheme uses the DX register contents.to address the I/O port.The DX register contents directly corresponds as the address the port and again appear on the address bus for as the port address.

The various forms of IN and OUT Instructions are listed belows.

Assembly		Operation
IN	AL, port 1	8 bits of data inputted from port 1
IN	AX, port 1	16 bits of data inputted from port 1
IN	AL, DX	8 bits of data inputted from port DX
IN	AX, DX	16 bits of data inputted from port DX
OUT	Port 1, AL	8bits of data outputted to port 1
OUT	Outport 1, AX	16 bits of data outputted to port 1
OUT	DX, AL	8 bits of data outputted to port DX
OUT	DX, AL	16 bits of data outputted to port DX

TABLE 6.6. Form of IN and OUT Instructions

* for the 8088 the pin is the which will be opposite to the above example. This pin would be driven high by the IN or OUT instruction when executing on the 8088.

The figure 6.12. illustrates the execution of OUT lOH, AX instructionwhere the contents of AX are transferred to I/O port with address lOH.

Notice that the port address 10H is zero extended to OOlOH before being sent on the address bus.

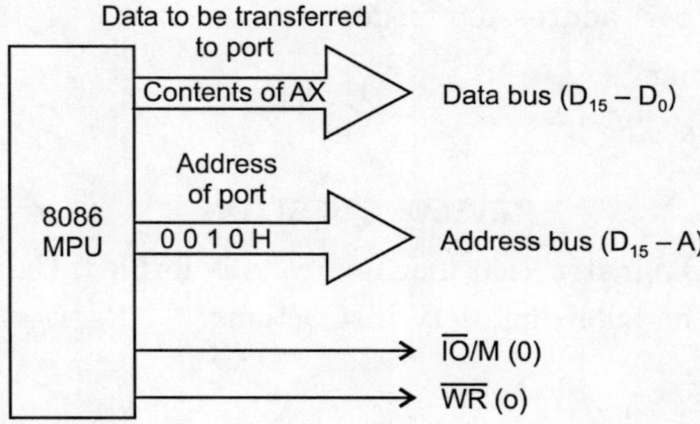

Figure 6.12. Operation of OUT 1 OH, AX instruction

6.6. KEY POINTS TO REMEMBER

1. Data movement instructions are instructions that are used to transfer (move) data from one location to another.
2. The MOV instruction is the most versatile data movement instruction. However, move instruction cannot be used to move data from memory to memory locations or between two segment registers.
3. The PUSH and POP instructions are the basic data movement instructions present in the 8086 instruction that allows for data transfer to and from to the stack memory.
4. Load effective address instruction loads the 16 bit specified register with the offset address of data specified by the addressing mode of the instruction.
5. The LDS and LES instruction load a 16 bit specified register with the offset address of data specified by the addressing mode as well as a segment register with new segment base address.
6. String data transfer instructions are dependent on the direction flag and use the index registers (SI or DI) to hold the offset address of the memory location to be accessed.
7. The DF = 1 selects auto decrement while DF = 0 selects autoincrement mode for string instructions.
8. MOVS instruction is the only 8086 instruction which allows for memory to memory data transfer.
9. XCHG instruction exchanges the contents of the specified registers. SAHF and LAHF are used to store AH onto the flag register or load the AH register with contents of flag register

respectively.

10. XLAT is the translate instruction which adds the AL register to BX register to produce a memory offset address for data. It is used in code conversion.
11. IN and OUT instructions are the only 8086 input and output to port instructions. They can use two types of port addressing modes
    - fixed port addressing
    - variable port addressing

REVIEW QUESTION

6.1. Explain how a MOV instruction machine code is formed. Using Appendix A find the machine code for the following MOV instructions
- MOVAX, BX
- MOV CX, [DI]
- MOV [SI], BX
- MOV DX, SS : [Dr]
- MOV AX, SS

6.2. How would the D bits and W bits of figure 3.9 affect the operation of a move instruction.

6.3. Explain the operation of PUSH and POP instructions in detail. Assume the SS register to hold the value of 0200H while the SP register hold 0100H. What would be the effect on SS and SP register if the POP CX were to execute? Explain the same if a PUSH AX where to occur?

6.4. A programmer wants to save the state of the flag register before entering a subroutine. What would the likely instructions that he would require to use so as to be able to do so ?

6.5. Explain the operation of the LEA instruction.

6.6. What is the difference between LDS, LES and LEA instructions?

6.7. Suppose that initial state of an 8085 based system is DI = 2121H, DS = 0100 H, AX = 2121H. The memory location states are given below

Address	Contents
031211H	22H
03122H	33H
03123H	44H
03124H	55H
03125H	55H

What would be the state of the processor after the execution of

(a) LEA, AX, [DI]

(b) LDS BX, [DI]

6.8. Explain the direction flag and its use in string data transfer instructions. Which instructilns would be required to select auto increment and auto decrement mode for string data instructions?

6.9. Write a program to transfer 150 bytes of data from memory location starting at 21000H to memory addressed by 31000H ?

(Hint: Use MOVS instruction)

6.10. Explain the use of SI and DI register in addressing for string operation. What are the default segment registers used by each?

6.11. What are the differences between

(a) LODSB (b) LODSW (c) STOSB (d) STOSW

Graphically explain the working of each?

6.12. Design a BCD to excess three code converter program. (Hint: Use XLAT instruction).

6.13. What are the differences between LAHF instruction and POP Flag reg instruction? Explain.

●●●

Chapter-7

ARITHMETIC AND LOGIC INSTRUCTIONS

INTRODUCTION

A very important class of Instructions are the Arithmetic and LOGIC instructions. As their name suggests these instructions operate on data and execute some arithmetic or logic operations. In this chapter we shall thoroughly discuss all the 8086/8088 arithmetic and logic instructions.

7.1. LOGIC INSTRUCTIONS

Logic instructions are a basic requirement for computer systems as they provide a mechanism for binary bit manipulation. These instructions allow specific bits to be set, cleared or complemented. All logic instructions affect the flag bits. However the carry and overflow flags are always reset (cleared) during logic operations. It does not make any sense for a logic operation to generate a carry or an overflow.

The basic logic instructions include AND, OR, Exclusive OR (XOR) and NOT logic operation. Here we explain these instructions along with other instructions such TEST (a form of AND) and NEG instruction.

7.1.1. And Instruction

The AND instruction obeys the basic logic multiplication truth table shown below in figure 7.1.

A	B	T
0	0	0
0	1	0
1	0	0
1	1	1

Figure 7.1. Truth table for AND operation

Suppose we have two eight bit numbers as shown below. The logical AND of these would be as shown in fig 7.1

```
  10011100
  10000101
  --------
  10000100
```

Figure 7.2. ANDing of two binary numbers.

The basic utilization of the AND instruction is in the process called masking. Masking is basically the process of clearing unrequired bits in a number while retaining the bits that are required in the program. Figure 7.3 shows the process of masking.

7 6 5 4	3 2 1 0	bit no.
X X X X	X X X X	unknown no.
0 0 1 1	0 0 1 1	Mask
0 0 X X	0 0 X X	Result

Figure 7.3. Process of masking

The above operation clearly masks out bit 2, 3, 6, 7, (as the AND of any bit with zero is zero). However the bit no. 0, 1, 4, 5 are retained. This process can be used to convert ASCII 30H to 39H (ASCII equivalent of 0 to 9) into a equivalent BCD number. (Example 7.1)

Example 7.1:

```
0000 | C6 C3  35 | MOV  BL,   3 5H ; Load ASCII Number
0003 | 80  E3  0F | AND, B L  O FH ; Mask BL
```

The AND instruction can operate in all addressing modes except memory to memory addressing mode and segment register addressing mode. Corresponding flags are set (except carry and overflow flags which are reset).

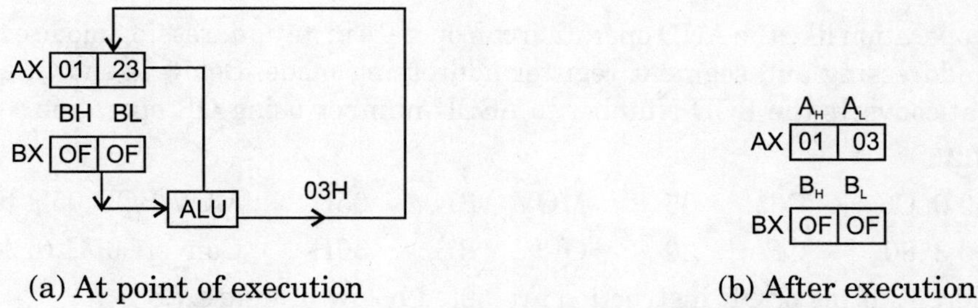

(a) At point of execution (b) After execution

Figure 7.4. Operation of AND AL, BL, instruction

Notice the above instruction AND AL, BL. We observe that the contents of AL and BL are ANDED by the 8086 MPU ALU to give 03H. After fhis value is stored in the AL register. The different forms of AND instruction are listed in figure 7.5.

Assembly	Operation
AND AL, BL	AL = AL AND BL
AND CX, DX	CX = CX AND DX
AND CL, 33H	CL = CL AND 33H
AND SI, 4242 H	SI = SI AND 4242H
AND BX, [D I]	BX = BX AND contents of address DS: [DI]
AND BX, [B P]+[D I]	BX = BX AND contents of address SS: [BP]+[DI]

Figure 7.5. Forms of AND Instruction

7.1.2. OR Instruction

The OR instruction follows the truth table for logical addition as shown below in figure 7.6. Here a zero will appear only when both bits ,are zero while a one will appear when any of the bits are one. An OR instruction can be also used as a mask. However here unlike the AND masking process unrequired bits will acquire a value of 1. The figure 7.7 shows the OR maskrig process.

A	B	Y
0	0	0
1	0	1
1	1	1

Figure 7.6. OR Operation Truth table

7	6	5	4	3	2	1	0	bit No
X	X	X	X	X	X	X	X	unknown No.
0	0	1	1	1	1	1	1	mask
X	X	1	1	1	1	1	1	Result

Figure 7.7. OR masking process

We observe in figure 7.7 that bit 0 to 5 are masked and thus remain set at 1. The bits 6 and 7 of the number will thus be transmitted unchanged to the Result.

The OR operation like the AND operation can operate in any addressing mode except memory to memory addressing and segment register addressing mode. Below is shown a portion of a program that converts the BCD Number to ASCII number using OR operation.

Example 7.2 :

```
0000 C6    C3    05    MOV   BL,    05H    ; MOV BCD 05H to BL
0003 80    CB    30    OR    BL,    30H    ; Convert 05H to ASCII B5H
```

The different forms of OR instruction are listed below in figure 7.8 .

Assembly Language		Operation
OR	DH, CL	DH = DH OR CL
OR	DX, CX	DX = DX OR CX
OR	BH,OIH	BX = BH OR OIH
OR	BP,0004H	BP = BP OR 0004H
OR	AX, [DI]	AX = AX OR contents ofDS: [DI]
OR	[DI], BX	Contents of address DS: [DI] = Contents of address DS: [DI] OR BX

Figure 7.B. Forms of OR Instruction

7.1.3 XOR (Exclusive Or) Instruction

As studied in digital electronics the exclusive OR operation provides an output that is logically if both variables are exclusive (different). The figure 7.9 shows the XOR truth table.

X	Y	Z
0	0	0
0	1	1
1	0	1
1	1	0

Figure 7.9. XOR truth table

The XOR gate behaves as a comparator because if both inputs are same the gate gives an output of zero logic. However if the two inputs are different the XOR gate provides an output of one.

The XOR instruction is of vital importance to us when there is a requirement for us to invert specific bits of a number while retaining some bits. The figure 7.10 shows the powerful qualities of the XOR gate.

	7	6	5	4	3	2	1	0	bit position
	X	X	X	X	X	X	X	X	Unknown Number
+	1	1	1	1	0	0	0	0	Mask
	X	$\overline{X}$	$\overline{X}$	$\overline{X}$	$\overline{X}$	X	X	X	

Figure 7.10. XOR operation on a number to Invert some bit while retaining some

Observing figure 7.10 we notice that bit 0 to bit 2, and bit 7 will remain same. However bit 3 - bit 6 are inverted.

Another very important use of an XOR instruction is to clear a register or memory location. The XOR instruction provides the fastest way as well as uses a less amount of system memory to do the job. The Example below shows the use of XOR instruction to clear a register.

Example 7.3.

```
0000    XOR    CX, CX    ;    Clear CX register.
```

The XOR instruction for clearing a register would require just 2 bytes of memory. However if we decided to use MOV CX, OOH (which would also accomplish the same operation) we would require 3 bytes of memory. Thus for efficient programming the XOR instruction should be used to clear a register or memory.

Logic instructions AND, OR and XOR allow a programmer complete control over the state of any bit stored in a register or memory location. These instructions are of vital importance when we interface control equipment (such as motors, drives etc.), to the microprocessor which are required to be turned on (state 1), turned off (0) or toggled from on to off or vice-versa. A simple example is a LED controller interfaced to the 8086 .

Example 7.4.

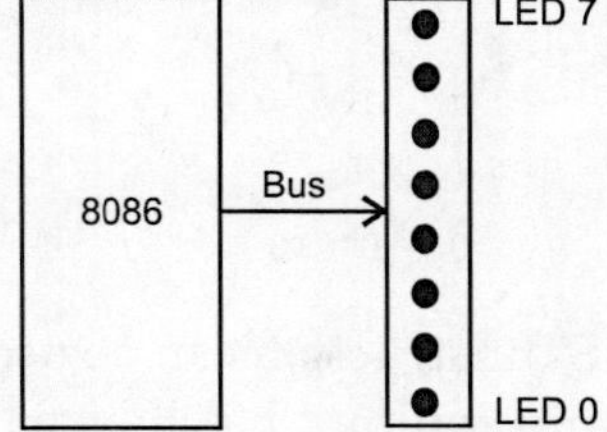

8 LED Panel (port address 0020H)

Figure 7.11. Application of Logic Instructions for interfacing circuitry

PROGRAM

XOR	AX,AX	:	Clear AX register
OR	AX,OOOFH	:	Set lower 4 bits at logic high
OUT	AL,0020H	:	Switch on LED 0 to LED 3
AND	AX,OOOAh	:	Switch off LED 0 and LED 2
OUT	AL,0020H		
XOR	AX,OOOFH	:	Toggle LED 0 to ·LED 3
OUT	AL,0020H		

On running the program we observe that the first instruction will clear AX register to OOOOH. The OR instruction will cause AX to store value OOOFH where AH = OOH and AL = OFH. On outputting this to LED panel (whose port address we are assuming at the moment is 0020H) it will switch LED 0 to LED 3 to logic on. After this the AND instruction will mask all bits except bit 1 and bit 3. On outputting AL again to Port 0020H, LED 1 and LED 3 will glow. The XOR instruction is used to toggle the lower four LED's i.e., on the execution of last instruction (OUT AL, 0020H) we find LED 1 and LED 3 off and LED 0 and LED 2 On.

7.1.4. Test Instruction

The test instruction is as the name suggests used for TEST purposes. It performs the AND operation. However unlike the AND operation which changes the destination operand, the TEST instruction does not do so. The TEST instruction only affects the condition of flag register.

The example is

TEST AH,OIH

In this instruction AH is Anded with 01. Consider AH to be OFH. Thus the TEST intruction would in effect AND OFH and 01 H.

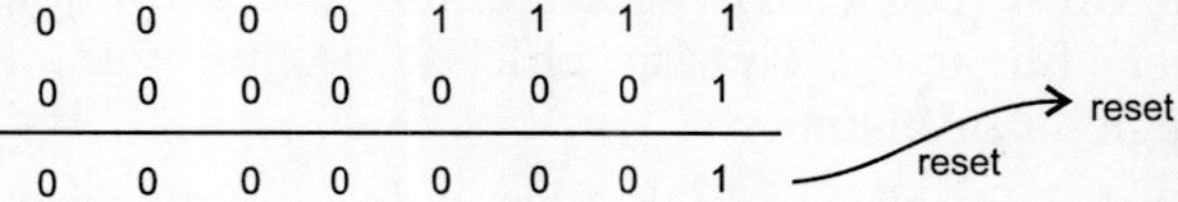

Since the bit number zero is not zero the zero flag would be reset i.e., ZF = 0 to show a state of non zero result.

However if AH is OEH then the TEST instruction would result in

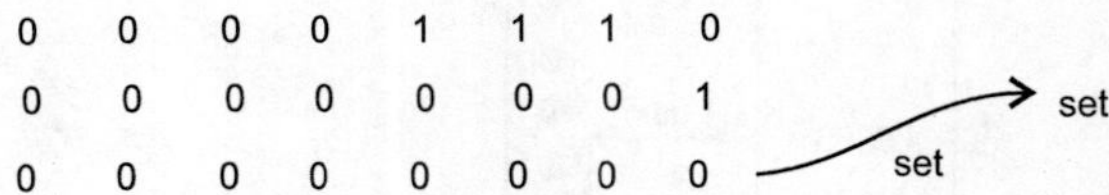

Since the result is zero the ZF = 1. TEST instruction can be used for checking multiple bits. Note however in the above case AH contents will not be changed after TEST instructon. The figure 7.12 shows the TEST instruction execution for the instruction TEST AH, AL. (The temporary register is shown only for understanding the execution).

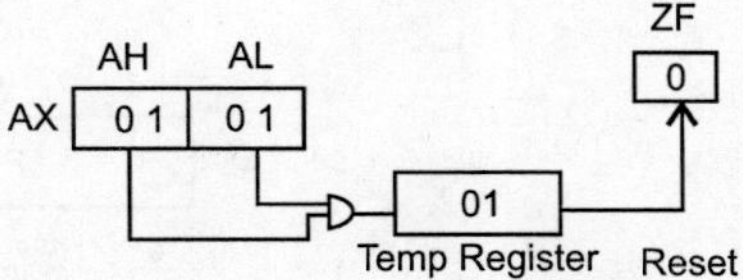

Figure 7.12(a). TEST AH, AL Operation. Note AH and AL donot change

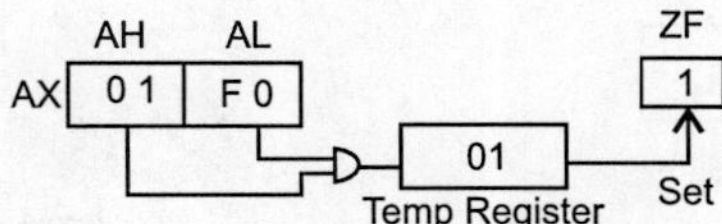

Figure 7.12(b). TEST Operation forTEST AH, AL when ZF = 1

The TEST instruction uses the same addressing modes as AND operation. The different forms of TEST instruction are listed in figure 7.13.

Assembly	**Language**	**Operation**
TEST	DL,DH	DL anded with DH. Flags are affected
TEST	CX,AX	ex anded with AX. Flags are affected
TEST	AH,4H	AH anded with 4H. Flags affected
TEST	CX, [DI]	ex anded with contents of memory location DS [DI]. flags affected.

Figure 7.13. Forms otTEST Instruction

7.1.5. NOT And NEG Instructions

The NOT operation or instruction on a register or memory location implies to complement to its contents (L's complement). NEG instruction when executed on a register or memory location simply finds its two's complement and stores it. These instruction have only one operand (It does not more any sense to apply a inversion or two's complement on two number's). These instructions can use any addressing mode except segment register addresing mode. Figure 7.14 shows the operation of NOT and NEG instructions·

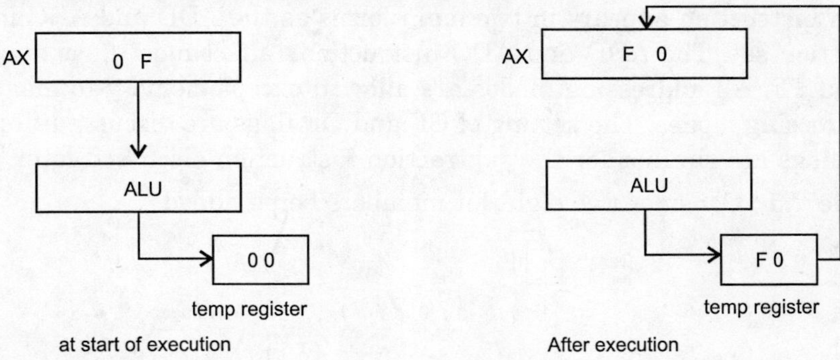

(a) Operation of NOT AX instruction (Note: The not instruction compleme -ts AX)

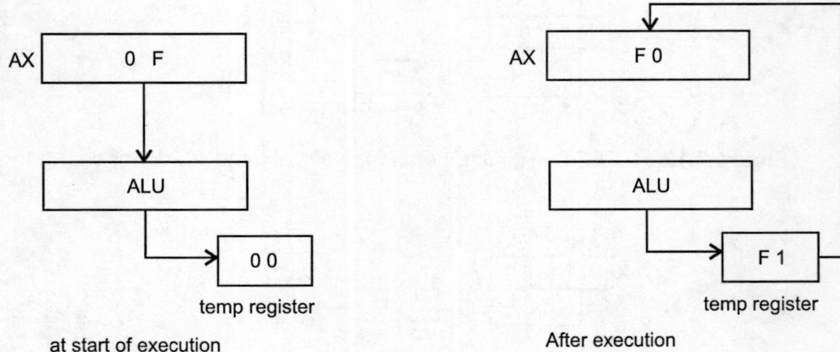

(b) Operation of NEG AX instruction (Note: the neg. instruction finds the 2's complement of AX)

Figure 7.14. Operation of NOT and NEG instructions

Figure 7.15 shows the different types of NOT and NEG instructions

Assembly	Language	Operation
NOT	AL	AL is complemented
NOT	CX	CX is complemented
NEG	AL	AL is 2's complemented
NEG	AX	AX is 2's complemented
NEG	[DI]	DS:[DI] memory address contents are two complemented

Figure 7.15. NOT and NEG instruction

7.2. ARITHMETIC INSTRUCTIONS

As of now we have discussed the logical instructions found in the 8086/8088 instructions. The other types of instructions that are present in the instruction set are the arithmetic instructions which include ADD, SUB, DIV, MUL etc.

7.2.1. Addition Instructions

The addition instruction appears in two main forms as the ADD and ADC instructions in the 8086 instruction set. The ADD and ADC instructions affect most flags such as SF, ZF, CF, ACF, OF and PF. All addressing modes are allowed except memory to memory and segment register addressing modes. The setting of CF and OF flags are discussed first. The changes of CF and OF flags are similiar for the subtraction instruction discussed later.

Example 7.5 : Consider two eight bit members being added

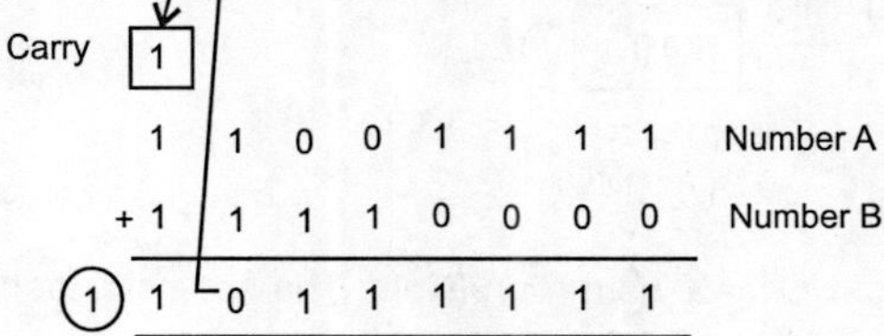

When these two numbers are added we observe there are two carries generated, one when the first seven bits are added and one when the MSB's added. We also observe that there is a final carry generated. When two 8 bit numbers or two 16 bit numbers are added and a final carry is generated this effectively sets the carry flag. Thus for the above number the final carry C (shown circled) is reflected by the carry flag being set. However when two 8 bit signed numbers are added such as -127 and - 1 then the final answer should be -128. Let us see the working below.

```
          [1] 1 1 1 1 1 1      →  Carry
          (1) 1 1 1 1 1 1 1    →  – 127
          (1) 0 0 0 0 0 0 1    →  – 1
Carry (1)  1  0 0 0 0 0 0 0
```

Observe above two numbers which we are assuming at the moment are signed numbers. The MSB (circled) reflects the sign ofthe remaining 7 bit number. If the circled bit is 1 it is a negative binary number while if the circled bit is zero the number is assumed positive. Now when these two numbers are added the result requires 8 bits to be stored, but for signed numbers only (n-1) bits i.e., 7 bits are present for storage of the magnitude. Such a condition is said to be an overflow. Thus the carry that is transmitted to the MSB (or sign bit) shown as square encircled sets the overflow flag so as to show that the two addition have overflowed their bit space (i.e. the result cannot be accomodated in the magnitude bits which is 7 bits for an eight bit number).

Now let in consider the two addition instructions in detail.

7.2.1.1 Add Instruction

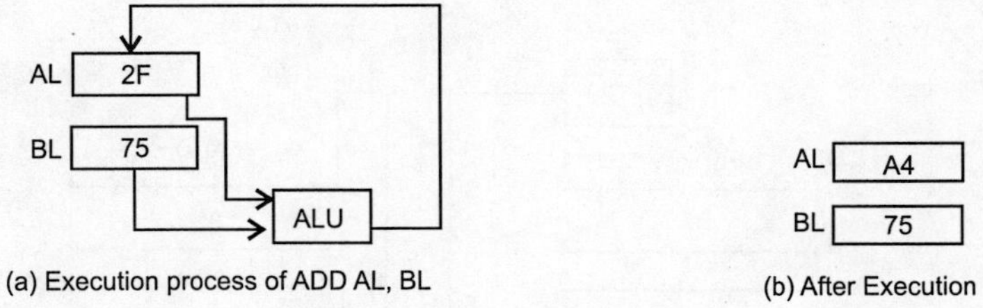

(a) Execution process of ADD AL, BL (b) After Execution

Figure 7.16. Operation of ADD AL,BL

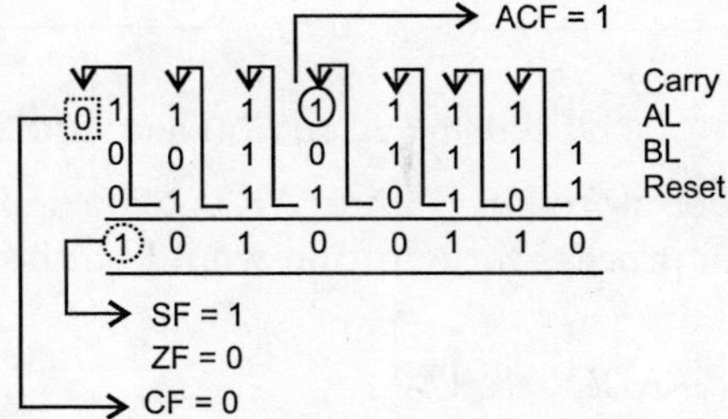

Figure 7.17. Flag changes due to ADD AL, BL

Observe figure 7.17 where in 2FH is added to 75H. Now ZF = 0 (zero flag) as the result of the operation is not zero. The SF is set because the MSB of the result is One. The CF is reset as no final carry generated. ACF = 1 as there is a half carry or carry after bit 3. (Similarly the ACF is set if there is a half carry or carry after bit 7 for 16 bit additions). The different forms of the addition instruction are listed in figure 7.1B.

Assembly	Language	Operation
ADD	AL, BL	AL = AL+BL
ADD	AX,DX	AX = AX+DX
ADD	BP,40H	BP = BP+40H
ADB	BX, [BP]	BX = BX +Contents of ss : [SP]
ADD	[DI], CX	Contents ofDS: [DI]
		= Contents of DS : [DI] + BP

Figure 7.18. Forms of ADD instruction

7.1.2. ADC Instruction

The ADC or addition with carry instruction is similar to the ADD instruction. Here the carry flag is also added to the operand. The ADC instruction is of vital important if we are to execute a 32-bit addition. Since the 8086 operates only on 16 bits of data thus the ADC is of vital importance to operate on data of lengths greater than 16 bits as an addition over 16 bits would require the carry bit to be added. Figure 7.19 shows a technique to add two 32 bits numbers. Notice that in the figure 7.9 we first add AX and CX registers to produce a 16 bits result and the appropriate carry. Now the ADC instruction will Add BX and DX registers and this carry to produce the corresponding 32-bit result.

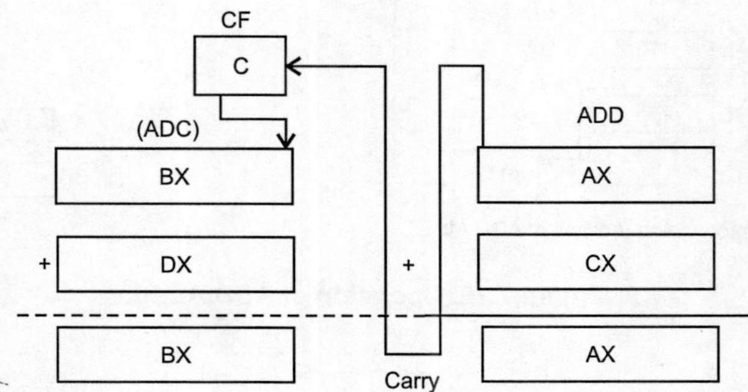

Figure 7.19. 32 bit addition using ADD and ADC instructions

Suppose that 32 bit number stored in BX and AX is to be added to a 32 bit number stored in DX and CX registers then the process for addition would be the program for figure 7.19 would be as

```
0000   03   C1   ADD AX, CX
0002   13   DA   ADC BX, DX
```

The some different forms of ADC instruction appear as shown below a in figure 7.20.

Assembly	Language	Operation
AD C	AL, AH	AL = AL+AL+CF
ADC	BX,DI	BX=BX+DI+CF
AD C	[DI], CX	contents of DS: [DI]
		= contents of DS : [DI] + CX + CF
AD C	DH, [BX]	DH = Contents of DS : [BX] + DH + CF

Figure 7.20. ADC instruction types

It should at this point be understood that the figure 7.20 shows only a few forms of the ADD instructions. Since there are a wide variety of ADD, ADC (and other instruction) forms we can only show some of their variations.

7.2.2. Subtraction Instructions

Subtraction instructions in the 8086 have two main forms, the SUB and SBB instruction. These are called the Subtract and Subtract with Borrow Intruction. Both these instructions can use any addressing mode except memory to memory addressing mode and segment register addressing mode. First let us consider the process of subtraction.

Example 7.6: Subtract 3H and 2H

	1	1	binary for 3H
–	1	0	binary for 2H.
	0	1	Result (01H)

Now let us consider the process of a subtraction where a larger number is subtracted from a small number.

Example 7.7: Subtract 2H and 3H

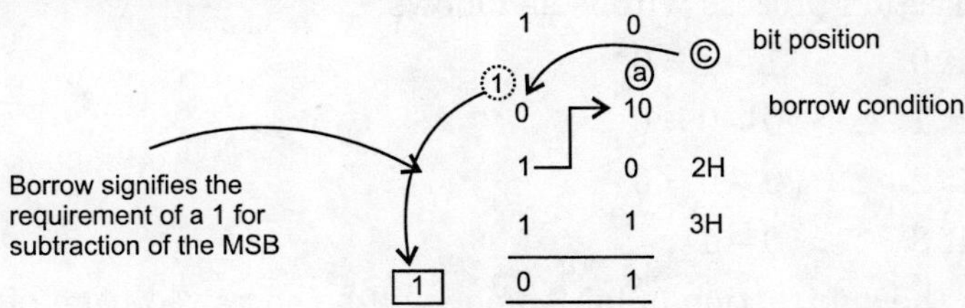

(*a*) Observe in above that at bit position 0, zero is to be subtracted from one. Thus we borrow 1 from bit position 1 (of 2H). This results in 10 and thus 1 is subtracted from 10 to give 1 at the LSB.

(*b*) The borrow of 1 from bit position 1 results in a zero. Again we have to subtract 1 from 0 at the MSB position.

(*c*) To accomplish this we have to borrow L But since this is the MSB, i.e., no other bits from where we can take a borrow, we say a final borrow has been generated. Thus we take a 1 to MSB position and get 10. Now 1 subtracted from 10 to give 1 (this final borrow is reflected by the carry flag in the 8086 microprocessor). We now extend the above example for 8 bits as used in the 8086 microprocessor. (Note: the process will be same for 16 bits except that now the two numbers will be 16 bits wide).

Example 7.8 : Subtract 21H and 39 H.

Answer: Consider the two numbers.

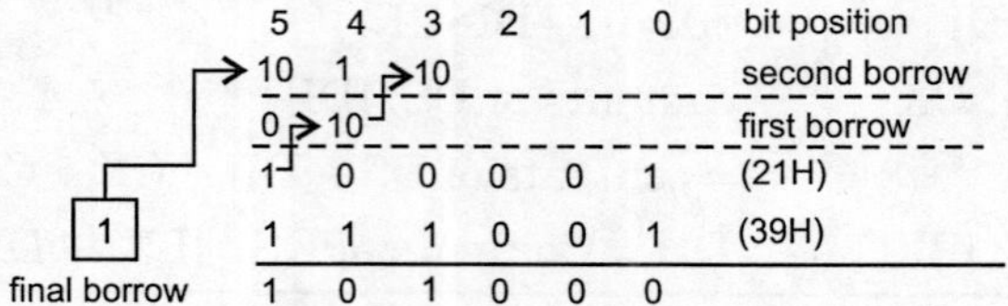

Thus subtraction would yield 28 H with a borrow of L

Example 7.9: 8 bit subtraction of 21 Hand 39H.

We now extend the above example for 8 bits as used by the 8086 microprocessor (the only difference will be that the numbers of the above example will now be treated as 8 bit numbers). (Note: the process will be same for 16 bits except that now the two numbers will be 16 bits wide.)

Answer: The subtraction here will differ from the above process because 21H now is a 8 bit number and so is 39 H. Thus there will be two more bit positions to be considered.

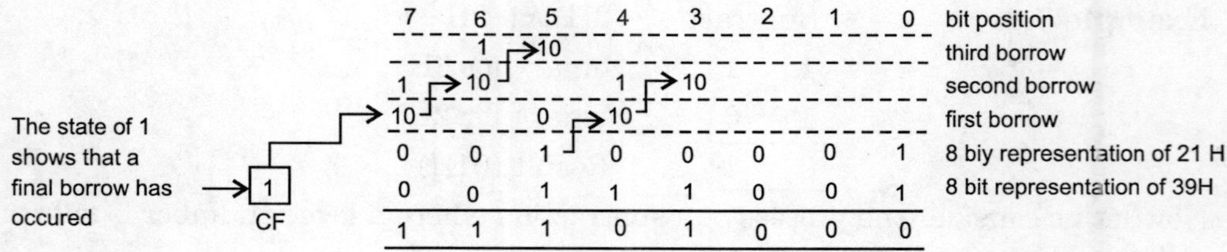

Here the subtraction process will be as follows:

bit 0 : 1 – 1 = 0

bit 1 : 0 – 0 = 0

bit 2 : 0 – 0 = 0

bit 3 : 0 – 1

To accomplish the subtraction of bit 3 a one will be borrowed from bit 5. This will result in 0 at bit 5. This borrowed one will result in 10 at bit 4. Now 1 will again be borrowed from bit 4 to bit 3. Thus 1 will be borrowed from bit 4 leaving one at bit 4 (10 - 1 = 1) and a 10 at bit 3.

Thus

bit 3 : 10 – 1 = 1

bit 4 : 1 – 1 = 0

bit 5 : 0 – 1

Now again a borrow is required at bit 5. But since bit 6 and bit'7 are both zero, thus a final borrow of 1 will be required. This will result in the carry flag being set i.e CF =1 to show the borrow. Now bit 7 will become 10 (due to borrow ,5). Now 1· will be borrowed to bit 6, thus bit 7,will become (10 - 1) ='1. Again I will be borrowed from bit f to bit 5 leaving bit 6 equal to 1 and bit 5 equal to 10.

Thus,

bit 5 : 10 – 1 = 1
bit 6 : 1 – 0 = 1
bit 7 : 1 – 0 = 1

Thus the result will be 11101000 or E8H with a borrow of 1 (which will be in the state of carry flag shown) or the final answer will be a shown below:

Ans = CF	1	E 8
	↓	↓
	borrow	result

The subtraction instruction will always always follow the above process.

We now have a final borrow which is reflected in a carry of one following the subtraction. Though here it shall be reflected by setting the carry flag, "the borrow should not be mistaken for a carry generated by the addition process,

7.2.2.1. SUB Instruction

The SUB or subtract instruction can operate on, both 8 and 16 bit data. It uses all the addressing modes 'that appear for the 'ADD instruction. The Figure 7.21 lists the different forms of subtract instruction.

Assembly	Operation
SUBCL, BL	CL= CL- BL
SUB CX, AX	CX= CX-AX
SUB DX, 0016H	DX = DX -0016 H
SUB CX, [DI]	CX = CX -contents of address DS : [DI]
SUB [BP] AH	Content of SS :[BP]
	= Content of SS : [BP] - AH

Figure 7.21. Different forms of SUB instruction

A simple program to subtract two eight bit numbers is listed below:

Example 5:	0 0 0 0	B5	22	MOV	CH, 22H
	00 0 2	80	ED 75H	SUB	CH, 75H

The SUB instruction affect the same flags as ADD which are ZF, CF, SF, ACF, PF and OF. However here the carry flag (CF) and Auxiliary carry Flag (ACF) hold borrows instead of carnes.

7.2.2.2. SBB Instruction

A subtract with borrow (SBB) operates similarly to the SUB instruction. However here the CF (carry Flag) which is assumed to hold borrow is also subtracted. The flags affected are same as SUB instruction.

Example 7.10: Discuss operation of SBB AX, CX

Answer: The SBB instruction will subtract AX and CX. The carry flag also will be subtracted thus.

AX = AX – CX – CF (CF is assumed to hold a borrow)

Thus if AX = 2000 H

CF = 1232 H

CF = 1

then A X = 0 DC D H

The flags will hold

CF = 0 → i.e., no final borrow

ZF = 0 → result non-zero

ACF = 1 → half borrow

SF = 0 → MSB = 0

OF = 0 → no overflow

PF = 1 → even parrity

The SBB is a vital instruction if we require 32 bit subtraction in a 8086/8088 program. A simple program is given below

Example 7.11: Program for 32 bit subtraction

```
0 0 0 0  2B    C7    SUB    AX, DX
0 0 0 2  1B    DE    SBB BX, CX
```

Here it is assumed the 32 bit number is stored in BX - AX register while the second 32 bit number stored in CX - DX register pair. The process of execution is shown below in figure 7.22

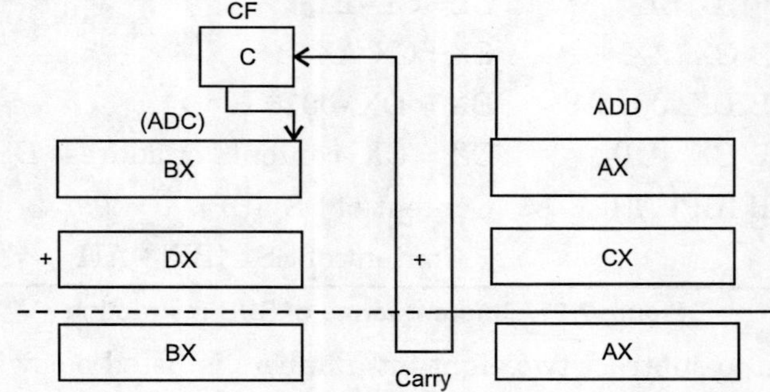

Figure 7.22. 32 bit subtraction

Different forms of SBB instruction are listed below in table 7.23.

Assembly		Operation
SBB	AL,BL	AL = AL - BL - CF
SBB	AX,DI	AX = AX - DI - CF
SBB	CX,2222H	ex = cx - 2222H - CF
SBB	AX,[BP]	AX = AX - contents of SS : [BP] - CF
SBB	[DI], SI	Contents of DS : [DI] = contents of DS : [DI] - SI - CF

Figure 7.23. Forms of SBB instructions

7.2.3. INC And DEC Instructions

A special form of ADD and SUB instructions also appear in the 8086/8088 instruction set. These are INC (increment) and DEC (decrement) instrutions. These instruction work exactly as their name suggests. Both these instructions can operate on memory or registers. However the segment register cannot be operated by either of these two.

INC Instruction: The increment instruction adds one to either a memory or register contents. Both these instructions can operate on 8 or 16 bit data. However for assembler detail when we are incrementing the contents of a memory location such as in. INC [D I], we specify whether the increment is on byte or word data. Thus, when we do assembly language programming we shall specify the word length when INC or DEC instructions refer to memory locations. Figure 7.24 lists different forms of INC instruction

Assembly	Language	Operation
INC	BL	BL = BL + 1
INC	SP	SP = SP + l.
INC	DI	DI = DI + 1
INC	BYTE [BX]	byte content of DS : [BX]
INC	WORD [BP]	= byte contents of DS : [BX] + 1
		word contents of SS : [BP]
		= word contents of SS : [Bp] + 1

Figure 7.24. Forms of INC instruction

INC affects all flags just as the ADD instruction except however for the carry flag. The INC instruction is of vital use when Arrays of data are to be operated on. Consider the program

```
MOV   AX, 2000H
MOV   DS,AX              ; load DS = 2000 H
MOV   AX, OOOOH
MOV   DI,AX              ; load off set adress
ADD   AL, [DI]           ; DI = 0000 H
INC   DI
ADD   AL, [DI]           ; Add
INC   DI                 ; DI = 0002 H
ADD   AL, [DI]
```

The above program adds three 8 bit numbers. Here we load DS = 2000H, AX = 0006H and DI = OOOOH. During the first ADD we add AL + DS : [DI] i.e., AL + contents of 20000H. Thus using the data contents of an arbitary memory as listed below AL = OOH + 22H or AL = 22H. Now the increment instruction will cause DI = 0001H. Thus ADD AL, [DI] will cause AL = AL contents of DS : [DI] = AL + contents of 20001 H or AL = 22 + 27 = 49 H

20000H	22H	
20001 H	27H	Memory space
20002H	92H	

thirdly the last INC and ADD will cause

AL = AL + contents of 20002 H

= 49 + 92 = DBH

AL = DBH

DEC Instruction: The decrement instruction operates in the same way except that the register or memory location contents are decremented by one. All flags except for the Carry flag are affected. The different forms of DEC are listed below in table 7.25 ..

Assembly		Operation
DEC	BL	BL = BL – 1
DEC	BP	BP = BP – 1
DEC	BYTE [DI]	byte conents ofDS :[DI]
		= byte content of DB: [DI] – 1
DEC	WORD [BP]	word contents of SS:[BP]
		= word contents of ss:[BP] – 1
DEC	BYTE NAME	byte contents DS : offset of label NAME
		= [byte contents of DS : offset of label NAME] – 1

Figure 7.25. Forms of DEC instruction

7.2.4. Multiplication Instructions

The 8086 family of processor have in built multiply instructions in their instruction set. This was a major advancement over earlier 8bit processors which did not multiply but had to use a combination of addition and shifting to achieve the multiplication operation. In 'the 8086 multiplication is performed on 8 bit and 16 bit data. The product is always double the size of the operands registers (or memory locations). That is a 8 bit multiplication (i.e., on two 8 bit number) will result in a 16 bit result. A 16 bit multiplication will generate a 32 bit result. There are two multiplication instruction in the 8086 processor: The MUL or unsigned multiplication instruction and the IMUL or signed multiplication instruction. Flag bits such as CF, ACF, OF, PF, SF etc, change. When the MUL or IMUL instruction executes but all flags except OF and CF have unpredictable result and are thus not used.

8 Bit Multiplication

In 8 bit multiplication the AL register always contains the multiplicand. The multiplier can be any 8 bit register or memory location. Immediate multiplication is not allowed. The result is always stored in the AX (16 bit) register. If the most significant 8 bits of the result are zero the CF and OF both are cleared (i.e., CF = OF = 0). If the result is 8 bit wide then CF = 0 and if the result is 16 bits wides then CF = 1. When signed multiplication is used (IMUL instruction) the product is in true binary form when positive and in two's complement form when negative. Table 7.26 bits the different forms of multiplication instruction :

Assembly	Language	Operation
MUL	CL	AX=AL × CL
IMUL	DH	AX=AL × DH
IMUL	[DI]	AX = AL × contents of DS : [D1]
MUL	[SI]	AX = AL × contents of DS : [S1]

Figure 7.26. Multiplication Instructions

Figure 7.27 illustrate Multiplication execution.

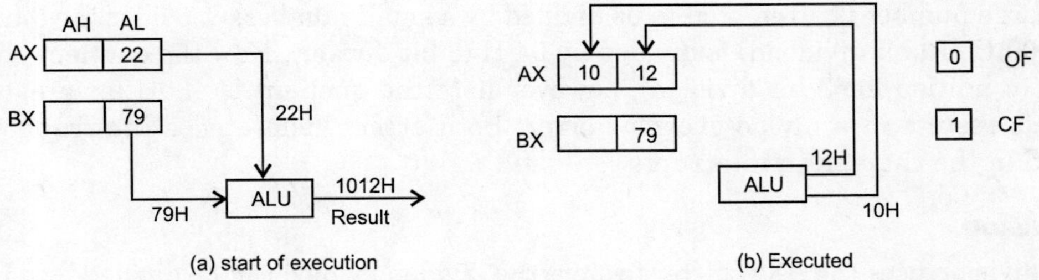

(a) start of execution (b) Executed

FIG. 7.27. Execution of MUL BL Instruction

16 bit Multiplication

Word multiplication is exactly same as 8 bit multiplication except that data operands are now 16 bit wide. Now AX stores the multiplicand while the multiplier can be any 16 bit regis- ter or memory location. Here, the CF = 0 and OF = 0 if the most significant 16 bits are zero. If the result is 16 bits wide then CF = 0 and if result 32 bits wide. the CF = 1. The result is always stored in DX-AX register where DX holds most significant 16 bits. For signed multiplication the DX-AX register hold the number in true binary form if positive or the result in DX-AX is in 2's complement form if negative. Figure 7.28 shows 16 bit signed multiplication while figure 7.29 shows different forms of 16 bit multiplication instructions.

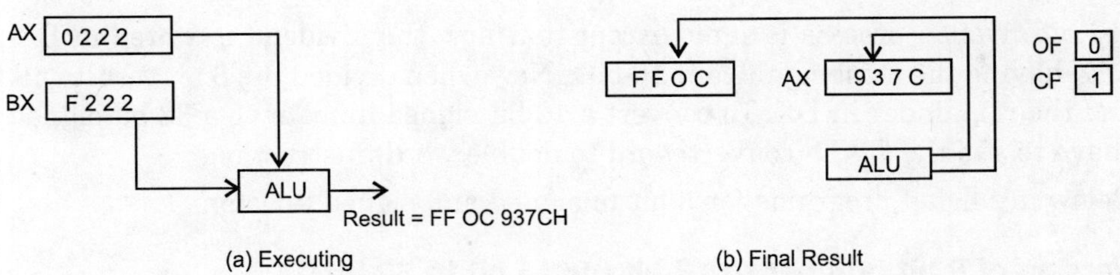

(a) Executing (b) Final Result

Figure 7.28. Execution of MUL BX instruction

Assembly	Language	Operation
MUL	CX	DX-AX = AX x CX
MUL	DI	DX-AX =AX x DI
MUL	[BX]	DX-AX= AX x contents of D8 : [DI]

Figure 7.29. Forms of Multiplication Instruction

7.2.5. Division Instruction

There are two division instructions present in the 8086 instruction set namely: the unsigned division or DIV instruction and signed division IDIV instruction. Both operate in 8 bit division mode and 16 bit division mode. In the 8 bit division mode a 16 bit number is divided by a 8 bit number to give a result 8 bits wide. In the 16 bit division mode a 32 bit number is divided by a 16 bit number to give a result 16 bits wide.

No flag is considered to respond predictably during the division process and therefore should not be used to control program flow. The divisicn process in microprocessors can result in two types of errors, one is the divide by zero error and the second the divide overflow error. A divide by zero error occurs when a number is attempted to be divided by zero. A divide overflow occurs when a large number is attempted to be divided by a small number. Consider the 8 bit division when 1200 H (16 bit dividend) is divided by 04 H (8 bit divisor). Now the quotient should be 8 bit wide (condition for 8 bit division), however here the quotient is 480H (or greater than 8 bits). This results in a divide overflow error. Both errors generated a interrupt which are discussed in the chapter on interrupts.

8 bit division

An 8 bit division uses the AX register to store the dividend which can be divided by a register or memory location. Usually when we work using the processor we generally use 8 bit numbers. Thus this number must be first converted to a 16 bit number. This is accomplished by the CBW (convert byte to word) instruction. The CBW sign extends a 8 bit number into a 16 bit number. Now the process of converting a byte to word will obviously vary for signed and unsigned numbers.

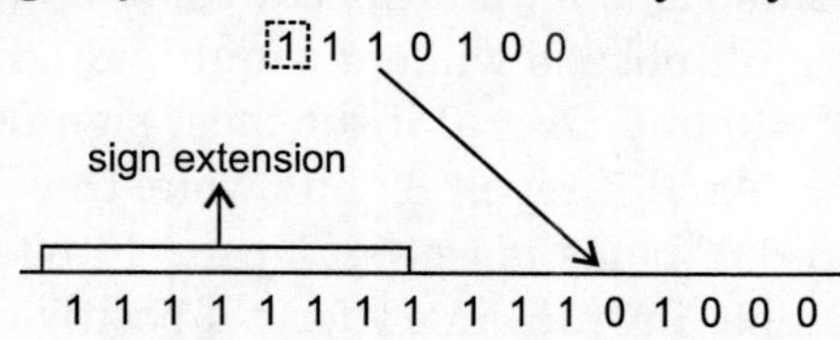

Figure 7.30. CBW operation on signed numbers

16 bit division

The 16 bit division process is same except that now the dividend is stored in DX-AX register with DX having the most significant 16 bits. Now when divided the 8 bit result will be stored in AX and the remainder in DX. To convert a 16 bit signed number to a 32 bit signed number we now have to use the CWD (convert word to double word) instruction.

Below are listed programs for 8 bit unsigned and signed division.

Conversion of 8 bit number to 16 bit (or 16 bit to 32 bit)

When converting a byte (or word) to word (or double word) the 8086/8088 microprocessor can be made to perform the operation differently depending on the type of number that we are using.

Zero extension: For unsigned numbers the programmer must clear the most significant 8 bit so as to convert a 8 bit number to a 16 bit unsigned number.

Sign extensions: However if we are using signed numbers then the process of conversions of a 8 bit numbers to 16 bit. We have to sign extend the number. The reason for this lies in the fact that the MSB is used to the sign of a signed number. Therefore, zero extension will result in loss of sign (ifMSB = 1) for a signed number. The 8086/8088 has two instructions for sign extending a number: The CBW and CWD instruciton.The CBW instruction (convert byte to word) sign extends a 8 bit signed numbers to a 16 bit signed number. The CWD instruction converts a 16 bit signed number to a 32 bit signed number.

Example 7.12 : 8 bit unsigned division

```
MOV     AL, 43 H
MOV     AH, OU H
MOV     BX, 0006H
DIV     BL
```

This will program divide 43 H by 6 H to give the answer OlOB H in AX where aBH is the quotient and 01H the remainder.

Example 7.13 : 8 bit signed division

```
MOV     AL,16H      ; move AL = 16H
MOV     BL, FDH     ; move BL = 3H
CBW
IDIV    BL.
```

This will result in AX = 01 FBH where the quotient is FBH (-5) and the remainder is + 1.

The process of converting 8 bit numbers to 16 bits depends on our usage of the numbers i.e., whether we are using unsigned* division process for unsigned numbers we simply clear the AH contents. However for converting a 8 bit signed number to a 16 bit signed number we have to use the CBW (convert to byte) instruction. The operation of DIV instruction is shown in figure 7.31.

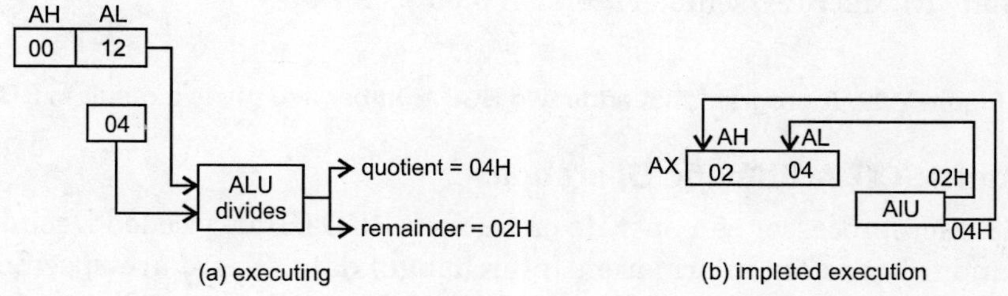

Figure 7.31. Shows the operation of a 8 bit division

It should be clearly understood that the 16 bit dividend is divided by the 8 bit divider to give a 13 bit quotient. The 8 bit quotient is stored in AL register while the remainder is stored in AH register.

```
Microsoft (R) Macro Assembler Version 5.00
                              • model small
    0000                      • data              ; data segment
    0000      1023            a dw 1023h          ; load packed bed
    0002      2345            b dw 2345h          ; numbers
    0004      ????            c dw?
    0000                      • code
    0000                      begin proc
    0000                      start
    0000      B8------R       mov ax, @data       ; load ds register
    0003      BE D8           mov ds, ax
    0005      33 D2           xordx,dx
    0007      8B 0E 000R      mov al, bl          ; move low byte of bcd to al
    0011      02 C1           add ai, cl          ; add both low byte bed no.s
    0013      27              daa                 ; adjust
    0014      8A DO           mov dl, al          ; store lower bed to al
    0014      8A C7           mov ai, bh          ; repeat above process
    0018      02 C5           adc ai, ch          ; for upper byte bed
    001A      27              daa                 ; adjust
    001B      8A F0           mov dh, al          ; store result bed in dh
    001D      d9 16 0004 R    mov c, ux           ; mov the resultant bed to memory
    0021      B4 4C           mov ah, 4ch
    0023      CD 21           int 2th
    0025            begin end
    0025            end start
  00 Microsoft (R) Macro Assembler Version 5.00    2/14/3
  Symbols-1
```

Figure 7.32. A program that adds two BCD Number and gives a result in BCD

7.2.6. BCD And ASCII Arithmetic Operations

The 8086/8088 microprocessor can operate on data on BCD (Binary coded Decimal] and ASCII* (American standard code for information Interchange) data. There are specific instruction ill the 8086/8088 instruction bit that operate on data of this kind. Let us discuss both these types of instructions in detail.

7.2.6.1. BCD Instruction

There are two types of arithmetic operation supported on 8086/8088 microprocessor for BCD data. This is the addition and subtraction, arithmetic operations. However, the 80861 8088

* Refer to chapter 1.

instruction set provides instructions to correct the result of a BCD arithmetic or subtraction operation. Both the instruction are only correcting instructions. BCD data appear in the packed BCD form where two BCD numbers are stored per byte.

The correcting or adjusting instructions operate on the AL register only. The BCD instruction that appear in the 8086/8088 instruction set are DAA (decimal adjust after addition) and DAS (decimal adjust after subtraction) instructions.

DAA Instruction: The decimal adjust after addition is used to adjust the result, after a ADD or ADC instructions that has operated on BCD data. Consider the program illustrated in figure 7.32 for addition of two 16 bit (4 digits per word or 4 BCD digits in 16 bit of data) BCD numbers.

The program add two 4 digit BCD numbers stored in memory and then stores the result again in memory. Also due to the DAA instructions limitation, that is, its operation on the AL registers the program first stores the two BCD numbers from memory (referred by A & B) inthe bx and ax registers. Then first BL and CL register are added and then adjusted to BCD. After this the result is stored in DL register. Following this the BH and CH registers are added along with the carry generated by the preceding step by the ADC instruction. The result is once again adjusted and then stored in DH register. Finally the DX register is stored back in memory. In the figure 7.32 The program adds 1023 to 2345 to get a sum of3368 and stores this at an offset address of 0004H in the data segment. You may note that only the four bits are added to get the result and this operaton is similar to decimal addition.

DAS Instruction: The decimal adjust after subtraction works similarly to the DAA instruction. However here the instruction is used to adjust the number after subtraction. Figure. 7.33 shows a program that subtracts two 4 digit BCD (16 bits) numbers. Notice theprogram is same as figure 7.33 except that the ADD and ADC instruction are replaced by SUB and SBB instructions respectively. Also the DAA instruction is replaced by DAS instruction .

```
❑❑ Microsoft (R) Marco Assembler Version 5.00 2/14/3
                                   Model small
0000                               data
0000      2344                     a dw 2344h
0002      0011                     b dw 0011h
0004      ???                      c dw ?
0000                               code
0000                               start proc
0000                               begin:
0000      B8........R              movax @data         ; load ds reg with
0005      88 1 E 000R              mov bx. a           ; in bx cx
0009      88 1 E 0002 R            mov cx. b           ; in bx ex
000D      8A C3                    mov ai, bl          ; mov lower bed byte to al
000F      2A C1                    sub ai, cl          ; subtract
0011      2F                       das                 ; adjust
```

```
0012    8A C7           mov dl, al          ; store lower result in dl
0014    8A C7           mov ai, bh          ; mov upper bcd byte to al
0016    2A C5           sbb ai, ch          ; subtract
0018    2F              das                 ; adjust
0019    8A FO           mov dl, al          ; store upper reslut in dh
0018    89 16 0004 R    move,dx             ; store in memory
001 F   844C            mov dh, 4ch
0021    CD 21           int 21h
0023            start end
0023            end begin
```

❑❑ Microsoft (R) MacroAssemblerVersion 5.002/14/3

Symbols-1

Figure 7.33. Program that subtracts two 8CD. numbers and give a result in 8CD

7.2.7. ASCII Instructions

ASCII Instructions operate on ASCII coded numbers. The ASCII code set is listed in the Chapter 1. The ASCII code ranges from 30H to 39H for the decimal numbers 0 to 9. The ASCII ASCII equivalent of decimal four ASCII equi valent of decimal seven arithmetic instructions in the 8086/8088 instruction, AAS (ASCII adjust after subtraction) AAM (ASCII adjust after Multiplication) and AAD (ASCII adjust before division) instructions. All ASCII instructions use the AX register as the source register for data and as the destination register for storing the result.

AAA Instruction

Direct addition operation on ASCII data results in false results. Consider the case when ASCII four (34H) and ASCII seven (37H) are to be added. This will result in a sum of 6 BH as given below,

```
  34  H   ASCII equivalent of decimal four
  37  H   ASCII equivalent of decimal seven
 ________
  6B H
```

However when we add 4 and 7 we would expect a result of 11. Similarly when we add 34H (ASCII four) and 37H (ASCII seven) we would expect the result to be ASCII equivalent of eleven (or would expect the result to be ASCII equivalent of eleven or 3131H). Tc ensure the proper addition of ASCII number the AAA instruction is used. Figure 7.34 shows a program that uses all the ASCII instructions. The first part of the program illustrates the method of adding ASCII number. In the example 34H (ASCII H) and 37 H (ASCII 7) are added. This is followed by the AAA instruction. The AAA instruction adjusts the result of addition and stores it in the unpacked BCD form. After execution the AX register would hold OlOIH. This however is not an ASCII number but can be converted to ASCII by adding 3030H. which would result in 3131H which is the ASCII equivalent of eleven. The AAA instruction during execution will always clear the AH register if the result is less than OAH* and add a 1 to AH if the result is greater than OAH.

AAD Instruction

In figure 7.34 the AAD instruction is used in the middle of the program. Observe that the AAD instruction is used before the division instruction instruction. However the AX register is required to hold the dividend in a two digit unpacked BCD form by the AAD (and not an ASCII value). On execution of the AAD instruction this unpacked BCD number is converted to its equivalent binary number. In figure 7.34 the dividend is 0405H (or decimal45H in unpacked** BCD form). On execution of the AAD instruction this number is converted to its binary equivalent 4DH (4DH is binary of 45 decimal). This is now divided by 6(unpacked BCD form). To generate a single digit quotient in AL and a remainder in AH. Now the AL register (i.e. quotient) can be converted back to ASCII by addition with 30 H.

However, AAD instruction can only convert unpacked BCD numbers the range 00 to 99 into their binary equivalent.

AAM Instruction

The AAM or ASCII adjust after multiplication instruction is similar to the AAD instruction. Both work on unpacked BCD data. Consider again the program in figure 7.34. Let us here see the use of the AAM instruction. We use the MUL instruction to multiply the two unpacked

BCD numbers stored in memory. Thus MUL will multiply 6 and 6 give the result of0024H*** in the AX register. Now when the AAM instruction will execute it will adjust AX register which would now hold 03 06H. This is unpacked BCD number equivalent to 0024H. Note here too if 3030H is added to the AX register it shall hold an ASCII result of 3336H (or an ASCII equivalent of decimal 36). The AAM instruction can only operate on data ranging from 000H to 0063H (held in the AX register). The AAM will adjust this range to an unpacked BCD number which will be held in AX register.

```
□□ Microsoft (R) Marco Assembler Version 5.00        2/16/3 ASCII operations
1-1
                          title ascii operations
                          PAGE 140, 100
                          .model small
0000                      .data
0000      0034                  four              dw0034h
0002      37                    seven             db 37h
0003      05                    div1              db 5
0004      0405                  dividend          dw 0405h
0006      06                    multiplicand      db 6
0007      06                    multiplier        db 6
0008      ????                  result_add        dw?
OOOA      ??                    result jiiv       db?
```

* OAH is decimal 10. Therefore if the result is less than 10 therefore there is no tens digit

** unpacked BCD stores one decimal digitpel byte.

*** 0024R is the equivalent hexadecimal of 36 decimal.

```
0008      ????          result_mul            dw?
0000          .code
0000          start proc
0000          begin:
0000      B8----R       movax,@data           ;load ds register with
0003      8E 08         mov ds, ax            ;data segment address
0023      A00006R                             ;load 8 bit multiplicand
0026      8A OE 0007R                         ; load 8 bit multiplier
002A      F6E1
002C      D40A
                                              ; ASCII addition process
0005      A1 00R        mov ax, four          ; moves ascii foru to al
0008      02 06 0002 R  add ai, seven         ; add ascii seven
000C      37            aaa                   ; adjust result to unpacked bcd
000D      05 3030       add ax, 3030h         ; convert to ascii
0010  A3  0008R         mov, result_add,ax    ; store ascii result
                                              ; ASCII Division Process
0013 8A IE 0003R        mov b1, div1          ; move 8 bit divisor to b1
0017 a 1 0004R          mov ax.dividend       ; move 16 bit dividend in ax
001 A 05 OA             div b1                ; divide
001 E 04 30             add ai, 30h           ; convert quotient to ascii
0020 A2 OOOA R          mov result_div, a1.store result
                                              ; ASCII Multiplication Process
002E 05 3030            add ax, 3030h         ; convert ax to ascii
0031 A3000BR            mov result jnut, ax   ; store result
0034          start endp
0034          end begin
□□ Microsoft (R) MacroAssemblerVersion 5.00
Symbols-1
2/16/3 ascii operations
```

Figure 7.34. The 8086 ASCII Instructions

AAS Instruction

The ASCII adjust after subtraction is used to adjust the AX register after a subtraction operation. Suppose 36H is subtracted from 39H to give a result of03H. Then the AAS instruction adjusts neither the AH nor the AL register. However if suppose 36H is subtracted from 34H, then after the AAS instruction execution AL will hold 08H and the AH register will be subtracted by one. This decrement of the AH register allows for multiple digit ASCII numbers to be subtracted from one another. (This is similar to use of carry flag and ADC for multiple addition of BCD numbers).

7.3. SHIFF AND ROTATE INSTRUCTION

Shift and rotate instructions are used for binary bit manipulation. These instructions are commonly encountered and used for controlling 110 devices. Let us discuss both these instructions in detail.

7.3.1. Shift Instructions

Shift Instructions move bits to the left or right in a number stored in a register or memory location. These instruction can operate on 16 bit or 8 bit data (the width selected by w bit of the instruction template*). There are four shift instructions of which two are logical shift instructions while the other two are arithmetic shift instructions. Remember from boolean algebra that a left shift is equivalent to multiplying the number by 211 (where n is no of times the number is shifted left). Similarly, a right shift is equivalent to multipling a number by 2-n (where n is number of times the original number is shifted right). Thus a single left shift is equivalent to multiplying the number by two and a single right shift is equivalent to dividing the number by two. The different shift instructions are SHL, SAL, SHR, SAR. Out ofthese four the SHL (shift logical left) and SHR (shift logical right) are logical shift instructions while SAL (shift arithmetic left) and SAR (shift arithmetic right) are arithmetic shift instructions. The logical shift instructions are designed to work on unsigned numbers while the arithmetic shift instructions are used to shift signed numbers. Thus logical shift instructions multiply or divide unsigned numbers while arithmetic shifts multiply or divide signed numbers. Figure 7.35 shows the operation of the four shift instructions.

All the shift instructions will shift the number in the specified direction and introduce a zero in the end except the SAR where the number is shifted right and the sign bit is again copied back to MSB. The shift instructions can be used on all registers (except segment register) and all memory addressing modes of the 8086/8088 instruction. Figure 7.36 shows a program to multiply and divide numbers.

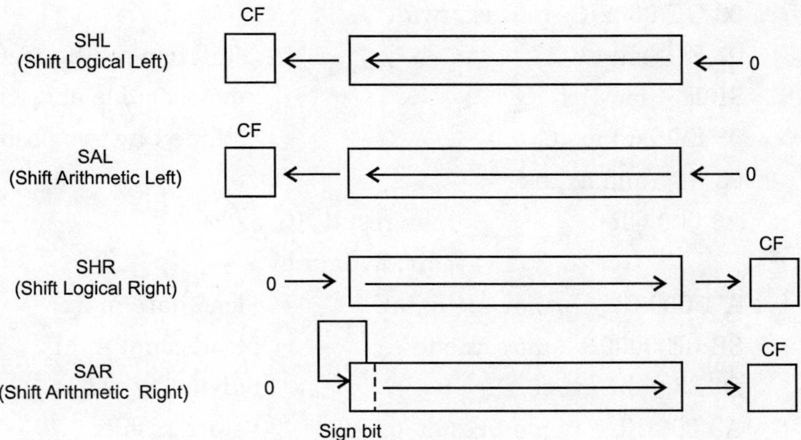

Figure 7.35. Operation of Shift Instruction

* figure 3.11

```
❑❑ Microsoft (R) Marco Assembler Version 5.00          2/16/3
Page 1-1
0064    .stack 64h
0000    .data
0000    0001   one dw 0001h            ; count data
0002    0002   two dw 0002h
0004    0003   three dw 0004h
0006    0004   four dw 0004h
0008    2367   dataO dw 2367h
OOOA    5443   data 1 dw 544h
OOOG    ????   result_ 10 dw ?
OOOE    ????   result_ 15 dw ?
0010    ????   result_ 15 dw ?
0012    ????   shiftleft32upperbyte dw ?
0014    ????   shiftleft3210werbyte dw ?
0016    ????   shiftright32upperbyte dw ?
0018    ????   shiftright3210werbyte dw ?

0000    .code
0000    begin:

                    ----- multiply data by 10 -------
0000    A 1 0008 R     mov ax, dataO
0003    88 OE 0002R    mov cx, two
0007    01 EO   shl ax, 1            ; shift one position i.e. multiple data X2
0009    8B08    mov bx, ax           ; move double of original data
000B    03 EO   shl ax, el           ; shift ax by two position i.e. data × 10
000D    03 C3   add ax, bx
000F    A3 000 CR         mov result_10, ax
                    ----- divide data by 2 -------
0012    A 1 0008R      mov ax, dataO         ; load data in ax
0015    8B 0E 0000R    mov cx.one            ; load count in c1
0019    D3E8    shr ax, cl                   ; divide by two
001B    A3 000ER       mov result_div_2ax    ; store results
                    ----- multiple data by 15 -------
001 E   88 OE 0006R    mov cx, four          ; load count = 4
0022 A 1 0008 R    mov ax, data 0            ; reload data in ax
```

```
0025 D3 EO          shl ax, cl                  ; multiply by 16
0027      28 06 0008 R   sub ax, dataO           ; subtract 16 ax-ax
002B      A3 0010R       mov result_15,ax        ; store ax times 15
          ; ----- shift a 32 bit number stored in ax-dx to
          ; left ----;
          ; ------ by one position ----------------
          ; ------ hence multiply 22 bit number by 2 ;
002E      A10008R              mov ax, data 0
0031      88 1 E 000AR         mov bx, data1
0035      01 E0   shl ax, 1
0037      0103    relbx,1
0039      A 1 0012R            mov ax, shiftleft 32 upperbyte
003C      88 1 E0014R          mov bx, shiftleft 32 lowerbyte
          ; ----- Shift a 32 bit number stored in ax-dx to
          ; right ----
          ; ------ by one position ----------------
0040      A1 0008R             mov ax, data0
0043      88 1E OOOAR          mov bx, data1
0047      01 E8                shr ax, 1
0049      01 03                rer bx, 1
0048      A1 0016R             mov ax, shiftright 32 upperbyte
004E      88 1 E0018R          mov bx, shiftright 32 lowerbyte

0052      end begin
    ❑❑ Microsoft (R) Macro Assembler Version 5.00        2.16.3
Symbols-1
```

Figure 7.36. Program showing use of 8086/8088 shift and rotate instruction for multiplication and division

Notethat the number of shifts' to be performed must be stated in the operand. We can use a immediate shift count with the instruction or use the CL register to hold the shift count. The. shift instruction uses the d bit of the instruction format to specify the count mode. Consider the SHL instruction format given in figure 7.37.

1 1 0 1 0 0	v	w	mod	100	r	m

Figure 7.37. SHL instruction format

The shift instruction use the v bit" as above in figure 7.37 to set the count mode. If v bit is 0 then the count = '1' and if v bit is 1 then shift count is assumed to be loaded in CL register. (Notice the v bit is same as d bit).

Consider the program in fig 7.36. Suppose we were to multiply AX by 10. This operation could be achieved by the MUL instruction or the SHL instruction could be used. We have shifted left the AX register by 1 position, effectively multiplying it by ,2. We then save this number which is double of data in the bx register. Next we shift the AX register by the count loaded in CL register (which is two). Thus AX get shifted left by 2 position. and effectively multiplies DATA by 23 (or 8). Now if we Add AX to BX we would get 8 x DATA + 2 x DATA = 10 times DATA. Shift instructions can be thus used for multiplication.

7.3.2. ROTATE INSTRUCTION

Rotate instructions of the 8086 instruction provide the programmer, the ability to rotate the data bits in a register or memory location. There are four available rotate instruction in the 8086 instruction whose execution is shown in figure 7.38.

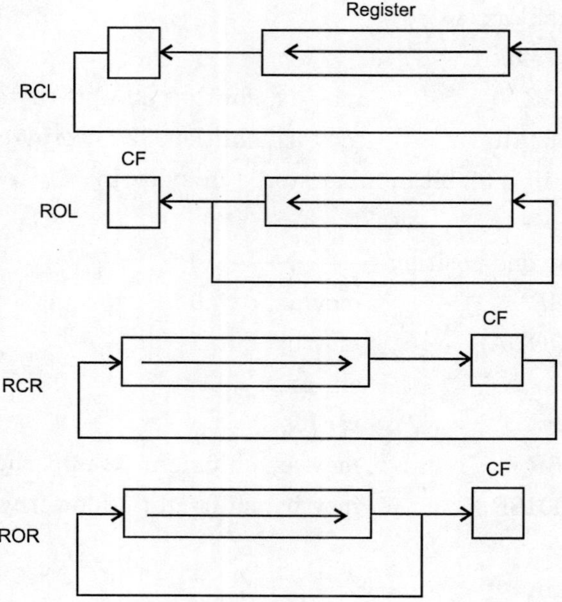

Figure 7.38. Rotate instructions

The four rotate instructions are the ROR, ROL, RCL and RCR. The ROR and ROL are the rotate right and rotate left instructions. These rotate the register as will as transmit the LSB or MSB into the carry flag as shown in figure 7.38. The RCL and RCR are rotate through carry instruction which rotate the register or memory location through the carry flag as shown in figure 7.38. Rotate instructions are often required when number wider than 16 bits have to be shifted. Rotate instructions use the same addressing modes as shift instruction. Also they can be programmed to rotate by a number stored in the CL register similar to the shift instruction. Figure 7.39 shows the RCL instruction format while figure 7.40 shown the various rotate instruction forms

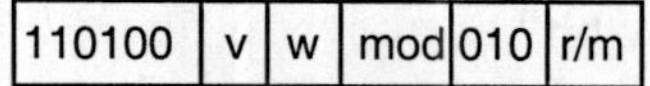

110100	v	w	mod	010	r/m

Figure 7.39. RCL instruction format

* the v hit is same as d hit.

Note v = 0, rotate count = 1

v = I, rotate count is loaded in CL register.

Assembly language	Operation
ROLCX,l	Rotate left CX by 1
RCLBL, CL	Rotate through carry left BL register by count loaded in CL register
ROR BYTE [DI], 1	Rotate right the contents of DS : [DI] by 1 {byte data)
RCR WORD [BP], 1	Rotate right through carry word contents of SS: [BP] by 1.

Figure 7.40. Rotate instructions forms

7.4. KEY POINTS TO REMEMBER

1. The 8086/8088 microprocessor contains a powerful set of arithmetic and logical instruction. Arithmetic and logical instructions are performed by the ALU ofthe 8086/8088 microprocessor.
2. The 8086/8088 instruction set contains a powerful set of logical instructions which include AND, OR, XOR, NOT and NEG instructions.
3. Logical instructions are powerful bit manipulation instructions allowing us control over bits that can be changed according to a programmers needs.
4. The TEST instruction is the only logical instruction which doesnot change the destination operands. The effect of this instruction is only on the flag register.
5. The 8086/8088 instruction set also contains a large number of arithmetic instruction which can operate on signed and imagined numbers. The arithmetic instructions include ADD, ADC, SUB, SBB, MUL, IMUL, DIV and IDIV instructions.
6. The arithmetic instructions INC, DEC are special forms of the addition and subtraction instructions.
7. During division operation the programmer may have to convert a 8 bit (or 16 bit) number to a 16 bit (as 32 bit) number. These are two processes for this
   - Zero extension: used for signed numbers
   - sign extension (using either CBW or CWD instructions) : used for unsigned numbers
8. The 8086/8088 instruction set also contain BCD and ASCII instructions such as AAA, AAS, AAM, AAD (ASCII instructions) and DAA, DAS (BCD instructions) which the result for code conversion back to ASCII or BCD form.
9. The shift and Rotate are available for bit manipulation. These include 4 shift and 4 rotate instructions. The shift instruction can also be used with the shift count loaded in CL register by specifying the v bit value of the shift instruction is "in figure 7.37.

REVIEW QUESTIONS

7.1. What are the restrictions on the addressing modes available to the 8086 add and subtract instructions?

7.2. Is it possible to Add two 64 bit numbers using the addition instructions? If so write the corresponding assembly language program?

7.3. Suppose AX = 2345H and DX = 2121H. What would be the state of the Flag register if the following instruction execute

(a) ADD AX. DX

(b) SUB AX. DX

7.4. Develop a program that adds 10 bytes of data from memory. Assume the first data location is 0000H. (Hint: use the INC instruction to hold the memory points and do continual addition with a memory location address which is changed by INC).

7.5. Write down the instruction that would perform the following operation.

(a) BX - CX

(b) DX -AX

(c) SI -EEH

(d) BX - Contents of DS : [SI]

(e)AX-contentsofDS: [BP]

7.6. What would be the state of the flag register and the AX register after execution of the following instruction .

(a) ADD AX. [BP]

(b) ADD AX. SI

(c) SUB AX. BX

(e) SUB AX. SI

Assume AX = 2222H, BX = 1111H. SI = 3333H.

7.7. Develop a program that would subtract two 64-bit number located in memory?

7.8. During 8bit multiplication which registers hold the multiplier and multiplicand? Also which register is used to hold the product? What is the size of the product?

7.9. Explain the state of flags during a multiplication instruction execution?

7.10. What is the difference between 8 bit and 16 bit number multiplication in context to the registers used?

7.11. What are the different errors generated during the division operation?

7.12. Explain the differences between

(a) MUL and IMUL instruction

(b) DIV and IDIV instruction

7.13. During an 8 bit division instruction operation which registers are used to hold the divider. dividend. quotient and remainder? What is the effect on the flag register?

7.14. What is divide by zero error and divide under flow errors?

7.15. Develop a program to divide two 8 bit numbers. where the number are initially stored in CL and DL register. Following division then multiply the result by 10 and store the result in AX register.

7.16. What are the different ASCII and BCD instructions found in the 8086/8088 instruction set.

7.17. Explain the operation of all the ASCII and BCD instructions ofthe 8086 instruction set.

7.18. Develop a program that adds three ASCII numbers and stores the result in AX register?

7.19. Develop a program for addition of three BCD numbers?

7.20. What is the primary use of the 8086 logic instructions? List and explain the opera tion of each?

7.21. What is the difference between the TEST and AND instructions?

7.22. Select an instruction which will perform the following opera tions.

(a) AND AX and BX and store result in BX

(b) AND AX and 23H

(c) OR CX and contents ofDS: [BP] and store result in ex

(d) OR CX and contents of DS: 1112H and store result in memory.

(e) XOR BX and DX and store result in DX.

7.23. Which instruction is the fastest way to clear a register or memory location.

7.24. List the differences between NEG and NOT instructions?

7.25. Develop a program that controls an 8-LED light panel (port 1000H). It is required that the LEDs switch on in the sequence that all even LED are an at one time followed by all odd LED's. Use the 8086 logic instruction to achieve this.

7.26. Write an 8086 program that uses the shift instructions to multiply a number by

(1) four

(2) five

(3) six

7.27. It is required to perform a division operation by using shift instructions. Write the corresponding program to divide the number stored in AX by

(1) two

(2) six

(3) eight

(4) nine

7.28. Explain the 8086 rota te instructions?

●●●

Chapter-8

PROGRAM CONTROL AND PROCESSOR CONTROL INSTRUCTIONS

INTRODUCTION

We have discussed data movement, logic and arithmetic instructions in the preceding two chapters. Now we shall discuss the program control and the processor control instructions. This chapter will be outlined with a lot of assembled programs so as to illustrate these instructions clearly. We shall also include a few programs using the instructions that have been discussed in the preceding chapters.

8.1. PROGRAM CONTROL INSTRUCTIONS

The program control instructions are responsible for control of program flow during program execution. These include the unconditional and conditional jump instructions, the LOOP and CALL instructions. First we shall discuss the jump instruction group which comprises the main group of program control instructions. In the jump instructions class we shall first discuss the unconditional jump instructions followed by the conditional jump instructions.

8.1.1. Unconditional Jump Instructions

The 8086 processor supports three types of unconditional jump instructions: the short jump, near jump and the far jump. The short jump is a two-byte instruction that allows change of program control to memory location within + 128 bytes from the address following the jump. The near jump allows change of program execution to memory locations within ±32 Kbytes from the jump instruction (i.e., any where in the current code segment). The third jump is the far jump which is a five byte instruction. The far jump allows a jump to any memory location within the 8086 memory system. The short and near jump are often called intrasegment* jump and far jump are called intersegment** jumps.

Short Jump: Short jumps are also referred to as relative jump instructions. The short jumps use only two bytes of memory and have the general format listed in figure 8.1.

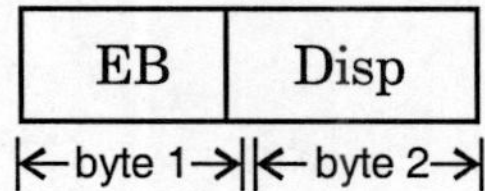

Figure 8.1. The short jump instruction

* These jumps only allow the program execution to be transferred in the current working code segment.

** These jumps allow program execution to relocated to any point in the 8086 system memory.

Here the second byte comprises the displacement of the jump location from the current instruction location. Since the displacement is stored in signed number format this the displacement can range from the values -128 to + 128. When the 8086/8088 microprocessor executes a short jump, the displacement is sign extended and added to the instruction pointer which effectively generates the jump address in the current code segment.

```
                              ;program to show short jump
                              ;model small
0064                          ;stack 64h
0000                          .data
0000     0023                 a dw 23h
0002     1221                 b dw 1221h
0004     ????                 c dw?

0000                                          .code
0000                                          begin proc
0000     33 DB                                x or bx, bx
0002     A1 0000R                    start:   mov ax, a
0005     03 C3                                add, ax, bx
0007     EB 13                                jmp short next ;short jump
0009     03 06 0002 R                         add ax,b
0000     A3 0004 R                            mov c, ax
0010     8B 1 E 0004 R                        mov bx, c
0014     03 1 E 0002 R                        add bx, b
0018     891E 0004 R                          movc, bx
001C     8B 08                       next:    mov bx, ax
001E     EB E2                                jmp start; short jump
0020                                          begin end P
0020                                          end begin
```

Figure 8.2. Program showing the short jump instruction

This jump address can be -128 byte below or + 127 bytes above the address that immediately follows following the jump instruction In figure 8.2. The first jump instruction uses the SHORT directive to inform MASM that the jump required is a short jump. However, the second jump doesnot use the short directive as MASM will automatically optimize and use the best jump type (short, near or far) depending on the final jump address. Observe the first jump statement, its address is 0007H. Also see the short jump sign extends the displacement 0013H and adds it to the address corresponding to the next instruction (0009H) to calculate the final jump address (OOlCH). [Refer figure 8.3]

displacement : 1 3 H

sign extended displacement : 0 0 1 3 H

address of next instruction : 0 0 0 9 H

jump address calculated : 0 0 1 CH

(*a*) jump using a positive displacement

displacement : E 2 H

sign extended displacement : F F E 2 H

address of next instruction : 0 0 2 0 H

jump address calculated : 0 0 0 3 H

(*b*) jump using a negative displacement (jump at address DOlE offig 8.2)

Figure 8.3. Calculation of jump address of the two jmp instructions of figure B.2.

Most assemblers require the jmp instructions to reference a label which specifies the jump address. As in figure 8.2. the first jump references the label next (which references the address 001CH). The assembler then calculates the displacement required using the jump address of the label.

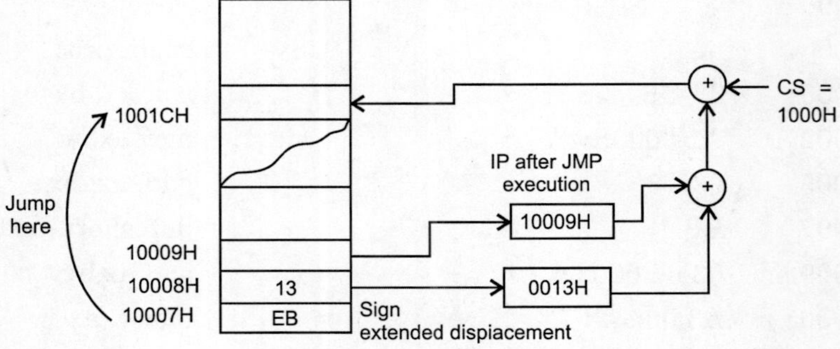

Figure 8.4. The Short Jump Execution

Near Jump: The near jump is used to jump to any location within the current code segment. This is a three byte instruction having the general format as in figure 8.5.

E9	Disp low	Disp high
←byte 1→	←byte 2→	←byte 3→

Figure 8.5. The near jump instruction

The two byte displacement contains the displacement of the location to which a jump is required. This is directly added into IP (instruction pointer) of the 8086/8088 microprocessor and added to the CS (code segment) register to produce the jump address. Since a 16 bit displacement follows the jump instruction the near jump can jump to any location in the current code segment. Figure 8.6 illustrates a program showing the use of the near jump instruction. The near jump like the short jump is relocatable as it is a relative jump. Thus if the code segment (i.e. program moved to a new location) the distance between the jump instruction and jump address will always remain constant. Thus program execution is not affected (or dependent) on the code segment. This is an important feature of the 8086 family of processors and one of the reasons for its extensive use in the personal computers (since programs can be written and loaded anywhere in the system memory and their execution is never affected due to the relative jumps and relocatable data segments).

```
                        ;program to show listing of near jum
                        .model small
0000                    .data
0000    0023            a dw 23h
0002    2123            b dw 2123h
0004    ????            c dw ?

0000                    .code
0000                    main proc far
0000    33 DB           xor bx, bx
0002    B80123          start:mov ax, 0123h
0005    03 C3           add,ax, bx
0007    E9 0100R        jmptwo;near jump
000A                    main endp
0100                    org 100h
0100                    next proc near
0100    8B D8           two: mov bx, ax
0102    89 1 E 0004R    movc, bx
0106    E9 0002R        mp start;near jump
0109                    next end p
0109                    end main
```

Figure 8.6. Program showing the near jump

Observe in figure 8.6 (and figure 8.2) any reference to the data segment is listed with a R. This R signifies relocatable data segments and is for use for the assembler internally. It is not present when the program is loaded in memory. Secondly, observe the jump instruction. It is listed as E9 OlOOR where E9 specifies the short jump, 0100H the jump address address and R stands for relocatable (or relative) jump address. (Note: the actual machine language instruction will assemble as E9 F600** and not as E9 0100 R).

Now let us observe how a jump address is calculated. Let us use the first jump instruction offigure 8.6.

displacement =	0 0 F 6 H
+ Address of instruction following jump instruction =	0 0 0 A H
offset address loaded to IP	0 1 0 0 H

Figure 8.7. Calculation of Jump Address of fig. 8.6. for the jump instruction at address 0007H

(*Note:* The MASM will never show the displacement but only the final jump address for a near jump. Care must be taken as one must not assume the near jump loads this offset into the IP. When observed in machine code the near jump of figure 8.6 will be listed as E9 F6 00. where the displacement will be 00F6H*. Similarly, all near jump instructions will be listed with the

* As when loaded in memory the displacement low byte shall precede the displacement high byte.

jump address, but when observed in machine code will be in the format as of figure 8.5 where the displacement will be distance between address of the next instruction and jump location address).

Figure 8.8 illustrates the method the 8086/8088 microprocessor uses for execution of a near jump instruction .

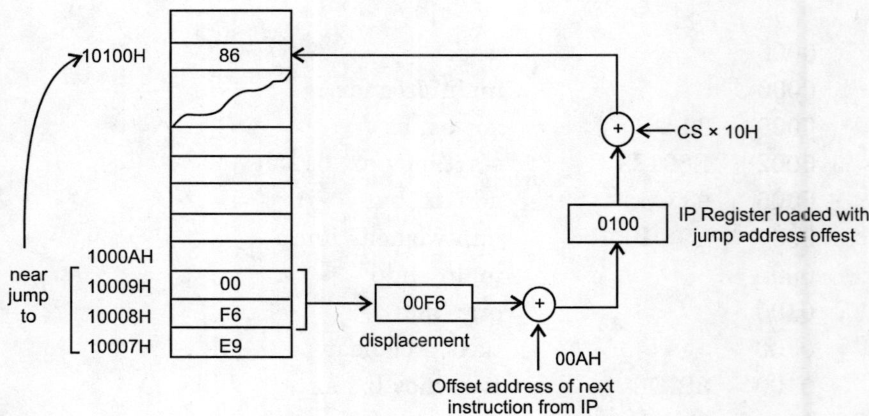

Figure 8.8. Near Jump (CS : 1000H) Execution

Far Jump: The far jump is a five byte instruction having the general format as shown in figure 8.9.

E9	Disp low	Disp high	CS low	CS high
←byte 1→	←byte 2→	←byte 3→	←byte 4→	←byte 5→

Figure 8.9. Far Jump instruction

The far jump can be used to jump to any location in the 8086/8088 system memory. (any address in its 1M Byte memory system).

The far jump format if you will observe has first two bytes following the opcode containing the displacement and the last two bytes contain the segment address of the jump location. The displacement constitutes the 16 bit offset address of the jump location. The address of the jump location is calculated as shown in fig. 8.10.

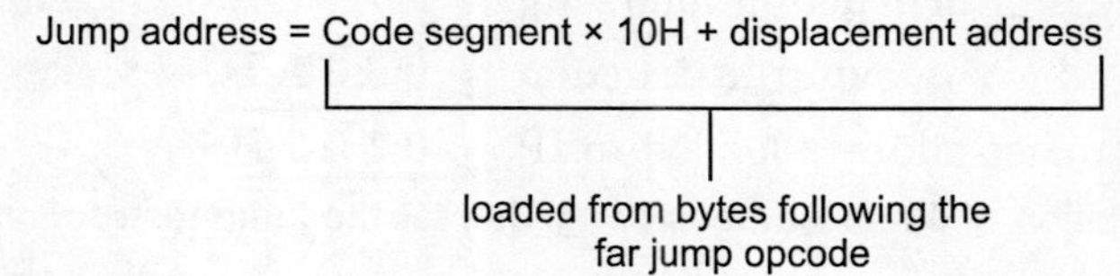

Figure 8.10. Jump address calculation

```
                .model small
0000            .data
0000   0023     a dw 23h
0002   2123     b dw 2123h
```

```
0004    ????         c dw ?

0000                 .code
0000                 main proc far
0000    33 DB        xor bx, bx
0002    B80123       start:mov ax, 0123h
0005    03 C3        add,ax, bx
0007    E9 0100R     jmptwo;near jump
000A                 main endp
0100                 org 100h
0100                 next proc near
0100    8B D8        two: mov bx, ax
0102    89 1 E 0004R movc, bx
0106    E9 0002R     mp start;near jump
010B                 next end p
010B                 end main
```

Figure 8.11. Program showing the far jump instruction

Figure 8.11. shows a program using a near jump and a far jump instruction. Note the far jump is initiated using the FAR PTR directive (which informs the assembler that the jump instruction is a far jump). You will also observe that the jump instruction is listed as EA 0002 ------ R. The code segment address is filled in when the .OBJ file is linked into a .EXE program by the Linker program. Figure 8.12. illustrates the far jump instruction.

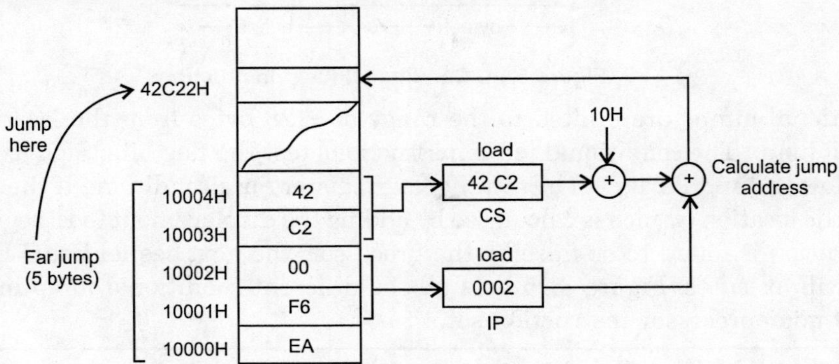

Figure 8.12. The far jump instruction execution

Unconditional Jumps using Registers Contents: These jumps are called the indirect jumps. An indirect jump instruction uses a register contents to specify the displacement of the jump address. However, these jump instructions donot add the displacement to instruction pointer, but transfer (copy) the displacement (i.e. the register) into the instruction pointer. Thus the displacement is the effective offset address of the jump location in such instructions. For example JMP CX copies the CX register contents to the instruction pointer (IP) when the jump instruction executes, to generate the jump location. Thus a indirect jump can be used to jump to any location in the current code segment. Figure 8.13 illustrates the jump instruction using register contents.

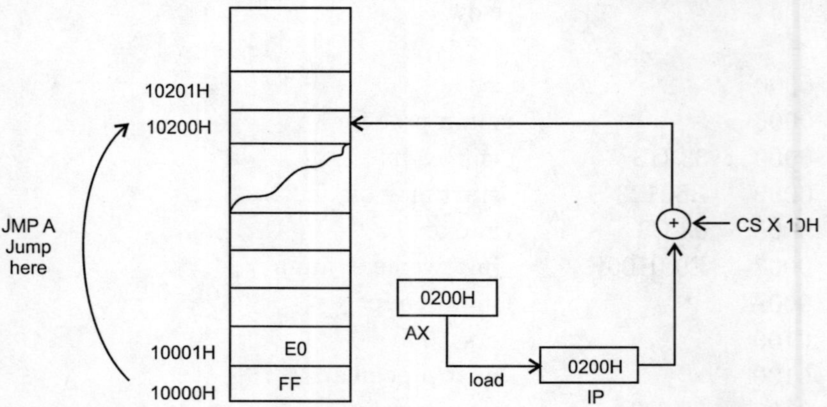

Figure 8.13. The jump instruction using register operands for execution of JMP AX Instruction assuming CS = 1000H

8.1.2. Conditional Jump Instructions

Conditional jumps are jump instructions that check the flag bit for a condition. If the condition tested is true the jump instruction will execute. However, if the condition tested is false, the next sequential instruction in the program shall execute. The 8086/8088 conditional jumps are always short jumps. Short jumps are only two bytes wide and have the general format as listed below in figure 8.16.

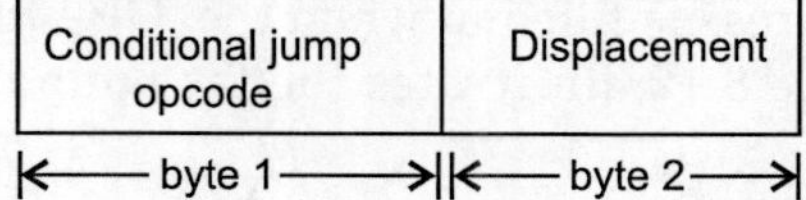

Figure 8.14. Conditional jump instructions

Conditional jumps are limited to the range of ±128 bytes from the location following the conditional jump. The conditional jump instructions test the flag bits: Sign (S), Carry (C), Zero (Z), Overflow (0), and Parity (P) bits in the flag register for a condition.· If the condition is true, a jump to the location (w hich is calculated by adding the displacement to IP register) is executed. If the condition is found to be false by the processor, the next sequential instruction in the program will executes. Figure 8.15 lists all the different conditional jump instructions of the 8086/8088 microprocessor instruction set.

Assembly	Condition under	Operation Name
JA	Z = 0 and C = 0	Jump if above
JAE	C = 0	Jump if above or equal
JB	C = 1	Jump if below
JBE	Z = 1 or c = 1	Jump if below or equal
JE/JZ	Z = 1	Jump when zero or equal
JNE/JNZ	Z =0	Jump when not zero
JG	Z = 0, S = 0	Jump when greater than
JGE	S = 0	Jump when greater or equal
JL	S = 1	Jump when lower

JLE	Z = 1 or S = 1	Jump when less or equal
JNC	C = O	Jump when no carry
JC	C = 1	Jump on carry
JNO	O = 0	Jump when no overflow
JO	O = 1	Jump on overflow
JPO	P = 0	Jump in odd parity
JPE	P = 1	Jump on even parity
JNS	S = 0	Jump when no sign
JS	S = 1	Jump of sign

Figure 8.15. Conditional Jump instruction

Since conditional jumps execute depending on the state of the flag bits they have important uses for magnitude comparison. Magnitude comparison refers to comparison of the magnitude of two numbers. When two numbers are compared the flag bits in the flag register would be set accordingly. These bits can then be used by conditional jump instructions for changing the program execution to a different location (depending on state of the flag bits).

We could simply illustrate a jump instruction execution as

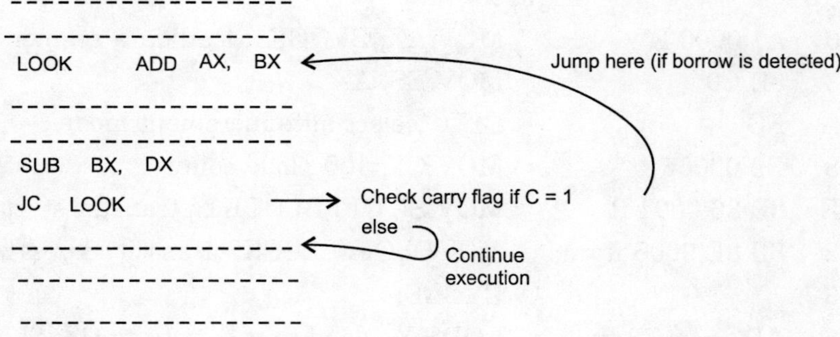

Figure 8.16. Execution of a Conditional Jump

When we are comparing two signed or unsigned number then depending on the resulting flag bits that are set we can induce a corresponding change of program execution by use of the correct jump instruction. For example, if we have the following two line of code.

```
SUB, BX, AX     ; Substract BX - AX
JA TOP          ; Jump to TOP if
                  BX greater then AX
```

This corresponding lines of code should only be used if BX and AX registers are holding unsigned number. However if BX and AX registers are holding signed numbers the JA instruction should be replaced by JG instruction.

Thus comparisons between two numbers can be made and depending on our requirement we can induce a corresponding jump if. However we must remember that

(1) JG, JL, JGE, JLE, JE and JNE instructions must be used when we are comparing

signed number. Observe in figure 8.15 that a jump instruction for signed number checks the sign flag bit for this reason.

(2) JA, JAE, JB, JBE, JE, JNE instructions are used for unsigned number comparison. For this reason these instructions check the carry flag bit.

However we shall not discuss each of these instructions in detail as the operation signified by these instructions is understood by the name of jump instruction itself.

An important conditional jump instruction that we shall however discuss is the JCXZ (JUMP when CX register zero) instruction. We can modify the data block movement program as listed in chapter 6. Here we shall use the JCXZ instruction as shown in figure 8.16 .

```
                        .model small
0000                    .data
0000   1000                BLOCK1 dw 1000h
0002   2000                8LOCK2 dw 2000h
0004   0000                OFFBLOCK1 dw OOOOh
0006   1111                OFFBLOCK2 dw 1111 h
0000                    .code
0000                    main proc
0000                    begin :
0000   A1 0000 R           MOV AX, BLOCK1; load data segment address
0003   8E C0               MOV ES, AX
000A   FC                  CLD  ;select auto increment mode
0008   B9 00064            MOV CX, 100 ; load count
000E   8B 36 0004 R        MOV Sl, OFFBLOCK1 ; transfer start address of data
0012   8B 3E 0006 R        MOV Dl, OFFBLOCK2 ; transfer address where data to be copied
0016                       CYCLE:
0016   AD                  LODSW  ;load first byte from DS: Sl
0017   49                  DEC CX  ;decrement CX
0018   AB                  STOSW  ;transfer data byte to ES: Dl
0019   E3 FB               JCXZ CYCLE
001B                    main end
001B                    end begin
```

Figure 8.17.

8.1.3. The Loop Instruction

The LOOP instruction is a powerful combination of a decrement CX and JNZ conditional jump. The program of 8.17 is modified by the LOOP instruction and shown in figure 8.18.

The LOOP instruction during each execution decrements the CX register and checks the state of the zero flag. If CX is not equal to zero it jumps to the jump location. If CX register becomes zero (CX = 0) the jump looping breaks down and the next sequential instruction is executed.

In the program listed in figure 7.20, we intend to transfer a block of memory whose address (segment base as well as offset address) is located in memory to a block of memory again whose address has been stored in memory. The first half of the program deals with loading the DS, ES, SI and DI with the address of source and destination blocks of memory.

The program following the label CYCLE is che part of the program that actually transfers the data. The LODSW instruction loads a data byte to the AX register from DS : SI while STOSW transfers it to ES : DI. The LOOP instruction decrements the CX register which is loaded with the block size and therefore the LOOP instructions executes 100 times till CX becomes zero. When CX becomes zero the program exits the loop .

```
                      .model small
0000                  .data
0000  1000                  BLOCK1 dw 1000h
0002  2000                  8LOCK2 dw 2000h
0004  0000                  OFFBLOCK1 dw OOOOh
0006  1111                  OFFBLOCK2 dw 1111 h
0000                  .code
0000                  main proc
0000                  begin :
0000  A1 0000 R             MOV AX, BLOCK1; load data segment address
0003  8E C0                 MOV ES, AX
000A  FC                    CLD ;select auto increment mode
0008  B9 00064              MOV CX, 100 ; load count
000E  8B 36 0004 R          MOV Sl, OFFBLOCK1 ; transfer start address of data
0012  8B 3E 0006 R          MOVDl, OFFBLOCK2 ; transfer address where data to be copied
0016                        CYCLE:
0016  AD                    LODSW   ;load first byte from DS: Sl
0017  49                    DEC CX  ;decrement CX
0018  AB                    STOSW   ;transfer data byte to ES: Dl
0019  E3 FB                 JCXZ CYCLE
001B                  main end
001B                  end begin
```

Figure 8.18

Conditional Loops

There are conditional loops present in the 8086 instruction set. The first is the LOOPZ/ LOOPE or loop while Zero/Equal. The LOOPE (or LOOPZ*). The LOOPE (loop while equal) "will continuously jump as long as CX > 0 or CX < 0 while an equality exists. It will exit the loop if the equality condition ceases to exist or the CX register decrements to zero, The LOOPNZ/ LOOPNE or loop while not zero/equal executes exactly opposite to the LOOPE/LOOPZ instruction.

8.1.4. Subroutines

Subroutines form an important part of optimized program coding. Though subroutines help save space as well as can be used by any number of programs they have their disadvantages. The first disadvantage lies in the complexity ofloading a subroutine while the second disadvantage is the extra time consumption that is required for transferring the program execution point.

The CALL instruction is used to change program execution to the first address of the subroutine while the RET (return) instruction is used to return back to the point in the program where we changed our program execution, CALL instruction when executed first pushes the address of the following** instruction onto the stack and then transfers execution to the subroutine. The RET instructions loads the address stored on the stack and forces the program to return to the meinory location that follows the CALL instruction.

While using assembly language programming, subroutine procedures may be stored with the use of PROC directive and ENDP directive as discussed in chapter 5. Following MASM ver. 6.0 procedures can be defined as NEAR and FAR types by using these directives. For storage of registers which might be altered the MASM 6.0 onwards allows them to be stored by the USES directive. The names of registers following uses would cause the assembler to automatically generate the code to push them onto the stack (when transferring control to the procedure) and pop then back from the stack (when returning from a procedure). Thus the statement PROCEDUREI PROC FAR USES AX BX would cause a far procedure to be defined. The assembler would also push the AX and BX registers when transferring main program execution to the procedure PROCEDURE 1 and pop them back when returning back to the main program. The assembler will also generate the corresponding RET opcode (whether inter segment or intrasegment return) from the last return instruction.

8.1.4.1. Call Instruction

The CALL instruction is needed to transfer the execution to the memory location of the procedure. The call instructions are of two types: the near (inter segment) call and the far (intra segment) calL

Inter Segment Call

The inter segment call or 'the near call has the general format as shown in figure 8.19(a). It is three bytes long and uses a 16 bits displacement. Like the near jump it can be used to transfer control to ± 32k bytes before or after the next instruction address. On execution the call instruction first transfers the address of the next instruction to the stack. It then adds the displacement to the IP (which holds the address of next instruction) to generate the memory address of the procedure. It then calculates the address thereby transferring control to the procedure.

E8	disp low	disp high

(a) Near Call

E8	offset low	offset high	seg low	seg high

(b) Near Call

Figure 8.19. Call Instruction format

Inter Segment Call

The far call like the far jump is a 5 byte instruction having the general format as shown in figure 8.19(b). The first byte stores the offset address while the next two bytes store the segment address of the procedure. On execution, the far call first pushes the address of the next instruction onto the stack. (The offset address and segment address are both pushed on the stack). Next the new offset address and segment base are both are loaded from memory. This allows the far call to access any procedure stored in memory. The execution process of the call instruction is same as figure 8.8 and 8.12 except that first the address of the next instruction is pushed onto the stack.

Indirect Calls

The third type of call instruction are similar to the indirect jump instruction. The IP register is pushed onto the stack. Next the microprocessor loads the AX register into the IP. Thus this would in effect cause a indirect jump to any location in the current code segment.

Figure 8.20 shows a program which effectively reads a key being typed by using the INT21H interrupt function 02. It then stores the AL register (which is loaded by the ASCII value of key) and processes it to find whether it is a number between 0 to 9 by the store procedure. If the key pressed is not between 0 to 9 decimal then it restarts the program. Otherwise the program stores the number in memory location referenced by num and then converts it to ASCII. The ASCII procedure then calls the display procedure which displays the number on the screen.

The call instruction which listed to read procedure is a far call (5 bytes) while the rest are near calls.

```
                        .model medium
0100                    .stack 100h
0000                    .data
0000    ??              datal db?
0001    ??              num db?
0000                    .code
0000                    store proc near     ; procedure to store data
0000    A2 0000 R       mov data 1, al
0003    2C 30           sub ai, 30h
0005    721E            jb start
0007    3C 09           cmp ai, 9
0009    77IA            ja start
0008    A20001 R        mov num, al
000E    C3              ret ;return if between zero to one
000F                    store endp
000F                    read proc far       ; procedure to read key
000F    8401            movah,l
0011    CO 21           INT 21H
0013    C8              ret
0014                    read endp
0014                    display proc near  ;procedure to display
```

```
0014    8402            mov ah, 2           ;load int and function number
0016    CD 21           int 21h
0018                    display endp
0018                    ascii proc near ;calculate ascii and then display
0018    AO 0001R        mov ai, number
0018    0430            add ai, 30h
0010    E8 0014 R       call display
0020                    ascii endp
0020                    main proc           ;main procedure '
0020                    begin:
0020    88 .... R       mov ax, number
0023    8E D8           mov ds, ax
0025                    start :
0025    9A 000F----R    call read
002A    E8 0000 R       call store
002A    E8 0018 R       call ascii
0030                    main endp
0030                    end begin
```

Figure 8.20. A Program showin the use of different call instructions.

8.1.4.2. Return Instruction

The return or RET instruction removes a two byte offset address (near return) from the stack and places it into the IP register. A far return removes 4 bytes and loads them into CS, IP registers so that the program execution can be transferred back to the next instruction following the CALL instruction. When using assemblers, the RET will be translated as a near or far RET depending upon the CALL executed automatically by the assembler.

The second type of RET instruction that is available on the 8086 instruction set is RET within segment while adding immediate data to SP register instruction. This instruction loads the offset address of the return location and then adds the immediate data to the SPa This instruction of use when certain locations of the stack location have to removed following a return as shown in the program in figure 8.21.

```
                    .model small
0064                .stack 64h
0000                .data
0000    000C                    data0 dw 12
0002    0002D                   data1 dw 45
0000                .code
0000                del proc near
0000    03 C3                   add ax, bx
```

```
0000    03 D0             add dx, ax
0004    C2 0003           ret 3
0007              del endp
0007              main proc
0007              begin:
0007    B8---R            mov ax, @data
000A    8E D8             mov dx, ax
000C    A1 0000 R mov ax, dataO
000F    8B 1E 0002 R                mov bx, data 1
0013    B9 007A           mov ex, 122
0016    50                push ax
0017    53                push bx
0018    51                push cx
0019    55                push bp
001A    E8 0000 R call del
001D    58                pop ax
001E    F4                hlt
001F              main endp
001F              end begin
```

Figure 8.21. A Program using the RET immediate

The program in fig. 8.21 shows the use of RET 3 instruction. The program is itself of no use but just a method to illustrate RET 3 use.

We are explaining the program from memory address 0016 H. The AX, BX, CX, BP register will be pushed onto the stack by the four push instructions. This is followed by the call to procedure del. del procedure performs some operations following which the RET3 instruction is encountered. The RET loads the offset address 001D (address of next instruction) and adds 3 to the SP. Now this is obviously the stack memory location address after PUSH AX instruction had executed. Therefore the contents of BX, C~, BP have been lost from stack. The RET instruction is used in such situations when certain portion of the stack have to be deleted following a procedure.

8.2. PROCESSOR CONTROL INSTRUCTIONS

The last type of the 8086/8088 microprocessor instructions are the processor control instructions group. These instructions operate on the flag register or are used to test the condition of the 8086/8088 microprocessor pins. Many of these instructions are used when interfacing hardware components such as the coprocessor or other support chips.

8.2.1. Flag Bit Conmol Instructions

CLC : Clear Carry flag instruction : This instruction clears the carry flag bit in the flag Register.

CMC : The Complement Carry flag instruction: This instruction causes the 8086/8088 microprocessor to complement the state ofthe carry flag bit.

STC : The set carry flag instructions causes the carry flag to be set.

CLD : The direction flag clear instruction causes the DF to be reset:

STD: This set Direction flag instruction does the opposite of the CLD instruction i.e. sets the direction flag bit.

CLI and STI : The clear interrupt flag and set interrupt flag instructions are used to control the state of IF flag. The STI places a 1 in the interrupt flag thereby allowing INTR pin to be used to interrupt the 8086/8088 microprocessor. The CLI places a 0 in the IF flag therefore disables the INTR pin.

8.2.2. Miscellaneous Processor Control Instructions

HLT : The halt instruction causes the 8086 microprocessor to cease all program execution. However an interrupt, DMA request or a processor reset will cause the 8086 to exit the halt state. HLT instructions are often used so as to place the microprocessor in a no operation state and therefore wait for external interrupts to restart the program execution.

WAIT: The wait instruction tests the condition of the 8086/8088 microprocessor $\overline{\text{TEST}}$ pin. If the $\overline{\text{TEST}}$ pin is high the microprocessor does nothing and proceeds with normal instruction execution. However if the $\overline{\text{TEST}}$ pin is low it causes the microprocessor to wait for the $\overline{\text{TEST}}$ pin to go high. The $\overline{\text{TEST}}$ pin of the 8086 microprocessor is connected to the BUSY pin of the 8087 coprocessor. This instruction is used to synchronize the 8086 with the 8087 and used to stop the 8086 microprocessor execution till the 8087 coprocessor has finished its instruction execution.

NOP : The no operation instruction causes the microprocessor to nothing. This instruction is used to introduce time delay within the program. Program Control and Processor Control Instructions

ESC (Escape to external Device) : The escape instruction is actually the 8087 coprocessor instruction. On execution of the ESC instruction the 8086 microprocessor sends the memory address of the next instruction. It is used for allowing the coprocessor instruction set to be used in conjunction with the 8086/8088 instruction set.

LOCK (Bus Lock Prefix) : The bus lock prefix is used before a normal 8086 instruction. It causes the $\overline{\text{LOCK}}$ pin to be driven low for the entire duration of the instruction before which the LOCK prefix has been used. The $\overline{\text{LOCK}}$ pin is gener-ally connected to the 8288 bus controller and (or) 8289 bus arbiter. When the $\overline{\text{LOCK}}$ pin is driven low, the 8288 and 8289 outputs are driven to their high impedance state thereby shutting of the system bus. After the execution of the instruction which has used the $\overline{\text{LOCK}}$ prefix, the $\overline{\text{LOCK}}$ pin goes high thereby enabling the 8288 and 8289 devices.

8.3. KEY POINTS TO REMEMBER

1. In this chapter we have discussed the program and processor control instructions.
2. Jump instructions are oftwo types: the conditional and unconditional jump instructions.
3. There are three unconditional jump instructions: the short jump, near jump and far jump. The near jump is a two byte instruction and uses a 8 bit displacement. It can be used to jump within ±128 bytes from the next sequential instruction. Near jumps permit a jump anywhere within the current code segment. They are three bytes long and use a

16 bit displacement. The far jump permits a jump any where within the 8086 memory.

4. The jump with register contents is an indirect jump and loads the contents of the specified register into the IP. These allow a jump anywhere within the current code segment.
5. Conditional jumps test conditions of the flag registers; if the condition is true a jump occurs. The flags bit checked are C, S, Z, 0, P bits. However, if the condition tested is false then the next sequential instruction is executed. All conditional jumps are short jumps.
6. The loop instruction is a combinational of JNZ and decrement CX registers. The LOOP instruction decrement CX and checks the ZF flag. IfZF = 0 the next sequential instruction executes else the jump occurs.
7. Other forms of LOOP are LOOPE and LOOPNE instructions.
8. Subroutines are important parts of optimized programming. They are implemented by the CALL and RET instruction. Subroutines can be called from any point in the program.
9. The CALL instruction pushes the address of next instruction into the stack. If then jumps to the desired procedure. Calls are of two types: the near and far calls.
10. Near Calls only place the offset address of next instruction onto the stack while far calls push both the CS and IP register contents of next instruction before executing the jump.
11. RET instruction is used to return to the program. It loads the address from the stack. RET instruction also has two forms: near RET which load only the IP; the far RET which loads both CS and IP registers from the stack.
12. The flag bit control instructions control the state of certain flag bits such as
    - CLl and STI → Clears or set interrupt flag
    - CLD and STD → Clears or set direction flag
    - CLC, CMC and STC → Clears, complements or set the carry flags.
13. The halt instruction is used to stop program execution. The processor can leave the halted states by any of the three actions listed below
    - microprocessor reset
    - interrupt signal or INTR
    - DMA request
14. Wait instruction checks the state of the $\overline{\text{TEST}}$ pin. If $\overline{\text{TEST}}$ = 1 the next instruction executes. If $\overline{\text{TEST}}$ = 0 the wait instruction causes the processor to wait till $\overline{\text{TEST}}$ goes high.
15. Lock instruction prefix causes the 8086 $\overline{\text{LOCK}}$ pin to be driven low for entire duration of the instruction that has been prefix by the $\overline{\text{LOCK}}$ instruction.

REVIEW QUESTIONS

8.1. Explain the operation of the following instructions graphically

- short jump

- far jump
- near call
- far return (RET)

8.2. Describe the execution of the near jump?

8.3. Differentiate between the near, short and far jumps?

8.4. The assembler will automatically optimize which jump in a program if the jump distance is

(*a*) 0010H

(*b*) 0256H

(*c*) 12345H

8.5. Explain the working of register indirect jumps ? How loes the JMP AX instruction execute?

8.6. Describe the conditional jumps that you can use ifthe numbers you are currently operating on are

(*a*) signed binary numbers

(*b*) unsigned binary numbers

8.7. Describe the different flag bits that may be tested by the different conditional jumps?

8.8. Explain the working of a LOOP instruction?

8.9. Suppose register AX and BX are compared by the eMP instruction? Which instruction would you use if you wanted a exit from the jump loop when

(*a*) AX >BX

(*b*)BX>AX

(*c*)AX=BX

Assume both numbers are unsigned.

8.10. Repeat the problem 9 when both registers are storing signed numbers.

8.11. Describe the operation ofthe LOOPE and LOOPNE instructions? How does it differ from the LOOP instruction?

8.12. Write a program to copy 500 bytes of data from memory location 12340H onwards to the memory location 55550H onwards?

8.13. It is required to load 100 bytes of memory following 1000H with FFH. Write a program to do the following? .

8.14. Explain by key differences between

(*a*) near and far call instruction

(*b*) near and far return instruction

8.15. Why are subroutines so useful?

8.16. Explain what would the RET12 instruction achieve?

8.17. Write a program to add four 64 bit numbers? Use procedures to do the above? Also store the carry flag by the changing the state of the DF (direction flag)

8.18. Explain operation of the wait instruction?

8.19. Which instruction change the following flag bits

(*a*) IF

(b) DF

(c) CF

8.20. Explain the difference between NOP and HLT instructions.

8.21. What does a LOCK prefix instaruaction accomplis? How can it be used in conjuction to the 8288 bus controller (Read Appendix B6).

8.22. Explain how a HLT instruction can be used to synchronize slower peripherals?

8.23. Write a program that prints your name by use of procedures?

●●●

Chapter-9

THE 8087 ARITHMETIC COPROCESSOR

INTRODUCTION

The intel designers formed a separate class of processors for high speed arithmetic data processing. We have observed till now that the 8086 has very limited arithmetic capability because mostly it can perform arithmetic operations such as addition and subtraction. Though multiplication and division instructions are included these are not very effective for large data (i.e. that is more than 16 bits wide). However these new processor chips that were introduced were capable of multiplication, division, addition, subtraction operations as well as were capable of finding square root, calculate the partial tangent, partial arc tangent and logarithms. These processors were capable of operating on 16 bit, 32 bit and 64 bit signed numbers, 18 digit BCD data and 32, 64 and 80 bit floating point numbers. The processors were aptly named co-processors as these were not stand alone* processors but required a CPU'like the 8086 to operate. The first of this family of processors was called the 8087 coprocessor and was designed for use with the 8086 processor. It must be clearly understood that coprocessors like the 8087 were designed for mathematical operations as discussed and thus were capable of higher speeds than the 8086. However for a deeper insight and understanding we shall now discuss the 8087 numeric coprocessor in greater detail. A note to all students that the 8087 cannot function as a stand alone processor but requires the 8086 microprocessor for functioning.

9.1. THE 8087 COPROCESSOR DATA TYPES

The intel 8087 uses data types that cannot be handled by the 8086 MPU. The data that can be handled by the 8087 numeric processor can be of the form of signed-integer, BCD data or floating point numbers. Though it can be recalled that the 8086 can operate on signed integer and BCD data it must be always -understood that the 8086 is limited to a data width of 16 bits. However as we discuss the 8087 coprocessor in greater detail we shall find it capable of better data handling capability (with respect to size of data and operations on data).

9.1.1. Signed Numbers

During the chapter on arithmetic and logic instructions, we discussed that the 8086 can operate on signed numbers of up to a maximum size of 16 bits. However this limits the 8086 MPU

* In computer terminology a processor is called a stand alone processor if it can perform all the operations of a CPU. However, coprocessor we capable (as we shall discuss) of only numeric operation hence are not called.standalone processors.

applicability because in many systems there can be a requirement of operation of numbers of larger.bit size. The 8087 coprocessor was designed to handle and operate on data (signed numbers) of size 16 bits (double byte or word). 32 bit (double word) or 64 bits. The conversion of decimal to signed binary numbers and vice-versa is same as earlier. However. we discuss signed numbers in great detail.

9.1.1.1. 16 Bit Signed Number

The 16 bit binary number is also referred to as word. The reason being the data bus of the 8086 is 16 bits wide (or of size of one word where word = 16 bits).

The data format for a 16 bit signed number is shown in figure 9.1

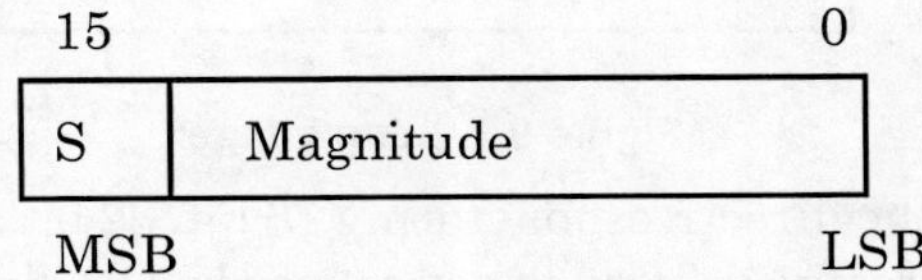

Figure 9.1. 16 bit signed binary number

Thus it can be observed that for a 16 bit number one bit (MSB) is reserved for sign denotion and 15 bits to store the magnitude of the number. Thus the maximum range of number storage will be -2^n to $2^n - 1$ where n = 15*. Thus the range will have limits

lower limit $= -2^n = -2^{15} = -32768$

upper limit $= +2^n - 1 = 32768 - 1 = 32767$

And the total no's that can be stored by 16 bit denotion is $2^{16} = 65536$ numbers (or 32767 + 32768 + 1)

Note: 1 is added as the number zero is to be included.

9.1.1.2. 32 bit Signed Number or Short Integer

The 32 bit signed number that the 8087 coprocessor can operate on is also called the short integer. The data format is shown in figure 9.2. Here the magnitude bits are a total of 31 bits while the MSB stores the sign bit of the number.

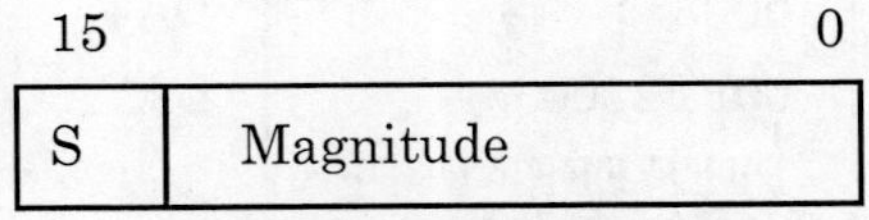

Figure 9.2. Short integer

The range for a short integer is similarly found as we did for a 16 bit, signed binary number.

lower limit $= -2^n = -2^{31} = -2147\ 483648$

$\cong 2 \times 10^9$

upper limit $= +2^n - 1 = +231 - 1 = 2147\ 483648$

$\cong 2 \times 10^9$

Thus range $= \pm 2 \times 10^9$

---

* *n* = 15 as this is the number of magnitude bits. The MSB or sign bit will only range the numbers from a negative to positive value.

9.1.1.3. 64 bit Signed Number or Long Integer

The 64 bit signed number or long integer has the data format similiar to the above discussed signed numbers only that here the number is 64 bits wide. Again the range is calculated.

lower limit = $-2^n = -2^{63} = -9.2 \times 10^{18}$

upper limit = $+\ 2^n - 1 = +\ 2^{63} - 1 \cong +9.2 \times 10^8$

Thus range = $\pm\ 9.2 \times 10^{18}$

The figure 9.4 shows the long integer

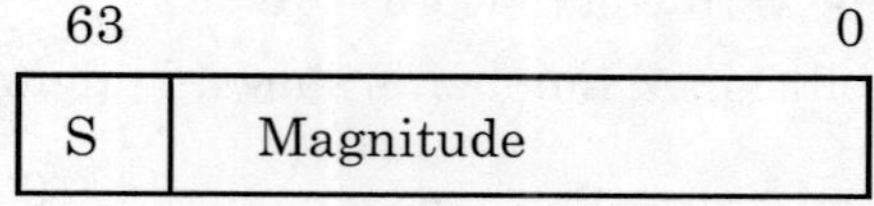

Figure 9.3. Long Integer.

In assembly language the directives for storing 16 bit, 32 bit and 64 bit signed numbers are same as earlier. The DW directive is used to define words, DD directive defines double words or short integers and DQ directive defines long integers as shown in figure 9.4.

Directive	Operation	Size
DW	define words	16 bit signed integer
DD	define double words	32 bit signed or short integers
DQ	define Quarternary words	64 bit signed number or long integer

Figure 9.4. Assembly language directives for signed numbers

Example 9.1. To show use of directives to store signed numbers in assembly language.

Answer:

Memory Address**	Data (Calculated)* by assembler	Name	Directive	Number
0000	0003	DAT	DW	+3
0002	00FF2300	DATA	DD	+16720640
0006	FFFF FFFF FFFF			
	FFEO	DATAl	DQ	–32
000A	FFEA	DAT2	DW	22
000C	FF FF FF E8	DATA 2	DD	–24

9.2. BINARY CODED DECIMAL (BCD) NUMBER

The 8087 can operate on BCD data. However here the BCD data for the coprocessor is 80 bits long. Thus a total of 10 bytes are used $\left(\frac{80}{8} = 10\text{bytes}\right)$. Here the first most byte is used to store

* This column will automatically be calculated by the assembler by use of directive.

sign. Thus there are a total of 9 bytes to store the BCD data. Now as earlier discussed a BCD number required 4 bits, thus two BCD numbers can be stored in one byte. Thus in 9 bytes 18 BCD numbers can be stored. Thus, the 8087 operates on 18 digit signed BCD numbers.

The DT directive is used to store BCD numbers in memory by assemblers.

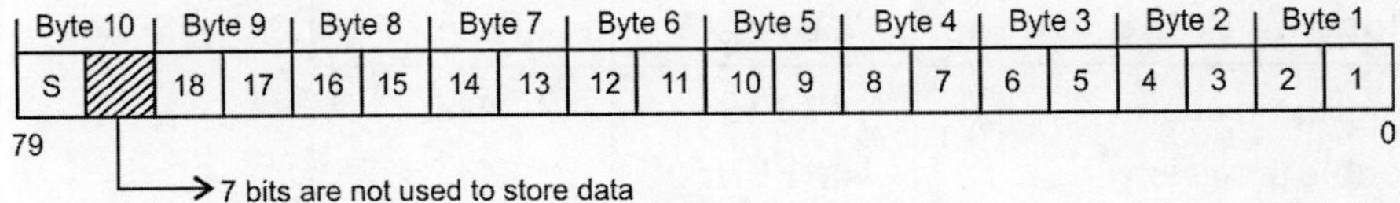

Figure 9.5. 8087 BCD format

9.3. FLOATING POINT NUMBERS

The floating point number representation is the most real form of expressing numbers. These can be used to store integers, fraction etc. Let us recall the real number system from senior school. It is of the form

$$24672.675$$

which can be represented as

Sign ⟶ (+) [2.46722675] × 10^4 ⟶ Exponent

Number ⟶ (2.46722675)

The binary floating point numbers are very similar to the decimal real numbers. The floating point numbers can be broken down into the three parts namely.

1. Sign bit
2. Exponent
3. Significant

Now, let us first consider the floating point number system in detail before we proceed to the 8087 coprocessor architecture.

9.3.1. Floating Point Number System

Consider the decimal number 250.25. Now let us perform (or convert) operations so as to convert this decimal number to binary.

Now the decimal 250 is converted to binary

$$(250)_{10} \longleftrightarrow (11111010)_2$$

Now let us convert 0.25 to binary

$$(0.25)_{10} \longleftrightarrow (0.01)_2$$

Thus

$$(250.25)_{10} \longleftrightarrow (11111010.01)_2$$

This corresponds to the step 1 which is conversion to binary. The next step is to normalize this binary number. i.e.,

$$1111\ 1010\ .\ 01 = \boxed{1.111101001} \times 2^7$$

(1.111101001 ⟶ Significand)

Thus we have

sign = 0

exponent = 111 (binary for 7)

significand = 111101001

Microprocessors however use a standardized representation to express floating point numbers for compatibility from one system to another.

The microprocessor use a biased exponent format (The IEEE 754 standard) where the exponent is biased by a number as shown in figure 9.6.

Floating Point	**Bias**	**Size**
Single precision	7 FH	32 bits
Double precision	3FFH	64 bits
Extended floating point	3FFF H	80 bits

Figure 9.6. Floating point number types and bias for exponents

Now let us first discuss the different types of floating point numbers, wherein we shall discuss conversion of decimal numbers to floating point and vice-versa.

9.3.2. 32 Bit Floating Point Numbers Or Single Precision Numbers

The data format for 32 bit floating point numbers is shown below in figure 9,7. These numbers use 32 bits where the MSB is the sign bit, bit 23 through 30 stores the exponent.

Now let us perform the inverse of the operation as discussed above. Suppose we are given the following bit pattern

1	1000111	10110001100000000000000,
↓	↓	↓
Sign	Exponent	Significand

Now first we will separate each of the three component.

Sign = 1 (thus negative number)

Exponent = 10000111

Significand = 1011 00011 0000000000000

Now let us calculate the true exponent by subtracting the bias from the biased exponent

true exponent = 10000111 – 01111111

= 1000

$=(8)_{10}$

Now the significand is 1011 00011

or $-\ 1.1011\ 000\ 11 \times 2^{s}$

Now we shall denormalize the above number to get

– 110110001.1

The final step is to convert to decimal which would be

$-(433.5)_{10}$

We shall extend this method to other floating point numbers by only changing the bias* and the size of the significand

* A biased exponent is the exponent added with a predetermined bias number such as 7FH for 32 bit floating point numbers.

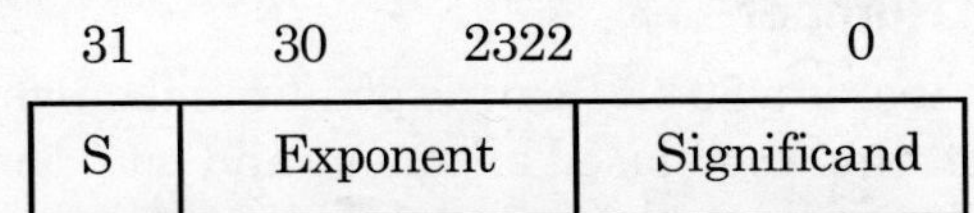

Figure 9.7. Single Precision Floating Point Numbers

As in figure 9.7 we observe a bias of 7 FH is used. Thus now let us convert 250.25 to floating point. Now observe 23 bits are used for storing the significand, 8 bits are used to store the biased exponent and 1 bit to store sign.

Example 9.1.

Input		Result
1. Conversion	250.25	11111010.01
2. Normalization	11111010.01	1.111101001×2^7
3. Biased exponent	111 + 0111111	10000110 (0111111 is binary of bias 7FH
4. Sign bit		0
exponent		10000110
Significand		1111,0100,1000,00000000000

Thus the final number as stored in memory would look like.

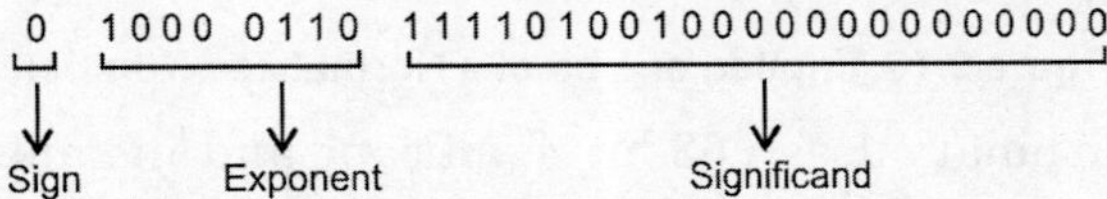

9.3.3. 64 Bit Or Double Precision Floating Point Numbers

A double precision floating point number is similiar to the single precision floating point number except for the fact that a double precision number takes 64 bits (8 bytes) of memory while a single precision number takes 32 bit (4 bytes) of memory space. The data format for a double precision number is given below in figure 9.8.

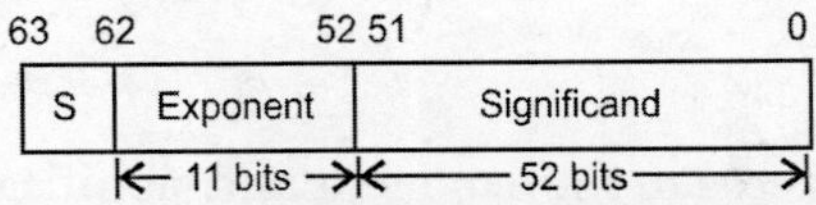

Figure 9.8. Double Precision Number

In the double precision floating point numbers we use a bias of 3FFH to bias the true exponent. Thus consider the following example of conversion of a decimal number to double precision floating point.

Example 9.2. Convert 225.25 to double precision binary

Result	Process
1. 225.25. = 11100001.01	Conversion of decimal to binary
2. 11100001.01 = 1.110000101×2^7	Normalization
3. 111 + 01111111111 = 100 0000 0110	Calculation of bias exponent by adding bias of 3FFFH
4. Sign = 0	
Exponent = 10000000110	
Significand = C280000000000H	
5. Number = 40 6C 28 00 00 00 00 00 H	

9.3.4. Extended Precision Number

The extended precision number is a 80 bit floating point number where the MSB is the sign bit, bit 64 through 78 is used for storing the biased exponent and bit 0 through 63 are used to store the significant. The data format is shown in figure 9.9.

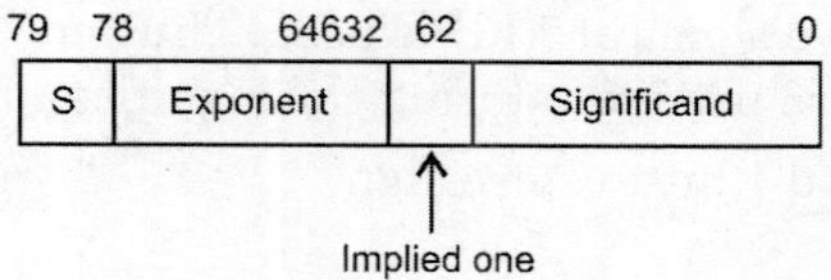

Figure 9.9. Extended precision binary

The extended precision binary uses a bias of 3FFFH to calculate the biased exponent. However the extended precision binary is the only format which includes the implied one bit of a significand. The significand of a normalized number is shown in figure 9.10.

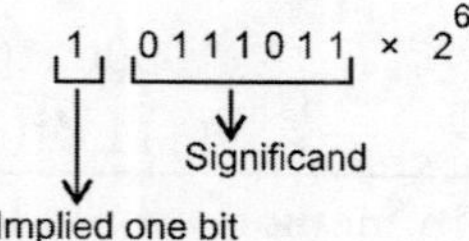

Figure 9.10. Implied one bit of a Normalized Number

The extended precision binary uses a 63-bit significand and an implied one bit as in figure 9.11.

9.4. SPECIAL CASES

There are three special cases of floating point numbers for storage of zero, infinity, negative and positive infinity and a non existing number. We only take the case for single precision numbers (can be similarly extended for double and extended precision numbers)

9.4.1. Zero

A zero is expressed by storing zeros in the exponent and significand portions of a single precision. The sign bit can zero or one as shown in figure 9.11.

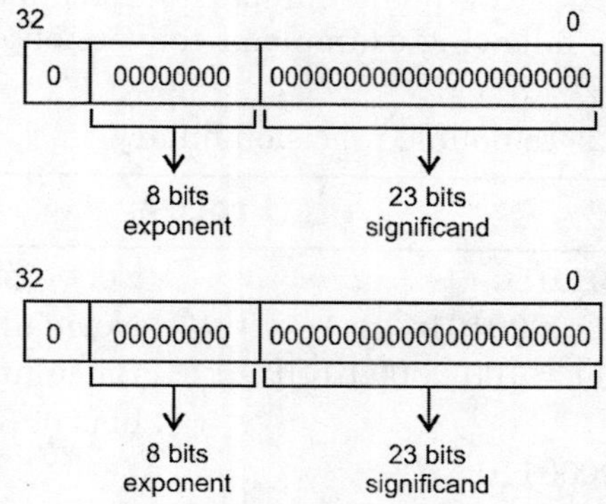

Figure 9.11. Data representation for Storage of zero for floating point numbers

9.4.2 Infinity

Infinity and negative infinity is stored when the exponent bits are all ones and the significant bits are zeros. If the sign bit is zero it represents infinity and if the sign bit is a one then the number stored is considered as negative infinity as shown in figure 9.12.

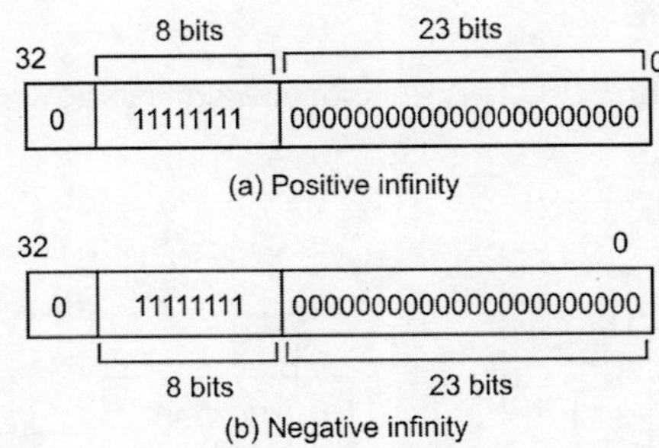

Figure 9.12. Data representation for storace of (a) positive (b) negative infinity

9.4.3. Invalid Floating Point Number

An invalid floating point number is consider to be stored when the exponent has all ones while the significand is not all zero.

9.4.4. Assembler Directives

To store floating point number by the use of assemblers we have the following directives as shown in figure 9.13

Directive	Floating point	Syntax in assembly
DD	single precision	NAME DD 99.42
DD	single precision	NAME 1 DD 22.462
DQ	double precision	NAME DQ3.2E + 10 ·
DT	extended precision	NAME 0 DT 3.425E + 22

Figure 9.13. Assembler directives and usage for storage of floating point numbers

9.5. 8087 INTERNAL ARCHITECTURE

The 8087 coprocessor is designed to work with the 8086 concurrently or simultaneously. Though since the 8087 can take large amount oftime to execute most of its instructions we thus prefer to operate the 8086 and 8087 concurrently. The 8087 is available in a 40 pin DIP package. The coprocessor is also called numeric processor (or math processor) as these processors are designed to efficiently execute arithmetic and transcendental operations.

When a 8086 microprocessor system that has a coprocessor (i.e. 8087) is prog-rammed, all we have to do is write down the instructions (either 8086 or 8087 instructions) straight into memory. When these instructions are executed both the 8086 and 8087 load these instructions into their respective queue. The 8086 decodes the instructions in its queue and executes them. On encountering an 8087 instruction the 8086 simply treates it as a NOP. Similarly when the

8087 decodes it instructions (i.e. in its queue) and finds a 8086 instruction it will treat it as a NOP. If you observe the 8087 instruction set in detail you will find that all 8087 instructions have the same five most significant bits in their first code byte that is 11011. Also on observing the 8086 set you will find this corresponds to the Ese instruction. The ESe instruction is in itself an instruction that tells the 8086 to do a NOP.

9.5.1. INTERNAL ARCHITECTURE

The internal block diagram of the 8087 numeric processor is shown in figure 9.14.

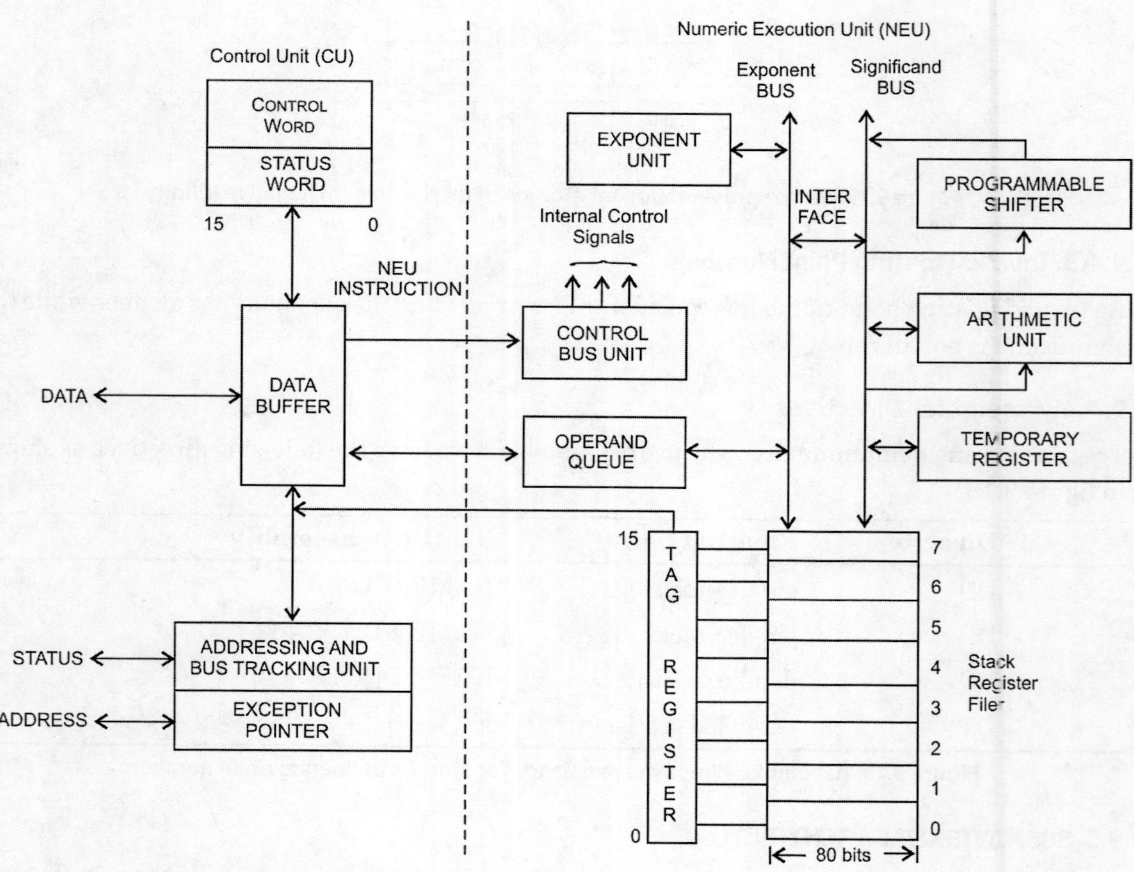

Figure 9.14.8087 Internal Block Diagram

The 8087 coprocessor can be considered to be made up of two sections

1. The Control Unit (CU)
2. The Numeric Arithmetic Unit (NED)

The pin diagram of 8087 coprocessor is shown in fig. 15. the details of the pins are discussed in following sections :

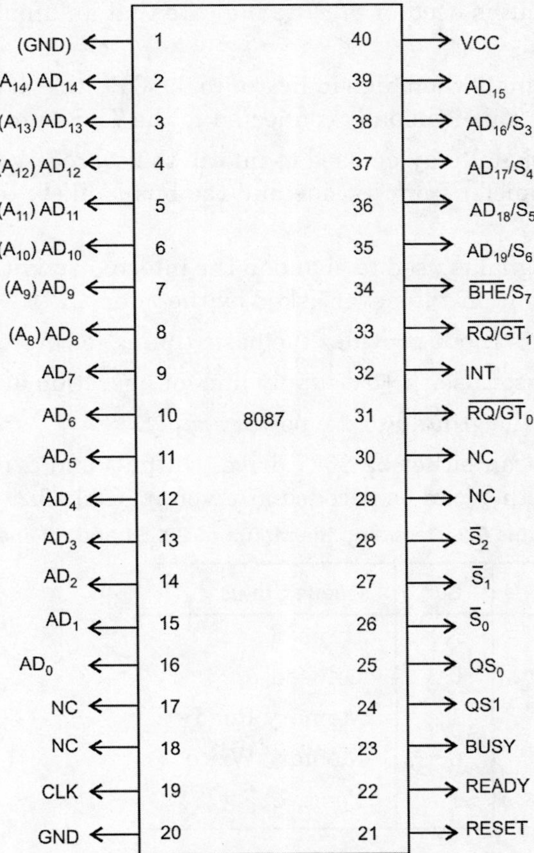

Figure 9.15 the pin diagram of 8087 corprocessor

$AD_0 - AD_{15}$ – These lines are the time multiplexed address/data lines. These lines carry address during T_1 cycle and hold data during T_2, T_3, T_w and T_4 states; Ao is used, whenever the transfer is on lower byte ($D_0 - D_7$) of data bus to derive the chip select.

$A_{19}/S_6 - A_{16}/S_3$ – These lines are the time multiplexed address/status lines. These lines functions in a similar to the corresponding pins of 8086 microprocessor. The S_6, S_4 and S_3 are remaining permanently high, while the S_5 remains permanently low.

$QS_1 - QS_6$ – The queue status input signal QS_1 and QS_6 enable 8087 to keep track of the instruction pre-fetch queue status of the CPU as shown in table 9.1

Table 9.1: Status of queue lines

QS_1	QS_2	Queue Status
0	0	No operation
0	1	First byte of op code from queue
1	0	Empty queue
1	1	Subsequent byte from queue

INT – The interrupt out is used by 8087 to indicate that an unmarked exception has been received during execution.

BUSY · This output signal, when high indicator to the CPU that it is busy with the execution of an allotted instruction. This is usually connected to the TEST pin of 8086 or 8088.

READY · This Input signal may be used to inform the coprocessor that the address device will complete the data transfer from its side and the bus is likely to be free for the next bus cycle.

RESET · This Input signal is used to abandon the internal activities of the coprocessor and prepare it for further execution whenever asked by the main CPU.

CLK – The CLK Inputs Input provides the basic timing for processor operation.

V_{CC} – The 8087 processor uses a +5V supply line for operation of the circuit.

GND – This provides the ground to the power supply.

$\overline{S}_2$, $\overline{S}_1$ and $\overline{S}_0$ – These can either be 8087 drive (Output) and drive (Input) by the CPU, if these are driven by 8087, they can be decoded as given in Table 9.2.

Table 9.2. Decoding the status of S2, S1 and S0 lines

S2	S1	S0	Queue Status
0	X	X	Unused
1	0	0	Unused
1	0	1	Memory Read
1	1	0	Memory Write
1	1	1	Passive

RQ/GT_0 – The Request Grant (RQ/GT_0) pin is used by the 8087 to gain control of the bus from the host 8086/8088 microprocessors for operand transfer, it must be connected to one of the request grant pins of the host microprocessor.

9.5.2. Control Unit Of 8081 Coprocessor

The control unit provides the interface between the coprocessor bus and the processor (8086) system bus. The coprocessor monitors and processes the system bus simultaneously with the 8086 MPU.On encountering an ESC sequence (Coprocessor instruction), the 8086 does nothing while the coprocessor executes. On the other hand, when the system bus contains a valid 8086 instruction the coprocessor ignores it while the 8086 microprocessor executes the respective instruction.

9.5.3. Numeric Execution Unit

The Numeric Execution unit is as the name suggest the equivalent of the 8086 Execution unit except here all instructions are numeric in nature. The 8087 coprocessor has eight 80-bit registers that work as a stack memory. This is referred to a stack register file. This stack register file holds all operands for the coprocessor instructions as well as is used to store results of an arithmetic operation. All coprocessor instructions either address this stack register file by accessing the data in a specific stack register or by using the push and pop method for retrieving and storing data on the top of the stack. All numbers that are stored onto the stack memory are automatically converted* to extended precision floating point numbers (80 bit) by the coprocessor when brought from the 8086 system memory. When numbers are stored from memory or 8086 system bus they are converted from the original state (i.e. signed integer, BCD, single precision or double precision formats) to an extended precision floating point number. The 8087 coprocessor internally only operates on 80 bit extended precision floating point numbers.

9.5.4.8087 REGISTERS

The 8087 coprocessor has three registers which are vital for operation and to decipher the results as they store status of result, control certain arithmetic properties of the result and show the status of the register stack file. The three registers are:

(1) Control Register or Control Word

(2) Status Register or Status Word

(3) Tag Register

9.5.4.1. Control Register

The control register of the 8087 provides control over arithmetic operations as well as provides control to the programmer so as to-mask exceptions** that may occur during (or after execution) of an arithmetic operation.

The control register format is shown in figure 9.15. The control register is used to select the precision control mode, rounding control mode, infinity control mode as well as to mask exceptions that may occur after an arithmetic operation. The FLDCW (Coprocessor load control word) instruction is used to load the state of the control register of the coprocessor. A brief summary of the bits of the control register is given below.

IE — Interrupt enable mask bit: Used to enable or disable the coprocessor interrupt signal.

rc — Infinity control: Selects the infinity mode Projective mode allows unsigned –(+∞) infinity while affine mode allows both positive and negative infinity. Refer to section 9.4.2.

RC — Rounding Control determines the rounding type as given in figure 9.15.

PC — The precision control sits the precision of the result i.e single, double or extended precision. This means that 00 will store a result which is effectively a single precision number.

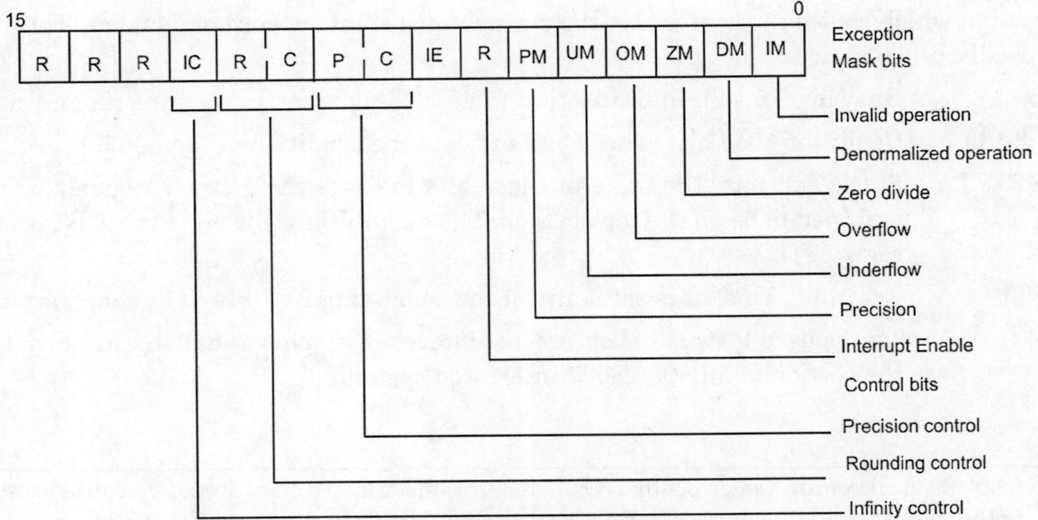

Figure 9.15. 8087 Control Register

Note: (1) **IE (Interrupt Enable Mask Bit)**

0 = interrupt enable

1 = Coprocessor interrupt enable

(2) PC (Precision Control Bit)

00 – 24 bits (i.e. single precision control)

01 – unused

10 – 53 bits (i.e. double precision control)

11 – 64 bits (i.e. extended precision control)

(3) RC (Rounding Control)

00 – Round to nearest or Even (Default after coprocessor initialization)

01 – Round downwards (towards $-\infty$)

10 – Round upwards

11 – Chop (truncate to zero)

(4) IC (Infinity control bit)

0 = Projective

1 = Affine

(5) R – Reserved for future bits

(6) All exception* make bits are enabled when 1 and disabled when 0.

Figure 9.15. 8087 Control Register

9.5.4.2. Status Register

The Status register at all times reflects the state that the coprocessor is in. There are a total of 15 valid status bits in the status register as shown in figure 9.16.

The status register can be checked by using the FSTSW (Coprocessor store status word) instruction which copies the contents of the status register into a word of memory. The status bits are discussed in detail below.

B : Busy bit. This bit indicates that the 8087 is currently executing an instruction.

C3-CO : Condition code bits. These bits indicate the condition of the 8087.

ST3,ST2 : Stack Top bits, the state of these bits indicate the current register which is assumed to be on the top of stack. The default (initialized) value of ST is 000 or register O.

ST1 : Error bit. This bit is set if any of the remaining five bits (if enabled) are set.

E : This is used to cause a coprocessor interrupt which we shall discuss in detail in the section of an 8086/8087 interfaced system.

* In the 8086 microprocessor errors that occur after arithmetic instructions are referred to as exceptions in the context of 8087 coprocessor, *e.g.* overflow error, divide by zero error (or exception).

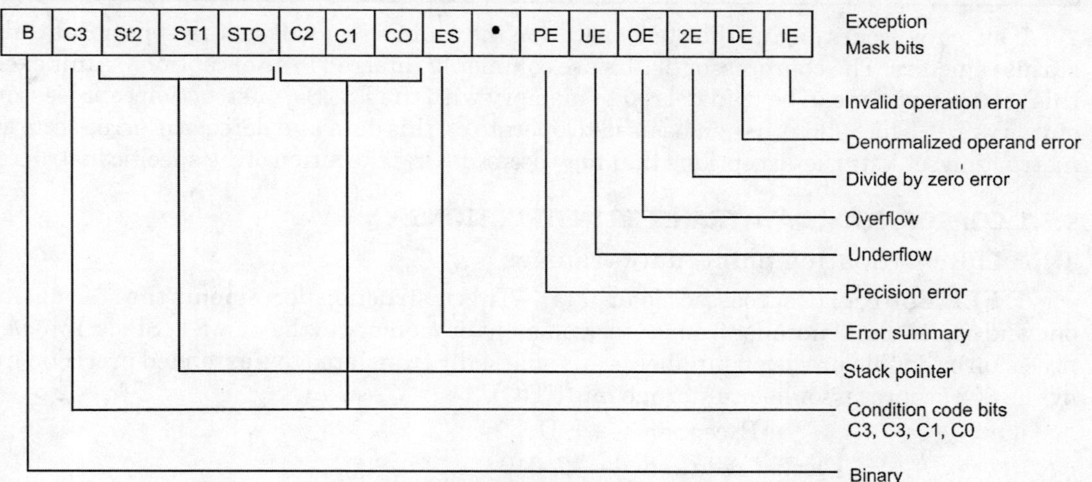

Figure 9.16. 8087 Status Register

PE : Precision error bit. This bit is set if the result or operand for an operation exceeds the precision set (as set in the control word)

UE : Underflow error bit. An underflow error is said to have resulted if the result is too small for the current precision level. This effectively set the UE bit.

OE : Overflow error bit. When the result is too large to be represented by the current precision level the overflow error is set.

ZE : Divide by zero error. Refer to section 7.2.5

DE : Denormalized error bit. If any of the operands is in denormalized state then this bit is set

IE : Any other error that maybe generated sets the IE bit.

9.5.4.3. Tag Register

The tag register maintains the status of all eight registers of the stack register file of the 8087. The tag is used to indicate whether the register holds valid data, zero, invalid (such as infinity) data or the register is empty. The tag register is shown in figure 9.17.

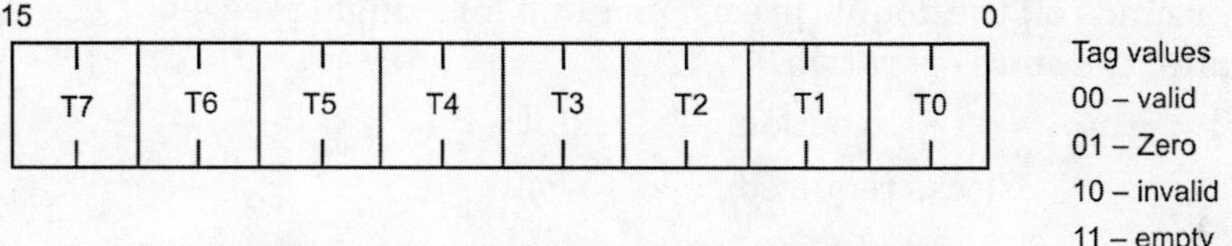

Figure 9.17. 8087 Coprocessor Tag Register

9.6. THE 8087 INSTRUCTION SET

The MASM assembler is used for coprocessor programming for a 8086 and 8087 based system. The use of assemblers is the best way to produce codes for such a system (as shown in figure 9.23).

All the coprocessor mnemonics begin with the letter F, which stands for floating point (i.e. the coprocessor) and is used so as to distinguish the 8086 and 8087 instruction.

The different types of the 8087 instructions are listed according to the operation group such as data transfer, arithmetic, compare, transcendental, constant and coprocessor control instruction. We shall now discuss each group in detail.

If an error occurs during the 8087 execution the coprocessor will set the appropriate bit in the status registers. The coprocessor errors are commonly referred to as exceptions. After execution this status register can be transferred to memory with the FSTSW instruction coprocessor store status word). The 8086 can then be used to operate on this data and detect the error occured. The instruction set lists the-exceptions that may occur during execution ofthe specific instruction.

9.6.1. COPROCESSOR DATA TRANSFER INSTRUCTIONS

Type I : Real (floating point) data transfer

1. FLD Source: (Coprocessor load data)—This instruction decrements the stack pointer by one and copies a real number from a stack element or memory to the new ST (Stack Top). A single precision or double precision number is automatically translated to a extended precision number by the 8087 coprocessor before storage into ST(O).

Exception — I, D
Clocks* required — 17–310

Examples:

FLD DATA— Loads a floating point number from location specified by DATA to stack top.
FLD ST(2)— Copies a floating point number from stack register 2 to stack top ST.

Formats:

1 1 0 1 1 0 0 1	0 0 0 0 0 m m m disp	32 bit memory load FLD
1 1 0 1 1 1 0 1	0 0 0 0 0 m m m disp	64 bit memory load FLD
1 1 0 1 1 0 1 1	0 0 1 0 1 m m m disp	80 bit memory load ,FLD
1 1 0 1 1 0 0 1	1 1 0 0 0 r r r	32 bit register load specified by r r r
1 1 0 1 1 1 0 1	1 1 0 0 0 r r r	64 bit register load specified by r r r
1 1 0 1 1 0 1 1	1 1 1 0 1 r r r	80 bit registerload specified by r r r

0 0 m m m → addressing equivalent to 8086 rIm mod fields (figure 3.11)

r r r → specifies register to stack register file.

2. FST destination: (Coprocessor store data)—This instruction copies the stack top (ST) to the specified stack register or memory location. If the number to be transferred is larger than memory location size then the number is automatically rounded off to that size. For e.g. if a single precision number is to be stored and the ST contains 'a double precision number the 8087 automatically rounds off the double precision number to single precision. The stack pointer is not incremented after the store operation.

Exception → I, 0, U, P
Clocks required → 15-540

Examples:

FST ST(6)— Copies 3T to ST(6). Stack pointer incremented.
FST DATA— Copies ST to memory referenced by DATA

Formats:

1 1 0 1 1 0 0 1	mod 0 1 0 r/m	disp	: 32 bit memory store (FST)
1 1 0 1 1 1 0 1	mod 0 1 0 r/m	disp	: 64 bit memory store
1 1 0 1 1 0 1 1	mod 0 0 0 r/m	disp	: 80 bit memory store
1 1 0 1 1 1 0 1	1 1 0 1 0 rrr		FST ST to rrr (stack register)

mod r/m → equivalent to 8086 mod and r/m bits for memory addressing.

rrr → specifies stack register.

3. FSTP destination: The equivalent of the FST instruction except that the stack pointer is incremented by 1 so as to point to the next location of the stack file. This may be considered as a POP operation preceded by memory storage of the stack top.

Exception → I, O, U, P

Clocks → 15—540

Examples:

FSTP DATA— Copies ST register to memory referenced by DATA and increment stack pointer

FSTP ST(2)— Copies ST to ST(2) and increments stack pointer.

Formats:

11 0 1 1 0 1 1	mod 0 1 1 rm	disp	: 32 bit memory FSTP
1 1 0 1 1 1 0 1	mod 011 rm	disp	: 64 bit memory FSTP
110 1 101 1	mod 111 rm	disp	: 80 bit memory FST
11011101	1 100 1 rrr		FST ST (rrr)

the 32 bit memory FSTP for example copies the ST (80 bit register) into 32 bits, i.e. it converts a extended precision number to a single precision number and stores it in 4 bytes (32 bits) of memory space. Thus, a 32 bit FSTP is a single precision store, a 64 bit FSTP is a double precision memory store and a 80 bit FSTP is a extended precision memory store.

4. FXCH Destination: The FXCH is a floating point exchange instruction. It exchanges the contents of ST with the specified stack register. If no destination is specified the ST and ST(l) are exchanged

Exception → I

Clocks → 10—15

Examples:

FXCH ST(2)— Contents of ST exchanged with contents of ST(2).

Formats:

1 1 0 1 1 0 1 1 1 1 0 0 1 r r r : FXCHI ST, ST (rrr)

Type II : Integer Transfers

1. FILD Source: (Coprocessor Integer load data instruction) The FILD copies a integer from memory and stores it on ST (stack top). The number is converted from integer format to extended precision automatically by the processor before storage on stack.

Exception → I

Clocks → 17—310

Examples:

FILD NAME: load a integer from memory referenced by NAME to ST

Formats:

1 1 0 1 1 1 1 1	mod 0 0 0 rm	disp	: word integer load (16 bit)
1 1 0 1 1 0 1 1	mod 0 0 0 rm	disp	: short integer load (32 bit)
1 1 0 1 1 1 1 1	mod 1 0 1rm	disp	: long integer load (64 bit)

2. FIST Source: The FIST instruction or the coprocessor integer store instruction stores the ST to a memory location. The coprocessor automatically converts the ST (a extended precision number) to integer format before storage

Exception → I, P
Clocks → 15—540

Examples:

FIST DATA : store ST to memory referenced by DATA. ST converted to integer prior to storage by coprocessor.

Formats:

1 1 0 1 1 1 1 1	mod 0 1 0 rm	disp	16 bit memory FIST
1 1 0 1 1 0 1 1	mod 0 1 0 rm	disp	32 bit memory FIST
1 1 0 1 1 0 1 1	mod 1 1 1 rm	disp	64 bit memory FIST

3. FISTP destination: The FISTP or coprocessor integer store and pop instruction is similar to the FIST instruction where a number from ST is converted first to an integer and then stored in memory, The stack pointer is incremented after storage

Exception → I, P
Clocks → 15—540

Examples:

FISTP DATA : Store ST to memory location referenced by DATA followed by stack point increment

Format:

1 1 0 1 1 1 1 1	mod 0 1 1 rm	disp	: word integer store and pop
1 1 0 1 1 0 1 1	mod 0 1 1 rm	disp	: short integer store and pop
1 1 0 1 1 1 1 1	mod 1 1 1 rm	disp	: long integer memory store and pop

Type III : BCD Data Transfer

1. FBLD Source: The FBLD or coprocessor BCD load data instruction loads a 80 bit BCD number from memory converts it to extended precision format and then stores it in ST. The stack pointer is decremented prior to storage on stack top.

Exception → I
Clocks → 17—310

Example:

FBLD DATA_BYTE : An 80 bit BCD loaded from memory referenced by DATA_BYTE

Format:

1 1 0 1 1 1 1 1 mod 1 00 rm disp 80 bit BCD memory load

2. FBSTP Destination: The coprocessor BCD store data and pop instruction converts the ST to BCD format and stores it in a memory location similar to a POP operation. The STis converted to a 10 byte BCD number. The stack pointer is then incremented by one.

Exception → I
Clocks → 15—540

Example:

FBSTP MONEY : ST converted to 80 bit BCD and stored on memory location ref- erenced by MONEY. Stack pointer is then incremented by one.

Format:

1 1 0 1 1 1 1 1 mod 1 0 0 rm disp 80 bit BCD memory load

9.6.2. COPROCESSOR ARITHMETIC INSTRUCTIONS

A point to be noted before we discuss these instructions is that the 8087 coprocessor never operates on data except the extended precision floating point number (80 bit). Since all numbers whether signed integers, BCD, single precision or double precision numbers are converted to the 80 bit extended precision floating point numbers before storage on the register stack or during an operation, the coprocessor effectively operates on 80 bit floating point numbers. It thus implies that internally the 8087 coprocessor operates on 80 bit floating point. The 8087 coprocessor again converts numbers that are integers, before storing in memory from the extended precision format. Thus the 8087 coprocessor may effectively do a BCD or an integer operation (such as ADD or SUB), and it may appear that the 8087 operates an BCD or integers but internally it only operates on extended precision numbers. Thus all discussed arithmetic instructions are the floating point arithmetic instructions for the extended precision format.

ADDITION INSTRUCTION

(1) FADD Destination Source: The coprocessor floating point addition instruction adds a source to the destination and stores it in the destination. The source can be a stack register or memory location while the destination has to be a stack register. If no source or destination is specified (i.e. default value) the ST(O) is added to ST(I) and stack pointer is incremented so as to point to ST(O). The source if a memory location must be a real number (either single precision, double or extended precision).

Exception → I, D, O, U, P

Clocks → 70—143

Example:

F ADD ST(7), ST(0)	:	ST(0) + ST(3). Store in ST(3)
FADD MEMORY	:	ST(0) + real number from Memory. Store in ST(0).
FADD	:	ST(0) + ST(I). Pop stack and store in ST*.

Format:

1 1 0 1 1 0 0 0	mod 0 0 0 rm	disp	32 bit FADD (from memory)
1 1 0 1 1 1 0 0	mod 0 0 0 rm	disp	64 bit memory FADD
1 1 0 1 1 0 0 0	1 1 0 0 0 reg		FADD ST, ST (reg)

The d bit is same as 8086 microprocessor's instruction direction bit. The mod (2 bits), r/m (3 bits) and reg (3 bits) perform same operation as in 8086 except that in the 8087 coprocessor the reg addresses the stack registers of 8087. When d = 0 the destination is ST (reg) while if $d = 1$ the destination for the result is ST(O).

(2) F ADDP source: The coprocessor real addition pop instruction adds the ST and specified ST(reg) register and stores the result ST. The operation is a pop type i.e. the stack pointer will hence be incremented by 1.

Exception → I, D, O, U, P

Clocks → 70—143

Example:

FADDP	ST(2)	:	Add ST(2) to ST and store result in ST and increment stack pointer so that ST(I) becomes stack top (ST).

Format:

1 1 0 1 1 1 1 0 1 1 0 0 0 : Add ST(reg) to ST(O) and increment stack pointer

3. FIADD Source: Add an integer from a memory location to the ST register. The integer will be automatically converted to an extended precision number prior to addition by the coprocessor. The result is stored in the ST register

Exception → I, D, O, P
Clocks → 70—143

Example:

FIADD memory : Add integer number referenced by memory to ST and store result in ST

Format:

1 1 0 1 1 1 1 0	mod 0 0 0 rm	word integer memory ADD (16 bit)
11 0 1 1 0 1 0	mod 0 0 0 rm	short integer memory ADD (32 bit)

SUBTRACTION INSTRUCTION

(1) FSUB Destination Source: The coprocessor real subtraction instruction subtracts the source from destination location (destination - source) and stores the result in the destination. The destination must at all times be. a stack register while the source can be a memory location or stack register. Thus a memory subtraction always subtracts a memory location contents from a stack register. The number loaded from the memory will also be a real number (i.e., must be stored a r-eal number, if you store a integer here the result obtained will be wrong).

Exception → I, D, O, U, P
Clocks → 70—143

Example:

FSUB	ST(4)*, ST :	ST (4) = ST(4) – ST(O)
FSUB	DATA	ST(0) = ST(0) – contents of memory referenced by DATA
FSUB		ST(0) = ST(0) – ST(l)

Format:

1 1 0 1 1 0 0 0	mod 100 rm	disp	: single precision memory FSUB
1 1 0 1 1 1 0 0	mod 100 rm,	disp	: double precision memory FSUB
1 1 0 1 1 0 0 0	1 1 101 reg		: FSUB ST, ST(reg)

The d bit is the same as in the FADD instruction and controls the flow *i.e.* to where the operation will take place. Thus if $d = 0$ the flow in from ST to ST (reg) in above example *i.e.* ST(reg) = ST(reg) – ST(0) and if $d = 1$ the flow towards ST or ST(0) = ST – ST (reg).

(2) FSUBP destination source: The coprocessor real subtraction and pop instruction works similar to FADDP instruction. Here the source is subtracted from the destination and the result stored in destination. The destination must be a stack register while the source may be a stack element or memory location. The stack pointer is then incremented by 1 (popped).

Exception → I, D, O, U, P
Clocks → 70—143

Example:

FSUBP ST(l), ST(2) : ST(l) = ST(l) – ST(2) and stack pointer increment by 1. (ST(l) becomes new ST)

FSUBP ST(l) : ST(l) = ST(l) – ST and stack pointer incremented by one (*i.e.* ST(l) becomes new stack top)

Format:

1101110 11101reg FSUBP ST, ST (reg)

(3) FISUB source: The coprocessor integer' subtraction instruction operates exactly as the FIADD instruction. Here a source* (a memory location contents) which is an integer is first converted to extended precision number by the coprocessor and then subtracted from the ST. The result is then stored in the ST register

Exception → I, D, O, P

Clocks → 70—143

Example:

FISUB DATA : ST = ST – integer from memory location referenced by DATA

Format:

11011110 mod 100rm disp : word integer FISUB.

11011010 mod 100rm disp : short integer FISUB

REVERSE SUBTRACTION INSTRUCTIONS

In addition to the subtraction instructions the 8087 has another type of subtraction called the reverse subtraction instructions (FSUBR, FSUBRP, FISUBR). These instructions operate in exact the same manner except that here the destination is subtracted from the source and stored into the destination.

i.e. destination = source – destination

However during execution of the subtraction instruction the operation performed is

destination = destination – source

The instructions become very important when a stack element must be subtracted from a memory location. Since in a subtraction instruction a memory location is always a source, thus to subtract a register from a memory location we would have to write first transfer the memory location to. the stack and then do a subtraction. However the reverse subtraction instruction here would save us upto 310 clocks if used as we would not have to store the memory location onto the stack. There are three reverse subtraction instructions.

FSUBR destination, source

FSUBRP destination, source

FISUBR source

All these instruction work exactly as their equivalent subtraction instructions except that the

destination = source – destination

MULTIPLICATION INSTRUCTION

The three multiplication instructions that appear on the 8087 instruction set are the FMUL (Coprocessor real multiply), FMULP (Coprocessor real multiply and pop) and the FIMUL (Coprocessor integer multiply) instructions. Now let us discuss the three in detail.

(1) FMUL destination source: A real multiplication takes place where the destination is multiplied by the source and the result is put back into the destination. The source can be a stack register or a memory location while the destination must be a stack register.

Exception → I, D, O, U, P

Clocks → 110—168

Examples:

FMUL DATA: ST = ST × contents of memory location referenced by DATA

FMUL ST(2), ST ST(2) = ST(2) × ST

Format:

11011000 mod 001 rm disp : 32 bit memory FMUL

11011100 mod 001 rm disp : 64 bit memory FMUL

11011000 i 1 001 reg : FMUL ST, ST(reg)

The d bit or direction bit sets direction as in F ADB and FSUB instructions.

(2) FMULP destination source: A real number from the specified source is multiplied to the specified destination, the result stored in the specified destination followed by a increment of stack pointer by 1.

Exception → I, D, O, U, P

Clocks → 110—168

Examples:

FMULP ST(I) = ST(1) × ST followed by increment of stack pointer

Format:

11011110 11001 reg → FMULP ST, ST (reg)

(3) FIMUL source: Convert the integer referenced by the source (i.e. memory) to extended precision and multiply to the ST and store result in ST

Exception → I, D, O, P

Clocks → 110—168

Examples:

FIMUL DTR multiply ST by the contents of memory location referenced by PTR i.e. ST = ST × contents of memory location referenced by PTR.

Format:

11011110 mod 001 rm disp : word integer FIMUL

11011010 mod 001 rm disp : short integer FIMUL

DIVISION INSTRUCTION

(1) FDIV destination, source : The coprocessor real divide instruction divides the destination by the source. The source can be a real number stored in memory or stack element. The destination will always be a stack element.

Exception → I, D, O, P

Clocks → 191—243

destination = destination / source

Examples:

FDIV MEM ST = ST/contents of memory location referenced by MEM

FDIV ST, ST(l) ST = ST/ST(l)

Format:

1 1 0 1 1 0 0 0 mod 110 rm disp 32 bit memory division
1 1 0 1 1 1 0 0 mod 100 rm disp 64 bit memory division
1 1 0 1 1 0 0 0 11 111 reg FDIV ST, ST(reg)

(2) FDIVP destination, source: Coprocessor division and pop instruction. Exactly same as FDIV except that stack pointer is incremented by one. See F ADDP or FSUBP instructions

Exception → I, D, Z, O, U, P
Clocks → 191—243

$$\text{destination} = \frac{\text{desination}}{\text{source}}$$

Format:

1 1 0 1 1 1 1 0 111 1 reg : FDIVP ST, ST (reg)

(3) FIDIV source: The coprocessor integer divide instruction divides the ST by an integer stored in memory referenced by the source. The result goes to ST register.

Exception → I, D, Z, O, U, P
Clocks → 191—243

Format:

1 1 0 1 1 1 1 0 mod 110rm disp : 16 bit memory FIDIV
1 1 0 1 1 0 1 0 mod 110rm disp : 32 bit memory FIDIV

REVERSED DIVISION INSTRUCTION

The reverse division instructions consist of three instructions namely FDIVR (the coprocessor real division reverse) instruction, FDIVRP (the coprocessor real division reverse pop) instruction and FIDIVR (the coprocessor integer division reverse) instruction. All these instructions operate exactly as the division instructions (FDIV, FDIVP and FIDIV). However here the source is divided by the destination and storedin the destination

Reverse division instructions : destination = source / destination

Normal division instructions : destination = destination / source

Exception → I, D, Z, O, U, P
Clocks → 191—243

Format:

1 1 0 1 1 0 0 0 mod 111 rm disp : 32 bit real FDIVR
1 1 0 1 1 1 0 0 mod 111 rm disp : 64 bit real FDIVR
1 1 0 1 1 2 0 0 11110 reg : FDIVR ST, ST (reg)
1 1 0 1 1 1 1 0 mod 111 rm disp : word integer memory FIDIVR
1 1 0 1 1 0 1 0 mod 111 rm disp : short integer memory FIDIVR
1 1 0 1 1 1 1 0 11 110 reg : FDIVRP ST, ST (reg)

OTHER ARITHMETIC INSTRUCTIONS

(1) FSQRT: The FSQRT or coprocessor square root instruction basically just calculates the square root of ST and stores it in the ST. Thus

$$ST = \sqrt[2]{ST}$$

Exception → I, D, P
Clocks → 180—186

Example: FSQRT

Format:

1 1 0 1 1 0 0 1 1111 1010

(2) FSCALE: The FSCALE instruction scales ST with respect to ST(l). Here the ST(l) register value is added to the exponent part of ST. Thus it can be said that the ST(l) which is converted first to a, signed integer and then added to the exponent of ST. This is short is equivalent to multiplying or dividing ST by powers of two.

Since the exponent part of an extended precision number is 16 bits.thus the integer value of ST(l) must range between + 2^{-15} and 2^{+15}.

Exception → I, O, U
Clocks → 32—38

Example:

FSCALE

Format:

1 1 0 1 1 1 0 1 1111 1101

(3) FPREM: The FPREM instruction is the coprocessor partial remainder instruction. The FPEM instruction divides the ST by ST(1) and stores the remainder in ST. The quotient here is rejected. Consider the figure 9.18 given below. The FREM instruction operates in a similar fashion.

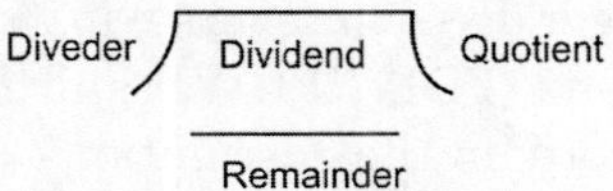

Figure 9.18. FREM Operation

5)27(⑤ → Ouotinet
25
②
Remainder

Example 9.2.

As in above example, 27 is divided by 5 to give a quotient of 5 and remainder. This remainder is what is stored onto the ST by the FREM instruction:

Exception → I, D, U
Clocks → 15 – 190

Example:

FPREM

Format:

1 1 0 1 1 0 0 1 1111 1000

(4) FRNDINT: This is the coprocessor round off register to integer instruction. The RC (Rounding control) bits in the control word are used to decide how the ST will be rounded off. For example if

- RC = 11 or chop mpde then a number such as 122.25 will be rounded as 122 .
- RC = 10 or round up (towards +∞) 122.25 will be rounded as 123.

Exception → I, P

Clocks → 16 – 50

Example:

FRNDINIT

Format:

1 1 0 1 1 0 0 1 1111 1100

(5) FABS : The coprocessor convert to absolute instruction converts the ST to its absolute value and stores this value in the ST. This instruction simply converts any signed number in the ST to a +ve value

Exception → I

Clocks → 10 – 17

Example: F ABS → ST = Abs (ST)

Format: 1 1 0 1 1 0 0 1 1111 0000

(6) FXTRACT : The FXTRACT instruction seperates (or extract) the exponential and significand bits of the extended precisioin number stored in ST. These two components are then stored as extended precision numbers onto the ST and ST(l) register where the significand is stored in ST and exponential portion in ST(l).

Exception → I

Clocks → 27 – 55

Example:

FXTRACT : ST = significand of ST

ST(1) = exponent of ST

Format:

1 1 0 1 1 0 0 1 1111 0100

(8) FCHS : This is the coprocessor sign complement instruction which complements the sign bit of the number stored in ST. This is equivalent to changing the sign from +ve to -ve and vice versa for a number

Exception → I

Clocks → 10 – 17

COMPARISON INSTRUCTIONS

The compare instructions is a useful class of instructions present in the 8087 coprocessor instruction set which are used for comparing two numbers. These numbers can be stack register contents or one a stack register and the other a memory location. On execution of the compare instructions the two numbers are compared and the condition code bits (in the status register) C3, C2 and CO are set. The status of different combination of C3, C2, CO are shown in figure 9.19. After comparing the FSTSW instruction of the 8087 coprocessor can be used to transfer the status register to memory the 8086 microprocessor can be used to determine the result of the compare operation.

C3	C2	CO	Inference
0	0	0	ST > source
0	0	1	ST < source
1	0	0	ST = source
1	1	1	uncomparable

Figure 9.19. Condition code bits for compare Instruction

Now let us study the comparison instruction that can be encountered during the programming of 8087.

(1) FCOM source: The coprocessor compare instruction compares the ST with a stack element or contents of memory. The code condition bits are then set depending on the two numbers.

Examples:

```
FCOM ST(6)   ; ST and ST(6) compared
FCOM         ; ST and ST(I) compared
FCOM DATA    ; ST and memory location referenced by DATA compared
```

Formats:

```
1 1 0 1 1 0 0 0   mod 010 rm   disp   ; 32 bit memory compare
1 1 0 1 1 1 0 0   mod 010 rm   disp   ; 64 bit memory compare
1 1 0 1 1 0 0 0   1 1 010 reg         ; FCOM ST,  ST(reg)
```

(2) FCOMP source: This instruction is identical to the FCOM instruction but the stack pointer is incremented by 1. Thus the old ST(1) becomes ST henceforth

Exception → I, D

Clocks → 40 – 93

Examples:

```
FCOM   ST(3)    ;   ST and ST(3) compared. Stack pointer incremented
FCOM   DATA     ;   ST and content of memory location referenced by DATA
                    compared. Stack pointer incremented
```

Formats:

```
1 1 0 1 1 0 0 0   mod 011 rm   disp   ; 32 bit memory FCOMP
1 1 0 1 1 1 0 0   mod 011 rm   disp   ; 64 bit memory compare
1 1 0 1 1 0 0 0   1 1  011 reg        ; FCOM ST, ST(reg)
```

(3) FCOMPP : The FCOMPP instruction compare ST with ST(l) and sets the condition code bits. The stack pointer is then incremented by two. Thus the old ST(2) becomes new ST after execution. Or simply the new ST is now above the two registers (ST and ST(l)) that we compared.

Exception → I, D

Clocks → 40 – 93

Formats:

1 1 0 1 1 1 1 0 11011001 FCOMPP

(4) FICOM source: This instruction compare ST with a word, short or long integer stored in memory. As usual the integer is converted to extended precision format before comparison

Exception → I, D

Clocks → 40 – 93

(5) FICOMP source: This instruction is exactly the FICOM instruction except for the fact the stack pointer is incremented by one after the comparison operation

Exception → I, D

Clocks → 40 – 93

Formats:

1 1 0 1 1 1 1 0	mod 010rm	disp	; 16 bit FICOM
1 1 0 1 1 0 1 0	mod 010 rm	disp	; 32 bit FICOM
1 1 0 1 1 1 1 0	mod 011 rm	disp	; 16 bit FICOMP
1 1 0 1 1 0 1 0	mod 011 rm	disp	; 32 bit FICOMP

(6) FTST : The FTST instruction compares ST with 0 and sets the condition code bits CO, C2, C3 as shown in figure 9.19.

Exception → I, D

Clocks → 38 – 48

Formats:

1 1 0 1 1 0 0 1 11100100 FTST

(7) FXAM : The coprocessor examine ST instruction does exactly what its name suggests. It examines ST for its status such as whether ST is normalized* unnormalized, zero, infinite, empty or whether ST is not a number. The condition code bits for FXAM are given below in figure 9.20.

C3	C2	C1	C0		Status of ST
0	0	0	0	→	+ unnormal
0	0	0	1	→	+ NAN
0	0	1	0	→	– unnormal
0	0	1	1	→	– NAN
0	1	0	0	→	+ normal
0	1	0	1	→	+ ∞
0	1	1	0	→	– normal
0	1	1	1	→	– ∞
1	0	0	0	→	– 0
1	0	0	1	→	Empty

1	0	1	0	→	– 0
1	0	1	1	→	Empty
1	1	0	0	→	+ denormal
1	1	0	1	→	Empty
1	1	1	0	→	– denormal
1	1	1	1	→	Empty

where: normal → normal floating point number

NAN → not a number (exponent all ones and significand not all zeros)

Empty → Tag register value of 11

unnormal → leading bits of significand are zero

denormal → exponent at most negative value

Figure 9.20. CO-C3 bits for FXAM

Exception → none

Clocks → 12 – 23

Formats:

1 1 0 1 1 0 0 1   1 1 1 0 0 1 0 0

9.6.3. Transcendental Instructions

The transcendental instructions constitute a powerful set of instructions that 8087 coprocessor has. These are trigonometric and exponential instructions which are totally unavailable on the 8086 microprocessor.

(1) FPTAN: The coprocessor tangential instruction calculates the radio Y/X for the number stored in 8T. *i.e.,*

$$\frac{Y}{X} = \tan ST$$

Now for correct operation of this instruction ST must be in radians. The FPTAN instruc- tions can only operate for an angle range of $0 < ST < \pi/4$. The Y component of the Y/X ratio replaces the angle (that is stored in ST) i.e., Y component is stored in ST, then the stack pointer is incremented. The ST(1) now becomes the new ST where X component of the Y/X ratio is stored.

Exception → I, P

Clocks → 30 – 450

Format:

1 1 0 1 1 0 0 1   1 1 1 1 0 0 1 0

(2) FPATAN: This is the coprocessor tangential angle instruction where tangential angle is calculated for n ratio ofY/X. i.e.

$$\text{angle} = \tan \frac{Y}{X}$$

Now X must be in ST while Y should be in ST(I)

The angle so calculated is stored in Y, i.e. ST(I). After the operation is completed the stack pointer is incremerited therefore the angle is. effective in ST.

Exception → U, P

Clocks → 250 – 800

Limits → $0 < Y < X <$ i.e., $0 < ST(1) < ST < \infty$

Format:

11011001 11110010

(3) F2XMl : (Coprocessor 2 to power X Minus 1 instruction). This instruction calculates the function $F = 2^x - 1$. X is stored in ST while the result of the operation is again stored in ST. Thus

$ST = 2^{ST} - 1$ F2 × M1 Instruction

Exception → P

Clocks → 310 – 630

Limits → $0 \le X \le 0.5$

Format:

11011001 11110000

(4) FYL2X : Coprocessor Y log base 2 X Instruction calculates $F = Y \log_2 X$ where Y must be in ST(l) and X in ST. The result in then stored in ST(I). The stack pointer is then incremented so that F is effectively the stack top. Thus

$F =$ (new) $ST = ST(1) \log_2 ST$

Exception → P

Clocks → 900 – 1100

Limits → $0 \le X \le \infty$

$-\infty < Y < +\infty$

(5) FYL2XP1 : (Coprocessor Y log 2 X plus 1 Instruction) This instruction calculates the function $F = Y \log_2[X + 1]$. Y must be in ST(1) while X in ST. The result F is stored in ST(I) and stack pointer increments so as to make ST(1) the new ST. Thus

$F =$ new $ST = ST(1) \log_2 ST + 1]$

Exception → P

Clocks → 700 – 1000

Limits → $0 \le X \le 1 - 2\sqrt{2}$

$-\infty Y < +\infty$

Format:

11011001 111111001

9.6.4. Constant Load Instructions

The instruction set of the 8087 coprocessor includes instructions which load constants on the top of the stack. The instruction are given below in the figure 9.21.

Instruction	Constant pushed onto top of stack
FLDZ	Zero
FLD1	+1
FDLP1	+π
FLD2T	$LOG_2 10$
FLD2E	$\log_2 e$
FLDLG2	$\log 10^2$
FLDLN2	$\log e^2$

Figure 9.21. Constant load instructions

9.6.5. 8087 Control Instruction

Now let us discuss the 8087 control instructions. These instructions perform no math- ematical operations but are required for synchronization of the 8087 coprocessor with 8086 main processor. In the next section we shall discuss a 8086 microprocessor interfaced to an 8087 coprocessor. Here we shall discuss the coprocessor control instructions.

(1) FINIT/FNINIT : This is the coprocessor initiallization instruction. This is equivalent to the 8086 reset operation. The coprocessor is effectively reset. During reset, the default conditions for the control word are set. These default conditions are IE Mask bit = 1, RC bits = 00, Infinity control = O. Also, the stack pointer is initialized to ST(O) and the tag bits for each stack register holds 11 (i.e. all stack registers are empty)

Clocks – 2 — 8

Format – 1 1 0 1 1 0 0 1 1 1 1 1 0 1 1 0

(2) FDISI/FNDISI : This coprocessor instruction disables the 8087 interrupt output pin so that it cannot cause an interrupt whenever an exception occurs. Since whenever an exception occurs, the coprocessor interrupt pin goes high so as to interrupt the 8086, this process can be bypassed by this instruction

Clocks – 2 — 8

Format – 1 1 0 1 1 0 1 1 1 1 1 0 0 0 0 1

(3) FENI/FNENI : Enables the 8087 coprocessor interrupt output so that the 808'7 coprocessor can cause an interrupt to the 8086 main processor when an exception occurs

Clocks – 2 — 8

Format – 1 1 0 1 1 0 1 1 1 1 1 0 0 0 0 0

(4) FLDCW Source: This instruction is used to load the control word with the contents of a memory location referenced by source. This instruction is used to initialize the 8087 coprocessor control word.

Clocks – 7 — 14

Format – 1 1 0 1 1 0 0 1 mod 101 rm disp

(5) FSTCW/FNSTCW : This instruction copies the 8087 coprocessor's control word to a memory location referenced by destination. This is then used so as to check various status bits of the 8087 coprocessor by the 8086 instructions.

Clocks – 12 — 18

Format – 1 1 0 1 1 0 0 1 mod 1 1 1 rm disp

(6) FSTSW/FNSTSW Destination: This the coprocessor store status word instruction. This instruction copies the status word to a named memory location. Again we can now use the 8086 instructions to take further action

Clocks – 12 — 18

Format – 11011101 mod 111 rm disp

(7) FCLEX/FN CLEX : This FCLEX instruction clears all exception bits in the 8087 coprocessor status register. It also unasserts (disables) the BUSY and INT output pins

Clocks – 2 — 8

Format – 1 1 0 1 1 0 1 1 1 1 1 0 0 0 1 0

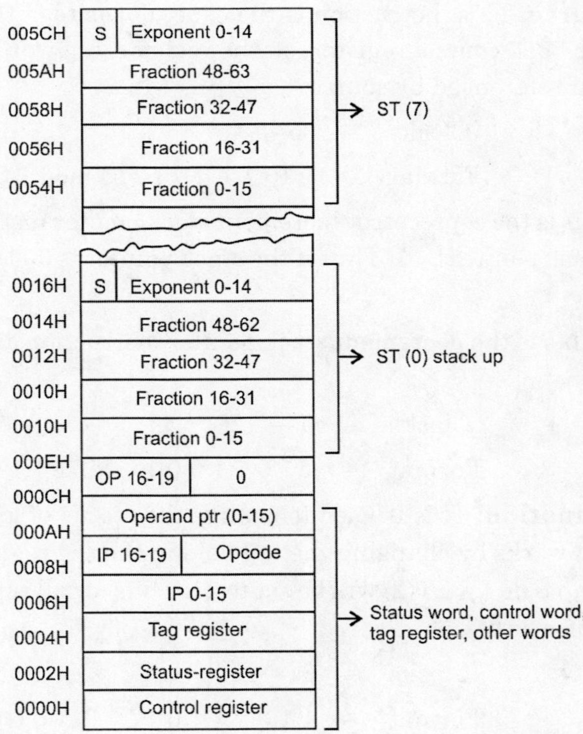

Figure 9.21. Memory format of 94 byte area for FSAVE instruction

(8) FSA VE/FNSA VE : The FSAVE instruction is used to copy the state of the 8087 coprocessor to memory. This instruction copies the 8087 status word, control word, pointers, tag register and the entire register stack to a name 94 byte area of memory. The memory format of this 94 byte area is shown in figure 9.21. After coping all this the 8087 is re-intialized like the FINIT instruction.

Clocks – 197 — 200

Format – 1 1 0 1 1 1 0 1

(9) FRSTORE Source: This is the coprocessor restore state instruction. This instruction copies a 94 byte area of memory into the 8087 registers. It is the exact opposite equivalent of FSAVE instruction.

Clocks – 197 — 207

Format – 1 1 0 1 1 1 0 1 mod 101 rm disp

(10) FSTENV/FNSTENV Destination: This instruction is similar to the FSA VE instruction except that here only the stack registers are not copied. Here only the control word, status word, tag register, exception pointers are copied to a destination memory location. Refer to figure 9.21 for the memory format. Only the portion from 000H to 000AH offset is copied to memory.

Clocks – 40 — 50

Format – 1 1 0 1 1 0 0 1 mod 110 rm disp

(11) FLDENV Source: This instruction is the exact opposite of the FSTENV destination instruction. It loads the 8087 control register, status register, tag word and exception pointers from a memory location referenced by source

Clocks – 35 — 45

Format – 1 10 1 1 0 0 1 mod 1 0 0 rm disp

(12) FINCSTP: This is the coprocessor increment stack pointer instruction. This instruction increments the 8087 stack pointer by 1. Thus if the stack pointer is 000 (or ST(O) it is increment to 001 (ST(1).

(13) FDECSTP : This is the decrement stack pointer instruction and decrements the stack pointer by 1

Clocks – 6 — 12

Format – 1 1 0 1 1 0 0 1 1 1 1 1 0 1 1 0

(14) FFREE Destination: This is used to change the status of a stack register to that of empty. This instruction works by changing the tag field in the tag register to 11 (empty). Thus for FFREE ST(2). the tap field for ST(2) will be set to 11. Thus for all coprocessor operations the ST(2) will be empty hencewith

Clocks – 9 — 16

Format – 1 1 0 1 1 1 0 1 1 1 0 0 0 reg

(15) FNOP : This is the coprocessor instruction which is equivalent to the 8000: microprocesor NOP instruction. It actually copies the ST register to ST register

Clocks – 10 — 16

Format – 1 1 0 1 1 0 0 1 1 1 0 1 0 0 0 0

(16) FW AIT : This instruction is basically to synchronize the 8086 and 8087. This instruction is the 8086 microprocessor's wait instruction and causes the 8086 microprocessor to enter a wait state until Its TEST pin goes low. Since when the 8087 coprocessor is executing an instruction, its BUSY pin is high. Now, busy pin is connected to the 8086 microprocessor TEST pin thus effectively causing the 8086 microprocessor to wait for the 8087 coprocessor to finish execution. When the 8087 coprocessor execution completes the BUSY pin will go low causing the TEST pin input to become true. Thus now the 8086 microprocessor will restart execution. This instruction must be inserted whenever the we require the 8087 coprocessor execution to be completed before 8086 microprocessor execution.

Clocks – 4

Format – 1 0 0 1 1 0 1 1

Note: All assemblers (MASM and TASM) include a wait instruction (9BH) of the 8086/ 8088 microprocessor before a coprocessor instructions. Thus after we write an assembly program in an assembler we will find that assembler will include the 9BH (or wait instruction) before each coprocessor instruction. This is done so as to synchronize the 8086 and 8087 processors. However some instructions such as the FNINT, FNDISI etc. where there is an N before, the main instruction function have the same function as FINIT and FDISI except that here a NOP will be introduced by the assembler. Thus we can say that an assembler will introduce 9BH before FDISI instruction while for FNDISI it will introduce 48H (NOP).

9.7. AN 8087 ASSEMBLY LANGUAGE PROGRAM

Let us now consider a simple program that requires a 8087 coprocessor. Consider the pythogoras theorem. We can easily recall from our geometry classes that the pythagoras theorem is written as

$$C^2 = A^2 + B^2$$

or simply

$$C = \sqrt{A^2 + B^2}$$

```
                    Page 60, 120
                    ; pythagoras theorem example
                    ; a program to show 8087 insturction
                    .8087
                    ;------------DATA SEGMENT------------
0000  00004041          DATASG SEGMEMENT PARA 'data'
0000  00007041      A DO 12.0       ; side a
0004  0000A040      B DD 5.0        ; side b
0008  ?????????     C DD            ? hypotenuese
000C  0000          CONTROL_WD OW 0
000E  0000          STATUS_WD OW 0
0010                DATASG ENDS
                    ; ------------CODE S EGM ENT------------
```

```
0000                  COOESG SEGMENT PARA 'code'
                      ASSUME CS : COOESG, DS:DATASG
0000  B8----R         BEGIN: MOV AX, DATASG ; load ds with
0003  8E D8           MOV DS, AX                  ; data seg address
0005  9B D8 E3        FINIT                       ; initialize 8087
0008  C7 06 000C  R 03FF MOV CONTROL_WO, 03FFH ; set round control to even
000E  9BD9 2E 000C R FLDCW CONTROL_WO            ; load 8087 control word
0013  9B D9 06 0000R FLD A                 ; load a to st
0018  9B D8 C8        FMUL ST, ST(0)              ; square a
0018  9B D9 06 0004 R FLO 8                ; load b to st
0020  9B D8 C8        FMUL ST, ST(0)              ; square b
0023  9B D9 C1        FADO ST, ST(1)              ;add a² and b²
0026  9B D9 FA        FSQRT                       ; calculate the hypotenuese
0029  9B DD 3E 000E R FSTSW STATUS_WD             ; store status word to memory
002E  A1000E R        MOV AX, STATUS_WD           ; load ax ax status word
0031  24 BF           AND AL, 08FH                ; check to see bits
0033  75 05           JNZ ERROR                   ; jump if error
0035  9B D9 1 E 0008 R FSTP C                     ; copy hypotenuese
003A  90              ERROR: NOP
003B                  CODESG ENDS
                      END BEGIN
Segments and Groups:
Name                  Length      Align       Combine     Class
CODESG .........      003B        PARA        NONE        'CODE'
DATASG ..........     0010        PARA        NONE        'DATA'
Symbols
Name                  Type        Value       Attr
A ............        LDWORD      0000        DATASG
B ............        LD WORD     0004        DATASG
BEGIN .......         L NEAR      0000        CODESG
C ............        LDWORD      0008        DATASG
CONTROL_WD            LWORD       000C        DATASG
ERROR          LINEAR      003A        CODESG
STATUS_WO             LWORD       000E        DATASG
@FILENAME             TEXT PYTHA
          36 Source Lines
          36 Total Lines
```

13 symbols
50938 + 398838 Bytes symbol space free
0 Warning Errors
0 Severe Errors

9.8 ARCHITECTURE OF 8089 MICROPROCESSOR

An 8089 I/O processor is supposed to take care of all the system I/O activities, once initiated by the host CPU, the IID processors receives request from the system peripherals. It tracks the peripherals and it's the operation at issue commands to system. The 8089 may be operated in tightly coupled or loosely coupled configurations. The $\overline{RQ}/\overline{GT}_0$ pin of 8089 may be used to communicate with other 8089 that will be treated as a slave by the former 8089. The internal architecture of 8089 is shown fig.23. The 8089 microprocessor has two Internal I/O channels and both channel share the common ALU. The main control unit derives the control signals required for the operation of the I/O channel. The I/O channels also share the common control unit. The bus Interface and control unit handles all the bus activities, under the control of the control unit.

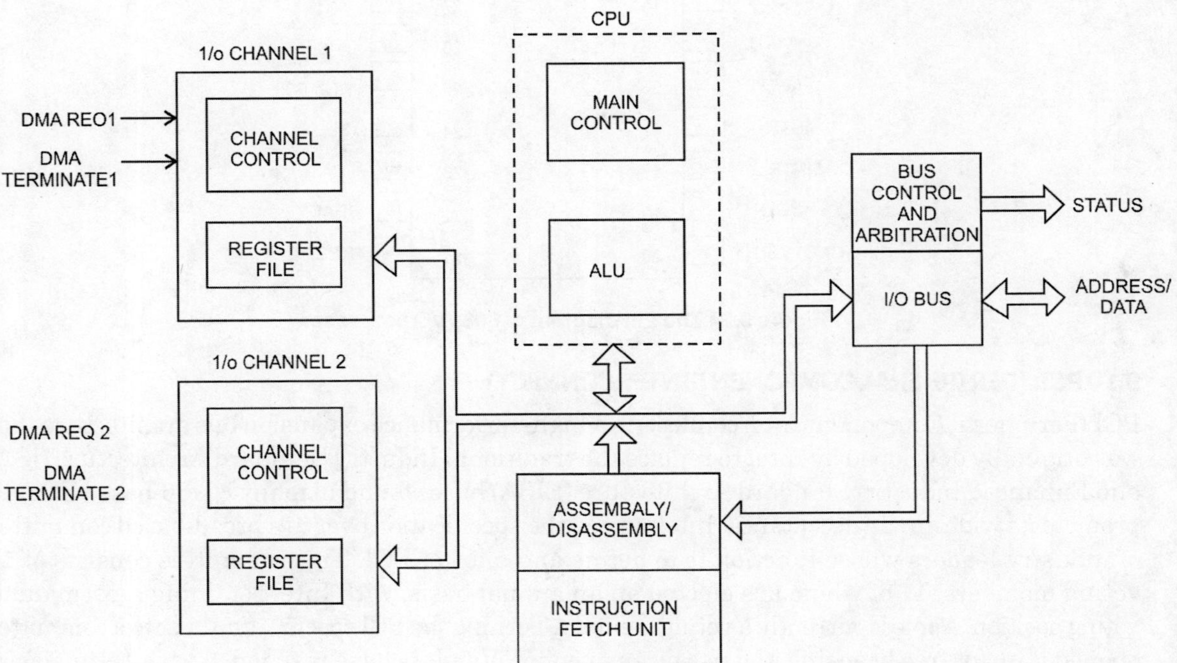

Figure 9.23 Architecture of 8089110 processor

9.9 PIN DIAGRAM OF 8089 MICROPROCESSOR

It used the same programming technique as 8087for input/output operations, such as transfer of data from memory to a peripheral device, and so reducing the load on the CPU. Figure 9.24 shows the pin diagram of the 8089 microprocessor and pin description is given in table 9.1.

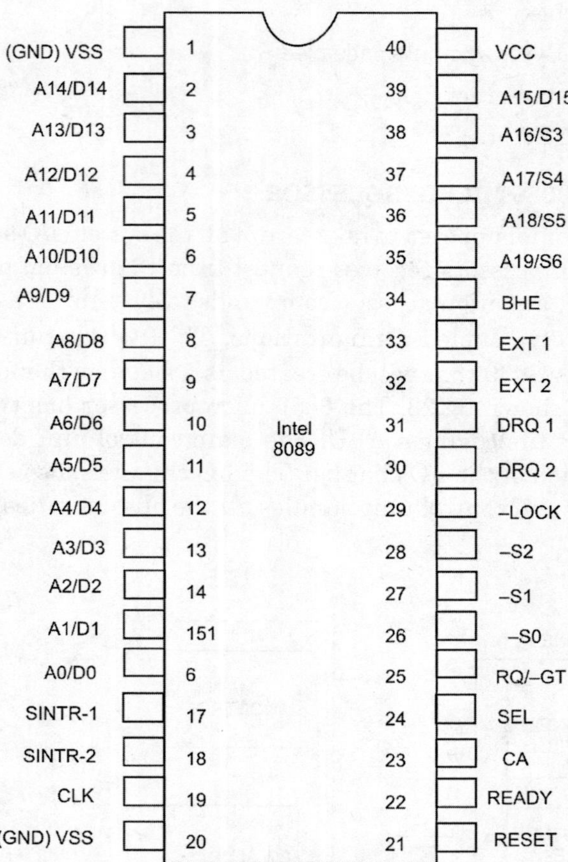

Figure 9.24 The Pin diagram of 8089 I/O processor

9.10 PCI: (PERIPHERAL COMPONENT INTERCONNECT)

PCI (Peripheral Component Interconnect) is a high-performance expansion bus architecture that was originally developed by Intel to replace the traditional Industry Standard Architecture (ISA) and Enhanced Industry Standard Architecture (EISA) buses found in many 80x86-based PCs. To promote its widespread acceptance, Intel turned the specification over to a broad-based committee of industry leaders whose function is to define and market PCI. This committee consists of 12 voting members, 11 of whom are elected on an annual basis, with Intel retaining a permanent voting position. Apple is currently a voting member. Turning the PCI specification over to a committee opened it up for free licensing and use by any vendor. Vendors have responded with enthusiasm. Currently, more than 200 vendors are producing PCI cards, and more than 500 cards are available for this bus architecture.

The majority of personal computers based on 80486 or Pentium processors already incorporate the PCI architecture. The comparison of different buses is given in table below:

Table 1 – PIN DESCRIPTION

Symbol	Type	Name and Function
AO–A15/D0–D15	I/O	**Multiplexed Address and Data Bus:** The functions of these lines are defined by the state of $\overline{SO}$, $\overline{S1}$ and $\overline{S2}$ lines. The pins are floated after reset ad when the bus is not acquired. A8–A15 are stable on transfers to a physical 8-bit date bus (same bus as MBL 8088), and are multiplexed with data on transfers to a 16-bit physical bus.
A16–A19/S3–S6	O	**Address and Status:** Multiplexed most significant address lines and status information. The address lines are active and are wise, the status line are active and are encoded as shown below. The pins are floated after reset and when the bus is not acquired. S6 S5 S4 S3 1 1 0 0 DMA cycle on CH1 1 1 0 1 DMA cycle on CH2 1 1 1 0 Non-DMA cycle on CH1 1 1 1 1 Non-DMA cycle on CH2
$\overline{BHE}$	O	**Bus High Enable:** The Bus High Enable is used to enable data operations on the most significant half of the data bus (DB–D15). The signal is active low when a byte is to be transferred on the upper half of the data bus. The pin is floated after reset and when the bus is not acquired. $\overline{BHE}$ does not have to be latched.
$\overline{SO}$, $\overline{S1}$, $\overline{S2}$	O	**Status:** These are the status pins that define the LOP activity during any given cycle. They are encoded as shown below: $\overline{SO}$ $\overline{S1}$ $\overline{S2}$ 0 0 0 Instruction fetch; I/O space 0 0 1 Data fetch; I/O space 0 1 0 Data store; I/O space 1 0 0 Instruction fetch; System memory 1 0 1 Data fetch; System Memory 1 1 1 Passive The status lines are utilized by the bus controller and bus arbiter to generate all memory and I/O control signals. The signals change during T4 if a new cycle is to be entered while the return to passive state in T3 or T_w indicates the end of a cycle. The pins are floated after system reset and when the bus is not acquired.
Ready	I	**Ready:** The ready signal received from the addressed device indicated that the device is ready for data transfer. The signal is active high and is synchronized by the MBI 8284A clock generator.

$\overline{\text{LOCK}}$	O	**Lock:** The lock output signal indicates to the bus arbiter that the bus is needed for more than one continuous cycle. It is set via the channel control register, and during the TSL instruction. The pin floats after reset and when the bus is not acquired. This output is active low.
RESET	I	**Reset:** The receipt of a reset signal cause the IOP to suspend all its activities and enter an idle state until a channel attention is received. The signal must be active for at least four clock cycles.
CLK	I	**Clock:** Clock provides all timing needed for internal IOP operation.
CA	I	**Channel Attention**: Gets the attention of the IOP. Upon the falling edge of this signal, the SEL input pin is examined to determine Master/Slave or CH1/CH2 information. This input is active high.
SEL	I	**Select:** The first CA received after system reset informs the IOP via the SEL line. Whether it is a Master or Slave (0/1 for Master/Slave respectively) and starts the initialization sequence. During any other CA the SEL line signifies the selection of CH1/CH2. (0/1 respectively.)
DRO1-2	I	**Data Request:** DMA request inputs which signal the IOP that a peripheral is ready to transfer/receive data using channels 1 or 2 respectively. The signals must be held active high until the appropriate fetch/stroke is initiated.
$\overline{\text{RO}}/\overline{\text{GT}}$	I/O	Request Grant Request Grant implements the communication dialogue required to arbitrate the use of the system bus (between IOP and CPU, LOCAL mode) or I/O bus when two IOPs share the same bus (RE-MOTE mode). The $\overline{\text{RO}}/\overline{\text{GT}}$ signal is active low. An internal pull-up permits $\overline{\text{RO}}/\overline{\text{GT}}$ to be left floating if not used.
SINTRI1-2	I	**Signal Interrupt:** Signal interrupt outputs from channels 1 and 2 respectively. The interrupts may be sent directly to the CPU or through the MBL 8259A interrupt controller. They are used to indicate to the system the occurrence of user defined events.
EXT1-2 V_{CC}	I	**External Terminate:** External terminate inputs for channels 1 and 2 respectively. The EXT signals will cause the termination of the current DMA transfer operation if the channel is so programmed by the channel control register. The signal must be held active high until termination is complete. **Voltage:** +5 volt power input.
V_{SS}		**Ground.**

Table: Comparison of different buses with PCI

Bus Type	Bus Width	Bus Speed	MB/sec	Advantages	Disadvantages
ISA	16 bits	8MHz	16 MBps	low cost, compatibility, widely used	Low speed, Jumpers & DIP switches. becominq obsolete
PCI	64 bits	133 MHz	1 GBps	very high speed, Plug & Play, dominant board-level bus	incompatible with older systems, can cost more
Compact PCI	64 bits	33 MHz	132 MBps	designed for industrial use, hot swapping/Plug & Play, ideal for embedded systems	lower speed than PCI, need adapter for PC use, incompatible with older systems

The following table provides a feature-by-feature comparison between the PCI and NuBus specifications:

FEATURE	PCI	NuBus
Bus Speed	33 MHz	10 MHz
Theoretical Maximum Throughput	123 MB per second	40 MB per second
Maximum Number of Slots	4 per bridge; 256 nridges	6 per bridge; 14 bridges
Real Throughput		
writes	90 MB per second writes	32 MB per second
reads	45 MB per second reads	20 MB per second
Software	Name Registry Open Firmware	Slot Manager
Form Factor	12.283 in. and 6.875 in.	12 in. and 7 in.

One point that is especially significant is the change in software that the move to PCI represents. All Apple computers with PCI will have built-in Open Firmware software, which is a standard (IEEE 1275) for booting PCI cards in an OS-independent environment. Using Open Firmware gives Apple flexibility in its choice of future operating systems. Finally, the runtime environment for PCI cards is also being rewritten to allow for growth in the future. The new expansion manager will allow PowerPC and 80 × 86 drivers to coexist in the PCI cards' ROM. This ability to accommodate multiple drivers permits the performance of I/O activities conducted via cards to improve significantly in the future.

9.10.1 PCI address spaces

PCI has three address spaces: memory I/O address, and configuration. Memory addresses are 32 bits (optionally 64 bits) in size, support caching and can be burst transactions. I/O addresses are for compatibility with the Intel × 86 architecture's 110 port address space. Although the PCI bus specification allows burst transactions in any address space, most devices only support it for memory addresses and not I/0. Finally, PCI configuration space provides access to 256 bytes of special configuration registers per PCI device. Each PCI slot gets its own configuration space address range.

The registers are used to configure devices memory and I/0 address ranges they should respond to from transaction initiators. When a computer is first turned on, all PCI devices respond only to their configuration space accesses.

9.10.2 PCI command codes

There are 16 possible 4-bit command codes, and 12 of them are assigned. With the exception of the unique dual address cycle, the least significant bit of the command code indicates whether the following data phases are a read (data sent from target to initiator) or a write (data sent from an initiator to target). PCI targets must examine the command code as well as the address and not respond to address phases which specify an unsupported command code. The commands that refer to cache lines depend on the PCI configuration space cache line size register being set up properly; they may not be used until that has been done.

0000

Interrupt Acknowledge

This is a special form of read cycle implicitly addressed to the interrupt controller, which returns an interrupt vector. The 32-bit address field is ignored. One possible implementation is to generate an interrupt acknowledge cycle on an ISA bus using a PCI/ISA bus bridge. This command is for IBM PC compatibility; if there is no Intel 8259 style interrupt controller on the PCI bus, this cycle need never be used.

0001

Special Cycle

This cycle is a special broadcast write of system events that PCI card may be interested in. The address field of a special cycle is ignored, but it is followed by a data phase containing a payload message. The currently defined messages announce that the processor is stopping for some reason (e.g. to save power). No device ever

9.11 USB (UNIVERSAL SERIAL BUS):

Universal Serial Bus (USB) is an industry standard developed in the mid-1990s that defines the cables, connectors and communications protocols used in a bus for connection, communication and power supply between computers and electronic devices. USB was designed to standardize the connection of computer peripherals, such as keyboards, pointing devices, digital cameras, printers, portable media players, disk drives network adapters to personal computers, both to communicate and to supply electric power. It has become commonplace on other devices, such as smartphones, PDAs and video game consoles. USB has effectively replaced a variety of earlier interfaces, such as serial and parallel ports, as well as separate power chargers for portable devices.

As of 2008, approximately 6 billion USB ports and interfaces were in the global marketplace, and about 2 billion were being sold each year

9.11.1 USB mass storage:

USB implements connections to storage devices using a set of standards called the USB mass storage device class (MSC or UMS). This was at first intended for traditional magnetic and optical drives, but has been extended to support a wide variety of devices, particularly flash drives, because many systems can be controlled with the familiar metaphor of file manipulation within directories. The process of making a novel device look like a familiar device is also known as extension. The ability to boot a write-locked SD card with a USB adapter is particularly advantageous for maintaining the integrity and non-corruptible, pristine state of the booting medium.

Though most post-2005 computers are capable of booting from USB mass storage devices, USB is not intended to be a primary bus for a computer's internal storage: buses such as Parallel ATA (PAT A or IDE), Serial ATA (SATA), or SCSI fulfill that role in PC class computers. However, USB has one important advantage in that it is possible to install and remove devices without rebooting the computer (hot-swapping), making it useful for mobile peripherals, including drives of various kinds. Originally conceived and still used today for optical storage devices (CD-RW drives, DVD drives and so on), several manufacturers offer external portable USB hard disk drives, or empty enclosures for disk drive

9.11.2 Pin diagram of USB 3.0

Connectors are backwards compatible with USB 1.1/2.0. The original 4 pins are still there as given in table below even up to expected transfer rate of 4.7 GB/s.

Pin	Name	Description
1	VBUS	+5VDC
2	D-	Data -
3	D+	Data +
4	GND	Ground
5	USB3_RX	USB 3.0 Data Receive (differential)
6	USB3_RX	USB 3.0 Data Receive (differential)
7	GND	Ground
8	USB3_TX	USB 3.0 Data Transmit (differential)
9	USB3_TX	USB 3.0 Data Transmit (differential)

9.11.3 Technical specs Features:

- True Plug'n'Play.
- Hot plug and unplug
- Low cost
- Easy of use
- 127 physical devices
- Low cost cables and connectors

9.11.4 Bandwidth:

- High speed: 480 Mbps speed (in USB 2.0 and above)
- Full speed: 12 Mbps speed (requires shielded cable)
- Low speed: 1.5 Mbps speed (non-shielded cable)

9.12. KEY POINTS TO REMEMBER

1. The 8087 numeric processor is a processor designed for high speed arithmetic data processing.

2. Coprocessor's are not stand alone processors. The 8087 coprocessor requires the 8086 microprocessor for proper operation and cannot function alone. An 8086 microprocessor must be operated in maximum mode so as to interface to the 8087 coprocessor.
3. The coprocessor can operate on signed numbers, 80 bit BCD numbers and floating point numbers.
4. Internally the coprocessor operates only to 80 bit extended precision numbers. Signed integers, single precision and double precision numbers are converted to 80 bit floating point number automatically by the coprocessor before it operates on them.
5. The 8087 coprocessor internally can be broken into two functional units
   (*i*) The Control Unit
   (*ii*) The Numeric Execution Unit
6. The 8087 coprocessor has three important register which are required to control the 8087 coprocessor and inform the programmer of the state of the 8087 coprocessor. These three registers are
   (*i*) Control Register
   (*ii*) Status Register
   (*iii*) Tag Register
7. The control register of the 8087 coprocessor provides a programmer control over the coprocessor and allows him to enable or disable the interrupting of the 8087 as well as Precision controt rounding control and Infinity control. This register is also used to enable or disable the 8087 coprocessor exceptions.
8. The 8087 coprocessor status register contains the status bits such as condition code bits, Busy bit, and various error bits. It also contains the status bits indicating which of the stack register is presently the stack.
9. The Tag register contains the fields for all the coprocessor stack registers. The Tag bits inform the coprocessor whether the stack register hold valid data or zero data or is empty.
10. The 8087 instruction set contains instructions such as data movement instructions, arithmetic, compare, transcendental, constant and control instructions.

REVIEW QUESTIONS

9.1. Explain briefly the 8087 internal architecture?

9.2. What is the difference between a coprocessor such as the 8087 and a microprocessor such as the 8086?

9.3. What is the use of the 8087 control register ?

9.4. Write a program to calculate the resonant frequency of an LC network?

9.5. Define the state of an 8087 coprocessor after it has been reset?

9.6. Develop a program to calculate the surface area of a cylinder whose side is R and height is *h*.

9.7. What is the use of the 8087 status register ?

9.8. What happens if the Tag register holds a value of FFFFH ? What is the use of the 8087 Tag register?

9.9. What is the difference between the 8087 division and reverse division ? What are the reasons that reverse division instructi ns are present in the 8087 coprocessor instruction set?

9.10. Convert the following decimal numbers to single precision floating point numbers

(*a*) 22.25

(*b*) –73.65

(*c*) –0.675

(*d*) 0.0

(*e*) 226

9.11. How is data stored inside an 8087 coprocessor?

9.12. How does an 8086 microprocessor synchronize with the 8087 coprocessor during program execution?

9.13. How is the rounding control and infinity control modes selected for an 8087 coprocessor? Explain the difference between affine and projective infinity control?

9.14. How are the constants + 1 and L loaded into a coprocessor stack register?

9.15. Which bit controls whether an 8087 can be interrupted or not?

9.16. What are the different exceptions that can occur during an 8087 coprocessor instruction execution?

9.17. What is the use of the coprocessor condition code bits?

9.18. Which instruction can be used to save the state of the 8087 coprocessor before it is reset?

9.19. What is the difference between

(*a*) FSTP and FST

(*b*) FFUJD and FADDP

(*c*) FIST and FST instructions

9.20. Which coprocessor instruction can be used to find the square of a number? What are the condition for using this instruction?

9.21. Which instructions is used to round an extended precision floating point number to an integer?
How is the rounding controlled?

9.22. What are limitations imposed on the 8087 coprocessor transcendental instructions?

9.23. What is the difference between

(*a*) FINIT anbd FNINIT

(*b*) FDISI and FNDISI

9.24. Which instruction perform the following operation

(*a*) store the 8087 control word to memory

(*b*) load the 8087 control word from memory

(*c*) store the 8087 status word in memory

(*d*) enable and disable the 8087 interrupt

(*e*) Used to clear all exception bits in the 8087 status word

9.25. What is the difference between the 8086 and 8087 instruction sets? How are these designed so as to synchronize both to one another.

●●●

Chapter-10

SUPPORT PERIPHERAL DEVICES

INTRODUCTION

Several. programmable devices from the intel family such as the 8255A peripheral interface, the 8254 programmable interval timer, the 8259 A programmable interrupt controller, the 8237 DMA controller, the 8155/8156 I/O and timer, and the 8279 keyboard/Display interface are discussed in this chapter. The various input/output functions are perform by a programmable interface device. This device can be set up to perform specific functions by writing an instruction (or instructions) in its internal register, called the control register. The functions can be changed at any time during execution of the program by writing a new instruction in the control register. These devices are widely used in microprocessor based products because they are flexible versatile and economical.

Since the 8255 A and the 8254 are compatible with any microprocessor, therefore, they are widely used general purpose programmable devices. The 8255 A consists three programmable ports, one of which can be used for bidirectional data transfer. The 8254 timer is similar to the 8155 timer except that it has three 16 bit independent timers with various modes.

In programmable devices, functions are determined through software instructions. Programmable interface devices are multiple I/O devices, but can also perform many other function such as time delays, counting, and interrupts. Generally it consists of many devices on a single chip, inter connected through a common bus.

10.1 THE 8251 PROGRAMMABLE COMMUNICATION INTERFACE

The 8251A programmable communication interface is a universal synchronous asynchronous recover and transmitter compatible with Intel processors. This may be programmed to operate in any of the serial communication modes built into it. This chip converts the parallel data into a serial stream of bits suitable for serial transmission, it is also able to receive a serial stream of bits and convert it into parallel data bytes to be read by microprocessor. The architecture of 8251 programmable communication interface is shown in figure 10.1.

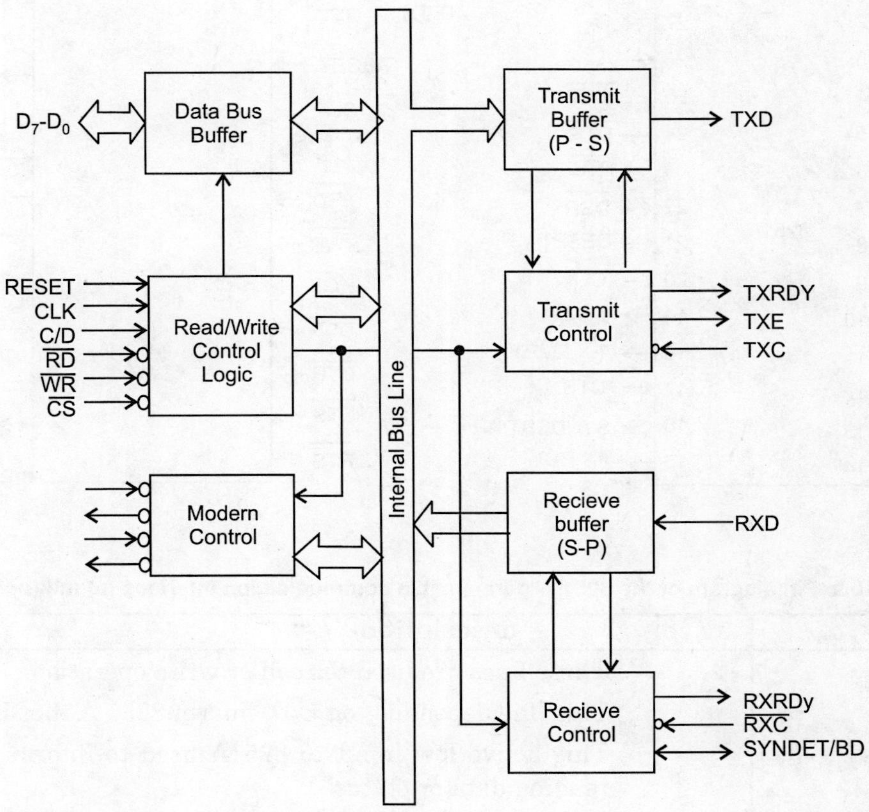

Figure 10.1: Internal Architecture of 8251 programmable communication interface

The data buffer interfaces with the internal bus of the circuit using system bus. The read write control logic controls the operation of the peripheral depending upon the operation initiated by the CPU. This unit also selects one of the two internal addresses those are control address and data address at the request of the C/f5 signal. The modem control unit handles the modem handshake signal to coordinate the communication between the modem and the USART. The transmit control unit transmit the data byte received by the data buffer from the CPU for further serial communication. This also decides the transmission rate which is controlled by the TXC input frequency.

10.1.1. THE PIN DESCRIPTION OF 8251 PROGRAMMABLE COMMUNICATION INTERFACE

Fig 10.2 shows the pin diagram of 8251 programmable communication interface and pin description is given in Table

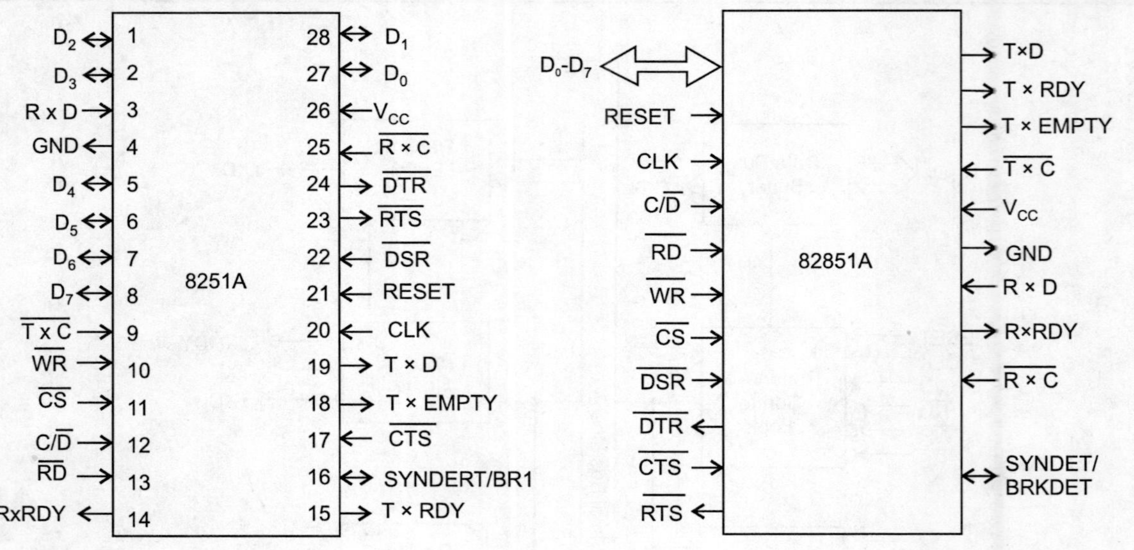

Figure 10.2. Pin diagram of the 8251 programmable communication interface (a) full (b) simplified.

Pin	description
D_0-D_7	These lines are used to read or write operation.
C/D	This line depending on C/D, inform 8251 A that inputs.
$\overline{RD}$	This active low input to 8251A used to inform the CPU is reading data or status information from its internal reqister.
$\overline{WR}$	This active low input to S251A used to inform the GPU is writing data control word to 8251 A.
$\overline{CS}$	This active low input selects the 8251A for read or write operation.
ClK	This input is used to generate internal devices timing and connected to clock generator output.
RESET	This active high input force 8251A to an idle state and remain idle till this signal goes low.
$\overline{T x C}$	This transmitter clock input controls the rate at which the character is to be transmitted. The baud rate is equal to the TxCfrequency in synchronous transmission mode.
T × D	This output pin carries serial stream of the transmitted data bits along with other information like start bit, stop bit

R × RDY	Data on this pin shows that receiver is ready
T × RDY	Data on this pin shows that transmitter is ready
$\overline{DSR}$	low input on this pin shows that data set is ready
$\overline{DTR}$	low input on this pin shows that data terminal is ready
SYNDET/BRKDET	input on this pin detect the synchronous or break detect
$\overline{RTS}$	low input at this line shows request to send data
$\overline{CTS}$	low input at this line clears to send data
TxEMPTY	It shows the transmitter is empty
V_{CC}	5V V_{CC}
GND	OV ground

Table 10.1 : Pin discription of 8251

10.2. THE 8255 A PROGRAMMABLE PERIPHERAL INTERFACE

The 8255 A is a widely used programmable parallel input/output (I/O) device. It can be programmed to transfer data under various conditions, from simple I/0*to interrupt I/O. However, it is flexible, versatile and economical (when multiple I/O ports are required), but some what complex. It is an important general-purpose I/O device that can be used with almost any microprocessor.

The 8255 A has 24 I/O pins. These pins are grouped in three 8-bit ports: A, B and remaining eight bits as port C. Further the eight bits of port C can be subdivided into two 4-bit ports: C_{UPPER} (C_U) and C_{LOWER} (C_L) as shown in figure 10.3. The functions ofthese ports are defined by writing a control word in the control register.

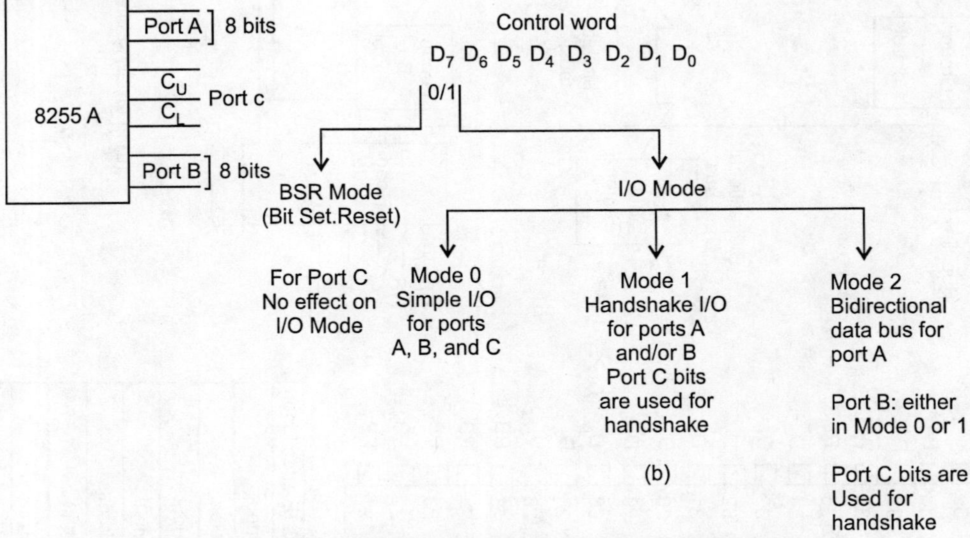

Figure 10.3. 8255A 1/0 Ports (a) and Their Modes(b)

All the functions of the 8255 A are classified according to two modes the Bit Set/Reset (BSR) mode and the I/O mode. The 8255A can enter either one of the two modes depending an the state of D_7 of the 8255A control registers. We use BSR mode to s,et or reset the bits in port C. The I/O mode is further divided into three modes. Mode 0, mode 1, and mode 2. All ports operating in mode 0 are simple I/O ports. Mode 1 is a handshake mode. The handshake' signal

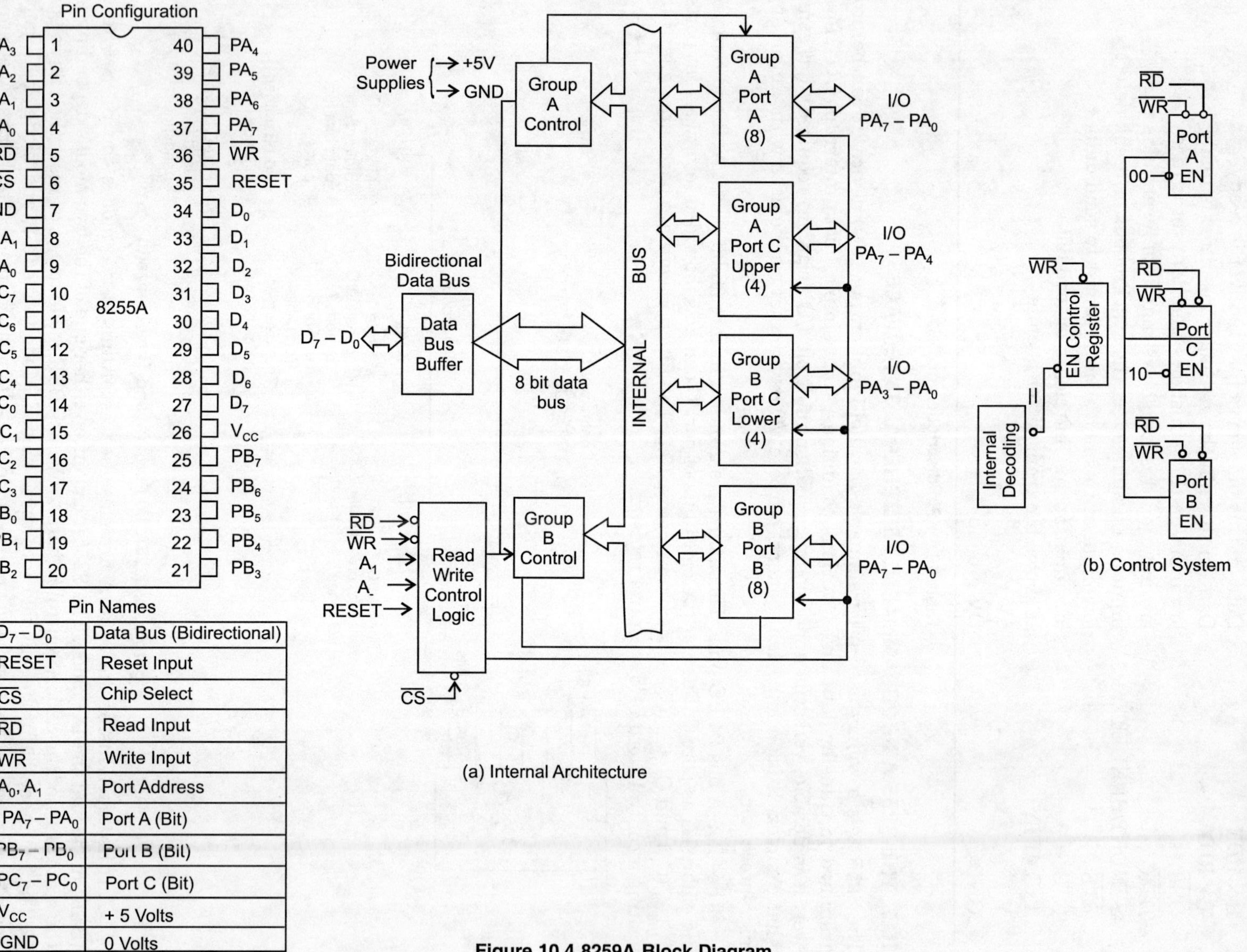

Pin Names

$D_7 - D_0$	Data Bus (Bidirectional)
RESET	Reset Input
$\overline{CS}$	Chip Select
$\overline{RD}$	Read Input
$\overline{WR}$	Write Input
A_0, A_1	Port Address
$PA_7 - PA_0$	Port A (Bit)
$PB_7 - PB_0$	Port B (Bit)
$PC_7 - PC_0$	Port C (Bit)
V_{CC}	+ 5 Volts
GND	0 Volts

(a) Internal Architecture

Figure 10.4 8259A Block Diagram

for port A and/or B is taken from port C. In the hand shaking mode, two types of I/O data transfer can be implemented namely: status check and interrupt. In mode 2, port A can be set up for bidirectional data transfer using handshaking signals from port C, and port B can be set up either in mode 0 or mode 1.

10.2.1. BLOCK DIAGRAM OF THE 8255 A

The block diagram of figure 10.4(a) shows two 8 bit ports (A and B), two 4 bit ports (C_U and C_L) the data bus buffer, and control logic. A simplified but expanded version of the internal structure including control register is shown in figure 10.4(b). This block diagram includes all elements that are required by a programmable device: port C performs functions similar to that of the status register in addition to providing handshake signals.

Control logic : The control section has six lines. We will discussed their functions and connections in following sections.

- **$\overline{RD}$ (Read) :** This control signal enables the Read operations. The 8086 reads data from a selected I/O port of the 8255 A when the signal is low.
- **$\overline{WR}$ (Write) :** This control signal enables the write operation. The 8086 write into a selected I/O port or the control register when the signal goes low.
- **RESET (reset) :** This is an active high signal; it clears the control register and sets all ports in the input mode.

$\overline{CS}$, A_0 AND A_1 : These are known as device select signals. $\overline{CS}$ is connected to a decoded address, and A_0 and A_1 are generally connected to the 8086 address line A_0 and A_1 respectively. The $\overline{CS}$ signal is master clip select, and A_0 and A_1 specify one of the I/O ports or the control register as shown in figure 10.4.

Figure 10.2. The 8255A le pinout, pin assignments and architecture

$\overline{CS}$	A_1	A_0	Selected
0	0	0	PortA
0	0	1	Port B
0	1	0	Port C
0	1	1	control Register
1	X	X	8255 A is not detected

figure 10.5 : truth table for 8255A functions due to CS, A_0 and A_1 pins.

For example the port addresses in figure 10.6(a) are determined by CS. A_0. and A_1 lines. When A_7 = 1 and A_6 through A_2 are at logic 0, the CS line goes low. The port addresses range from 80 H to 83 H when these signals are combined with A_0 and A_1 as shown in figure 10.4(b).

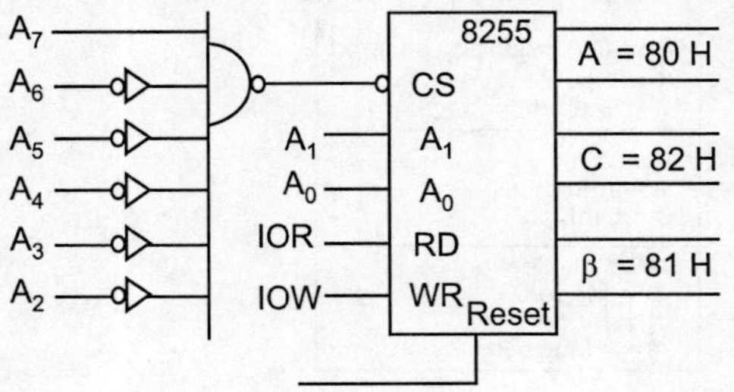

CS								Hex address	Port
A_7	A_6	A_5	A_4	A_3	A_2	A_1	A_0		
1	0	0	0	0	0	0	0	= 801 H	A
						0	1	= 811 H	B
						1	0	= 821 H	C
						1	1	= 831 H	Control Register

(b)

Figure 10.6. (a) 8255 A Chip Select Logic (b) I/O Port Addresses

CONTROL WORD

A register called the control register is shown in figure 10.7 (b). The contents of this register is called the control word, and is used to specify the I/O function for each port. As mentioned previously, this register can be accessed to write a control word when A_0 and A_1 are at logic 1. The register is not accessible for a Read operation.

The I/O function mode or the Bit Set/Reset function mode is specified by the state of bit D_7 of the control register as classified in figure 10.8(b). Bits $D_6 - D_0$ determine I/O functions in varIOUS modes if bit $D_7 = 1$ as shown in figure 10.8. Port C operates in the Bit Set/Reset (BSR) mode. If bit D_7 = O. The BSR control word does not affect the functions of ports A and B.

There are three necessary steps to communicate with peripherals through the 8255 A:

1. Determine the addresses of ports A, B, and C and of the control register according to the chip Select logic and address lines A_0 and A_1
2. Write a control word in the control register.
3. Write I/O instructions to communicate with peripherals ports. A, B, and C.

10.2.2. MODE 0 (SIMPLE INPUT OR OUTPUT)

Port A and B are used as two simple 8 bit I/O ports and port C as two 4-bit ports in this mode. Each port (or half-port, in case of C) can be programmed to function simply as an input port or as an output port. The input/output features in Mode 0 are as follows:

1. Outputs that appear on the ports are latched.
2. Input that appear on the ports are not latched.
3. Port do not have handshake or interrupt capability.

10.2.3. BSR (BIT SET/RESET) MODE

The BSR mode is concerned only with the eight bit port C. The output that shall appear on port C can be set or reset by writing an appropriate control word in the control register. A control word with bit $D_7 = 0$ is recognized as a BSR control word, and it does not alter any

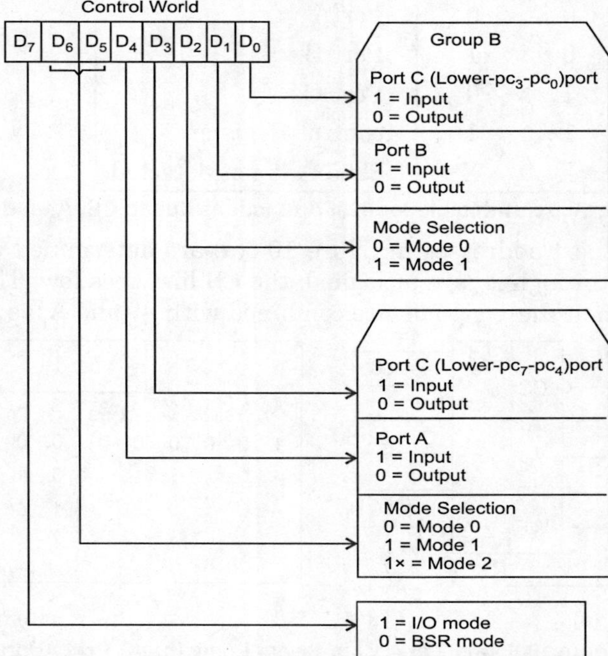

Figure 10.7. 8255A Control Word format for 1/0 mode

previously transmitted control word with bit $D_7 = 1$; hence the I/0 operations of ports A and B are not affected by BSR control word. In the BSR mode, individual bits of port C can be used for applications such as an on/off switch.

BSR CONTROL WORD

Figure 10.8 shows this control word, when written in the control register, sets or resets one bit of the output port at any time thereby allowing control over all high bits that are present in the output port at a time.

MODE 1: Input or Output with Handshake: Handshake signals in mode 1 are exchanged between the 8086 and peripherals prior to data transfer. The features that are available in this mode include the following:

1. Two port (A and B) function as 8-bit I/O ports. They can be configured either as input or output ports.
2. Each port uses three lines from port C as handshake signals. The remaining two lines of port C can be used for simple I/O function.
3. Input and output data are latched.
4. Interrupt logic is supported.

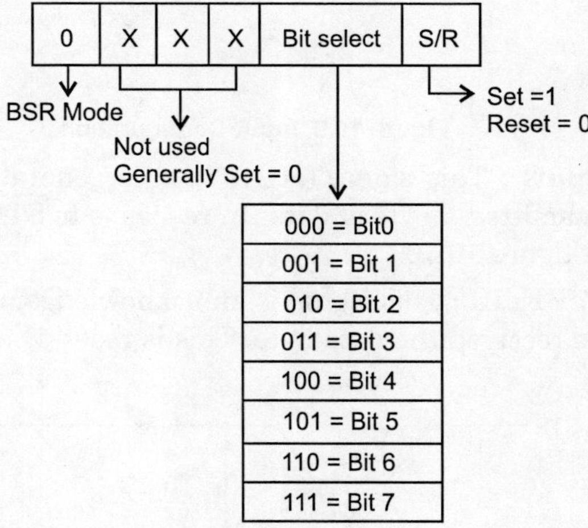

Figure 10.8. 8255A Control Word Format in the BSR mode.

MODE 1 : Input control signals: The associated control signals used for hand shaking when port A and B are configured as input ports shown in figure 10.9. Port A uses the upper three signals: PC_3, PC_4, and PC_5, port B uses the lower three signals: PC_2, PC_1 and PC_0 The functions of these signals one as follows:

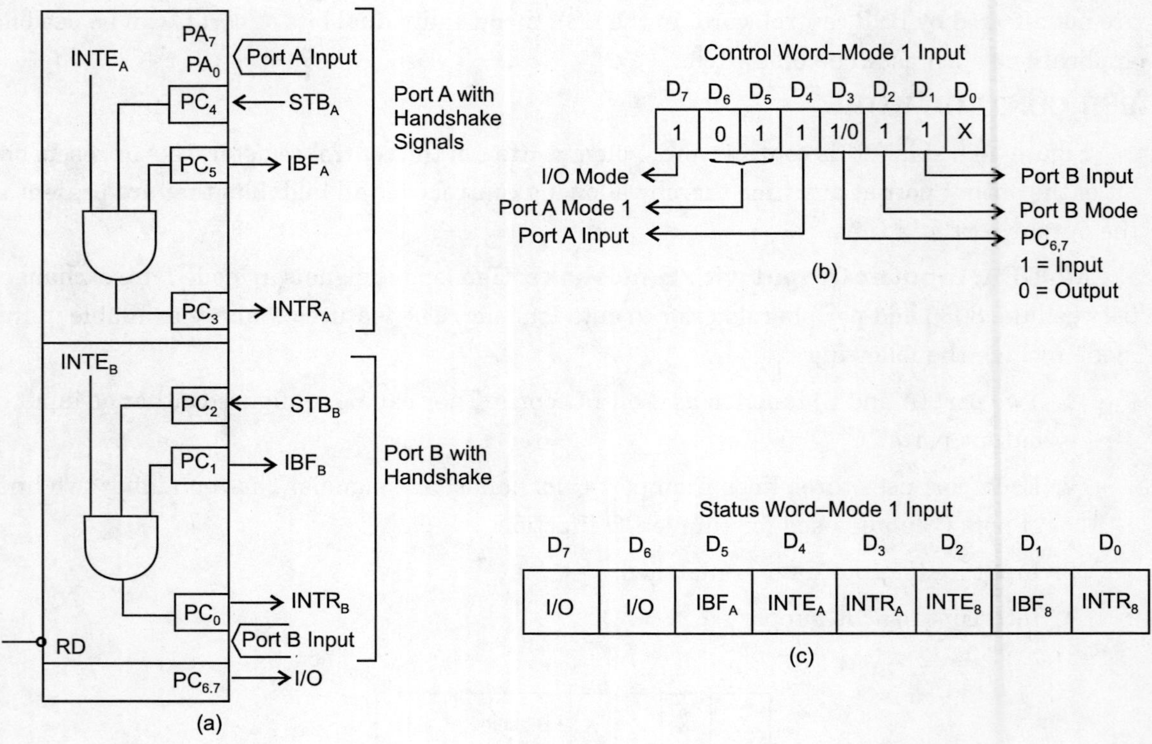

Figure 10.9. Input Configuration.

• **$\overline{\text{STB}}$ (Strobe Input) :** This signal (active low) is generated by a peripheral device to indicate that it has transmitted a byte of data. In response to $\overline{\text{STB}}$, the 8255 A generates 1 BF and INTR, as shown in figure 10.10.

• **IBF (Input Buffer Full) :** This signal is an acknowledgement by the 8255 A to indicate that the input latch has received the data byte. This is reset when the MPU reads the data:

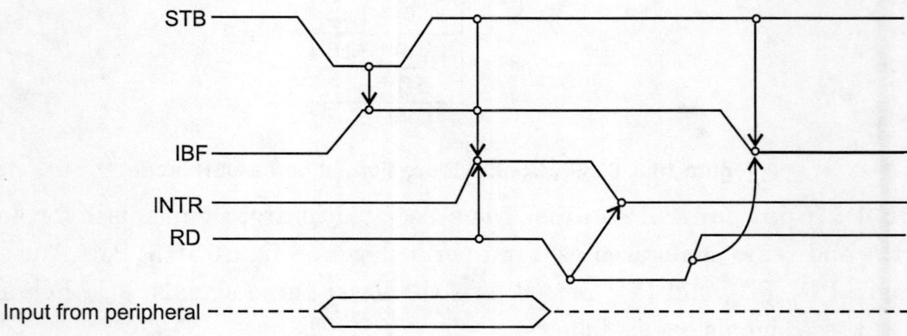

Figure 10.10. 8255A Mode 1: Timing waveforms for strobed Input (Handshake Mode).

• **INTR :** This is an output signal that may be used to interrupt the 8086. If $\overline{STB}$, IBF. and INTE (internal flip flop) are all at logic 1 then this signal is generated. This is reset by the falling edge of the $\overline{RD}$ signal.

• **INTE (Interrupt Enable):** This is an internal flip-flop used to enable or disable the generation of the INTR signal. The two flip-flops. $INTE_A$ and $INTE_B$ are set/reset using the BSR mode. (The $INTE_A$, is enabled or disabled through PC_4 and $INTE_B$ is enabled or disabled through PC_2).

CONTROL AND STATUS WORDS

Figure 10.7(b) uses control words derived from figure 10.4 to set up port A and port B as input port in Mode 1. Figure 10.7 (c) also shows the status word; which will be placed in the accumulator if port C is read.

PROGRAMMING THE 8255 A IN MODE 1

The 8255 A can be programmed to function using either status check I/O or interrupt I/O. (A flowchart for the status check I/O is shown in figure 10.11(a). In this flow chart. the 8086 continues to check data status through the IBF line until it goes high. This is a simplified flow chart; however. it does not show how to handle data transfer if two ports are being used. The technique is similar to that of mode 0 combined with the BSR mode. Since the 8086 is tied up in the loop, thus this is the disadvantage of the status check I/O with hand- shake.

Consider that vectored interrupts are available then the flow chart in figure 10.11 (b) shows the steps required for the interrupt I/O. Set INTE either for port A or port B is the confusing step in the interrupt I/O. Figure 10.9 (a) shows that the $\overline{STB}$ signal is connected to pin PC_4 and the $INTE_A$ is also controlled by the Pin PC_4 However; the $INTE_A$ is set or reset in the BSR mode and the BSR control word has no effect when port A and B are set in Mode 1.

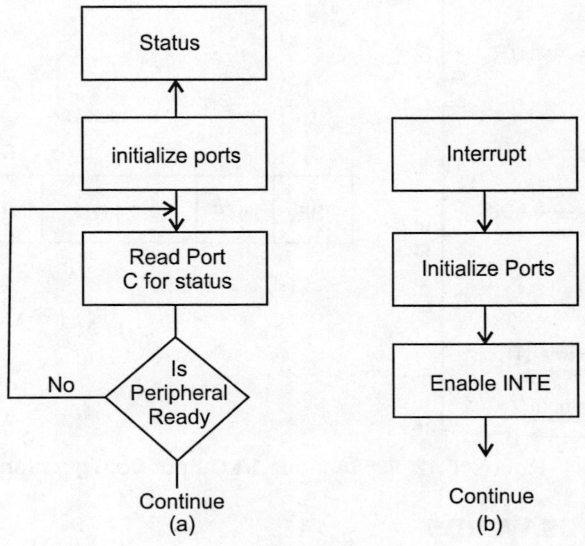

Figure. 10.11. Flow chart; (a) status check I/O (b)

MODE 1 : OUTPUT CO~TROL SIGNAL

The control signals when port A and B are configured as output ports are shown in figure 10.12. These signals are defined as follows:

• **$\overline{OBF}$ (Output Buffer Full) :** This is an output signal that goes low when the 8086 write data into the output latch of the 8255 A. The signal indicates to an output peripheral that new data is ready to be read (10-11). After the 8255 A receives an $\overline{ACK}$ from the peripheral, it goes high.

• **$\overline{ACK}$(Acknowledge) :** This is an input signal from a peripheral that must output a low signal when the peripheral receives the data from the 8255 A ports (figure 10.13).

• **INTR (Interrupt Request) :** This is an output signal, and it is set by the rising edge of the $\overline{ACK}$ signal. This signal can be used to interrupt the 8086 to request the next data byte for output. The INTR is set when $\overline{OBF}$, $\overline{ACK}$ and INTE are all one (figure 10.13). and reset by the falling edge of $\overline{WR}$.

• **$\overline{INT}$ (Interrupt Enable) :** This is an internal flip-flop and needs to be set to so as allow the 8255 to generate the INTR signal. PC_6 and PC_2 bit control the two flip-flop INTEA and $INTE_B$ through the BSR mode.

• **$PC_{4,5}$:** These two lines can be set up either as input or output.

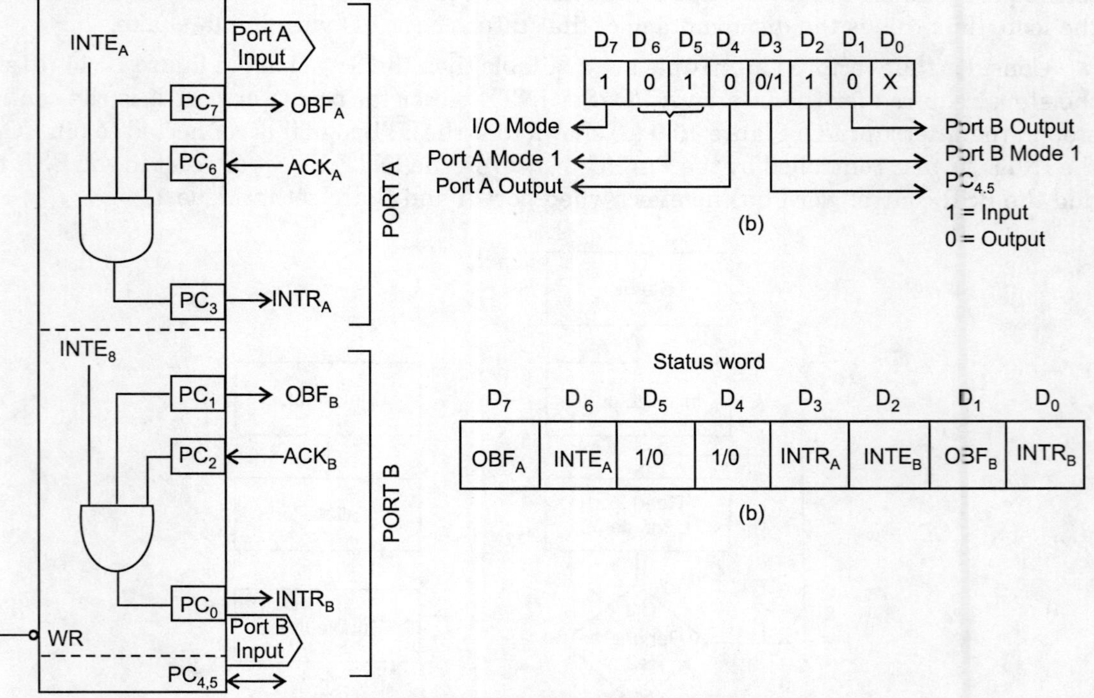

Figure 10.12. 8255A Mode 1 : Output Configuration

CONTROL AND STATUS WORDS

The control word is used to set up ports A and B as an output port in Mode 1 is shown in figure 10.12 (b). Similarly, the status words which will be placed in the accumulator if port C is read shown in figure 10.12(c).

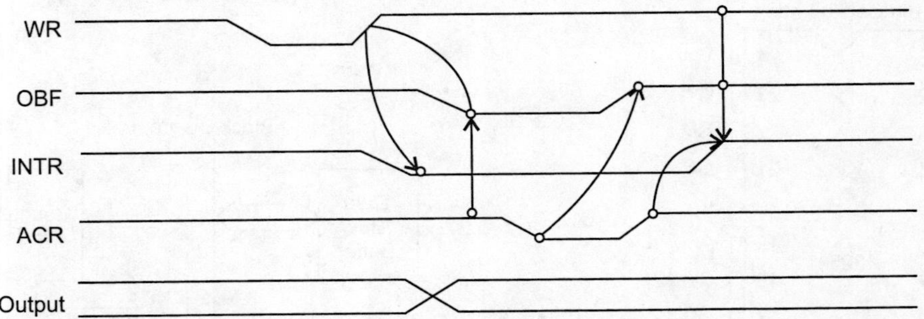

Figure 10.13. 8255 Mode 1 : Timing Waveforms for Strobed (with Handshake) Output

10.3. THE 8254 (8253) PROGRAMMABLE INTERVAL TIMER

The 8254 programmable interval timer/counter is functionally similar to the software-designed counters and timers which can be formed on the 8086. It generates accurate time delays and can be used for applications such as a real time clock, an event counter, a digital one shot, a square wave generator and a complex waveform generator.

The 8254 consists three identical 16 bit counters. The 8254 can operate independently in anyone of the six modes. It is packaged in a 24-pin DIP and requires a single + 5V power supply. A 16 bit count is loaded in the register to operate a counter and on command, begins to decrement the count until it reaches O. It generates a pulse at the end of the count that can be used to interrupt the 8086. The counter can count either in binary or BCD. In addition, a count can be read by the 8086 while the counter is decrementing.

The 8254 is an upgraded version of the 8253, and they are pin-compatible. The both device have almost the identical features except that:

- the 8254 can operate with higher clock frequency range (DC to 8 MHz and 10 MHz for 8254 - 2,) and the 8253 can operate with clock frequency from DC to 2MHz.
- the 8254 contains a status Read-Back command that can latch the count and the status registers of the counters.

10.3.1 : BLOCK DIAGRAM OF THE 8254

The 8254 consists of three counters (0,1 and 2), a data bus buffer, Read/write control logic, and a control register as shown in figure 10.12. Each counter has two input signals: clock (CLK) and GATE and one output signal: OUT.

DATA BUS BUFFER: This tri-state, 8-Bit bidirectional buffer is connected to the data bus of the 8086.

CONTROL LOGIC: The control section consists five signals. $\overline{RD}$ (Read), $\overline{WR}$ (write), $\overline{CS}$(Chip Select), and the address lines A_0 and A_1. The $\overline{RD}$ and $\overline{WR}$ signal in the peripheral I/O mode are connected to $\overline{IOR}$ and $\overline{IOW}$, respectively. While in memory mapped I/O, these are connected to $\overline{MEMR}$ (Memory Read) and $\overline{MEMW}$ (Memory Write). Address lines A_0 and A_1 of the 8086 are usually connected to lines A_0 and A_1 of the 8254 and $\overline{CS}$ is connected to a decoded address.

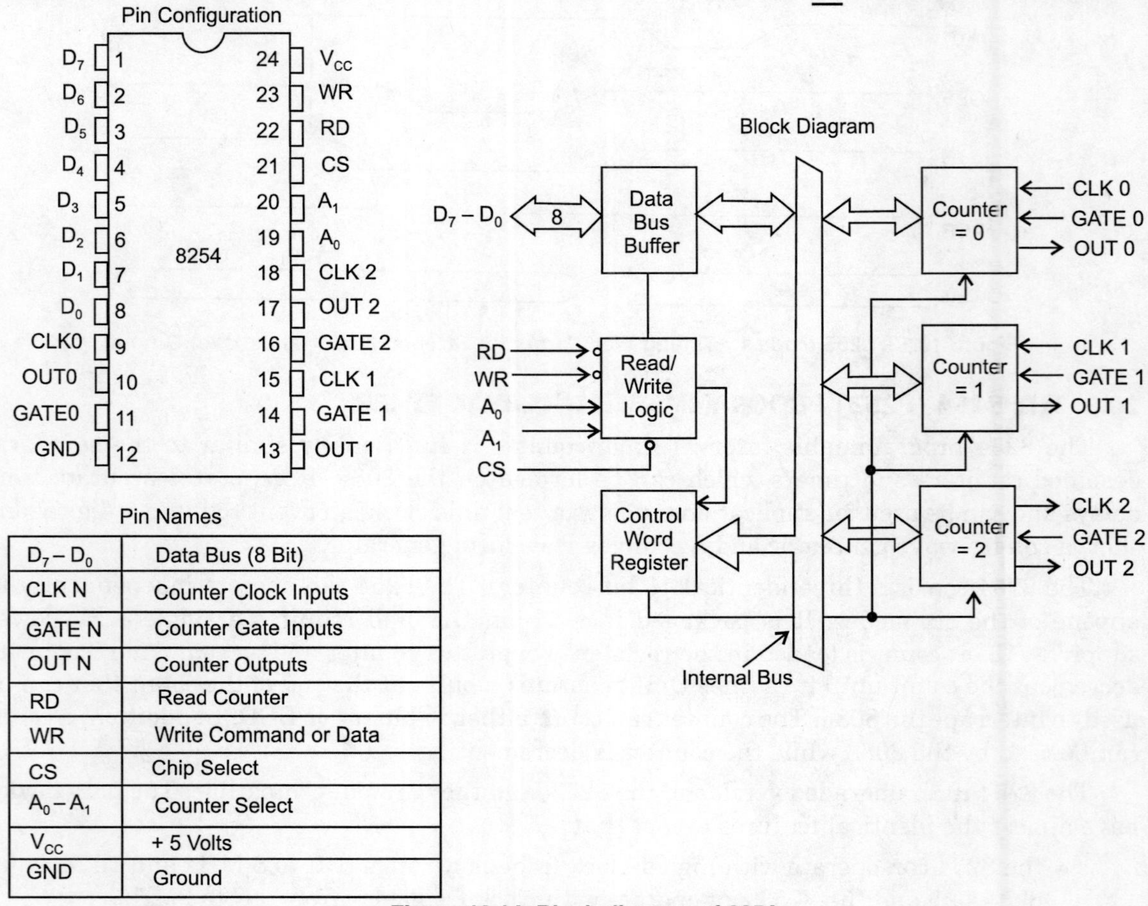

$D_7 - D_0$	Data Bus (8 Bit)
CLK N	Counter Clock Inputs
GATE N	Counter Gate Inputs
OUT N	Counter Outputs
RD	Read Counter
WR	Write Command or Data
CS	Chip Select
$A_0 - A_1$	Counter Select
V_{CC}	+ 5 Volts
GND	Ground

Figure 10.14. Block diagram of 8254

The control word register and counters are selected according to the signals an lines Ao and A_1, as follows:

A_1	A_0	Selection
0	0	Counter 0
0	1	Counter 1
1	0	Counter 2
1	1	Counter Register.

CONTROL WORD REGISTER: When lines A_0 and A_1 are at logic 1, then the control register is accessed. It is used to write a command word which specifies the counter to be used; its mode. and whether the current operation is a Read or a write operation. Figure 10.13 shows the control word format.

MODE: There are six different modes in which the 8254 can operate. The gate of a counter is used either to disable or enable counting as shown in figure 10.15. However, to maintain clarity, only one mode (Mode 0) is discussed first and the remaining modes are illustrated later on.

In mode 0, after the count is written and if the gate is high the count is decremented every clock cycle. The output goes high and remains high until a new count or control word is loaded when the count reaches zero.

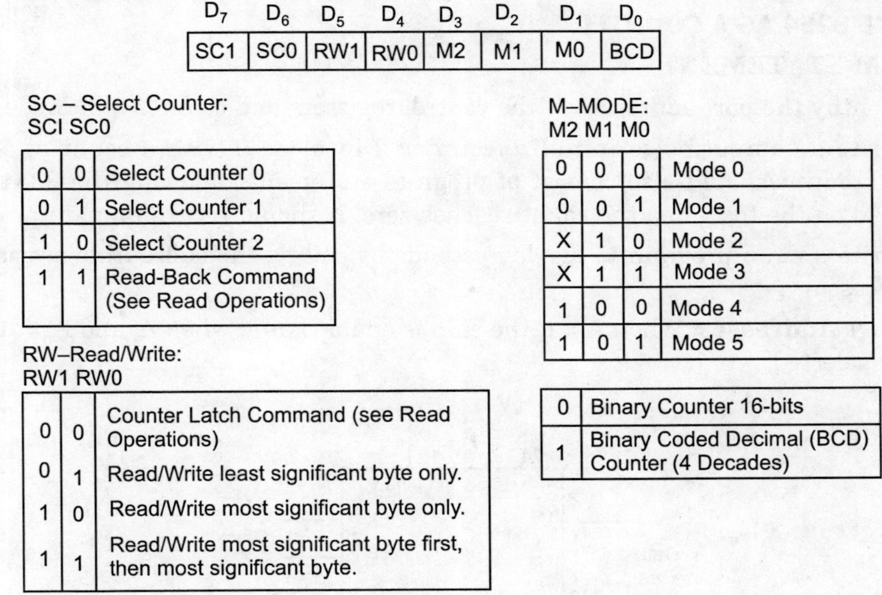

Figure 10.15. Gate settings of a Counter

10.3.2. PROGRAMMING THE 8254

The 8254 can be programmed to provide various types of output modes. The 8254 can also be programmed to check a count while counting by Read operations. The details of these operations are given below.

Write Operations: The following steps are necessary to initialize a counter.

1. Write a control word into the control register.
2. Load the low-order byte of a count in the counter register.
3. Load the high order byte of a count in the counter register.

The above steps should start the counter and provide appropriate output according to the control word, proper clock and an appropriate gate signal to one of the counters.

READ OPERATIONS: In event counters, it is necessary to read the value of the counter whose execution in progress. This can be done by any of following two methods. One method involves reading a count after inhibiting (stopping) the counter which is to be read. The second method involves reading a count while the count is in progress (Known as reading on the fly).

In first method, controlling the gate input or the clock input of the selected (or inhibited) counter and two I/O read operations required to be performed by the 8086. The first I/O operation reads the low-order byte, and the second I/O operation reads the high order byte.

In the second method, to latch a count in the output latch an appropriate control word is written into the control register and two I/O read operations are performed by the 8086. These Read/Write operations are shown below.

10.3.3 THE 8254 AS A COUNTER

PROBLEM STATEMENT

1. Identify the port addresses of the control registers and counter 2 in Mode 0.
2. Write a subroutine to initialize counter 2 in mode 0 with a count of $50{,}000_{10}$. The subroutine/should also consist of program statements allowing the 8254 count to be read on the fly; when the count reaches zero, it should return to main prɔgram.
3. Write a main program to display seconds by calling the subroutine as many times as necessary.

1. Port Addresses: When A_1=1, the chip is enabled, and when A_1 and A_0 = 1, the control register is selected.

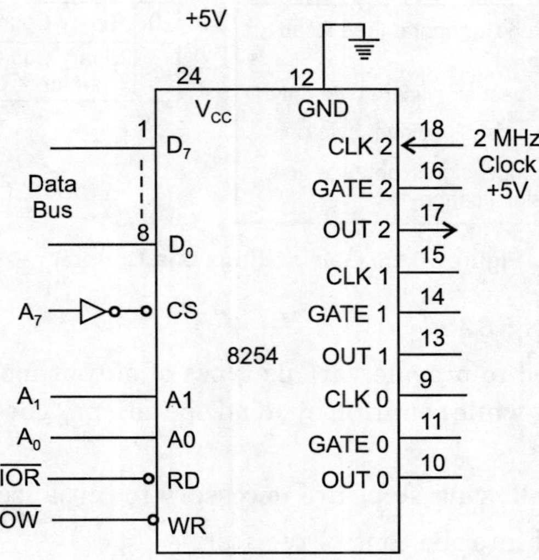

Figure 10.16. Schematic: interfacing the 8254.

In the same way, when A_1 = 1 and A_0 = 0 the counter 2 is selected. Consider that the un used address line A_6 to A_2 are at logic 0, the port address will be as follows:

Control register = 83 H

counter 2 = 82 H

2. Subroutine Counter: The following control word is necessary to initialize the 8254 for counter 2 in Mode 0.

Control word (Load operation)

D_7	D_6	D_4	D_5	D_3	D_2	D_1	D_0	
1	0	1	1	0	0	0	0	= BOH
Select Counter 2		Load 16-bit count		Mode 0			Count in binary	

Control Word for Latching: Bets D_5 and D_4 should be = 80 H subroutine.

COUNTER: MOV AL, BOH ; Control word to initialize counter 2

```
                OUT 83 H   ;  Write to control register
          MOV AL, LO BYTE  ;  Lows-order byte of count 50000
                OUT 82 H   ;  load counter 2
          MOV AL, HI BYTE  ;  higher order byte of count 5000
                OUT 82 H   ;  load counter 2
      Read : MOV AL, 80 H  ;  Control word to latch a count
                OUT 83 H   ;  Write in the control register
                 IN 82 H   ;  Read high-order byte
              MOV DL, AL   ;  Store low-orde byte in register D
                 IN 82 H   ;  Read high-order byte
               OR AL, DL   ;  Or low and high order bytes to set Z flag
              JNZ = READ   ;  If counter = 0 go back to read next count
                     RET
```

3. Main Program: The subroutine COUNTER provides 25 ms (50000 × 0.5 μs clock) delay; if this routine is called 40 times, the total delay will be one second.

Program

```
               MOVSP, STACK  ;  Initialize stack pointer
                MOV BL,OOH   ;  Clear register B to save number of seconds
SECOND:        MOV CL, 28H   ;  Set up register C to counter 40 (base 10)
WAIT:         CALL COUNTER   ;  Wait for 25 ms
                 DEC CX, 1
                  JNZ WAIT   ;  is this one second if not go back and wait
                 XORAL, AL
                ADDAL,01     ;  Add one second
                       DAA
                OUT PORT 1
                 MOVDX,AX    ;  Save seconds
                JMP SECOND   ;  Go back and start counting the next second
```

10.3.4. MODES

Now, we will describe briefly various modes of the 8254.

MODE 0 : INTERRUPT ON TERMINAL COUNT

Initially the OUT pin is low in this mode. The counter is decremented after every cycle after a count has been loaded in the register. The OUT goes high when the count reach zero. This can be used as an interrupt. The OUT remains high until a new count or a command word is loaded. The counting (m = 5) is temporarily stopped when the gate is disabled (G = 0), and continued again when the gate is at logic 1 as shown in figure 10.17.

MODE 1 : HARDWARE-RETRICCERABLE ONE-SHOT

In this mode, the OUT is initially high. The OUT goes low when the gate is triggered and the OUT goes high again at the end of the count, therefore generating a one-shot pulse as shown in figure 10.17.

MODE 2: RATE GENERATOR

This mode is used to generate a pulse equal to the system clock period at a given interval. The OUT stays high until the count reaches 1. When a count is loaded, and then the OUT goes low for one clock period. Since the count is reloaded automatically, the pulse is generated continuously. In this mode, the count = 1 is illegal.

MODE 3 : SQUARE-WAVE GENERATOR

In this mode, the OUT is high when a count is loaded. The count is decremented by two at every clock cycle and when it reaches zero, the OUT goes low, and the count is reloaded again. Since this is repeated continuously, therefore, a continuous square wave with period equal to the period of the count is generate. Or we can say that the frequency of the square wave is equal to the frequency of the clock divided by the count. This means that if the count (N) is odu, the pulse, stay high for (N + 1) /2 clock cycles and stays low for $\left(\frac{N-1}{2}\right)$ clock cycles.

Mode 0: Interupt on Terminal Count

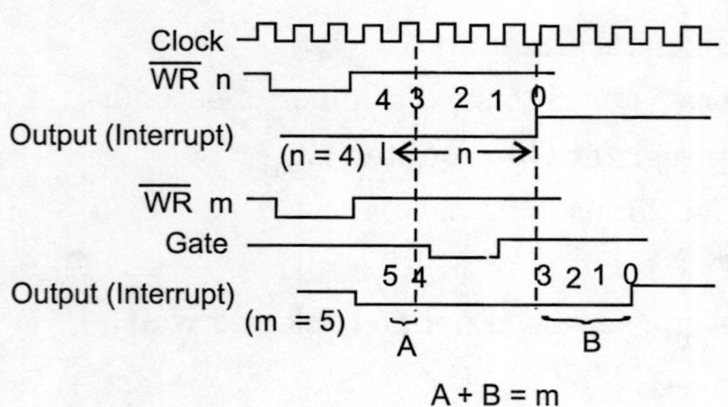

Mode 1: Programmable One-Shot

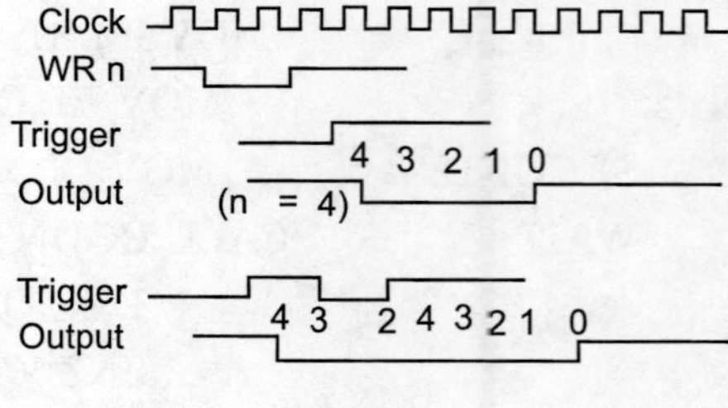

Mode 2: Rate Generator Clock

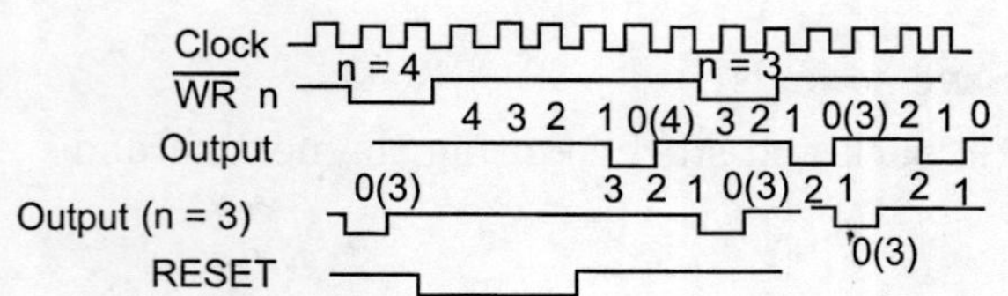

Mode 3: Square Wave Generator

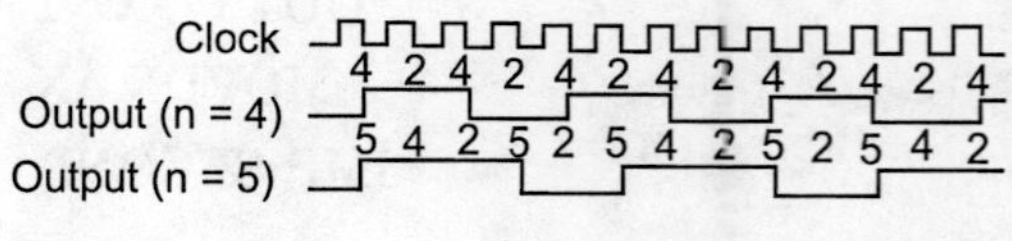

Mode 4: Software Triggered Strobe

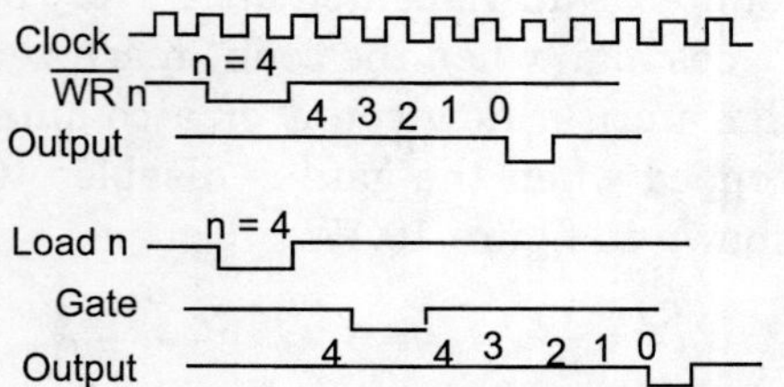

Mode 5: Hardware Triggered Strobe

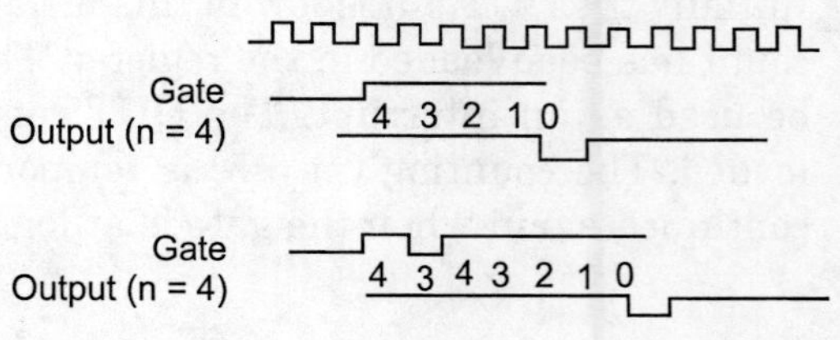

Figure 10.17. Six Modes of the 8254

MODE 4. SOFTWARE-TRIGGERED STROBE

In this mode, the OUT is initially high; it goes low for one clock period at the, end of the count. The count must be reloaded for subsequent outputs.

MODE 5 : HARDWARE-TRIGGERED STROBE

This is similar to Mode 4 except that it is triggered by the rising pulse at the gate. Initially, the OUT is low and when the Gate pulse is triggered from low to high, the count begins. The OUT goes low for one clock period at the end of the count.

READ-BACK COMMAND

The 8254 allows the Read-Back Command to read the count and the status of the counter; this command is not available in the 8253. Figure 10.17(a) shows the format of the command.

The command is written in the control register and if $\overline{\text{COUNT}}$ (bit D5) is 0, the count of the specified counter (s) can be latched. A counter or a combination of counters is specified by keeping the respective CNT bits (D_1, D_2 and D_3) high. For example, the control word 11010110 (D6H) when written in the register will latch the counts of counter 0 and counter 1, and these counts can be obtained by reading respective counter port addresses. The latched count are held until they are read or the counters are reprogrammed. The Read-Back command eliminates the need of writing separate counter-latch command for different counters.

If $\overline{\text{STATUS}}$ bit (D4) of the Read-Back command is low than the status of the counter (s) can be read. The format of the status byte is shown in figure 10.16(b).

(a) Read-Back Command Format

D_7	D_6	D_5	D_4	D_3	D_2	D_1	D_0
1	1	$\overline{\text{COUNT}}$	$\overline{\text{STATUS}}$	CNT 2	CNT 1	CNT 0	0

D_5 : 0 = Latch Count of Selected Counter (s)
D_4 : 0 = Latch Status of Selected Counter (s)
D_3 : 1 = Select Counter 2
D_2 : 1 = Select Counter 1
D_1 : 1 = Select Counter 0
D_0 : Reserved for Future Expansion; Must Be 0:

A_0, A_1 = 11
$\overline{\text{CS}}$ = 0
$\overline{\text{RD}}$ = 1
$\overline{\text{WR}}$ = 0

(b) Status Byte

D_7	D_6	D_5	D_4	D_3	D_2	D_1	D_0
OUT PUT	NULL COUNT	RW1	RW0	M2	M1	M0	BCD

D7 : 1 = Out Pin is 1
: 0 = Out Pin is 0
D6 : 1 = Null Count
: 0 = Count Available for Reading
D5 – D0 : Counter Programmed Mode

Figure 10.18. Read-Back Command Format(a) and Status Byte (b)

Example 10.3. *Write instructions to generate a pulse every 25 μs from counter 0.*

Solution: It should be initialized in Mode 2 so as t.o generate a pulse every 25 μs. from counter 0, and Gate 0 should also be held high.

D_7	D_6	D_4	D_5	D_3	D_2	D_1	D_0	
0	0	0	1	0	1	0	0	= 14H
Select Counter 0		Load 8-bit count		Mode 2			Count in binary	

Figure 10.19. Read command format

Count: In mode 2, the count is decremented every clock period, and at the last count, the counter generates a pulse equivalent to the clock period of the timer. Here the clock frequency of the 8254 is 2MHz (0.5 μs clock period) and the pulse should be generated every 25μs. Hence the count is

$$\text{Count} = \frac{25\times10^{-6}}{0.5\times10^{-6}} = 50 = 32\text{H}$$

This can also be calculated by dividing the clock frequency by the frequency cf the pulse (2 MHz / 40 KHz = 50) as frequency of the pulse is 40 KHz (1/25 μs).

Instruction:

```
PULSE: MOV AL, 14H   ; Control word Mode 2 and counter O.
       OUT 83 H      ; Write in 8254 control register.
       MOVAL, 32 H   ; Low-order byte of the count
       OUT80H        ; Load counter 0 with low-order byte
       HLT
```

Example 10.4. *Write instructions to generate a 2 KHz square events from counter 1 .*

Solution: To generate a square wave from counter 1, it should be initialized in Mode 3.

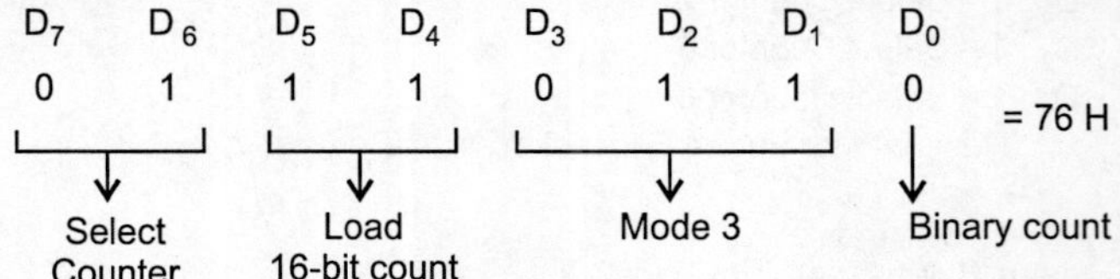

Figure 10.20.

Count: In Mode 3, the count is decremented by two for every clock period. 4N/2 clock pulse provides the upper half of the square wave if the count is N. The count is loaded again and N/ 2 clock pulses again provide the lower half.

In this example, the clock frequency of the 8254 is 2MHz (0.5 J..IS clock period) and square wave frequency is 2KHz (0.5 ms clock period). Thus

$$\text{Count} = \frac{0.5\times10^{-3}}{0.5\times10^{-6}} = 1000 = 03\text{F8H}$$

Instruction

```
SQWAVE : MOV AL, 764   ; Control word Mode 3 and counter 1
```

```
          OUT 83 H   ;  Write m 8254 control register
        MOV AL, F8H  ;  Low order byte of the count
          OUT 81 H   ;  Load counter/with low order byte
        MOV AL, 03H  ;  High order byte of the count
          OUT 81 H   ;  Load counter with high order byte
              HLT
```

Note: the gate of that counter must be tied high to run counter 1; otherwise the counter action is inhibited.

Example 10.5. *Write a subroutine to generate an interrupt every 2 sec. See figure 10.26 for counter addresses.*

Solution: Since the clock frequency shown in figure 10.13 and 10.16 is 2 MHz, therefore; a count is decremented every 0.5 μs. The count should be (2 sec/0.5 × 10^{-6}) 4 × 10^6. This count is too large for one 16 bit counter. We can divide this count, as an example, 50,000 for counter 1 and 80 for second counter (50k × 80 = 4 Meg). Ifwe set up counter 1 in Mode 2 with 50,000 as a count, it will generate a pulse every 25 millisec. The output of counter 1 can be used as a clock input to counter 2. This means that the count in counter 2 will be decremented every 25 millisec.If counter 2 is also set up in Mode 2 with the count = 80, the output pulse from counter 2 is generated every second that can be used to interrupt the processor.

Control word :

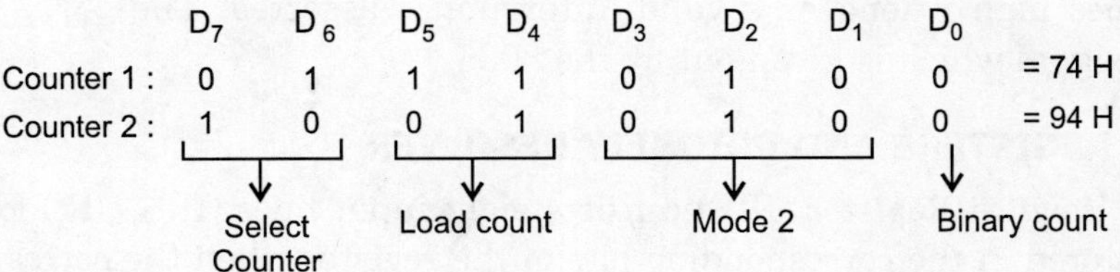

Instructions: The following subroutine is an initialization for 8254 timer. It uses two counters to generate an interrupt everyone second.

```
CNTIL0      EQU         50H
CNTIH 1     EQU         C3H
COUNT2      EQU         80
SECOND :
MOV    AL, 74H      ;    Control word.Mode 2, counter
OUT    83H          ;    Write in 8254 control register
MVI    AL, 96 H     ;    control word: mode 2, counter 2
OUT    83 H         ;    Write in 8254 control register
MVI    AL, CNTILO   ;    Low-order byte of count 50,000,
OUT    81 H         ;    Load counter 1 with low order byte
MVI    AL, CNTILO   ;    High-order byte of count 50,000
```

```
OUT    81 H         ;    Load counter 1 with high-order byte
MVI    AL, COUNT 2  ;    Count for counter 2
OUT    82 H         ;    Load counter 2.
RET                 ;
```

10.4. THE 8259 A PROGRAMMABLE INTERRUPT CONTROLLER

The 8259 is programmable interrupt controller designed to work with the intel microprocessor 8085,8086 and 8088. The 8259 A is upward compatible with its predecessor, the 8259. The 8259 A can be used only with intel's 8086/88 16 bit microprocessor, this is the main difference between two. It consists of additional features such as the level-triggered mode, buffered mode and automatic-end-of-interrupt mode.

Block Diagram of the 8259 A : The internal block diagram of 8259 A is shown in figure 10.22. The 8259 has eight blocks performing different functions such as: Control logic, Read Write logic, Data bus buffer, three registers (IRR, ISR and IMR.), priority resolver, and the cascade buffer. All the elements of a programmable device, plus additional blocks are Shown in this diagram.

READ/WRITE LOGIC: This is a typical Read/Write control logic. The controller is selected to write a command or read a status word when the address line Ao is at logic O. The chip select logic and A_0 determine the port address of the controller.

CONTROL LOGIC: This block includes two pins: INT (interrupt) as an output, and $\overline{INTA}$ (interrupt Acknowledge) as an input. The INT is connected to the interrupt pin of the 8086. This signal goes high whenever a valid, interrupt is asserted. The $\overline{INTA}$ is the interrupt Acknowledge signal which is given out by the 8086.

INTERR UPT REGISTERS AND PRIORITY RESOLVER

The interrupt Request Register (IRR) consists eight input lines (IR_0 – lR_1) for interrupts. The requests are stored in the corresponding bits of the register when the corresponding line goes high. The In-Service Register (ISR) stores all the levels that are currently being serviced and the interrupt Mask Register (IMR) stores the masking bits of the interrupt lines that are to be masked. The priority Resolver (PR) examines these three register and determines whether INT should be sent to the 8086 microprocessor when an interrupt is about to be serviced.

CASCADE BUFFER/COMPARATOR: This is used to expand the number of interrupt levels by cascading two or more 8259 As. .

Interrupt Operation: The interrupt enable flip-flop in the microprocessor can be enabled by writing the EI instruction, and the 8259 A should be initialized by writing appropriate control words in the control register. The 8259 requires two types of control words: initialization command word (ICWs) and Operational Command Words (OCWs). The ICWS are used to set up the proper conditions and specify RST vector addresses. The OCW s are used to perform functions such as masking interrupts, setting up status-read operation etc.

Priority Modes and other Features: Under software control in the 8259 A many types of priority modes are available and they can be changed dynamically during the program by writing appropriate command words. Following are the commonly used priority modes:

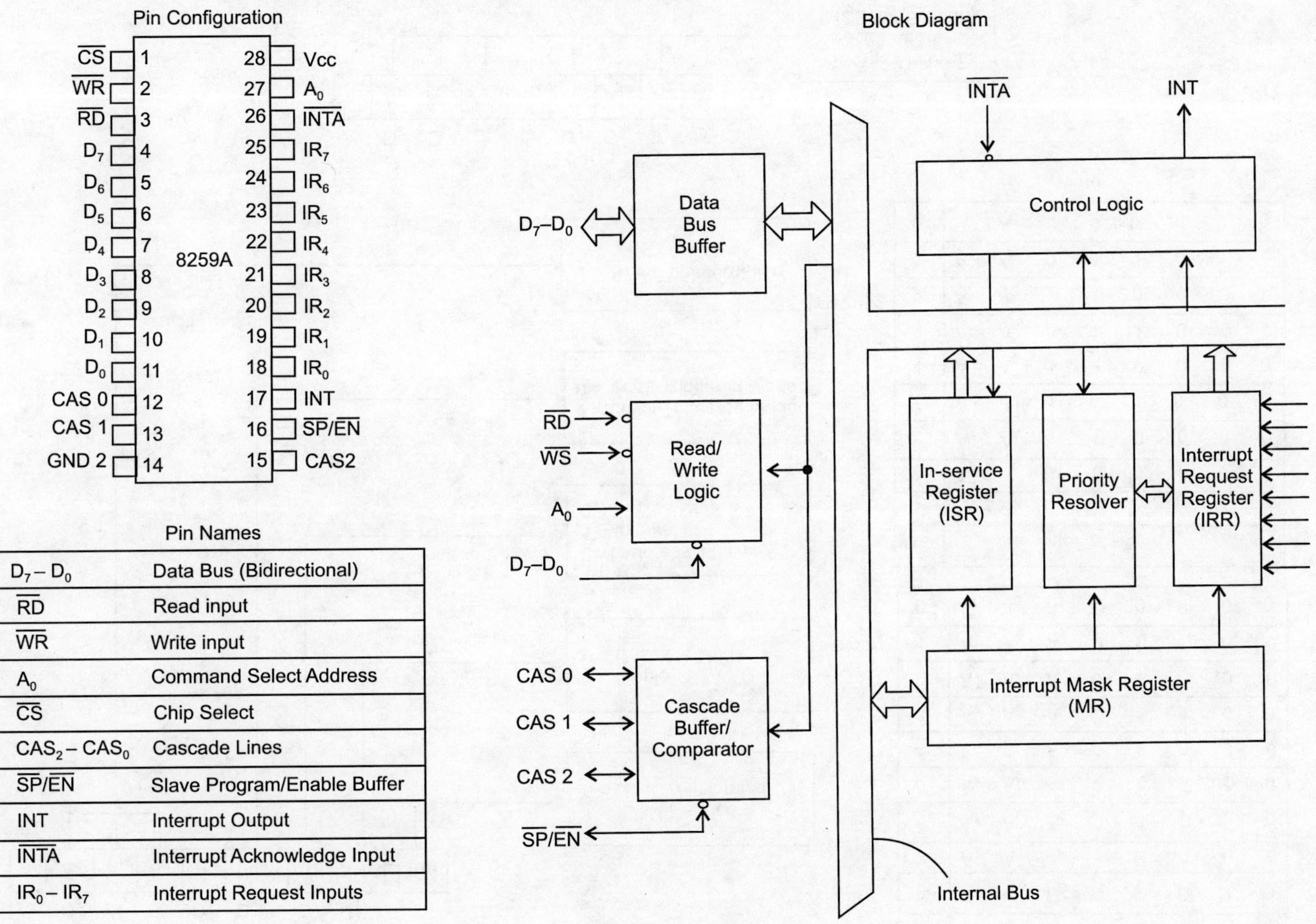

Figure 10.22 The 8259A Block Diagram

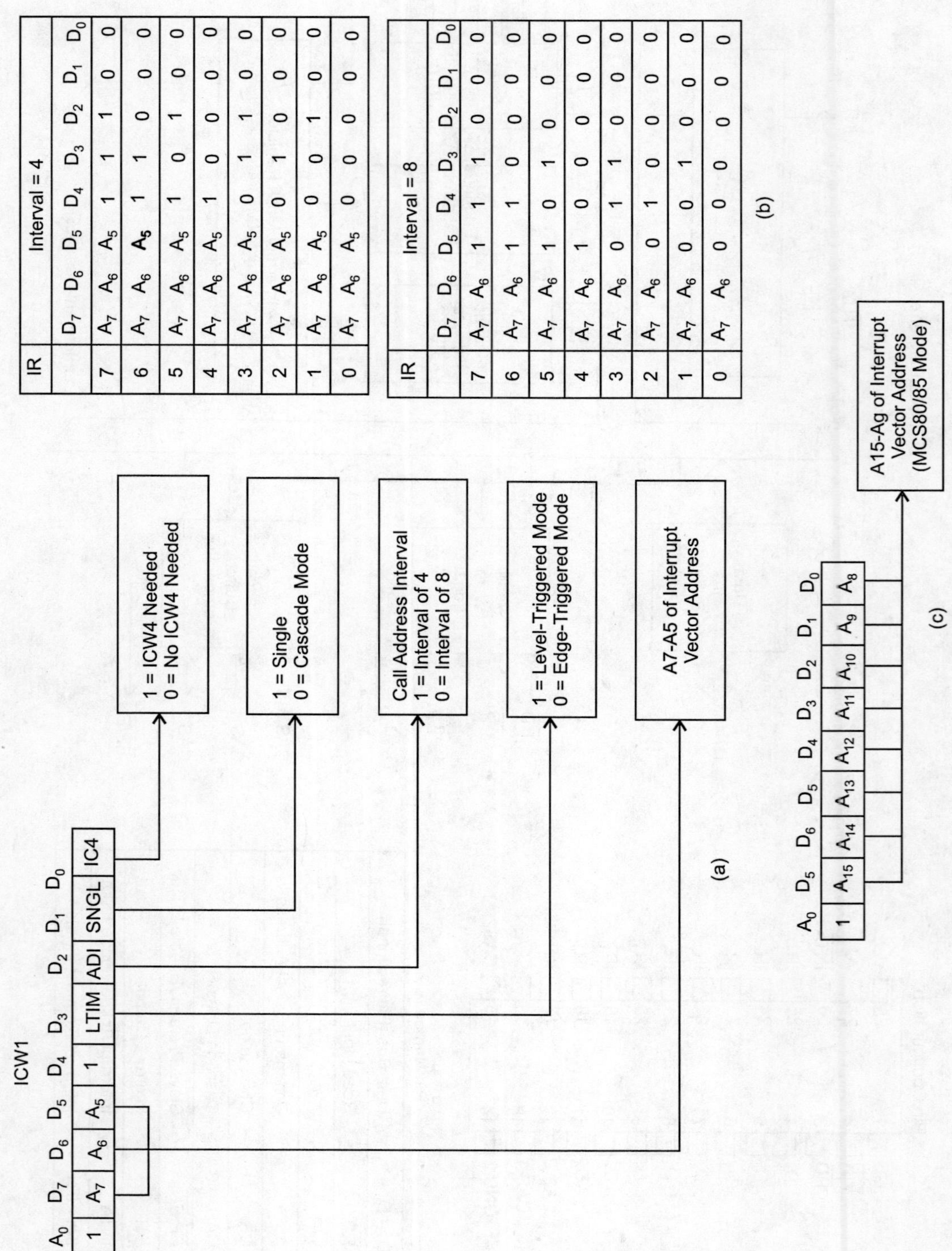

IR	Interval = 4							
	D_7	D_6	D_5	D_4	D_3	D_2	D_1	D_0
7	A_7	A_6	A_5	1	1	1	0	0
6	A_7	A_6	A_5	1	1	0	0	0
5	A_7	A_6	A_5	1	0	1	0	0
4	A_7	A_6	A_5	1	0	0	0	0
3	A_7	A_6	A_5	0	1	1	0	0
2	A_7	A_6	A_5	0	1	0	0	0
1	A_7	A_6	A_5	0	0	1	0	0
0	A_7	A_6	A_5	0	0	0	0	0

IR	Interval = 8							
	D_7	D_6	D_5	D_4	D_3	D_2	D_1	D_0
7	A_7	A_6	1	1	1	0	0	0
6	A_7	A_6	1	1	0	0	0	0
5	A_7	A_6	1	0	1	0	0	0
4	A_7	A_6	1	0	0	0	0	0
3	A_7	A_6	0	1	1	0	0	0
2	A_7	A_6	0	1	0	0	0	0
1	A_7	A_6	0	0	1	0	0	0
0	A_7	A_6	0	0	0	0	0	0

(b)

Figure 10.23 Initialization Command Words for the 8259A

1. Fully Nested Mode: This is a general-purpose mode in which all IR's (interrupt Requests) are arranged in a priority level starting from highest to lowest, with IR_0 having highest and IR_1 has the lowest priority.

In addition, any IR can be assigned the highest priority in this mode; the priority sequence will then begin at that IR.

2. Automatic Rotation Mode: In this mod~ devices that have been serviced, will then be assigned the lowest priority.

3. Specific Rotation Mode: This mode is similar to the automatic rotation mode, except that the user can select any IR for lowest priority, thus fixing all other priorities.

Programming the 8259 A : As earlier, the 8259 A requires two types of command words: Initiations command words (ICWs) and Operational Commond Words (OCWs). The 8259 A can be initialized with four ICW's: the first two are essential, and the other two are optional based on the modes being used. These words must be issued in a given sequence. Once initialized, the 8259 A can be setup to operate in various modes using three different OCW's. However, they no longer need to be issued in a specific sequence.

Figure 10.23 Shows the bit specification of the first two ICWs. The I_{cw1}, shown in the figure specifies:

1. Single or multiple 8259 A's are present in the system.
2. 4 or 8 bit interval between the interrupt vector locations
3. The address bits A, - A5 of the CALL instruction, the rest are supplied by the 8259 A, as shown in figure 10.21. The ICW_2 of figure 10.21 specifies the high order byte of the CALL instruction.

10.5 8155: MULTIPURPOSE PROGRAMMABLE DEVICE

The 8155 is specially designed for operation with the 8088 microprocessor and is a multipurpose programmable device. This device is connected directly to the ALE, IO/$\overline{M}$ $\overline{RD}$ and $\overline{WR}$signal from 8088. This eliminates the need for external demultiplexing of the low-order bus AD_I–AD_0 and generation of the control signals such as $\overline{MEMR}$, $\overline{MEMW}$, $\overline{IOR}$ and $\overline{IOW}$.

The 8155 includes in built 256 byte Of R/W memory, three I/O ports and a timer. The programmable I/O sections of this device are shown in the following sections.

1. The 8155 Programmable I/O Ports and Timer: The 8155 comprises of two main sections: the first is 256 bytes ofR/W memory and the second is programmable I/O circuitry. Functionally, these two section can be viewed as two independent chips. The I/O section consists of two 8-bit parallel I/O ports (A and B), one 6 bit port C and a timer as shown in figure 10.22 All the ports can be configured simply as input/output ports. Port A and B also can be programmed in the handshake mode, where each port uses three signals for handshaking which are taken from Port C. The timer is a 14 bit down counter and has four modes. Pins P_A; P_B and P_C shown in figure 10.22. Correspond to port A, Band C.

CONTROL LOGIC: The control logic of the 8155 is specifically designed to eliminate the need for external demultiplexing lines. AD_7 – AD_0 and generating separate control signals for memory and I/O. Five control signals: all except the chip Enable ($\overline{CE}$) are input signals directly generated by the 8088 as shown in figure 10.22.

• **$\overline{CE}$ - Chip Enable:** This is a master chip select signal which is connected to the decoded high order bus.

• **10/$\overline{M}$:** The memory section is selected when this signal is driven low. The I/O section (including timer) is selected when it is high.

• **ALE - Address Latch Enable:** This signal latches the low order address $AD_7 - AD_0$, $\overline{CE}$ and 101M into the chip.

• **$\overline{RD}$ and $\overline{WR}$:** These are the control signals which are used to read from and write into the chip registers and memory.

• **RESET:** This is connected to RESET OUT pin of the 8088 and this resets the chip and initialized I/O ports as input.

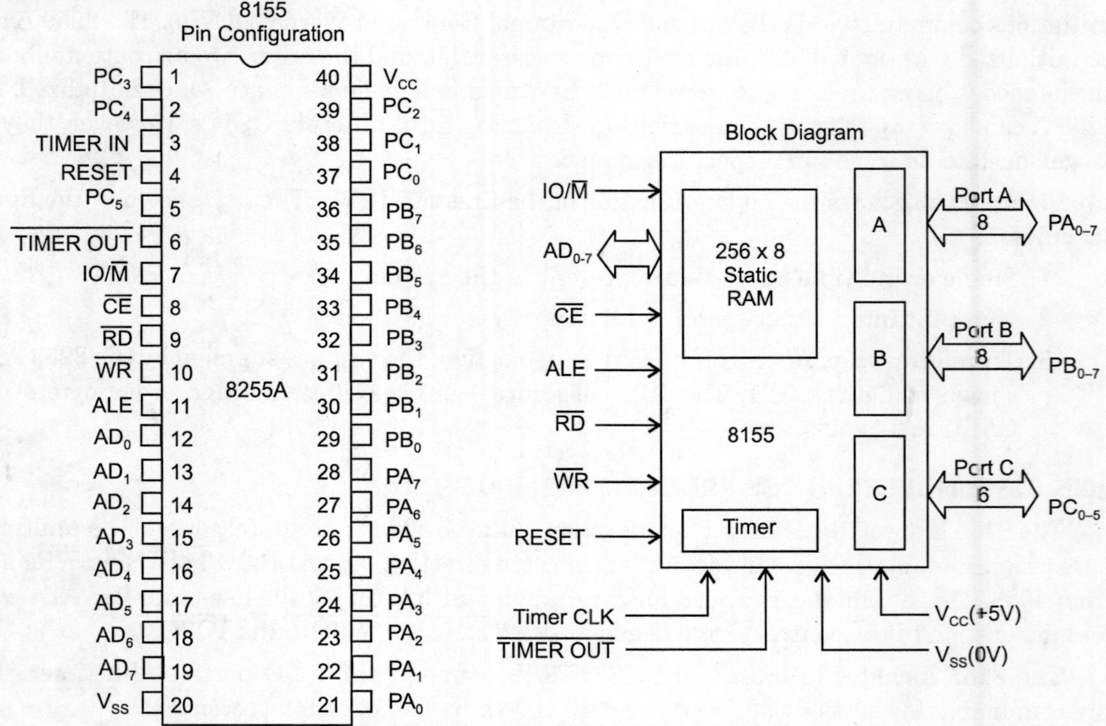

Figure 10.24. 8155 Pin configuration and Block Diagram

The 8155 110 PORTS: The I/O section of the 8185 internally includes a control register, three I/O ports, two registers for the timer as shown in figure 10.25. The block diagram of the I/O section represents a typical programmable I/O configuration.

The 8155 I/O section requires three address lines - AD_2 to AD_0 ($A_2 - A_0$ internally) - andthe chip enable logic to specify one of the seven registers, In addition, two control signals, $\overline{RD}$ and $\overline{WR}$, are necessary to read from and write into these I/O registers. The following steps are necessary to communicate with peripherals through the 8155.

1. Determine the address (port number of the registers and I/Os) based on the chip enable logic and address lines AD_0, AD_1 and AD_2.
2. Write a control word in the control register to specify I/O functions of the ports and the timer character-istics.

3. Write I/O instruction to port addresses to communicate with peripherals.
4. Read the status register, if necessary, to verify the current status of the I/O ports and the timer. In simple applications; this step is not necessary.

CHIP ENABLE LOGIC AND PORT/ADDRESSES

Address lines AD_2, AD_1 and AD_0 are shown as A_2, A_1, A_0 after internal demultiplexing are used to select one of the registers as shown in figure 10.25(b). Address lines (A_3–A_7) are don't care lines: however, the logic levels on the corresponding high order lines, (A_{11} – A_{15}) will be duplicated on lines A_3–A_7.

CONTROL WORD: The I/0 ports and the timer can be configured by writing a control word in the control register. The control register bits are defined as shown in figure 10.24.

In this control word, outputs are defined with logic 1 and input with logic O. I/0 functions of ports A and B are determined by the first two LSB's : D_0 and D_1

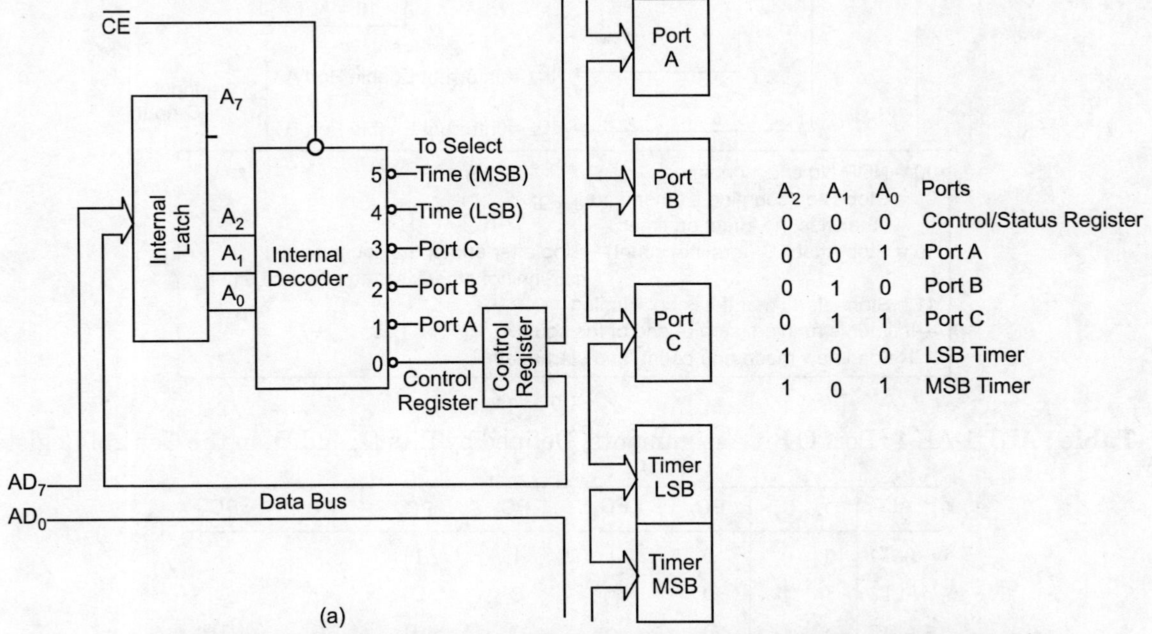

Figure 10.25. Expanded block diagram of the 8155(a) and its I/O Address: Selection (b)

Timer functions are determined by the MSB's, D_7-D_6 The functions of port C is deter- mined by bits D_2 and D_3; their combination specifies one of the four alternatives, from simple I/O to interrupt I/0 as shown in figure 1O.24(b). To enable or disable internal flip flops of the 8155, bits D_4 and D_0 are used only in the interrupt mode, These bit do not have any effect on the interrupt enable flip-flop (INTE)

10.5.1. INTERFACING SEVEN -SEGMENT LED OUTPUT PORTS USING THE 8155 PROBLEM STATEMENT

1. Design two seven-segment LED displays using port A and B of the 8155.
2. Write initialization instructions and display data bytes at each port.

HARDWARE DESCRIPTION: Two seven-segment output ports: Port A with the Hewlett-Packard HP 508217340, and port B with the 9370 Hexdecoders and common-anode seven segment LED's shown in figure 10.25. The HP 5082 consists of an internal decoder/driver. Both the 7340 and 5082 are functionally similar; however a seven-segment display with an internal built in decoder/driver is more expensive.

The port address are as follows

Control Register = 20H

Port A = 21 H

Port B = 22 H

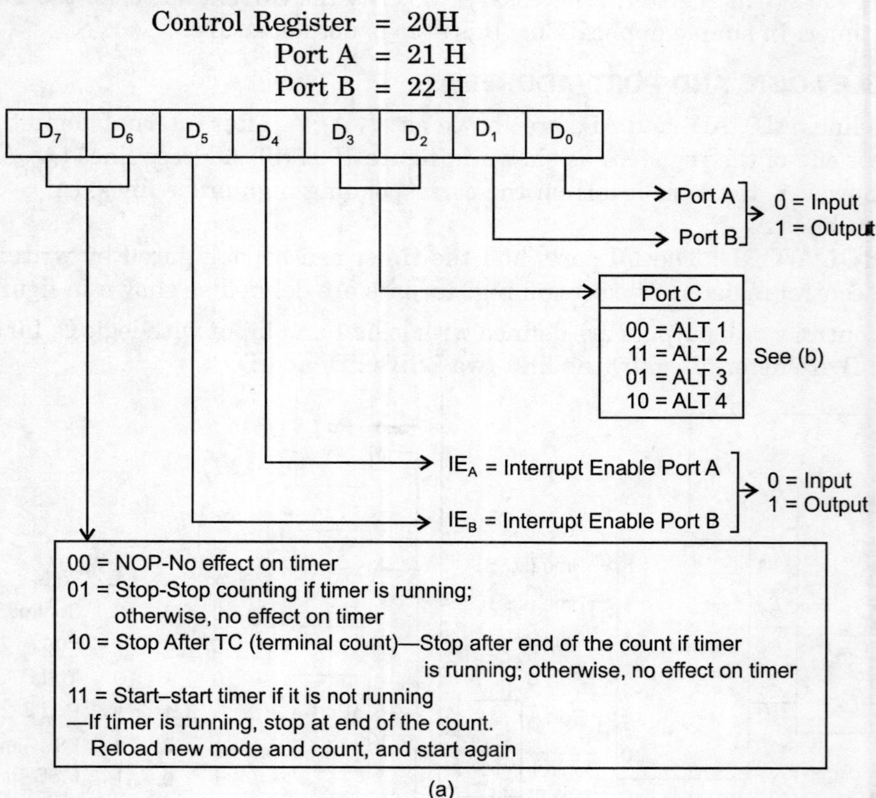

(a)

Table : ALT 1-AL 4 : Port C Bit Assignments, Defined by Bits D_3 and D_2 in the Control Register

ALT	D_3	D_2	PC_5	PC_4	PC_3	PC_2	PC_1	PC_0
ALT1	0	0	I	I	I	I	I	I
ALT2	1	1	O	O	O	O	O	O
ALT3	0	1	O	O	O	$\overline{STB_A}$	BF_A	$INTR_A$
ALT4	1	0	$\overline{STB_B}$	BF_B	$INTR_B$	$\overline{STB_A}$	BF_A	$INTR_A$

I = input, STB = Strrobe, INTR = Interrupt Request

O = Output, BF = Buffer Full, Subscript A = Port A

B = Port B

(b)

Figure 10.24. Control Word Definition in the 8155 (a) and Table of the Port C bit Assignments (b)

CONTROL WORD : The control word to configure port A and B as output is a follow:

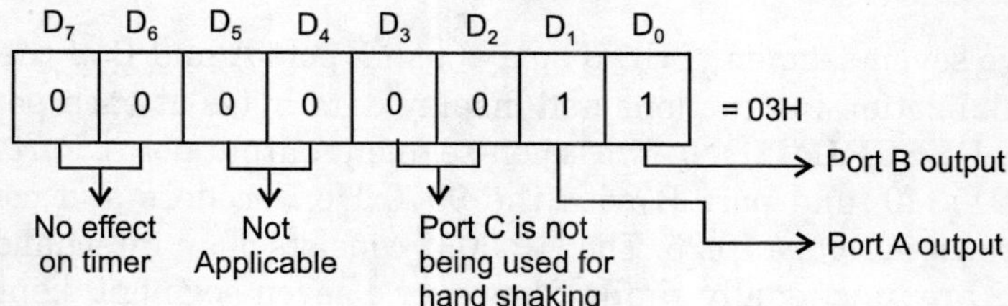

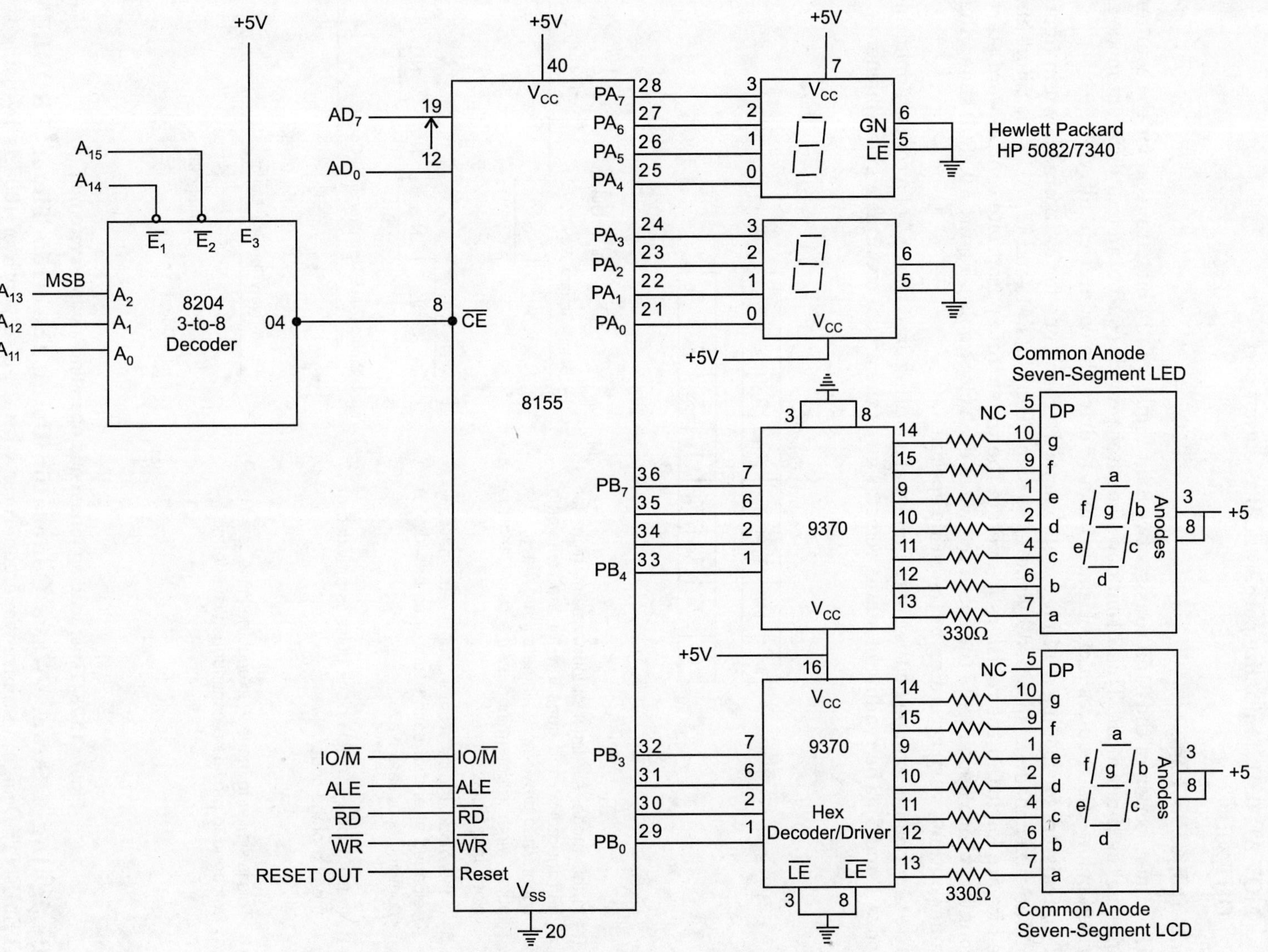

Figure 10.27 Interfacing 8155 I/O Ports with Seven-Segment LEDs

PROGRAM:

```
MOV AL, 03H  : Initialize port A and B as output port.
OUT 20H
```

10.5.2. THE 8155 TIMER

The timer section of the 8155 consists two 8-bit registers ; 14 bits are used for the counter, two bits for the timer mode. This 14-bit down-counter provides output in four different mode as described below and requires a clock pulse input to drive the counter circuitry.

Two registers for loading a 14 bit count, one for LSB and one for MSB are shown in figure 10.25. The timer mode-is specified by the most significant bits M_2 and M_1 A 14 bit count and mode bits are loaded in the registers to operate the timer. An appropriate control word starts the counter, which decrements the count by two at each clock pulse. Figure 10.28(a) shows that the timer outputs vary according to the mode specified.

The timer can be stopped either in the midst of counting or at the end of a count (applicable to modes 1 and 3). The reading of status register gives the actual count at a given moments.

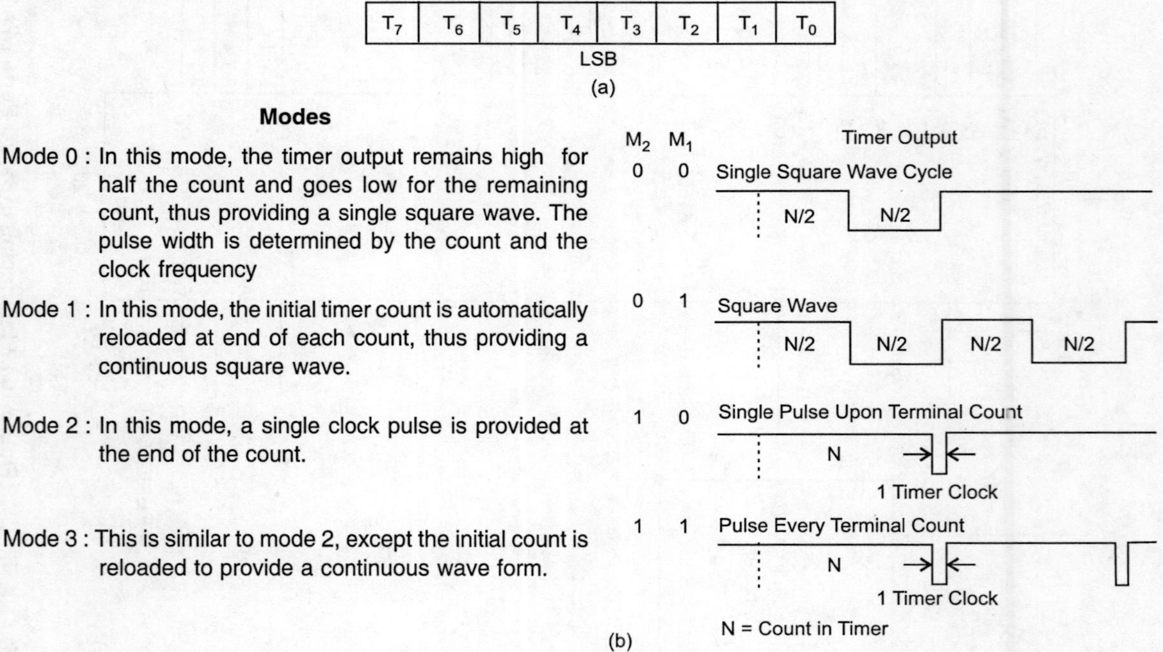

Figure 10.28. Timer loading format and operating mode waveforms

Designing a Square-Wave Generator Using the 8155 Timer PROBLEM STATEMENT: Design a square-wave generator with a pulse width of 100J.1s by using the

8155 timer. We setup the timer in Mode 1 an use a clock frequency of 3 MHz.

PROBLEM ANALYSIS

Timer count: The pulse width required is 100 μs, therefore, the count should be calculated for the time period of 200 μs. Since the timer output stays high for only half the count.

$$\text{Clock period} = \frac{1}{f} = \frac{1}{3\times10^6} = 330\text{ns.}$$

$$\text{Timer count} = \frac{\text{Pulse Period}}{\text{Clock period}} = \frac{200\times10^{-6}}{300\times10^{-9}} = 606$$

Count = 025H

Assuming the same decoder logic for the 8155 Chip enable line as in the previous example the port addresses for the timer registers are

Timer LSB = 24 H

Timer MSB = 25 H

The least significant byte, 5EH (of the count EH) should be loaded in the timer register with address 24H. MSB is determined as follows:

M_2	M_2	T_{13}	T_{12}	T_{11}	T_{10}	T_9	T_8	
0	1	0	0	0	0	1	0	= 42 H
Timer mode 1		MSB (25H)						

Figure 10.27. Timer MSB byte for above example

Thus, 42 H should be loaded in the timer register with the address 25 H.

Control word: Considering the same configuration for ports A and B as before, only bits D_7 and D_6 should be set to 1 to start the counter.

Therefore, control word: 11000011 = C3H

Initialization Instruction:

```
MOVAL,5EH    ; LSB of the count
OUT 24 H     ; Load the LSB timer register
MOVAL,42H    ; MSB of the count
OUT 25 H     ; Load the MSB timer register
MOVAL, C3H
OUT 20 H     ; Start the timer
HLT
```

10.5.3. THE 8155 I/O PORTS IN HANDSHAKE MODE

Data transfer in the handshake mode occurs between the 8086 and peripherals by using control signals called handshake signals. Two I/O ports of the 8155, A and B, can be configured in the handshake mode; each uses three signals from port C as 8086 a control signal as shown in figure 10.29. The details of configuring port A and B in the handshake mode by using the pins of port is given below.

Control Signals in Handshake Mode: PortA uses the lower three signals of port (PC_0' PCl and PC_2), and port B uses the upper three signals (PC_3, PC_4 and PC_5), when both port A and B are configured in the handshake mode as shown in figure. The functions of these signals are as follows :

- **$\overline{STB}$ (strobe input):** This is an input handshake signal from a peripheral to the 8155. The low signal on this pin informs the 8155 that data are being strobed into the input port.
- **BF (Buffer Full) :** This is an active high signal, indicating the presence of a data byte in the port.
- **INTR (interrupt Request) :** If the interrupt flip-flop (INTE) is enabled this signal is generated during the rising edge of the $\overline{STB}$ signal. This signal can be used to interrupt the 8086.

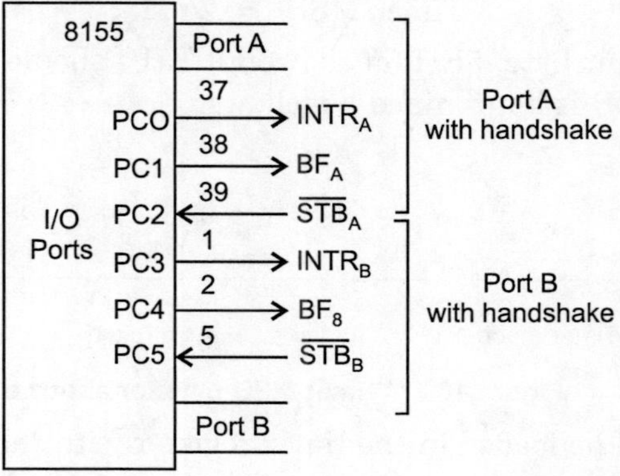

Figure 10.29. 8155 with Handshake Mode.

- **INTE (Interrupt Enable):** This is an internal flip-flop used to enable or disable the interrupt capability of the 8155. Bits D_4 and D_5 in the control register controls the interrupts for port A and port B respectively.

These control signals can be used to implement either of the I/O modes: Interrupt I/O or status check I/O

Input: The sequence of events and timing in data input to the 8155 shown in figure 10.30; they can be described as follows:

1. An external peripheral places data in the input port and inforn~s the 8155 by causing the $\overline{STB}$ signal to go low.
2. The falling edge of the $\overline{STB}$ sets signal BF (Buffer Full) high informing the peripheral to wait.
3. When the $\overline{STB}$ goes high, the rising edge of the $\overline{STB}$ can generate signal INTR if'the internal interrupt flip-flop INTE is set. The interrupt flips-flop are set or reset by the control word.

4. Data transfer from the 8155 input port to the 80861S the last step. This can be done either by interrupting the 8086 with the INTR signal or by checking the status of signal BE. Status register read operation can be checked by the 8086. The INTR andBF signals are resent when the 8086 reads data. It informs the peripheral that the port is empty and the device is ready for the next byte when the BF signal goes low.

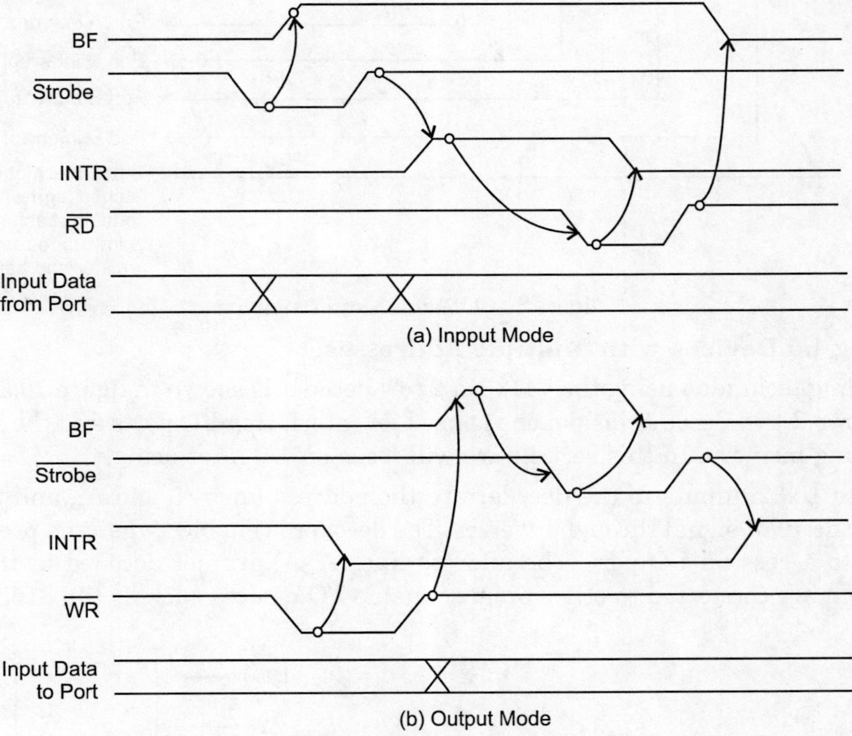

Figure 10.30. Timing waveforms for 8155 ports in Handshake mode

Output: The sequence of events and timing in data output from the 8155 port to a peripheral are as follows as shown in figure 10.30(a).

1. The MPU write in the port when the output is empty.
2. The falling edge of the WR signal resets the INTR signal and the rising edge sets the BF (Buffer full) signal high informing the peripheral that a byte is available in the port.
3. The peripheral acknowledges by sending the $\overline{\text{STB}}$ signal (active low) after receiving the data byte.
4. The $\overline{\text{STB}}$ signal resets the BF signal low and generates the interrupt request by setting INTR high. Now the 8086 can be informed by the interrupt signal to send the next byte or the 8086 can sense that the port is empty through a status check.

STATUS WORD: The 8086 can read the status register to check the status of the ports or the timer. Since the control register and the status register have the same port address. therefore, they are differentiated only by the $\overline{\text{RD}}$ and $\overline{\text{WR}}$ signals. Figure 10.31 defines defined the status register bits.

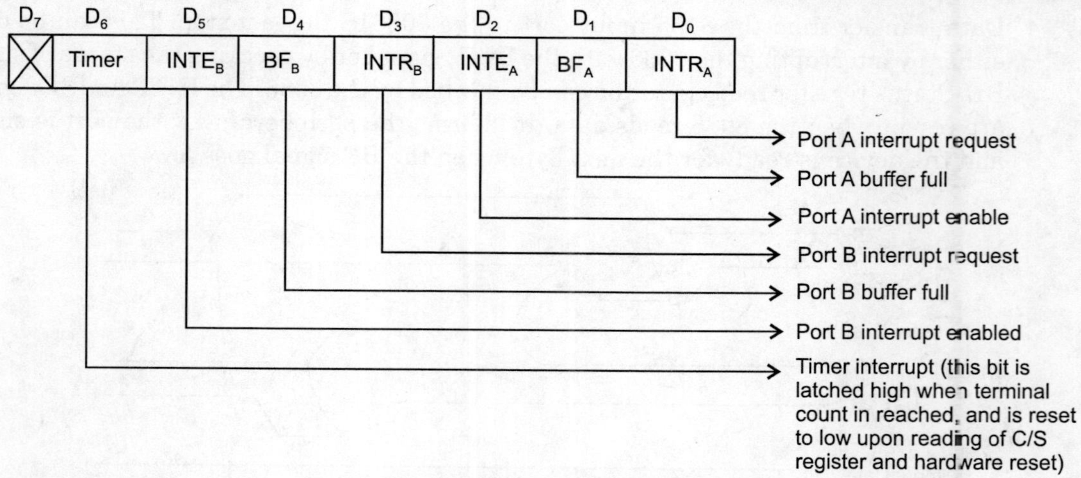

Figure 10.31. Status word Definition

Interfacing 1/0 Devices with Multiple Addresses

A decoding technique using the 74139, a 2 to 4 decoder is shown in figure 10.30. This device consists of two 2-to-4 decoder inside; one is used for interfacing I/O ports and the second is used for interfacing memory. In this section, we will focus on 1/0 interfacing.

In figure 10.32, inputs to the decoder are the address lines A_7 and A_6, and the decoder is enabled by the 10/M signal through inverter. The decoder is enabled when the processor asserts 10/M high to access an I/O port. The address lines A_5-A_6 are not decoded in this schematic; some of them are connected to other programmable I/O devices such as the 8155 and 8279.

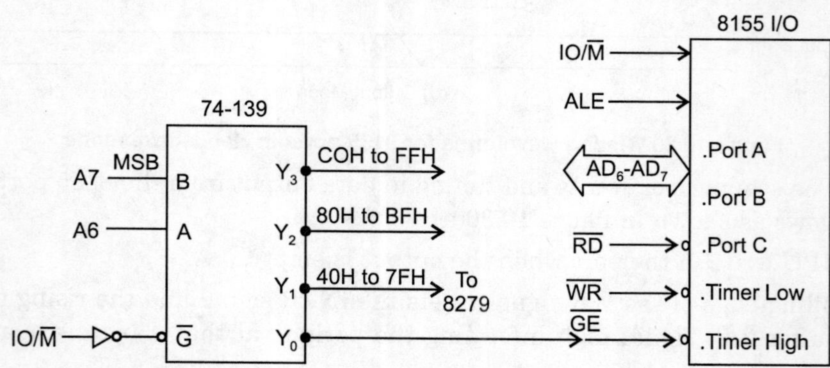

Figure 10.32. Interfacing 81551/0 ports and 8279 with Multiple Addresses.

ADDRESSES FOR 8155 110 PORTS

8155110 Addresses

A_7	A_6	A_5	A_4	A_3	A_2	A_1	A_0		
0	0	X	X	X	0	0	0	=	0 0control or status retgister
					0	0	1	=	0 1 Port A
					0	1	0	=	0 2 PortB
					0	1	1	=	0 3PortC
					1	0	0	=	0 4 Timer LSB

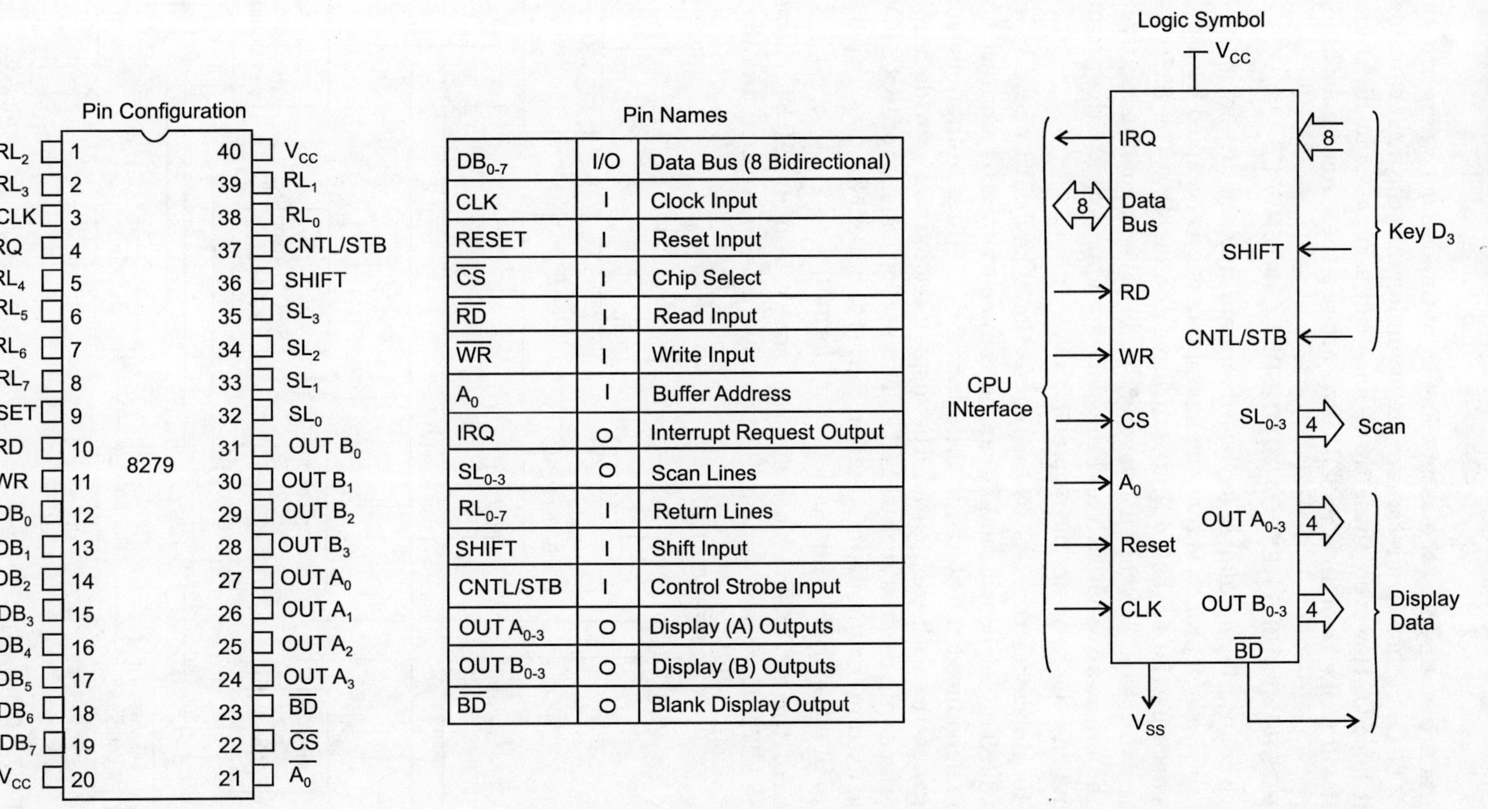

Pin Names

DB_{0-7}	I/O	Data Bus (8 Bidirectional)
CLK	I	Clock Input
RESET	I	Reset Input
$\overline{CS}$	I	Chip Select
$\overline{RD}$	I	Read Input
$\overline{WR}$	I	Write Input
A_0	I	Buffer Address
IRQ	O	Interrupt Request Output
SL_{0-3}	O	Scan Lines
RL_{0-7}	I	Return Lines
SHIFT	I	Shift Input
CNTL/STB	I	Control Strobe Input
OUT A_{0-3}	O	Display (A) Outputs
OUT B_{0-3}	O	Display (B) Outputs
$\overline{BD}$	O	Blank Display Output

Figure 10.33. The 8279 Logic Printout

1 0 1 = 0 5 Timer MSB

The 8155110 ports are accessed by the output line Y_0 of the decoder, therefore the address lines A_7 and AG must be O. In these addresses we are assuming that the don't care address lines A_5, A_4 and A_3 are at logic O. However, these three lines can assume eight different combina-tion's. The addresses will range from 38 H to 3DH if we assume these lines to be at logic 1.

10.6. THE 8279 PROGRAMMABLE KEYBOARD/DISPLAY INTERFACE

A hardware approach to interfacing matrix keyboard and multiplexed display is the 8279. The disadvantage of the software approach is that the microprocessor is occupied for a considerable amount of time in checking the keyboard and refreshing the display. The 8279 relieves the processor from these two tasks. The cost increase is the disadvantage of using the 8279. The trade offs between the hardware approach and the software approach are the production cost Vs the processor time and the software development cost.

The 8279 is a 40 pin device consists two major sections: keyboard and display sections shown in figure 10.33. The keyboard section can be connected to a 64 Contact key matrix. Keyboard entries are debounced and stored in the internal FIFO (first-in-first-out) memory, an interrupt signal is generated with each entry. The display section can provide a 16-character scanned display interface with devices such as LEDs. This section consists of 16 × 8 R/W memory (RAM), which can be used to read/write information that is to be sent for display purposes. The display can be set up in either right-entry or left entry format.

Block diagram of the 8279: The block diagram shown in figure 10.34. shows the four major sections of the 8279: Keyboard section, Scan section, Display section and the processor interface.

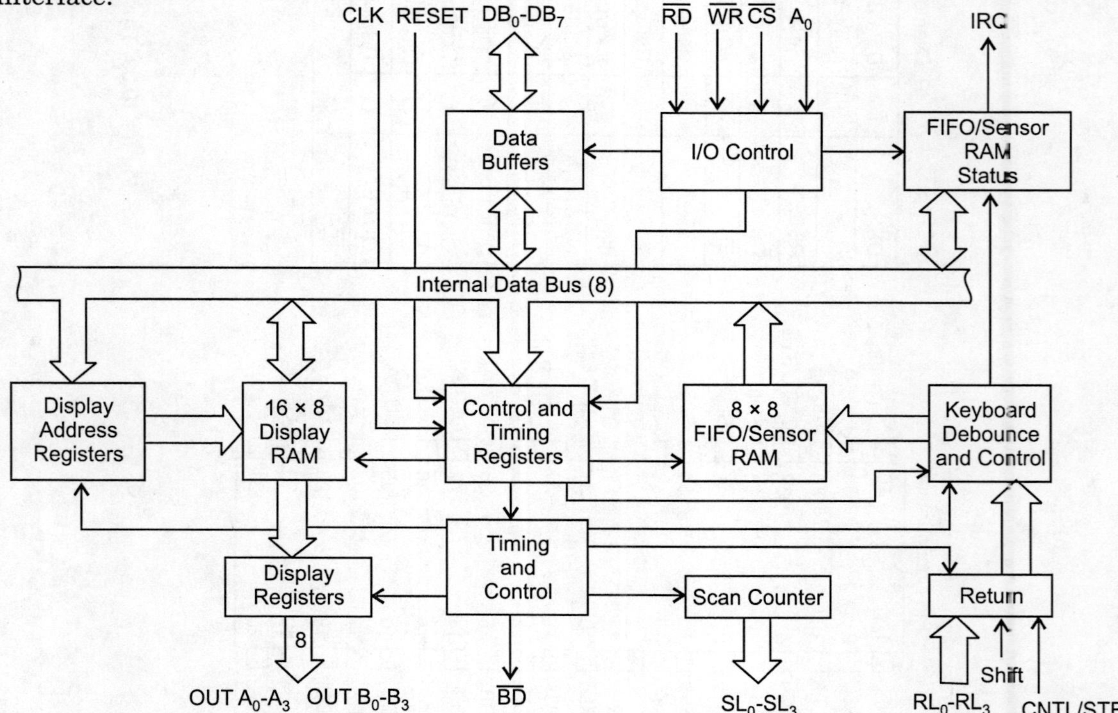

Figure 10.34. the 8279 Logic Block Diagram

KEYBOARD SECTION: This section consists of eight lines (RL_0 - RL_7) that can be connected to eight columns of a keyboard, plus two additional lines as well as to shift and CNTL/ 8TB (control/strobe) keys. The status of the SHIFT key and the control keys can be stored along with a key closure. The key pressed are automatically debounced, and the keyboard can operate in two modes two-key lock out or N-key rollover. If two keys in the two key lock out mode are pressed almost simultaneously. only the first key is recognized. In the N-key, rollover mode, simultaneous key are recognized and their codes are stored in the internal buffer. The 8279 can also be set up so that no key is recognized until only one key remains pressed.

The keyboard section also contains 8 x 8 FIFO (first-in-First-out) RAM. The FIFO RAM includes eight registers that can store eight keyboard entries; each is then read in the order of entries. The status logic keeps track of the number of entries and provides an IRQ (Interrupt Request) signal when the FIFO contains atleast one valid entry.

SCAN SECTION: The scan section includes a scan counter and four scan lines (SL_0- SL_3). These four scan lines can be decoded using a 4. to 16 decoder to generate 16 lines for scanning. These lines can be connected to the rows of a matrix keyboard and to the digit drivers of a multiplexed display.

DISPLAY SECTION: The display section consists of eight output lines divided into two groups A_0 - A_3 and B_0 - B_3. These lines can be used either as a group of eight lines or as two groups offour, in conjunction with the scan lines for a multiplexed display. By using the $\overline{BD}$ line. the display can be blanked. This section also has 16 x 8 display RAM to store data for display. The 8086 can read from or write into any ofthese registers.

PROCESSOR INTERFACE SECTION: This section has eight bidirectional data lines (DB_0 - DB_7). one Interrupt Request line (IRQ), and six lines for interfacing, including the buffer address lines:

Input signals are interpreted as control word or status word when A_0 is high while input signals are interpreted as data when A_0 is low. Whenever data entries are stored in the FIFO the IRQ lines goes high and therefore indicates the availability of data. This signal is used to interrupt the processor.

PROGRAMMING THE 8279 : The 8279 is a complex device that can accept eight different commands to perform various functions. The initialization com mands can specify the following

1. Left or right entry and key rollover.
2. Clock frequency prescalar.
3. Starting address and incrementing mode of the FIFORAM.
4. RAM address to read and write data and incrementing mode.
5. Blanking format.

8279 I/0 Addresses: In the keyboard / display device, the address line A_0 is connected to 8279. When A_0 is 0, it accesses the internal data port and when it is 1, it accesses the internal command port. The output line Y_1 of the decoder accesses 8279; therefore, the address lines A_7 and A_6 must be 0 and 1. respectively. Hence the range. of port addresses of the 8279 is as follows.

8279 I/O address

A_7	A_6	A_5	A_4	A_3	A_2	A_1	A_0	
0	1	X	X	X	X	X	0	= 4 0 H Command\staus port
							1	= 4 1 H Data port

The possible logic combination of, these addresses can range from 40H to 7FH if we assume the don't care address line A_5 through A_1 to be equal to logic '0'.

Multiple 1/0 address: The 8279 needs only two addresses, but due to the five don't care lines, it occupies the space of 64 I/O addresses. In the case of the 8155, three addresses lines are don't care, therefore, it has eight sets of address ranges for its 1/0 ports and the timer. The advantage of such a technique is in saving costs and space on a printed circuit board. We can decode only two address lines by using one two-input decoder. We need additional chips and space on the board to decode the remaining line in small systems, the space of the printed circuit called real estate and is of prime importance. However, we pay the price for this technique due to assignment of multiple addresses. However, if we do not need all the I/O space in a given system, wasting I/O addresses for a given device to reduce cost is of no consequence.

10.7. DIRECT MEMORY ACCESS (DMA) AND THE 8237 DMA CONTROLLER

Direct memory Access is an I/O technique commonly used for high-speed data transfer. For example for data transfer between system memory and a floppy disk. Data transfer in status check I/O and interrupt I/O is relatively slow because each instruction needs to be fetched and executed. In DMA, the processor relinquishes the control of the buses to a device called a DMA. controller. The controller manages data transfer between memory and a peripheral under its control, thus bypassing the processor. It introduces two new signals available on the 8086 : HOLD and HLDA (hold Acknowledge).

If a peripheral requests the processor for the control of buses then a DMA controller uses these signals. The processor communicates with the controller by using the chip select line, buses, and control signals. However, once the controller has gained control, it plays the role of a processor for data transfer. To perform this function the DMA controller should have:

1. A data bus
2. An address bus
3. Read/write control signal and
4. Control signal to disable its role as a peripheral and to enable its role as a processor.

This process is called switching from the slave mode to the master mode.

For all practical purposes, the DMA controller is a processor capable of only of copying data at high speed from one location to another location.

10.7.1 Internal Architecture of 8257

The Internal architecture of 8257 is shown in fig. 10.35. The chip supports the four DMA channel i.e. four peripheral device can independently request for DMA data transfer though these channel at a time. The DMA controller has 8-bit internal data a ReadlWrite Unit, a Control Unit, a priority resolving Unit along with a set of register.

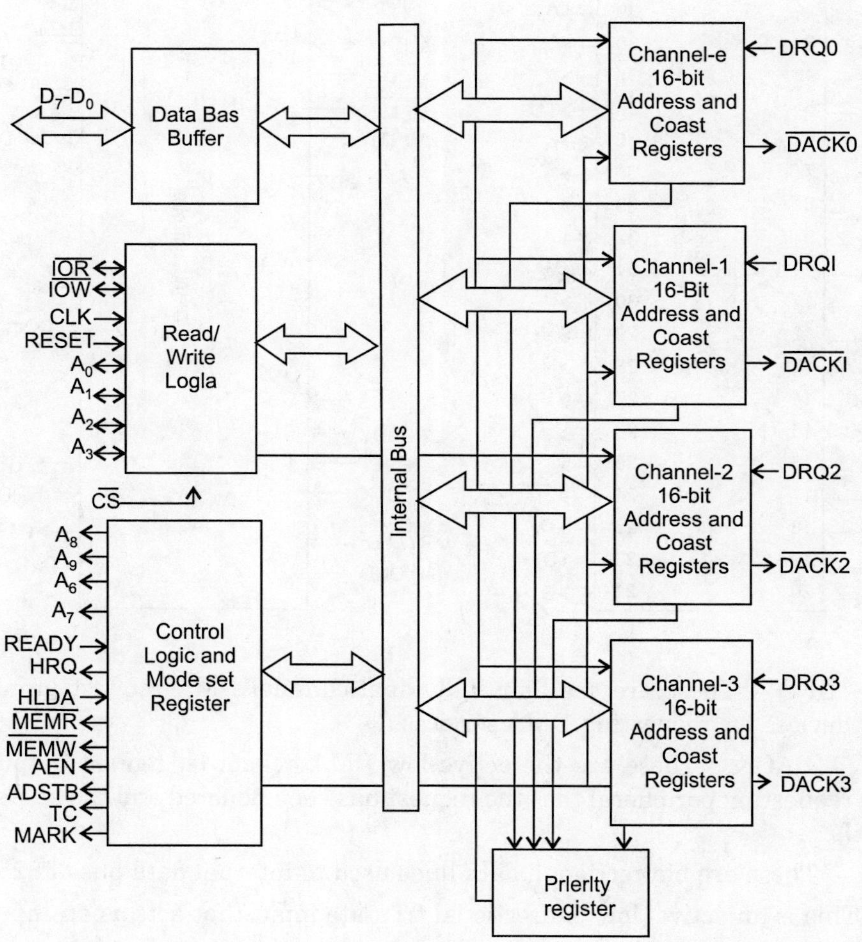

Figure 10.35 Architecture of DMA 8257

Register Organization of 8257 : The 8257 performs the DMA operation over four independent DMA channels. Each of the four channel of 8257 has a pair of two 16- bit DMA Address Register and terminal count Register. The CPU can select one of these ten register using address line A_0-A_3.

Terminal Count Register : As in the previous case, each of the four DMA channels of 8257 has one terminal count Register (TC). This 16- bit Register is used for ascertaining the data transfer though a DMA channel cease or stops after the required number of DMA cycle.

10.7.2. Pin Diagram of 8257

Figure shows the pin diagram of 8257 DMA controller and pin description is given below:

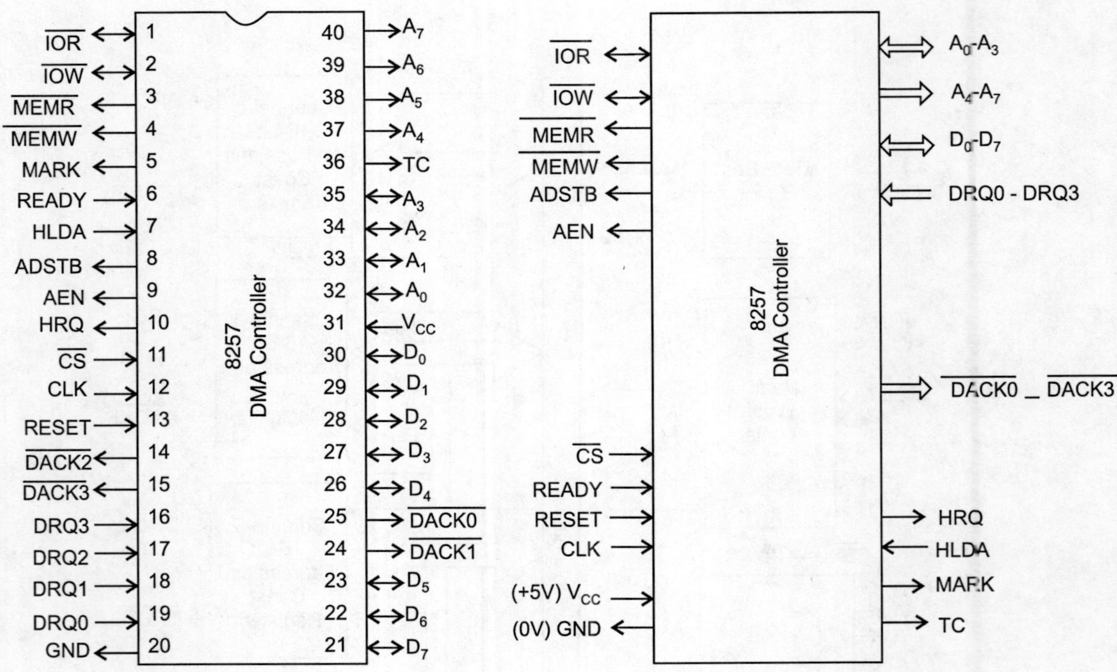

$DRQ_0 - DRQ_3$ – These are the four Individual channel DMA request input used by the peripheral devices for requesting DMA services.

$DACK_0 - DACK_3$: These are the active low DMA acknowledgement output lines which inform the requesting peripheral that the request has been honored and the bus is relinquished by the CPU.

$D_0 - Q_7$ – These are bidirectional, data lines used to internal data bus of 8257.

$\overline{IOR}$ – This is an active- low bidirectional tri state input that acts as an Input in the slave mode.

$\overline{IOW}$ This is an active- low bidirectional tri state line that acts as input in slave mode to load the contents of data bus to the 8- bit mode Register.

CLK – This is a clock frequency input required to derive basic system timing for the internal operation of 8257.

$A_0 - A_3$ – These are the four least significant address lines.

$A_4 - A_7$ – This is the higher nibble of the lower byte address generated by 8257 during the master mode of DMA operation.

ADSTB – This output from 8257 stroble the system data bus and the control the driven by the CPU.

AEN – This output pin is used to disable the system data bus and the bus driver is the control by the CPU.

10.7.3 Internal Architecture of 8237

The Internal block diagram of 8237 is shown in fig. 10.34. The 8237 contains three basic blocks of its operational logic. The timing and control block generates the internal timing and External control signals. The priority Encoder block resolves priority between the DMA channels requesting the services simultaneously.

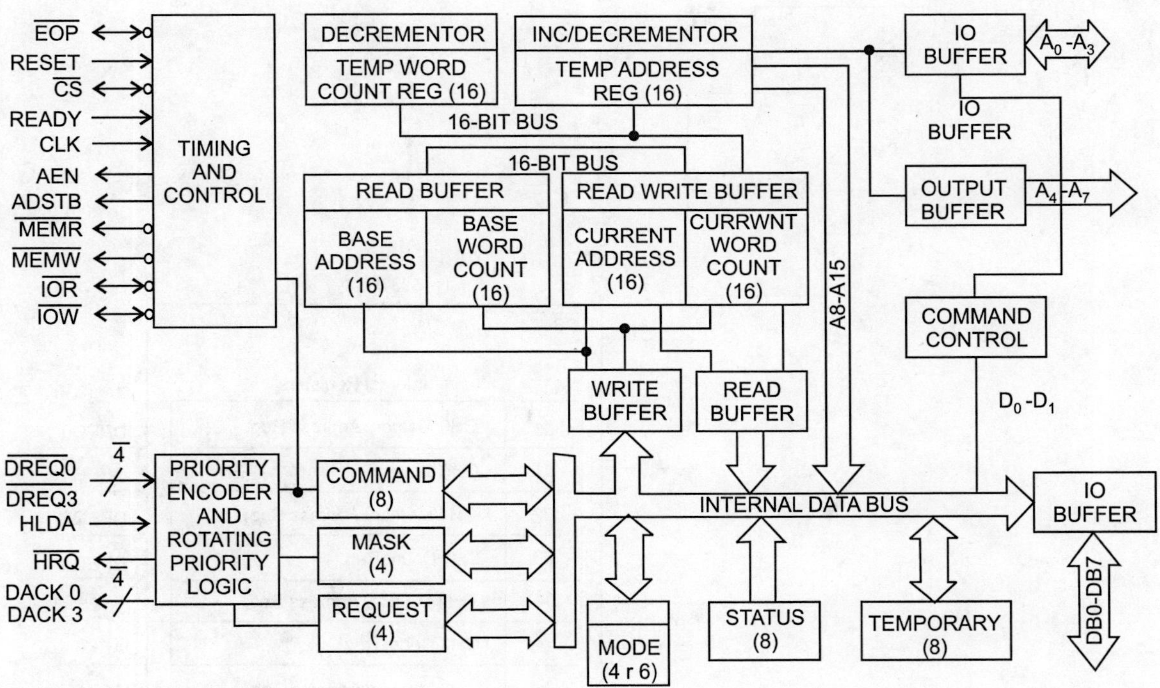

Figure 10.36. Internal architecture of DMA 8237

Current Address Register – Each of the four DMA channels of 8257 has a 16- bit current address register that holds the current memory address.

Current Word Register – Each channel has a 16- bit current word Register that holds the number of (counts) data byte transfer to the carried out.

Base Address and Base word count Register – Each channel has a pair of these register. These maintain an original copy of the respective initial current address Register and current word Register.

These cannot be Read by the CPU – The contents of these Register are used internally for auto – initialization.

Command Register – The 8- bit Register controls the complete operation of 8237. This can be programmed by the CPU and cleared by a reset operation.

10.7.4. DMA CHANNELS AND INTERFACING

The pin out and Internal registers of the 8237 is shown in figure 10.33. It also shows the interface with the 8088 using a 3-to-8 decoder.

The 8237 includes four independent channels, CH_0 to CH_3. Internally, each DMA channel is associated to two 16 bit registers. One register is used to load the starting address of the byte to be copied and the second is used to load a count value that signifies the number of bytes to be copied. Eight such registers that can be accessed by the 8088 is shown in figure 10.33.

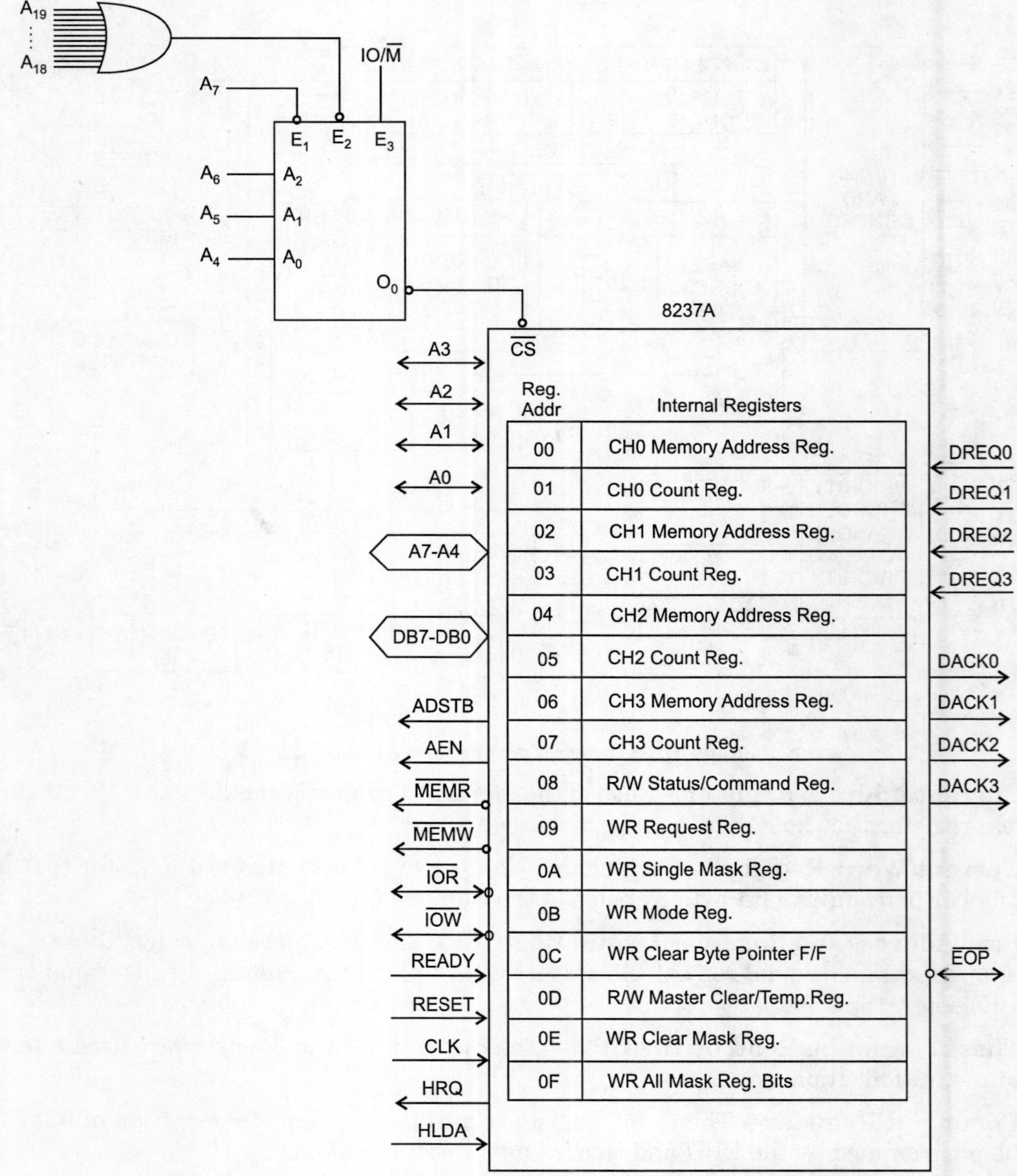

Figure 10.37. 8237A—DMA Controller with Internal Registers

Four address lines A_3 to A_0 and chip select (CS) signal determines the address of these registers. CH_0 Memory Address Register (MAR) is selected by address 0000 on lines A_3-A_0 and the address 0001 selects the next register; CH_0 count. Similarly, all the remaining registers are selected in sequential order. The last eight registers are used to' write commands. or read status as given below. The 8088 accesses the DMA controller by asserting the signal Y_0 of the decoder. Therefore, the addresses of these internal registers range from OOOOH to OOOFH as follows:

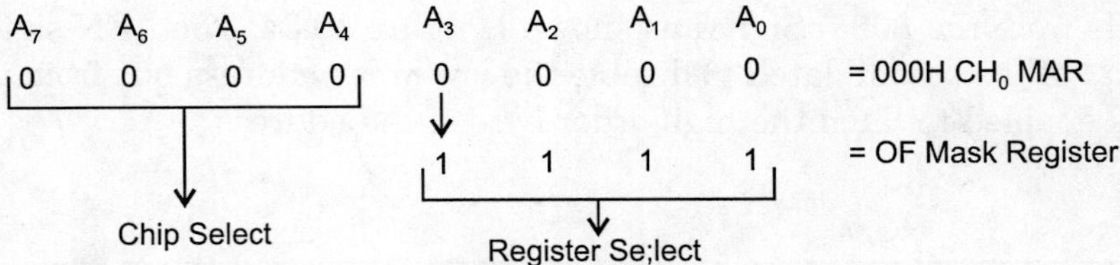

10.7.5. DMA SIGNALS

In figure 10.33, signals are divided into two groups: (1) One group of signals shown on the left of 8237 and are used for interfacing with the 8088. (2) The second group shown on the right-hand side of8237 is for communicating with peripherals. Some of these signals are bidi- rectional and their functions are determined by the DMA mode of operation (I/O or processor).

- **DREQO - DREQ3 :** DMA Request: There are four independent asynchronous input signals to the DMA channels which can be used by different peripherals such as floppy disks and hard disk drives. A request is generated by activating the DREQ lines of the channel to obtain DMA service.

- **DACK 0 - DACK 3:** DMA Acknowledge: These are output lines to inform the indi- vidual peripherals that a DMA has been granted. DREQ and DACK are equivalent to hand shake signals in I/0 devices.

- **AEN and ADSTB - Address Enable and Address stobe:** These are active high output signals that are used to latch a high-order address byte to generate a 16-bit address.

- **$\overline{\text{MEMR}}$ and $\overline{\text{MEMW}}$: Memory read and Memory Write Strobes:** These output signals used during the DMAcycle to write and read data bytes from memory.

- **A_3 - A_0 and A_7 - A_4 : Address:** A_3 - A_0 are bidirectional address lines. They are used as input to access different control registers. These lines are used as output lines during the DMA time cycle to generate a low-order address that is combined with the remaining address lines A_7 - A_4.

- **HRQ and HCDA:** Hold Request and Hold Acknowledge: HRQ is an output sigr.al used to request the 8088 for control of the system bus. The 8088 completes the bus cycle in process and issue the HLDA signal after receiving the HRQ.

10.7.5.1. System Interface

As previously stated that the DMA technique is used to transfer data bytes between I/O (such as a floppy disk) and system memory (or from memory to memory) at high speed. It consists of eight data lines, four control signals ($\overline{\text{lOR}}$, $\overline{\text{lOW}}$, $\overline{\text{MEMR}}$, and $\overline{\text{MEMW}}$), and eight address lines (A_7 - A_0). However, it needs 20 address lines to access 64 K bytes. Thus an additional eight lines must be generated as shown in figure 10.34.

The DMA places the low order byte on the address bus and the high-order byte on the data bus and asserts AEN (Address Enable) and ADSTB (Address strobe) when a transfer begins, these two signals are used to latch the high order byte from the data bus; thus, it places the 16 bit address on the system bus. The latch is updated when the lower byte generates a carry (or borrow) after the transfer of the first byte Two latches: One latch (373 # 2) to latch a high-order address from the data bus by using the AEN and ADSTB signal, and the second latch (373 #2) to demultiplex the 8085 bus and generate the low order address bus by using the ALE (Address Latch Enable from the 8085) Signal as shown is figure 10.34. The AEN signal is connected to the OE signal of the second latch to disable the low order address bus from the 8085 when the first latch is enabled to latch the high-order byte of the address.

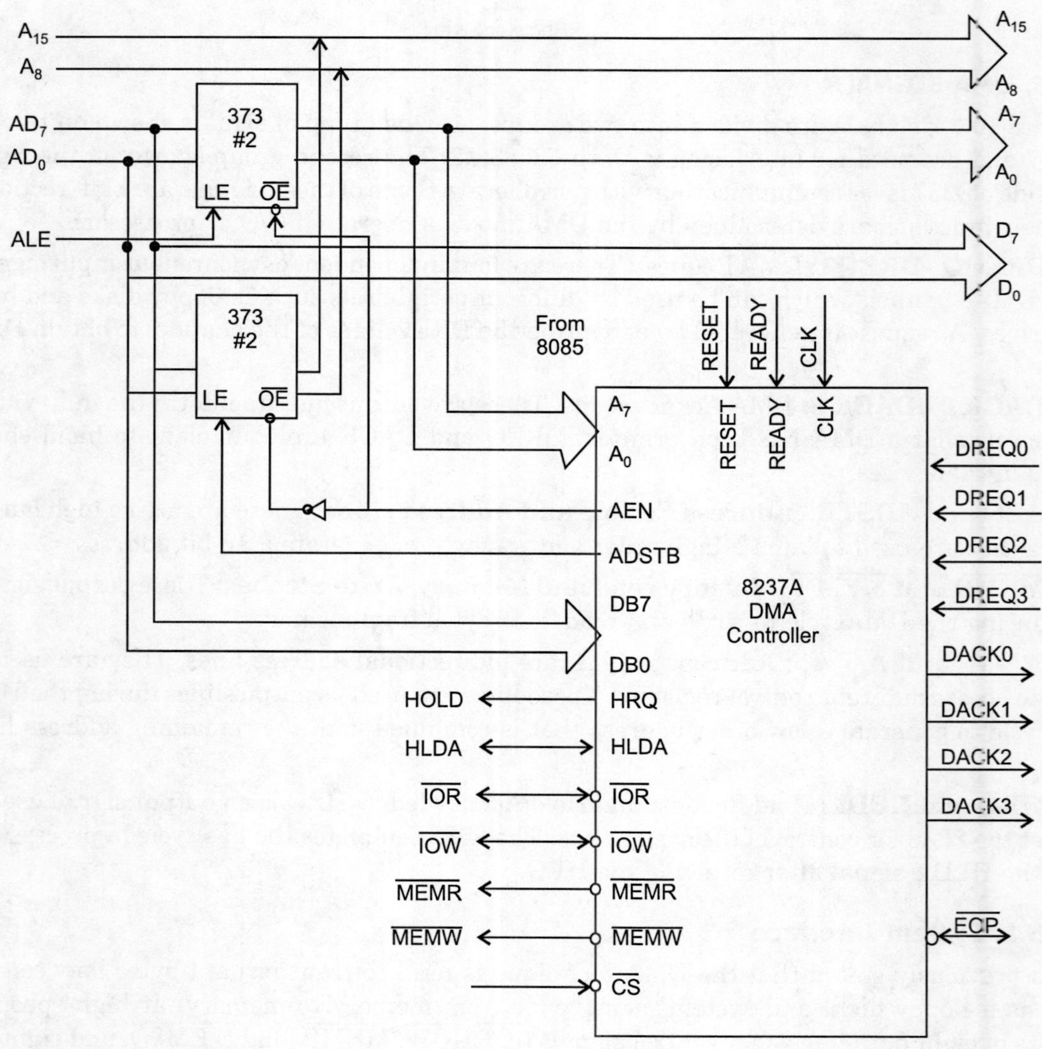

Figure 10.38. Interfacing 8237A-DMA Controller with the 8085

PROGRAMMING THE 8237 : The 8237 should be initialized by writing two various control registers to implement the DMA transfer. The following steps necessary to initialize the 8237.

1. Write a control word in the mode register that selects the channel and specifies the type of transfer (Read, Write, or Verify) and the DMA mode (block, single byte etc.)
2. Write a control word in the command register that specifies parameters such as priority among four channels, DREQ andDACK active levels, and timing and enables the 8237.
3. Write the starting address of the data block to be transferred in the channel Memory Address Register (MAR).
4. Write the count (the number of the bytes in the data block) in the channel count register.

10.7.5.2. DMA EXECUTION

We can classify the process of data transfer from the peripheral to the system memory under the DMA controller into two modes: the slave mode and the master mode.

Slave Mode: The DMA controller in the slave mode is treated as a peripheral; using the following steps:

1. The processor selects the DMA controller through the chip select
2. The processor writes the control word in channel registers and command/status registers by using control signal $\overline{IOW}$ and $\overline{IOR}$.

In this mod", the output signals of the 8237, such as $A_7 - A_4$, MEMW, and MEMR are in tristate.

Master Mode: The 8237 after the initialization in master mode keeps checking for a DMA request, and if requested steps in. Data transfer will then follow the following steps:

1. It sends a high signal to DRQ so as to prepare the peripheral is ready for data transfer.
2. The control logic sets HRQ high, when the DRQ has been received and the channel enabled.
3. In the next cycle, the processor relinquishes the buses and sends the HLDA signal to the 8237.
4. The DMA asserts AEN signal high after receiving the HLDA signal. The high AEN signal disables 373 latch #2 thus disconnecting the demultiplexed bus $A_7 - A_4$ of the processor and enables 373 Latch # 1 through an inverter. Next, the DMA asserts ADSTB (Address strobe high that is connected to Latch enable (LE) of 373 Latch # 1 and places the contents ofthe data bus, which is a high order byte of the starting address, on $A_{15} - A_8$. At the same time, the DMA also outputs the low order address $A_7 - A_0$ on the low address bus.
5. The DMA sends DACK to the peripheral when the entire address A_{15}-A_0 is available on the address bus.
6. The DMA controller continues the data transfer by asserting the necessary control signals ($\overline{IOR}$, $\overline{IOW}$, $\overline{MEMR}$ or $\overline{MEMW}$) until DACK remains high.
7. The DMA asserts $\overline{EOP}$ (End of process) signal low at the end of the/data transfer that can be used to inform the peripheral that the data transfer is complete. The DMA data transfer can also can be terminated by sending a low signal to $\overline{EOP}$ from an external source.

10.8. RS-232 INTERFACE

The RS- 232 standard defined a scheme for asynchronous communications, where there is a specified timing between data bits but no fixed timing between the characters that the bits form. In contrast, synchronous communication requires specific timing between bits and characters, and the clock for recovering data bits must be synchronized to or derived from the bit stream. Asynchronous systems are simpler but transmit data at much lower rates and are therefore less efficient than synchronous system.

Communication defined by RS·232 is serial data communication. There is a single wire, or link, for each direction of data flow, and the bits of the message are sent in sequence one at a time. Compared to parallel systems where multiple lines carry many data bits simultaneously, serial communication requires less circuitry at either end of the data link, is serial systems have lower through put than parallel systems, but in practice a parallel link is practical only over short distances of a few feet.

Although the RS-232 standard defines a serial system with just a single wire for each direction, the standard allows for other signal wires between the DTE and DCE as well. These additional signals are used to manage the data interface and to indicate communications status at any time to both the DTE device and the DCE device.

The RS-232 specification is intended to provide reliable communication up to a distance of soft, at rates up to 20,000 band. However, RS-232 interconnection should not be longer than soft, it has been successfully used for longer distance. Since the limiting factors of cable capacitance and noise are much less critical when lower band values are used; there are successful RS-232 links several hundred feet long, although this exceeds what the standard is designed to allow.

Finally, the phrase serial ASCII format is often used in conjunction with RS-232. The format standard itself does not specify at all what bit patterns should be used to represent the actual information being transmitted. Most commonly, ASCII representation is used. In special circumstances, however non-ASCII format are used, especially where only number are being sent. As we SRW in the discussion of ASCII, it is relatively in-efficient for numerical information since it requires a seven or eight bit field for each digit of the entire number. In contrast a straight binary format can express values up to 255 in an eight bit field. However, unless noted otherwise, we will assume that the RS-232 data are in ASCII format. RS-232 is primarily a signal voltage and timing, DTE·to·DCE interface standard. Just as it does not discuss how the data should be represented by the bits, it does not define the over all message format and protocol. The definitions of these areleft to the developers of the communications system. This means that RS-232 provides a great deal of flexibility, but also can have problems when signal levels and timing agree but message format, content and protocol differ between the two DTE devices that it inter connects in a complete system.

The RS-232 standard defined twenty five signal lines. This does not means that every DTE to-DCE mterface requires 25 signal wires- many RS-232 interfaces use just a few. The lines are divided into four groups: data, control, timing and specifies ground for protection and to make sure that the DCE and DTE chasis are at the same electric potential. In practice, the signal ground and chasis ground are often the same wire or the chasis ground is omitted, which are really violations of the standard's requirements; fortunately, this usually works, but it can cause operational (or even safely) problems.

The data lines are the most important signals. Received data and transmitted data lines permit full-duplex communication between the DTE and DCE. To make sure that there is no confusion on the direction of data flow on the received data and transmitted data lines, the RS-232 standard specifies data flow directions from the perspective of the DTE: Data go from the DTE to the DCE on the transmit data line and data go from the DCE to the DTE on the received data line from the perspective of the DCE, then data are transmitted on what is called the received data line and received on what is called the transmitted data line. This may seen confusing but it really avoids the more serious problem that will occur if both units try to transmit on the same line (signal clash) or receive on the same wire (there is no signal present, but there are two listeners on the same line and no signal source).

Line Pin	Description	EtA Destinations
1	Protective ground	AA
2	Transmitted data	BA
3	Received data	BB
4	Request to send (RTS)	
5	Clear to send (CTS)	CA
6	Data set ready (DSR)	CB
7	Signal ground (common return)	AB
8	Received line signal detector (data carrier detect)	CF
9	Reserved for data set testing	
10	Reserved for data set testing	
11	Unassigned	
12	Secondary received line signal detector	SCF
13	Secondary clear to send	SCB
14	Secondary transmitted data	SCA
15	Transmissional signal clement timing (CDE source)	DB
16	Secondary received data	SBB
17	Receiver signal element timing (DOE source)	DD
18	Unassinged	
19	Secondary request to send	SCA
20	Data terminal ready	CD
21	Signal quality detector	CG
22	Ring indicator (R)	CE
23	Data signal rate selector (DTEIDCE source)	CA/CI
24	Transmit signal element timing (DTE source)	DA
25	Unassigned	

Figure 10.39. Function of RS-232 lines by pin number

	Line	Description	Gnd	Data		Control		Timing	
	Pin		DCE	From DCE	To DCE	From DCE	To DCE	From DCE	To
Data	1	Protective ground	x						
	7	Signal ground/common return	x						
Control	2	Transmitted data		x					
	3	Received data	x						
	4	Request to send (RTS)				x			
	5	Clear to send (CTS)			x				
	6	Data set Ready (DSR)			x				
	20	Date-terminal ready (DTR) .				x			
	22	Ring detector			x				
	8	Received line signal detector or data carrier detect (DCD)			x				
	21	Signal quality detector			x				
	23	Data signal rate selector (DTE)				x			
	23	Data signal rate selector (DCE)			x				
Timing	24	Transmitter signal element timing (DTE)						x	
	15	Transmitter signal element timing (DCE)					x		
	17	Receiver signal element timing (DCE)					x		
Secon-	14	Secondary transmitted data		x					
dary	16	Secondary received data	x						
group	19	Secondary request to send				x			
	13	Secondary clear to send			x				
	12	Secondary received. time signal detector			x				

Figure 10.36. RS-232 lines, organized by functional grouping together with direction of signal flow

The control lines are next in importance are shown in figure 10.36. These lines are used for hand shaking, so either the DTE or the DCE can signal to the other that there are data to be transmitted and can indicate to the other if either DTE or DCE is ready to accept new data. Four control lines are used most frequently:

1. Request to send (RTS) from the DTE signals the DCE that the DTE has new data it would like to transfer.
2. Clear to send (CTS) from the DCE indicates to the DTE that the DCE can accept the new data.
3. Data terminal ready (DTR) from the DTE is another indication that the DTE is ready. Note that in the RS-232 standard the hand shake lines are asserted or active (binary value = 1) when they are in the space zone of +3 to 3 to +25V, which also corresponds to a binary 0 for data. This sometimes causes confusion when trouble shooting since an asserted control line is said to be in the binary 1 state, yet its signal voltage is the same as a data bit that is binary 0. As long as everyone follows the standard, however, there is no operating signal.

Using these hand-shaking lines, the complete data transmission sequence is a series of repeated requests to send new data, a handshake showing that the request can be honoured, sending the data, a handshake indication that new data cannot be accepted because either the DCE or the DTE is full or the transmission link is unavailable, then watching for the "OK to send" signal finally appearing and so on. The other control lines let the DqE indicate to the DTE that it has detected a ringing signal or that the communications line signal quality is acceptable among other function.

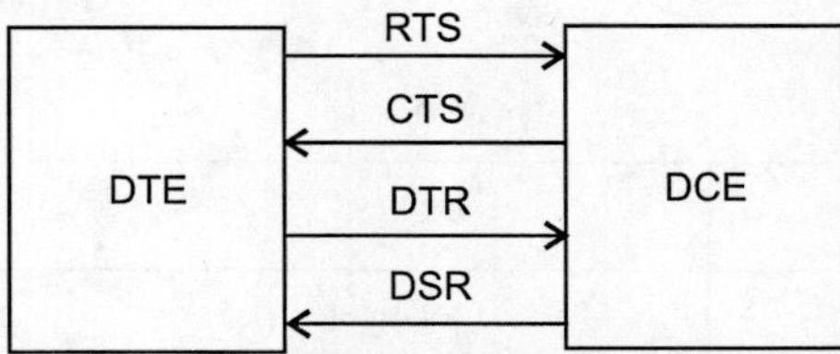

Figure 10.40. Four key control (handshaking) lines for DTE/OCE interfacing

10.8.1. RS-232 CONNECTORS

For many years, a 25-pin D-shaped connector with 13 pins. On one row and 12 on a second row was most common connector for RS-232 as shown in figure 10.41. However, since many of the 25 wire specified by the standard are not used, this is wasteful and costly in terms of wire terminations and circuit board space. Newer systems often use a smaller connector such as a nine pin D- shaped unit, or even a smaller type with just a few wires at the DTE end of the able However most DCE devices, such as printers, terminals or modems, continue to use the 25-pin D-shaped connector, although this is changing. The figure shows the pin assignments based on the 25-pin connector.

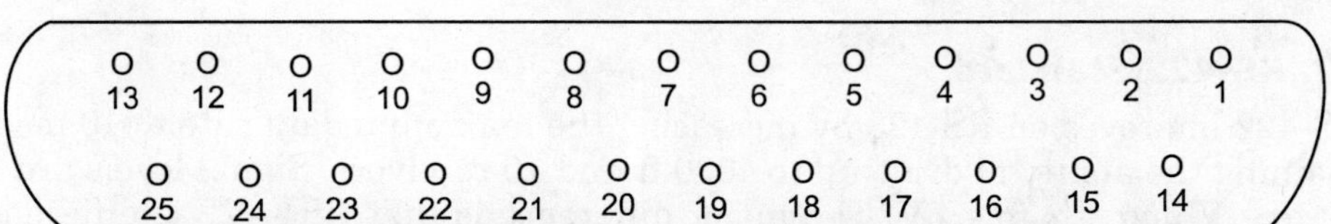

Figure 10.41. Pin diagram of 25 pin connector used for interfacing

10.9. OTHER COMMUNICATIONS INTERFACE STANDARD

The RS-232 interface standard is among the most common but it does have some technicalweaknesses. Most of these-except for low speed-arise not from the original application of connecting a DTE to its DCE device, but from the extended uses that have been added since the standard was developed. These include interfacing computer to printers, electronic instrumentation to other instruments, or using data systems in harsh electrical environments. Such as industrial plants. The summarized limitations ofRS-232 include:

(1) 50 ft is the maximum guaranteed distance from one end to the other.
(2) The 20 kiloband rate is too slow for many applications.
(3) The standard is designed for point to point links where there are just two users on the link.
(4) The single ground wire of RS-232 that is used as the OV reference for all signal can pick up noise and suffers voltage drops and radiated signals which is turn cause errors.
(5) The –25 to –3V and +3 to +25 signal levels are not compliable with most newer electronic equipment which has + 15V or +5V power supplies.

To overcome these weaknesses of RS-232 in some applications other E/A standards were developed. We will discuss these in the following section.

10.9.1. RS- 423 Standard

RS- 423 allows for distances to 4000ft, data rates to 100 K bits/s, and up to 10 line receivers driven by a single driver, as shown in figure 10.41. The standard achieve this distance and ability to support 10 receivers by using drive circuitry that produces more current along with line receiver input circuitry which draws less current.

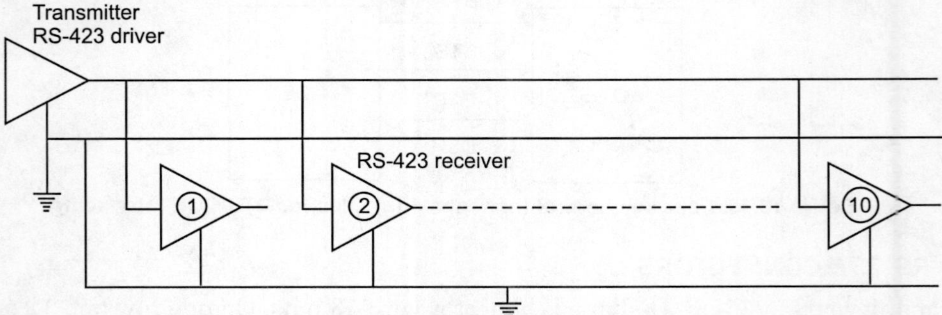

Figure 10.42. RS-423 with one driver, capable of driving to receiver

By lowering the voltages to - 3.6 to - 6V for a binary and +3.6 to + 6V for a binary 0, we can accomplished the increase in speed. This reduces the maximum slow rate requirement on the line driver by a factor of 4 compared to the RS - 232 standard (6V versions 25V) and are uses that the line driver output can transverse the entire maximum voltage span in less time (12V versons 50 swings), yielding faster data rates RS - 423 is a single ended specification like the RS- 232 standard, with each signal represented by a single wire and one ground wire common to all signals.

10.9.2. RS-422 Standard

RS-422 improves on RS-423 by increasing the maximum data rate to 10 mega bits/s while maintaining the ability to drive up to 4000 ft and 10 receivers. Signal levels are similar to RS-423 -2 to -6V and + 2 to + 6V. The major difference is that RS - 422 is differential standard. Each signal is represented by a pair of wires and the voltage difference across these wires is what is sensed at the receiver. This minimizes the effect of ground noise or the voltage drop along the signal leads. The signal arrives at the receiver with less noise since the ground wire noise is irrelevant and with larger useful amplitude (the voltage difference between the wires, not the voltage between a signal wire and ground).

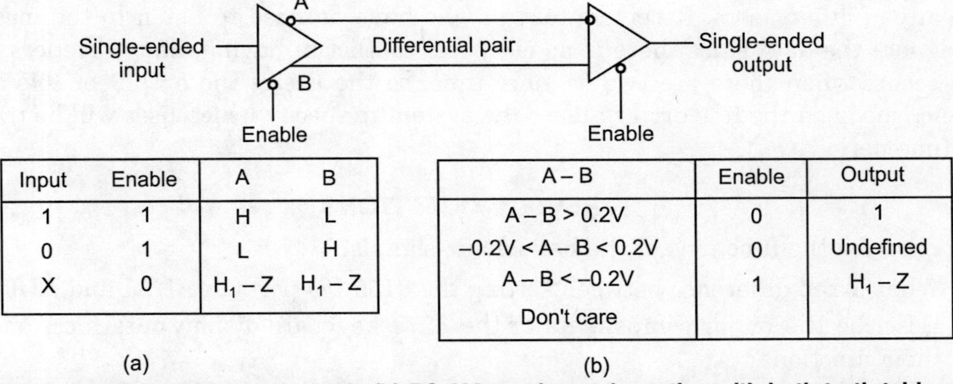

Input	Enable	A	B
1	1	H	L
0	1	L	H
X	0	$H_1 - Z$	$H_1 - Z$

A – B	Enable	Output
A – B > 0.2V	0	1
–0.2V < A – B < 0.2V	0	Undefined
A – B < –0.2V	1	$H_1 - Z$
Don't care		

Figure 10.43. (a) RS-422 driver schematic (b) RS-422 receiver schematic· with both truth table showing voltage levels and logic states.

The penalty for using RS -422 is that two wires are needed for each signal, in contrast to the one wire per signal used in RS -232 and RS -423 (plus a signal common ground for all). Line drivers for RS -422 take the single ended 5V signal with in the circuitry and produce a differential output on two wires as shown in figure 10.43(a). When the output is a binary 0, the A output is greater than B; the situation is reversed when the output is a binary 1. An RS - 422 line receiver as shown in figure 10.43(b) performs the reverse operation, taking the two differential wires and producing a 0 Vor IV single - ended output based ori the relative difference between input is a binary 1 at 5V, while difference is 0.2V, produces a binary 0 at OV.

10.9.3. RS-485 Standard

RS-485 is a standard that expand on RS - 422 by allowing the same speed and distance and up to 32 receivers, with signal levels of - 1.5 to - 6V and + 1.5 to + 6V. However, the major point of enhancement is that RS 485 allows multidrop operation, where more than one transmitter can be connected on a line - though not active - at a time as shown in figure 10041. Although only a single transmitter can be operated at any instant, the standard specifies that up to 32 transmitters can be connected on the differential wire pair at the same time. This is done by making sure that line drivers for RS -485 can be disabled, using a single control pin on the line driver LC, When disabled, the line driver goes to a high-impedance mode*, where it draws virtually no current and produces neither a binary o nor 1. As far as the system is concerned, these disabled drawers are not present on the line.

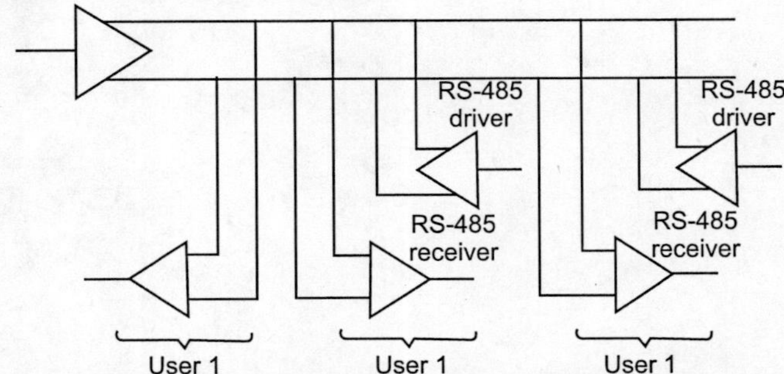

Figure 10.44. RS-485 multidrop standard can support multiple transmitters

The RS -485 standard is useful when many user points must be connected together, with a single user sending data at any time. To ensure that one and only one sender is active, the system circuitry must take special actions to instruct the users as to whose turn it is to become active. This forms the basis for the back bone of a network where messages can be sent bit any user to any of other users. Receivers (users) who have no need to listen to the message can simply ignore the data since there is no electrical conflict in having many listeners. The only conflict occurs when there are two or more time, so the use of the output enable and high-impedance mode on the line driver allows the system to specify which user will be transmitter at any time period.

REVIEW QUESTIONS

10.1. Describe the functions of the handshake signals?

10.2. What is the difference between setting the 8155 I/O ports in ALT I and ALT $_3$.

10.3. Describe the major components of the 8279 keyboard/display interface, and explain their functions?

10.4. Port B of the 8155 is setup in the handshake mode, and the reading of status word is 20H. Is the port setup for status check or Interrupt I/O ?

10.5. List the operating modes ofthe 8255 A Programmable peripheral interface.

10.6. Specify the handshake signals and their functions when port A of the 8255 A is setup as an output port in Mode 1.

10.7. Specify the two control words that are necessary to set bit peG (assume all other ports are in active).

10.8. List the necessary conditions to generate INTR when port A of the 8255 A is setup as an output port in mode 1.

10.9. Specify the major conditions required to the start the timer 8254.

10.10. Describe the major components of the 8259A interrupt controller and explain their functions?

10.11. Explain how the 8237 DMA controller transfers 64 Kb of data per channel with eight addresslines.

10.12. Setup the 8254 as a square wave generator with a I ms period, if the input frequency to the 8254 is 1 MHz.

10.13. Describe the internal architecture and working

- 8155
- 8279
- 8254
- 8255
- 8259

●●●

Chapter-11

8086 INTERRUPT STRUCTURE, MEMORY INTERFACING AND MODULES

INTRODUCTION

In this chapter we shall discuss the 8086 interrupt as well as, the memory module that may be interfaced to an 8086 or 8088 microprocessor.

11.1. 8086 INTERRUPTS

Here, we shall discuss the 8086 hardware and software interrupts in great detail. We shall also present you with the 8086 interrupt instructions. There are two types of interrupts that are used in the 8086 microprocessor external or hardware interrupts and internal or software interrupts. Both these interrupts can interrupt the current program execution and transfer control to an interrupt service routine. After execution of the interrupt service routine the program control is transferred back to the program which again starts its execution.

11.1.1. Interrupt Vector or Pointers

The interrupt vector table is a predefined memory area from 00000H to 003FFH which is used to store addresses of the interrupt service routine. The interrupt vector in itself a four byte number which contains the segment address and offset address of the location where the interrupt service routine or procedure is stored.

The memory locations 00000H to 0003FFH are equivalent to 1024 bytes of memory and the interrupt vector itself takes 4 bytes (2 bytes for IP and 2 bytes for CS) giving us a total of 256 different interrupts vectors. Each of these 256 interrupts vectors address are assigned with a code called the interrupt type code.

Having the interrupt type code itself is sufficient to load the address of the interrupt service routine. Taking an arbitrary interrupt INT m where m is the interrupt code we can calculate the address in the interrupt vector table where the IP and CS values are stored for interrupt m. The IP and CS values are calculated as given below:

$$IP = 4 \times m$$

$$CS = 4 \times m + 2$$

For example to service the 8086 non-maskable interrupt (from NMI pin), the 8086 assigns the interrupt type code 2. Therefore, to obtain the interrupt address vector for INT2* (Interrupt 2) the 8086 would automatically calculate:

* The NMI interrupt is assigned the interrupt code 2.

$$IP = 4 \times 2 = 00008H$$

$$CS = 4 \times 2 + 2 = 0000AH$$

The 8086 would then automatically load the IP with the contents oflocation 00008H and the CS with contents oflocation OOOOAH. Thus the user must load the corresponding valuesof IP and cs in these memory locations. The value of IP and CS must also correspond to the start address of the program that service the INT2. The general layout for the interrupt vector table for the 8086 is given figure 11.1.

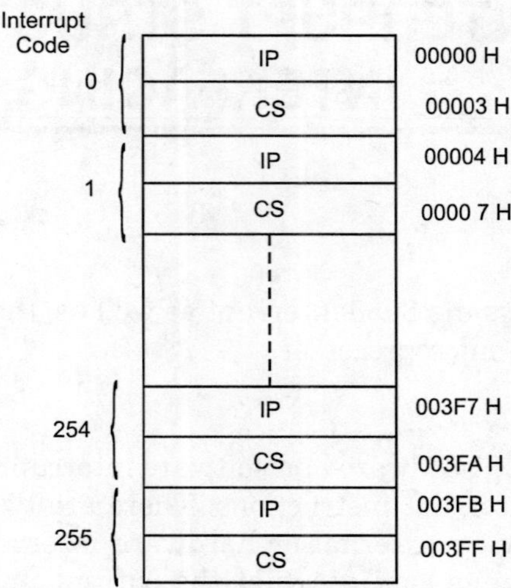

Figure 11.1. Interrupt vector table

However, on encountering an interrupt the 8086 will first push the flags, cs, IP register onto the stack and then reset the IF and TF flags. After this the CS and IP register will be loaded with the new values from the interrupt vector table. The interrupt service routine would now be executed (as the vector table contains the cs and IP values for the start of this routine). The interrupt service routine always ends with the IRET instruction which reloads (pops) back the IP, cs and Flag register from the stack, therefore, transferring control back to the main program. The IF and TF flag states are also restored. The 256 interrupt type codes are assigned as follows:

- Type 0 to 4 : predefined interrupts
- Type 5 to 31 : reserved for future purposes by intel
- Type 32 to 255 : user defined interrupts

11.1.2. Predefined Interrupts

The predefined interrupts are included for processor control and are not changable*. These five interrupts are:

1. Divide by Zero: type 0
2. Single Step Mode: type 1
3. Non-maskable Interrupt: type 2
4. Breakpoint: type 3
5. Interrupt on overflow: type 4

* These interrupts will always be assigned the above given interrupt codes.

The divide by zero interrupt automatically interrupts the 8086 when a divide error occurs. This interrupt will occur every time the divide instruction executes and a divide by zero is encountered.

The single step interrupt or type 1 interrupt occurs whenever the TF (trap flag) is set by an instruction. When the TF flag is set is causes the microprocessor to service the type 1 interrupt after execution of each instruction. This interrupt is of vital use during program debugging and also allows a programmer to write a service routine which allows him to check the processor state after every instruction.

The non-maskable interrupt or type 2 interrupt is initiated through the-8086 NMI pin. The NMI pin itself requires a positive edge signal which is active for atleast two processor clocks to cause the type 2 error. This is used for vital system functions such as power failures or external emergency shut down procedures. The 8086 automatically obtains the interrupt vector address from the vector table and start executing the service routine. This interrupt has the highest priority among all the interrupts. The type 3 interrupt or breakpoint is a very special single byte software interrupt and initiated by the instruction INT 3. This is generally used debugging software. The INT 3 interrupt causes the program execution to stop and the service routine to be serviced. This instruction can be put in any point of the program, where we want to check the state of the machine.

The interrupt on overflow is a type 4 interrupt and occurs when the OF flag is set and the INTO instruction executed. The overflow flag is affected by signed addition or subtraction operations. The INTO instruction can follow these instruction and can be used to detect the overflow condition and perform necessary correcting action if required.

11.1.3. Interrupt Instructions

The 8086 microprocessors have three different software interrupt instructions namely INTO, INT3 and INT. These instructions fetch a interrupt vector address from the interrupt vector table and then transfer control to the address located by the vector address. The interrupt call itself is similar to the far call because both place the return addresses (lP and CS) registers on the stack memory.

INT Instruction: The INT instruction has the general format listed below in figure 11.2. The first byte stores the opcode while the second byte stores the interrupt type code.

CD	Type
Opcode	Interrupt type no.

Figure 11.2. The INT instruction format

The interrupt vector address is obtained by multiplying the type number by 4. Therefore, INT 21H will begin at memory location 00084H.

On execution ofthe INT instruction the following operations take place:

1. Flag register is pushed onto stack
2. Trap Flag (TF) and Interrupt Flag (IF) are reset
3. Pushes the current CS onto stack.
4. Loads the new CS value from the interrupt vector address (obtained by 4 x Interrupt type code +2).

5. Pushes the IP onto the stack.
6. Loads the new IP value from interrupt vector table (from location 4X interrupt type number).
7. Transfers program control to the location addressed by the new CS and IP registers.

However, since the INT instruction when executed, disables the IF flag thereby disabling the INTR pin from requesting an interrupt. When the interrupt service routine is being serviced a programmer can allow INTR to interrupt the processor when servicing an ISR by -setting the IF in the start of ISR*.

IRET INSTRUCTION

The interrupt return instruction is only used with software or hardware interrupt instructions. The IRET is used to return from interrupt service routines. The IRET instruction execution is as follows:

1. Reload IP register from stack
2. Reload CS register from stack
3. Reload the flag register

Following this the program control is transferred back to the next instruction which follows the interrupt call. All interrupts are re-enabled as the states of IF and TF are restored back to original state due to the flag register reload from stack.

The other 8086 interrupt instructions are INTO and INT3 which have been discussed in section 11.1.2. These correspond to the interrupt on overflow (type 4) and breakpoint (type 3).

11.1.4. Hardware Interrupts

The 8086 software interrupts have been discussed uptill now. However. the 8086 is provided with two interrupt pins: NMI and INTR which can be used to externally interrupt the microprocessor. The NMI pin when activated causes a type 2 interrupt. The INTR pin when triggered can be used to cause any interrupt type whose interrupt type code must be applied to the 8086 data bus. The $\overline{\text{INTA}}$ or interrupt acknowledge pin is used to synchronize the reception of the interrupt type code on the data bus.

INTR Hardware Interrupt

The INTR or interrupt request pin responds to a high level input signal and interrupts the processor provided that the interrupt flag is enabled. If the above conditions are true the processor will complete the execution of the current instruction. Following instruction execution processor shall drive the $\overline{\text{INTA}}$ pin low for one bus cycle as shown in figure 11.3. It shall also drive the $\overline{\text{LOCK}}$ pin low so that a bus hold request will not be serviced until the next bus cycle. During the second bus cycle the interrupt type vector is fetched through the DATA BUS. This interrupt code is then multiplied by 4 and used as a pointer to the vector table. This effectively causes the service routine to be executed. Prior to this the processor would have pushed the flag Register. IP and CS registers on to the stack and disabled the IF and TF flags. Therefore. INTR cannot interrupt the program execution until the IF flag is set during the interrupt service routine.

A simple interrupt circuit interfaced to the 8086 which transfers the interrupt type 14 onto the data bus is shown below in figure 11.4.

---

* ISR = Interrupt Service Routine

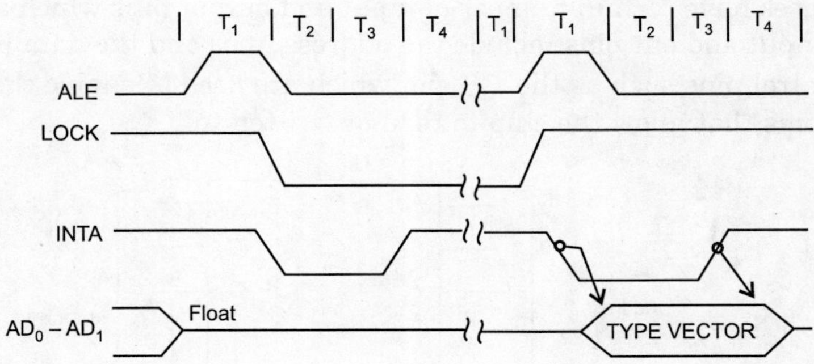

Figure 11.3. Interrupt Acknowledge Sequence

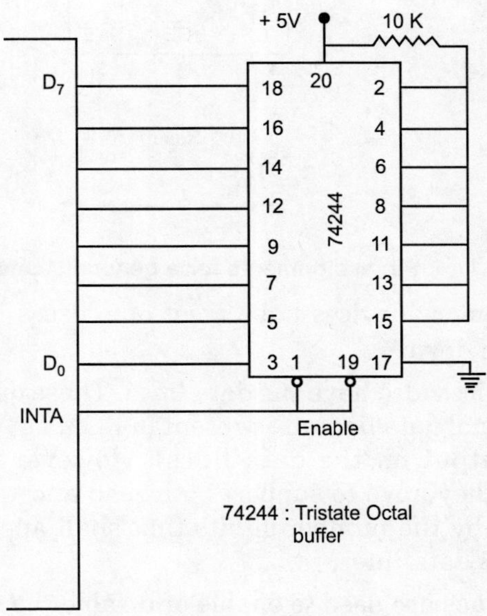

Figure 11.4. Circuit that loads INT FEH onto the DATA BUS

11.2. THE 8086 MEMORY MODULE

We here discuss the 8086 memory module with special consideration to the different memory devices available in the market today. We shall however provide you with some data sheets of readily available memory devices. First, we shall discuss SRAM and EEPROM devices. However, there are quite a few different types of memory chips available in the market such as in a EUROM's, SDRAM's, NVRAM's DRAM's and the latest memory chips: the flash memory device. However, we except the students to read the required data sheets before interfacing the memory device to the 8086 as we do not have sufficient space to give you all the different memory components available.

11.2.1. MEMORY PIN CONFIGURATION AND ASSIGNMENTS

All memory devices have certain number of input and output pins which are found on all such devices. These input and out pins include the address lines and the data lines. All devices also have certain control pins such as the $\overline{CS}$ pin which are used to enable the chip. Other control inputs include pins that allow the chip to read or written to.

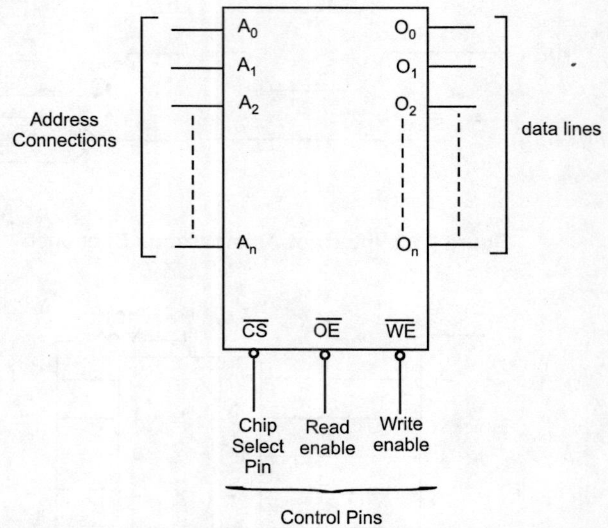

Figure 11.4. Pin assignments for a general Memory device

Address Iines: All memory devices have a set of address lines that are used to select a memory location within the device.

Data lines: Memory devices also have the data lines. These data lines may be unidirectional or bidirectional. Unidirectional data lines are present in ROM as they only allow a read operation (i.e., Data will only be output on the data lines). However RAM's and DRAMS contain bidirectional data lines as they have to support both read and write operation. Manufacturers typical label the data lines by the number of bits that shall appear. For example a 8 bit wide memory device shall have 8 data lines.

Selection lines: These lines are used to enable or disable the memory device. This input pin is often labeled as chip select ($\overline{CS}$) or chip enable ($\overline{CE}$). Many memory devices have more than one enable pins therefore requiring all to be active to select the device.

Control lines : Memory devices have a set of control inputs that is used to enable the devices to be read or write to. ROM's have one control input while RAM's have two control inputs. The control input on ROM is the output enable ($\overline{OE}$) pin. If the $\overline{OE}$ and $\overline{CS}$ pins of the ROM are both active the ROM places the data on its output lines. However if $\overline{OE}$; is inactive the output lines feet to high impedance state.

RAM memory devices contain two control inputs: One to enable the RAM to be read from while the other enables the RAM to be written to. Some RAM contain only one pin that controls both the function. In general the RAM pins are labeled $\overline{OE}$ and $\overline{WE}$. The $\overline{OE}$ pin enables the RAM to place data on the output lines while $\overline{WE}$ enables the data on data lines to be written to the RAM. However care must be taken that both pins donot become active at the same time.

11.2.2. The Rom Memory Device

The ROM or Read only memory is used for permanent storage of data. This is un-affected by the change in power supply. Since the ROM is permanently programmed so data is never lost, therefore, it is also called non-volatile memory. There are many types of ROM devices available such as PROM, EPROM, EEPROM etc.

PROM (Programmable Read only Memory)

PROM devices are for programmable devices that may be programmed in the field. However PROM's are one time programmable devices and therefore are not commonly used.

EPROM (Electrically Programmable Read Only Memory)

EPROM devices can be programmed by use of electricity. EPROM contents can be erased by exposure to UV light for about 20 minintes. However EPROM devices require a special device called EPROM programmer which is required for programming of the device.

EEPROM (Electrically Erasable PROM)

EEPROM devices are erasable and programmable by electric voltage. These/are the most widely used ROM devices.

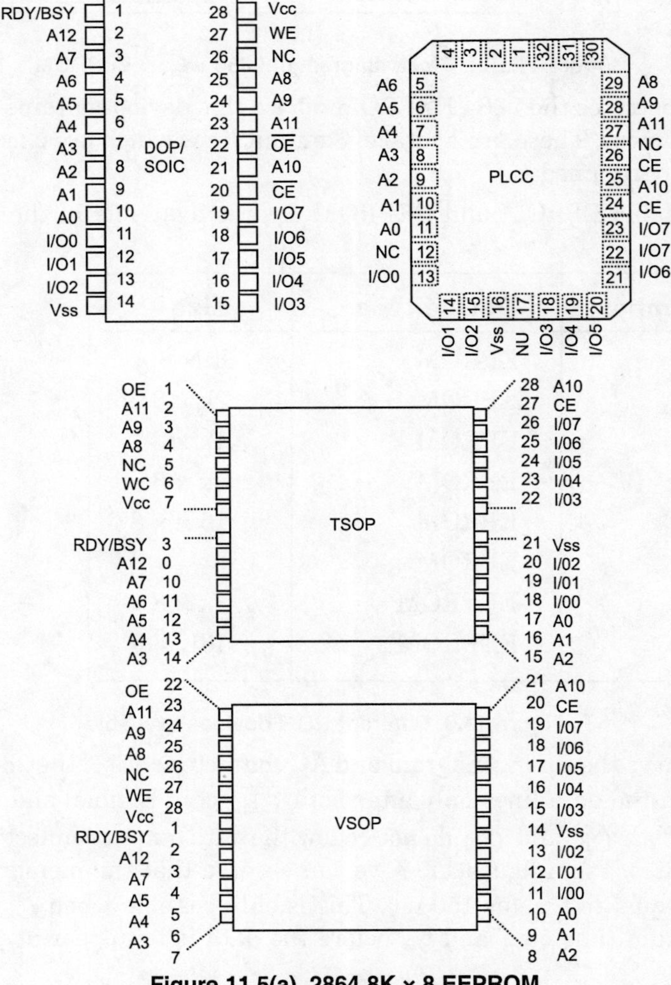

Figure 11.5(a). 2864 8K × 8 EEPROM

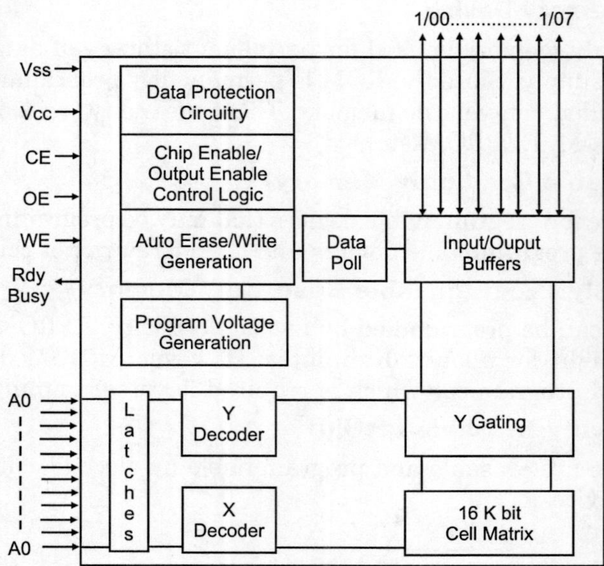

Figure 11.5(b). Block diagram of 2864 8K × 8 EEPROM

Figure 11.5. illustrates the 2864 EEPROM device this device contains 13 address lines and is a 8K x 8 ROM memory. These are a total of 8 output lines which are used as input lines when the EEPROM is programmed.

The other available EPROM and EEPROM devices available in the market are listed in figure 11.6.

DeviceNumber	Device type	Size
2708	EPROM	lk × 8
2716	EPROM	2k × 8
2732	EPROM	4k × 8
2764	EPROM	8k × 8
27128	EPROM	16 k × 8
27256	EPROM	
2864	EEPROM	8k × 8
28128	EEPROM	16K × 8

Figure 11.6. Different ROM devices available

Figure 11. 7. shows the timing diagram and AC characteristics of the 2864 EEPROM devices. Data will appear on the data lines only after both $\overline{CE}$ (chip Enable) and $\overline{OE}$ (Output Enable) go low. However if both $\overline{CE}$ and $\overline{OE}$ do not go low the data output connection will remain in the high impedance states. From figure 11. 7 we can observe that the memory access times of the 2864 EEPROM is equal to t_{ACC} (or 150 ns). This is only possible when $\overline{CE}$ and $\overline{OE}$ have gone low before a minimum time oft_{CE} and t_{OF} before the data is to appear at the output. Since the memory access time of the 2864 EEPROM device is much less that the 460 ns memory access time allowed by the 8086 microprocessors, there will be no problem in interfacing this device to the microprocessor.

Maximum Ratings*

Vcc and input voltages w.r.t. Vss........... –O.6V to + 6.25 V

Voltage on OE w.r.t. Vss........... –O.6V to + 13.5 V

Voltage on A9 w.r.t. Vss........... –O.6V to + 13.5V

Output Voltage w.r.t. Vss........... –O.6V to Vee + O.6V

Storage temperature........... –65°C to + 125° C

Ambient temp. with power applied........... –50°C to + 95°C

Name	**Function**
A0 – A 12	Address Inputs
$\overline{CE}$	Chip Enable
$\overline{OE}$	Output Enable
$\overline{WE}$	Write Enable
$I/O_0 - I/0_7$	Data Inputs/Outputs
RDY/ $\overline{Busy}$	Ready/ $\overline{Busy}$
Vcc	+ 5V Power Supply
Vss	Ground
NC	No Connect; No Internal Connection
NU	Not used; No External Connection is allowed

(A) Maximum Ratings And Pin Functions

Readiwrite Operation Dc Characteristic

V_{CC} = + 5V ± 10%
Commercial (C): T_{amb} = 0°C to + 70°C
Industrial (1): T_{amp} = –40° C to + 85°C

Parameter	Status	Symbol	Min	Max	Units	Conditions
Input Voltage	Logic '1'	V_{IH}	2.0	V_{CC} + 1	V	
	Logic '0' .	V_{IL}	–0.1	0.8	V	
Input Leakage	–	V_{IL}	–10	10	μA	V_{IN} =–O.lV To VCC + 1
Input Capacitance	–	CIN	-	10	pF	V_{1N} = OV; Tamb = 25°C; *f* = 1 MHz (Note 2)
Output Voltage	Logic '1'	VOH	2.4		V	I_{OH} = – 400 μA
	Logic '0'	VOL	0.45		V	I_{OL} = 2.1 mA
Output Leakage	–	ILO	–10	10	JlA	V_{out} = –O.lVto VCC + O.1V
Output Capacitanc	–	COUT	–	12	pF	V_{1N} = OV; Tamb = 25°C; *f* = 1 MHz (Note 2)

Power Supply Current, Active	TTLinput	I_{CC}	–	30	mA	f = 5 MHz (Note 1) V_{CC} = 5.5V
Power Supply Current, Stand by	TTL input	ICC(s) TTL	–	2	mA	CE = VIH (O°C to + 70°C)
	TTLinpul	ICC(s) TTL			3	mA CE= VIH (–40°C to + 85°C)
	CMOS input	ICC(s) CMOS		100	mA	CE = VCC–0.3 to VCC + 1

(b) Read/write DC characteristics

Figure 11.7(a). 2864 EEPROM Characteristics

(c) READ OPERATION AC CHARACTERISTICS

AC Testing Waveform;	V_{IH} = 2.4; V_{IL} = 0.45V; V_{OH} = 2.0V; V_{OL} = 0.8V		
Output Load:	1 TTL Load + 100 pF		
Input Rise and Fall Times	20 ns		
Ambient Temperature:	Commercial	(C)	Tamb = O°C to + 70°C
	Industrial	(1)	Tamb = –40°C to + 85°

Parameter	Symbol	28C64A-15		28C64A-20		28C64A-25		Units	Conditions
		Min	Max	Min	Max	Min	Max		
Address to Output Delay	t_{ACC}	–	150	–	200	–	250	ns	$\overline{OE} = \overline{CE}$ = VIL
$\overline{CE}$ to Output Delay	t_{CE}	–	150	–	200	–	250		$\overline{OE}$ =VIL
$\overline{OE}$ to Output Delay	t_{OE}	–	70	–	80	–	100	ns	$\overline{CE}$ =VIL
$\overline{CE}$ or $\overline{OE}$ High to Output Float	t_{OFF}	0	50	0	55	0	70	ns	Note 1
Output Hold from Address, $\overline{CE}$ or $\overline{OE}$, whichever occurs first,	t_{OH}	0	–	0	–	0	0	–	note 1

(D) READ WAVEFORMS

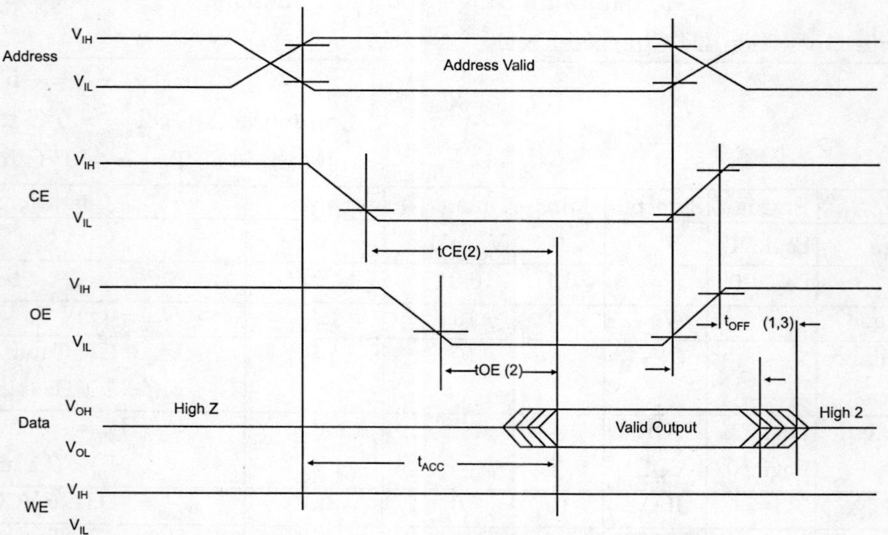

Figure 11.7 (b) Tuning diagram for Read Cycle

(Note the 8716 EPROM has a memory access time of 450 ns which makes doesnot allow it to be properly accessed by the 8086 systems which has been buffer. This is due to delays that occur due to the buffer and the minimal lines gap. To interface a 2716 we may have to introduce wait states to allows proper interfacing write the 8086.)

The 2864 write (programming) timing cycles have also been listed in figure 11.8. We will not explain these to you but expect you to understand these as an exercise.

AC Testing Waveform:	VIH = 2.4V; VIL = 0.45V; 2.0V; VOL == 0.8V
Output Load:	1 TTL Load + 100 pF
Input Rise/Fall Times:	20 ns
Ambient Temperatures:	Commercial (C) : Tamb = O°C to +70°C
	Industrial (I) : Tamb = –40°C to +85°C

Parameter	**Symbol**	**Min**	**Max**	**Units**	**Remarks**
Address Set-up Time	tAS	10	–	ns	
Address Hold Time	tAR	50	–	ns	
Data set-up Time	tDS	50	–	ns	I
Data Hold Time	tDH	10	–	ns	
Write Pulse Width	tWPL	100	–	ns	Note 1
Write Pulse High Time	tWPH	50	–	ns	
OE Hold Time	tOEH	10	–	ns	
OE Set-up Time	tOES	10	–	ns	
Data Valid Time	tDV	–	1000	ns	Note 2
Time to Device Busy	tDB	2	50	ns	
Write Cycle Time (28C64A)	tWC	1	µs		0.5 µs typical
Write Cycle Time (28C64AF)tWC	200	µs			100 s typical

Note1: A write cycle can be initiaited be CE or WE goinglow, whichever occurs last. The data is latched on the positive edge WE,whichever occurs first.

Note 2 :Data must be valid within 1000ns max. after a write cycle is initiated and must be stable at least until fDH after the positive edge of $\overline{WE}$ or $\overline{CE}$, whichever occurs first

Figure 11.8 (a) Byte Write AC Characteristics

11.2.3. Ram Devices

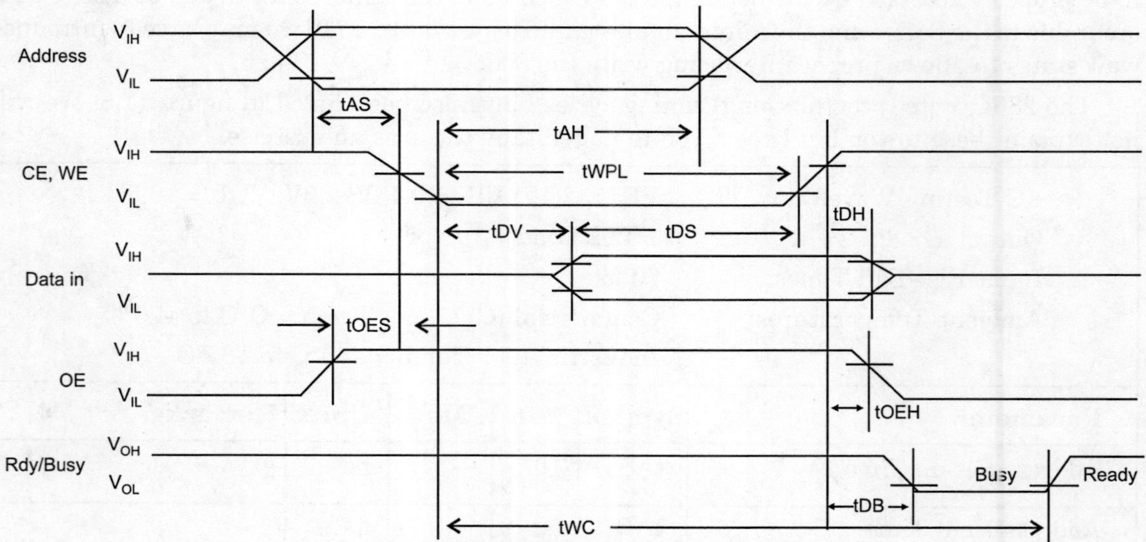

Figure 11.8. (b) 2864 Wrtte Cycle Tuning Diagram

RAM memory components are called Random Access Memory. RAM's are volatile memory devices because they cannot retain the data stored when the power has been switched off. These are two RAM devices of importance to us.

(*i*) SRAM, (*ii*) DRAM

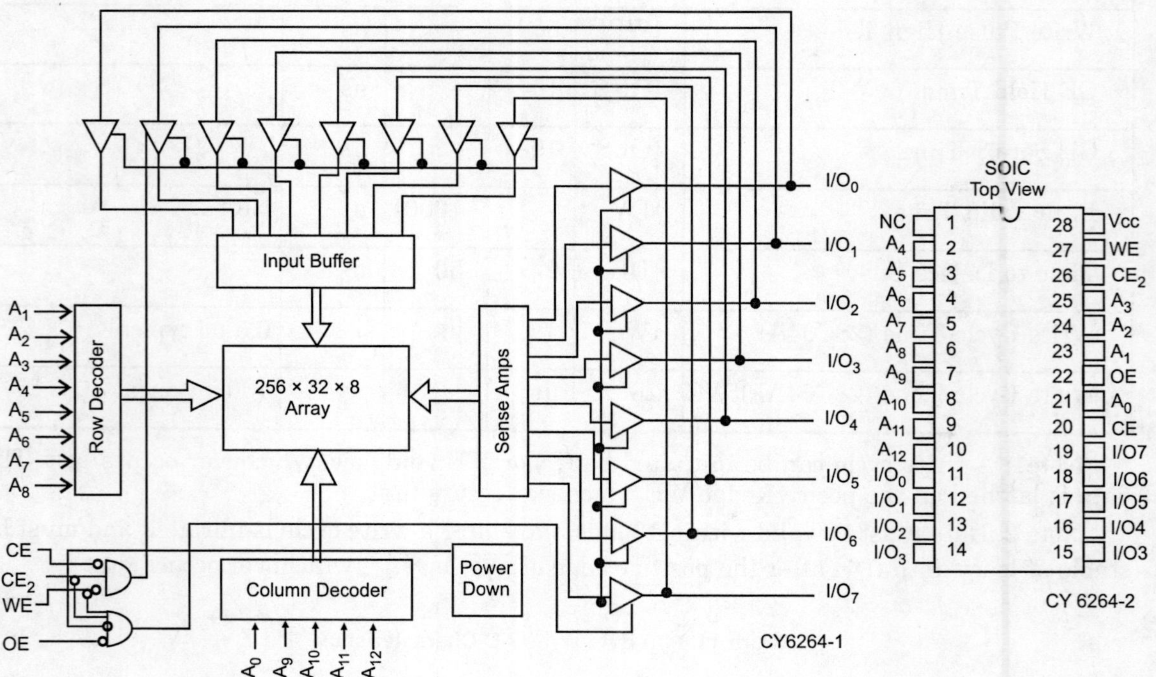

Figure 11.9(a) Cypress 6264 SRAM logic diagram and pin out

SRAM : SRAM devices are called static Random access memory devices as they retain the data as long as the device is supplied with power. The RAM devices differ from ROM devices as they can be read as well as written to. The RAM devices are used to store temporary data that the microprocessor is working on.

Figure 11.9(a) shows the cypress 6264 CMOS SRAM devices. This RM1 device is a 8K x 8 memory device. You may note that the 6264 has a total of 13 address lines and 8 data lines (labeled 1/00 to 1I0s). These data lines may be used for inputting data to the SRAM or as output lines to read 'data from the SRAM memory device.

$\overline{CE_1}$	$\overline{CE_2}$	$\overline{WE}$	$\overline{OE}$	Input/Output	Mode
H	X	X	X	High Z	Deselect/Power-Down
X	L	X	X	HighZ	Deselect
L	H	H	L	Data Out	Read
L	H	L	X	Data in	Write
L	H	H	H	HighZ	Deselect

Figure 11.9(b) Control signal truth table

Figure 11.9(b) shows the control signals truth take for the 6264 SRAM device. The 6264 has four control signals CEl, CE2 ,OE and WE. Figure 11.9(c) shows the electrical characteristics of the 6264 SRAM device.

Parameter	**Description**	**Test Conditions**	**6264-55**		**6264-70**		
			Min.	**Max.**	**Min.**	**Max.**	**Unit**
V_{OH}	Output HIGH Voltage	V_{CC}=Min., I_{OH} = – 4.0 mA	2.4		2.4		V
V_{OL}	Output LOW Voltage	V_{CC} = Min., I_{OL} =8.0 mA		0.2		0.2	V
V_{IH}	Input HIGH voltage		2.2	V_{CC}	2.2	V_{CC}	V
V_{IL}	Input LOW Voltage	–0.5	0.8	–0.5	0.8	–0.5	V
I_{IX}	Input Load Current	GND $\leq V_1 \leq V_{CC}$	–5	+5	–5	–5	μA
I_{OS}	Output leakage Current	GND $\leq V_1 \leq V_{CC}$	–5	+5	–5	+5	μA
I_{CC}	Output Short Circuit Current[2]	V_{CC} = Max., V_{OUT} = GND		–300		–300	mA
I_{CC}	V_{CC} Operating Supply Current	V_{CC} = Max., I_{OUT} = 0 mA		100		100	mA
I_{SBI}	Automatic $\overline{CE}_1$ Power-Down Current	Max. V_{CC}, $\overline{CE}_1 \geq V_{IH}$ Min. duty cycle=100%	20		20	mA	
I_{SB2}	Automatic $\overline{CE}_1$ Power-Down Current	Max. V_{CC}, $\overline{CE}_1 \geq V_{CC} - 0.3V$. $V_{IN} \geq V_{CC} - 0.3V$or $V_{IN} \leq 0.3V$		15	15		mA

Shaded area contains advanced information

1. Minimum voltage is equal to -3.0V for pulse durations less than 20 ns.
2. Not more than 1 output should be shorted at one time. Duration of the short circuit shouldexceed 30 seconds.

CAPACITANCE

Parameter	Description	Test Conditions	Max.	Unit
CIN	Input Capacitance	TA = 25°C,	f = 1 MHz, 7	pF
COUT	Output Capacitance	V_{CC} = 5.0V	7	pF

Figure 11.9(c) electrical Characteristics of the 6264 SRAM device

Figure 11.10 shows the read and write timing diagram for the 6264 SRAM device 1 read cycle timing diagram reveals that the memory access time is t_{AA} This time is abo 70ns for the 6264 - 70 SRAM device. This very low time allows the 8086 to be direct, connected to the 6264 -70 SRAM without requirement of wait states. The write time required for the 6264 SRAM is t_{SD} which is 35ns. Therefore the 6264 SRAM can be easily connected to the 8086 microprocessor without any compatibility problems.

Switching Characteristics Over the Operating Range[1] characteris-

Parameter	Description	6264-55		6264-70		Unit
		Min.	Max.	Min.	Max.	
READ CYCLE						
t_{RC}	Read Cycle Time	55		70		ns
t_{M}	Address to Data Valid		55		70	ns
t_{OHA}	Data Hold from Address Change	5		5		ns
t_{ACEI}	$\overline{CE_1}$ LOW to Data Valid		55		70	ns
t_{ACE2}	$\overline{CE_2}$ HIGH to Data Valid		40		70	ns
t_{DOE}	$\overline{OE}$ LOW to Data Valid		25		35	ns
t_{LZOE}	$\overline{OE}$ LOW to Low Z	3		5		ns
t_{HZOE}	$\overline{OE}$ HIGH to High [Z]		20		30	ns
t_{LZCEI}	$\overline{CE_1}$ LOW to Low Z	5		5		ns
t_{LZCE2}	$\overline{CE_2}$ HIGH to LOW Z	3		5		ns
t_{HZCE}	$\overline{CE_1}$ HI	0		0		mA
	$\overline{CE_2}$ LOW to high Z					
t_{pu}	$\overline{CE_1}$ LOW to Power-up	0		0		ns
t_{pD}	$\overline{CE_1}$ HIGH to Power-Down		25		30	ns

Write Cycle						
t_{wc}	Write Cycle Time	50		70		ns
t_{SCEI}	CEI LOW to Write End	40		60		ns
t_{SCE2}	CE2 HIGH to Write End	30		50		ns
t_{AW}	Address Set-up to Write End	40		55		ns
t_{HA}	Address Hold from Write End	0		0		ns
t_{SA}	Address Set-up to Write Start	0		0		ns
t_{PWE}	$\overline{WE}$ Pulse Width	25		4		ns
t_{SD}	Data Set-up to Write End	25		35		ns
t_{HD}	Data Hold from Write End	0		0		ns
t_{HZWE}	$\overline{WE}$ LOW to High Z[5]		20		30	ns
t_{ZLWE}	$\overline{WE}$ HIGH to Low Z	5		5		ns

Shaded area contains advanced information

Figure 11.10(a) Switching Characteristics

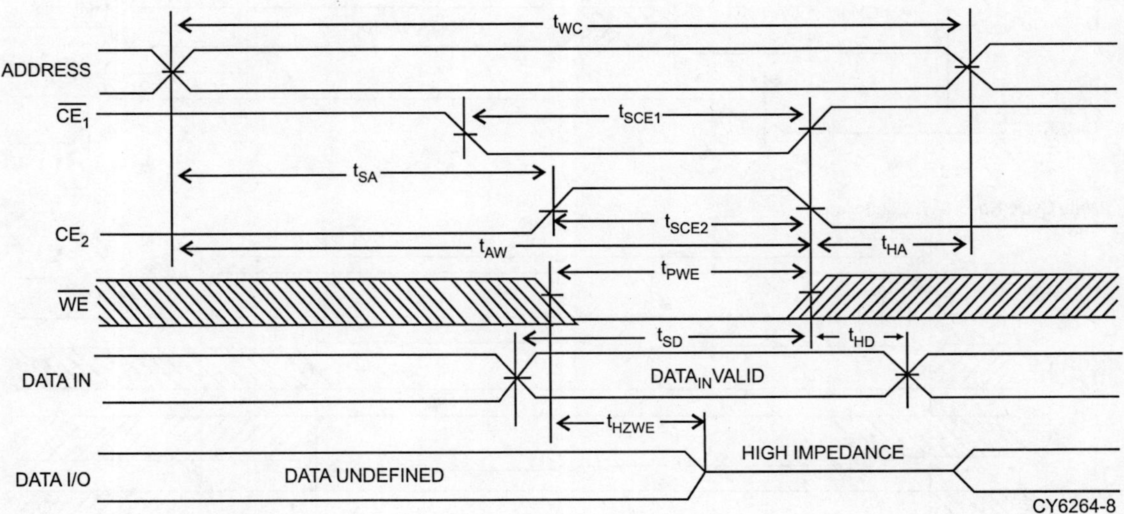

Figure 11.10(b) Read and Write Cycle Timing Diagrams for 6264 SRAM

11.2.4. ADDRESS DECODING

In order to access the memory components attached to the 8086/8088 microprocessor, we require to decode the addresses sent by the 8086 microprocessor. Decoding allows our system memory to be located between different memory sections. Decoding also allows our 8086 memory to be present from say between 10000H to 20000H and between 30000H to FFFFFH while the memory section between OOOOOH to OFFFFH to OFFFFH and 2000H to 2FFFFH to be not present at all.

Another reason for requiring address decoding is the fact that the microprocessor and memory devices may have different number of address lines. For example the 8086 microprocessor has 20 address lines while the 6264 SRAM has 13 address lines. This in effect means that

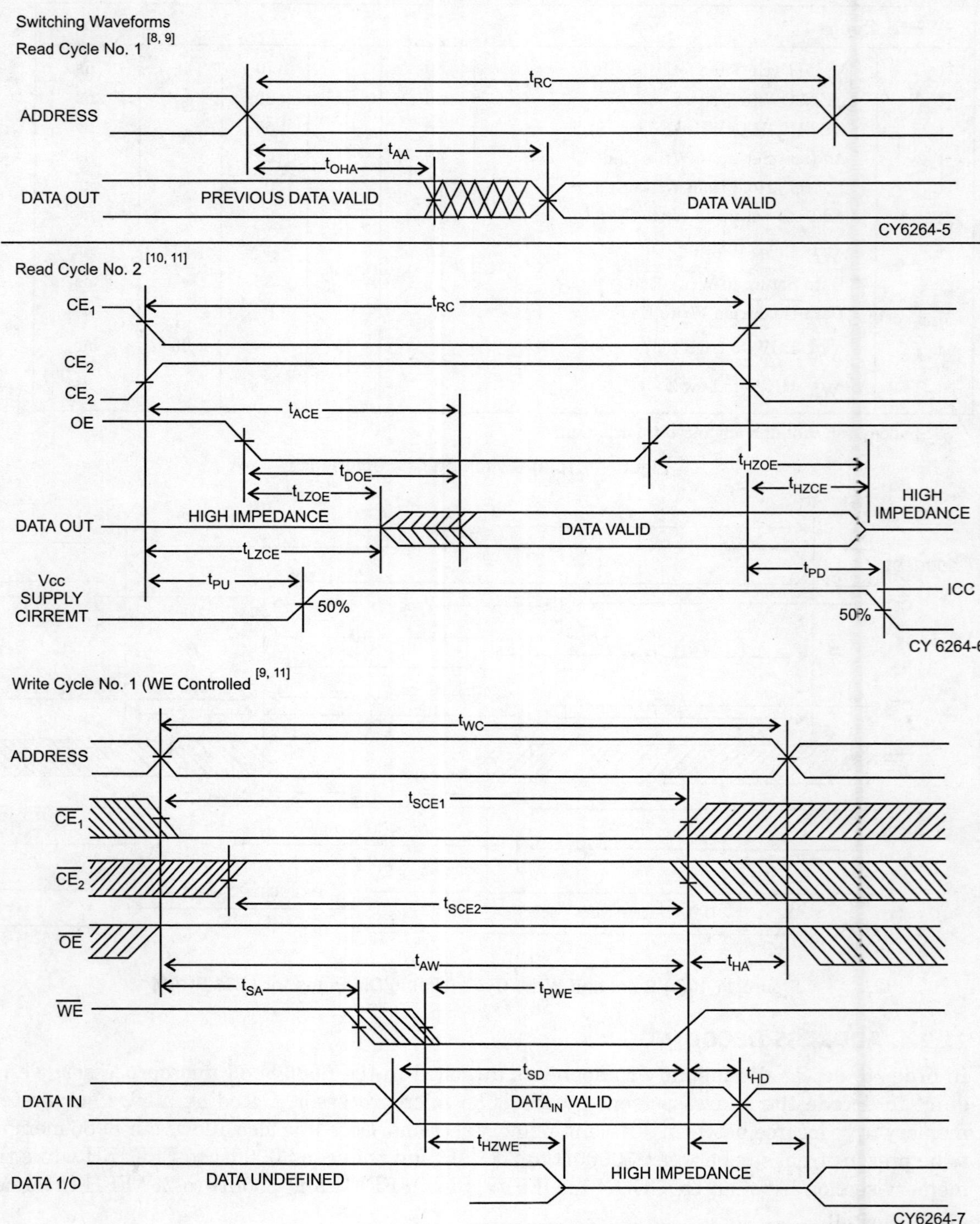

Figure 11.10(c) Read and Write Cycle Timing diagram

how would we access a memory device which has 13 address lines to a microprocessor having 20 address lines. Secondly if we connect only 13 address lines of the microprocessor to the SRAM device like microprocessor would then only be able to access 8 k bytes of SRAM instead of the 1M byte that the 8086 can access. To go around this problem we have to use address decoding circuitry which allows memory components of different sizes (e.g., 16 k, 61k) to be connected to the microprocessor.

11.2.4.1. Address Decoding circuits

Nand Gate Decoder

Suppose we were to connect the 2864 EEPROM to the 8088 microprocessor. This would require the Ao - A12 address inputs of the 2864 EEPROM to be connected to the Ao - A12 address lines of the 8088 microprocessor. The remaining 7 lines ($A_{13} - A_{19}$) of the 8086 are connected to a NAND Gate decoder. The 8086 M/$\overline{\text{IO}}$ pin is also connected to the NAND Gate there by allowing the EEPROM to be selected only when the 8086 wants to read a memory location and not when an I/O location with the same address with the same address. The circuit is shown in figure 11.11.

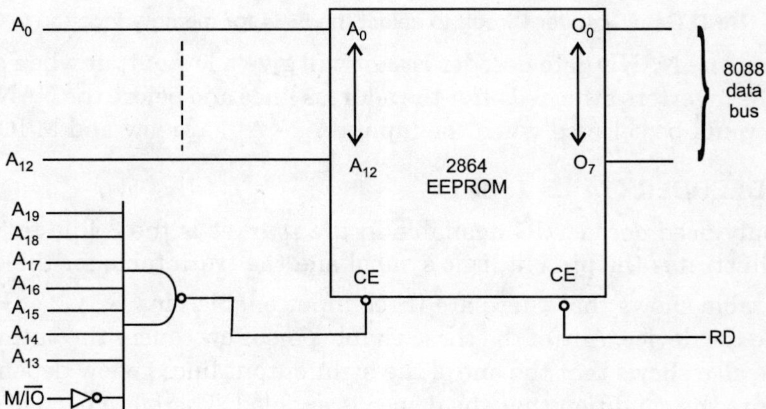

Figure 11.11 NAND decoder Circuit to select the 2864 for memory location FE000H to FFFFFH.

You may observe the NAND gate output would go low when all the input are high. This would in effect cause the EEPROM device to be accessed for memory address ranging from FEOOOH to FFFFFH. The NAND gate output when low would select the EEPROM device in the address of above as it is connected to the CE pin of the 2864. The address range can be calculated as:

Power address	1111	1110	0000	0000	0000	= FE000H
	1111	1111	1111	1111	1111	= FFFFFH
Higher address						

7 bits that go to NAND gate

If it is desired that the memory address starts from the memory location 00000H. The range would obviously be 8 k bytes (as the size of the EEROM is 8 k) therefore the last address would be 01FFFH. Figure 11.12 shows the NAND decoder circuit so that the EEPROM is accessed for the 8088 address 00000H to 01FFFH.

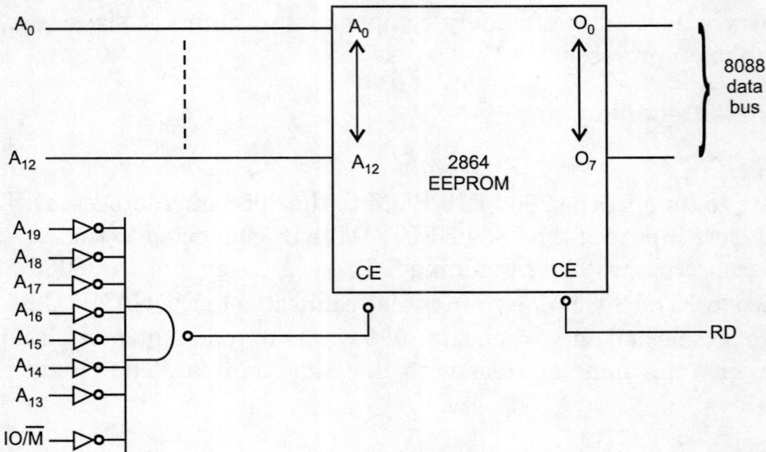

Figure 11.12. NAND Gate Dceoder Circuit to select the 8864 for memory location 00000K to 01FFFH.

Observe that the NAND gate decoder circuit will give a low output when all inputs are low. This is due to the inverters attached after the address lines and before the NAND gate. Therefore the EEPROM would be selected when the inputs $A_{13} - A_{19}$) are low and M/IO is high.

3-TO-8 LINE DECODER (74 LS 138)

A very commonly used decoder IC available in the market is the 3 line to 8 line decoder IC. Figure 11.13 illustrates the pin out logic symbol and the truth table for the 74138 I_C.

The truth table shows that there are three input enable pins G_1, $\underline{G2}_A$, $\underline{G2}_B$ which must be active to enable the device. Any of the three enable pins iflow causes the outputs to float high .o The truth table also shows that the one of the eight output lines go low depending on the input A. B. C subject to the condition that the device is enabled. Therefore each of these output pins can select a memory device thereby enabling us to select upto 8 memory devices by a singe IC. Now we connect a total of eight 2864 (8k x 8) memory devices to the 8088 microprocessor. The decoder is used to select each of these 8k devices allowing the 8088 to be able to access 64 k bytes of memory.

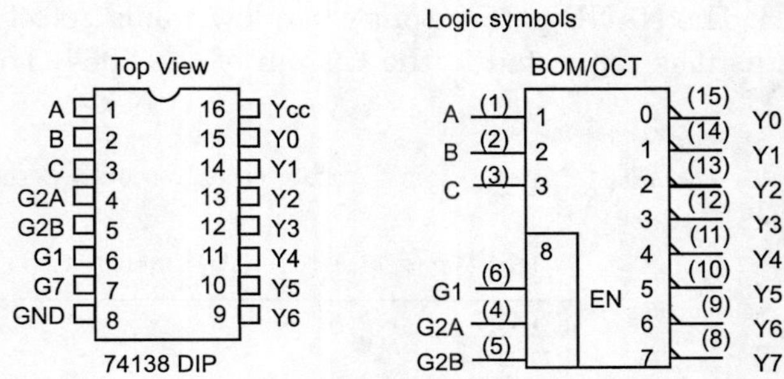

Figure 11.13(a) Pin out and Logic Diagram

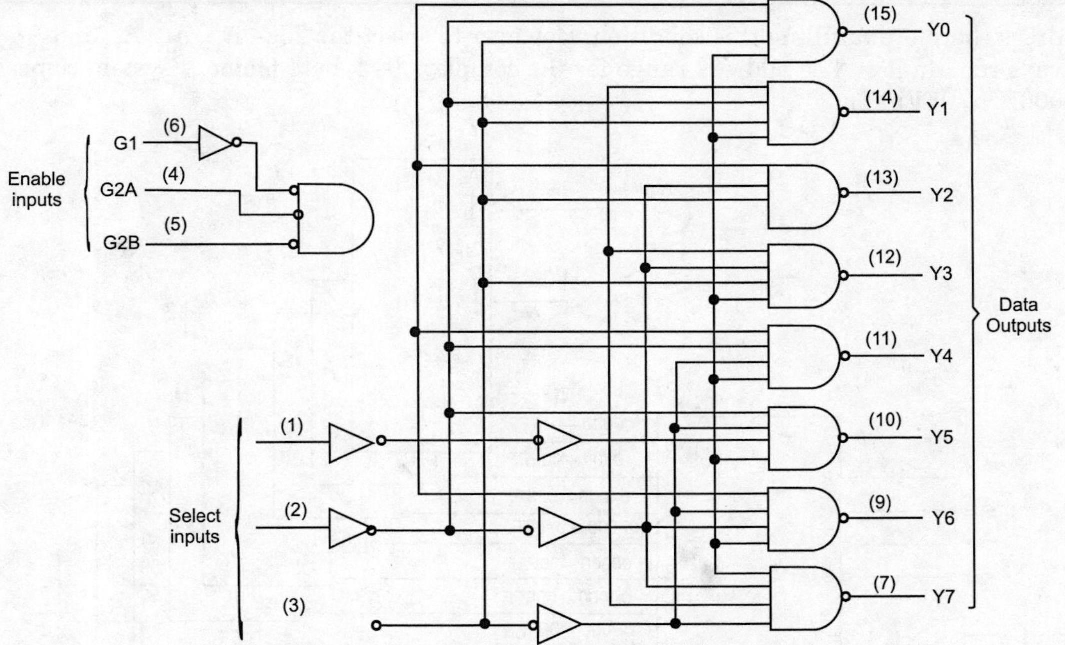

Figure 11.13(b) SN74 S138A LS138, SN 54S138

The figure 11.14 also shows the address range of the eight devices and the connections of control inputs. The 2864 has thirteen address lines leaving us with a total of 7 lines which must be connected to the decoder circuit.

You may observe that A16 and AI7 are connected to $\overline{G2}_B$ and $\overline{G2}_A$ pins of the 74138 respectively therefore to select the device Aw and An always remain zero. Similarly the A_{18}, A_{19} and $IO/\overline{M}$ lines are first inverted and then connected to an AND gate. Now G_1 must be high to select the device therefore requiring A_{18}, A_{19} and $IO/\overline{M}$ to be low (as they are first inverted before being fed to AND gate). Now in the 8088, the $IO/\overline{M}$ is low when the address is a memory

Inputs

Enable		Select			Outputs							
G1	G2	C	B	A	Y0	Y1	Y2	Y3	Y4	Y5	Y6	Y7
X	H	X	X	X	H	H	H	H	H	H	H	H
L	X	X	X	X	H	H	H	H	H	H	H	H
H	L	L	L	L	L	H	H	H	H	H	H	H
H	L	L	L	L	H	L	H	H	H	H	H	H
H	L	L	H	L	H	H	L	H	H	H	H	H
H	L	L	H	H	H	H	H	L	H	H	H	H
H	L	H	L	L	H	H	H	H	L	H	H	H
H	L	H	L	H	H	H	H	H	L	H	H	H
H	L	H	H	L	H	H	H	H	H	H	L	H
H	L	H	H	H	H	H	H	H	H	H	H	L

Figure 11.13(c) Function table and truth table for the 74138 IC

address therefore fulfilling this condition. However to select the 2864 A_{16}, A_{17}, A_{18} and A_{19} must always remain low. The address range for the complete 64 k byte memory system comes from 00000H to 0FFFFH.

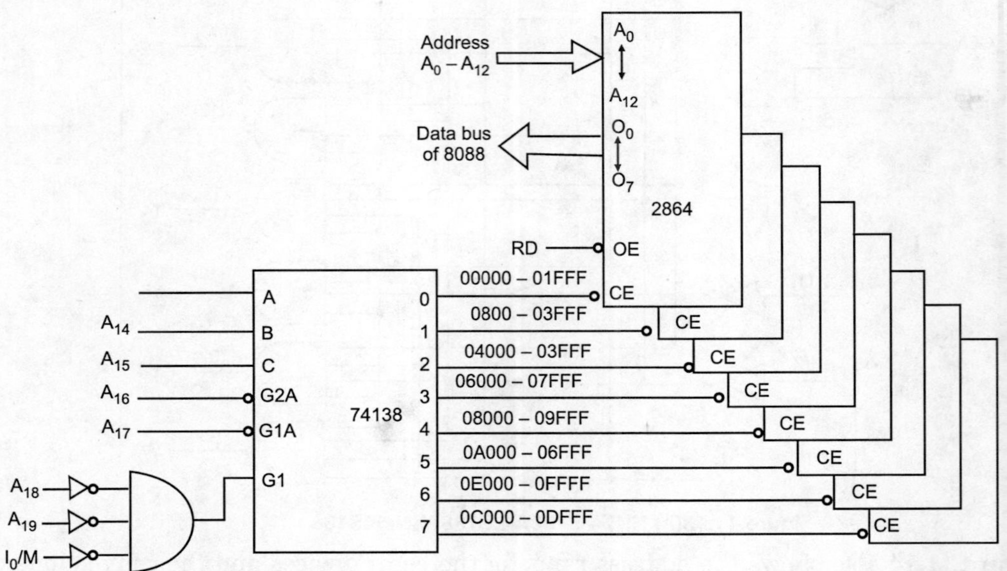

Figure 11.14. A circuit that uses a 74138 to select eight 2864 EEPROM for a 64K x 8 system memory

Address range of the circuit is 00000H to FFFFFH.

11.3 THE 8086 MEMORY INTERFACING

We have until now discussed only the 8088 memory interfacing. The 8088 memory interfacing is very easy as the 8088 has an 8 bit data bus and most memory devices available too are 8 bit devices. However interfacing the 8086 to the memory differs significantly due to three reasons.

(1) Data bus of the 8086 is 16 bit wide therefore requiring a 16 bit memory device.

(2) The $IO/\overline{M}$ pin of the 8088 is replaced by the $M/\overline{IO}$ pin the 8086.

(3) A new control signal $\overline{BHE}$ (bus high enable).

16 bit Memory interface

The 8086 microprocessor data bus is 16 bit wide. This therefore implies that the 8086 must be able to write data to any 16 bit memory location. This is further complicated by the fact the 8086 also writes to 8 bit data locations if required (for example a MOV M,AL where M is a memory location).

The memory interfaced to the 8086 must therefore be accessable as a16 bit wide memory or as a_8 bit wide memory. This problem is resolved by a memory interface technique that is used with a 8 bit the 8086 called memory banking. Memory banking uses two separate memory sections. each 8 bit wide where each section is called a bank. The 8086 microprocessor can write to both banks simultaneously (therefore a 16 bit memory device is viewed by the 8086) or can write to either one of the banks (which would therefore be viewed a 8 bit memory device by the 8086) at any time. Figure 11.15 shows the two banks of memory where one bank (odd or high bank) holds all odd numbered memory locations while the other bank (low or even bank) holds all even numbered memory address.

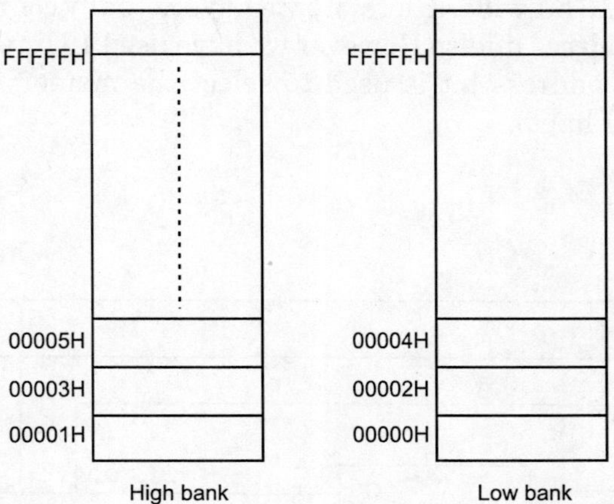

Figure 11.15. Dividing the 8086 memory into banks where each bank is 8 bit wide.

The, 8086 microprocessor then uses the $\overline{BHE}$ signal (bus high enable) and the AO address line to select one or both banks for data transfer. Figure 11.16 shows the truth table for these two signals and the bank chosen accordingly.

$\overline{BHE}$	A0	Function
0	0	both banks selected (i.e., 16 bit transfer)
0	1	High bank selected
1	0	Low bank selected
1	1	No bank selected

Figure 11.16. Truth table for BHE and AO and bank selected accordingly.

Bank selection logic can be designed in two ways:

(1) Seperate bank selection decoders for each bank.

(2) A seperate write selection logic technique.

Seperate Bank Selection Decoding Technique

This is the simplest way to select the banks. However this technique is an expensive method due to the large number of discrete IC's required to generate the corresponding decoding circuit. We shall show you this technique by interfacing two 62512 SRAM chips, where one chip will be the odd bank and the second an even bank. The 62512 is a 641{ x 8 SRAM IC. Figure 11.17 shows the bank selection by using different bank decoders.

Observe the figure you shall find the M/IO pin is used to select the 74138 IC's. Further the $\overline{G2}_B$ pin is grounded on both 74138s. However the $\overline{G2}_A$ of the upper 74138 is connected to the $\overline{BHE}$ while the same pin of the lower 74138 is connected to A_0. This allows the 62512 chips to be selected according to truth table of figure 11.16. Though you may find only single 62512 SRAM's connected to the decoder (74138) output this is only for clarity. You may if you desire, attach 8 62512 SRAM chips on each 74138 shown or attach other devices such as EPROM's.

The configuration offigure 11.17 allows a memory address range form OOOOOH to IFFFFH. The Ao pin is not connected to memory but is used to select one of two decodes. Though you may say that we have interfaced 64 I{ byte of SRAM to the 8086 this would be in correct as the memory attached now is 16 bit wide. Therefore we have actually connected 128k byte of memory. (Note: this requires 17 address line-so However we have used 16 lines (A_1-A_{16}) to directly address the SRAM while the A_0 address bit is used to select the memory device. Therefore we have actually used 17 address lines).

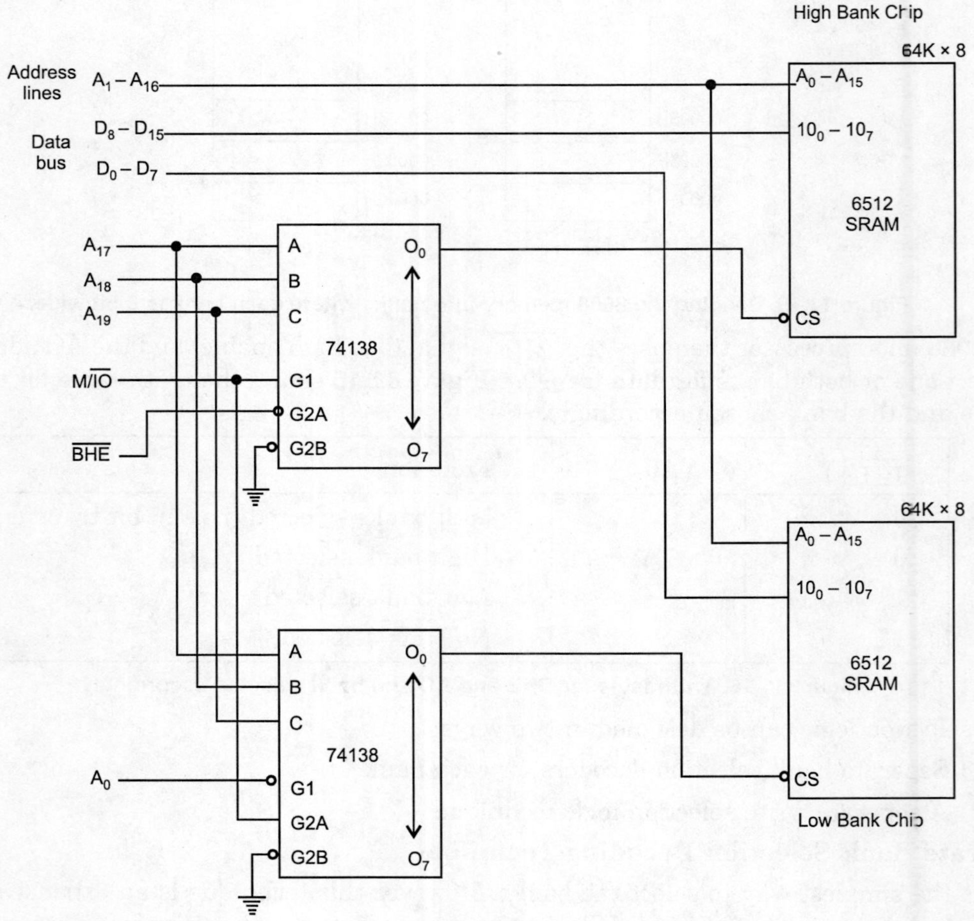

Figure 11.17. Bank selection logic to select banks where each bank is a 62512 64K x 8 SRAM. (Note: we have interfaced only one chip with both the 74138, through it can be further expanded to accommodate 8 chips through each decoder)

Seperate Bank write signal technique for Bank selection

The most cost effective technique to select the different banks is by generating of different bank write signals. This reduces the necessary components such as the decoders that we may require in the earlier technique.

However separate read strobe are not required because the 8086 always only reads the data byte that it requires from the data bus. If 16 bit of data is present on the data bus the 8086 will read either the data byte on D_0–D_{15} data lines or which is on $D_8 - D_{15}$ data lines, depending on the data byte that it requires, (The other byte isautomatically rejected). This, therefore, results that we only require the bank write signals. Figure 11.18 shows the generation of the separate bank write strobes for memory.

These bank write strobe are connected to SRAM or DRAM write input control pins. (ROM's do not require these signals). Thus depending on A_0 and $\overline{BHE}$ bits the bank write strobes allow the data transfer to be 8 bit or 16 bit.

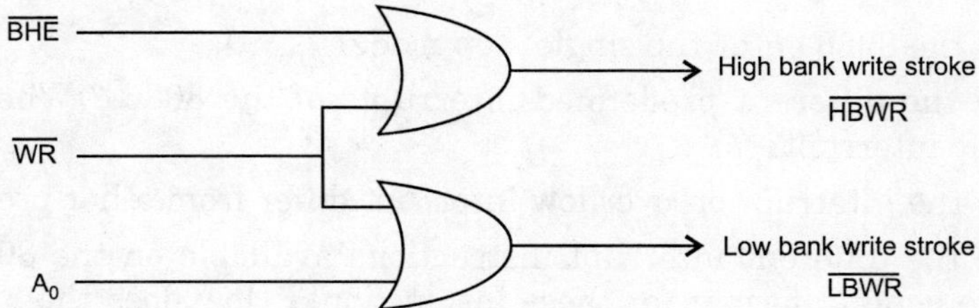

Figure 11.18. Using different bank write stores for bank selection.

11.4. KEY POINTS TO REMEMBER

1. The 8086 contains two types of interrupts: external or hardware interrupts and internal or software interrupts.
2. The interrupt vector table is used to store the address of the interrupt service routines. It occupies the first 1024 bytes of the 8086 system memory.
3. There are five predefined interrupts available on the 8086 microprocessor and a total of 224 user defined interrupts on the 8086 microprocessor.
4. The programmer can call interrupt service routines by the use of the lN T instruction.
5. When an interrupt is serviced the following sequence of operations occur.
   (*i*) Flag register is pushed onto stack
   (*ii*) TF and IF are reset
   (*iii*) CS register pushed onto the stack
   (*iv*) The new CS value is loaded from the interrupt vector table
   (*v*) IP pushed onto the stack
   (*vi*) The program control transferred to new CS : IP memory address.
6. The IRET instruction is used to return from an interrupt call.
7. There are many memory devices which can be interfaced to the 8086 microprocessor such as EPROM, EEPROM, SRAM etc.
8. Address decoding allows the 8086 to be interfaced to any memory chip. It also allows the designer to design systems only portions of the 1MB 8086 system memory exist.
9. The 8086 memory interfacing requires the use of memory banking. Memory banks are selected by the use of the $\overline{BHE}$ pin and the Ao address bit.

REVIEW QUESTIONS

11.1. What do you mean by the interrupt vector table? What are the uses of the interrupt vector table?

11.2. What would be memory addresses from where the CS and IP register values would be fetched for the following int errup ts ?

- type 2
- type 6
- type 8
- type 200

11.3. How does the 8086 enter the single step mode?

11.4. What are the different predefined interrupts of the 8086 ? What is the use of the breakpoint interrupt?

11.5. How does the interrupt on overflow interrupt differ from other predefined interrupts?

11.6. What are the different interrupt instructions available on the 8086 microprocessor? Define the general format for these instructions? How does the breakpoint interrupt format differ?

11.7. What are the sequence of operations that occur when the INT 21H interrupt instruction executes ?

11.8. What is the state of the TF and IF flags after an INT instruction is executed?

11.9. What is the purpose of the IRET instruction?

11.10. What are the sequence of operations that occur when an IRET instruction executes?

11.11. What are hardware interrupts? What are the different hardware interrupts of the 8086?

11.12. What is the NMI interrupt used for?

11.13. How can the INTR pin be used to interrupt the 8086 microprocessor? What are necessary conditions for the INTR to interrupt the 8086 ? How is the interrupt type code fetched?

11.14. Give a general pin out for a memory device? Explain the use of the following pins for these devices

(*i*) Address lines

(*ii*) Control inputs

(*iii*) Data lines

(*iv*) Selection lines

11.15. Repeat the above problem for the 6264 SRAM and 2864 EEPROM memory devices?

11.16. What are the different ROM devices available?

11.17. Suppose a EPROM device having a memory access time of 460ns is interfaced to the 8086. What are the problems that you are sure to face and why?

11.18. What are SRAM devices? Give examples?

11.19. What is address decoding and why is it necessary?

11.20. It is required to interface a 64k x 8 SRAM and a 128k × 8 EPROM device to the8088. The SRAM must have an address range from 00000H to OFFFFH while the EEPROM must have an address range of 20000H to 3FFFFH. Design the necessary interface?

11.21. Interface a 256 KB SRAM and a 512 KB EPROM to an 8086.

11.22. Due to bad economy, presently only 8 kbyte SRAM chips are available in the market? Design the necessary address decoder circuit so that you may interface 16 such chips to the 8088 where the first memory address must start from 20000H.

11.23. How does memory interfacing to 8086 differ from 8088 ?

11.24. What is memory banking? How are different memory banks selected?

11.25. What is the use of the $\overline{BHE}$ signal?

11.26. Design a memory interface to the 8086 microprocessor. Use 64 KB RAM devices and design the memory of the 8086 to be 1 M byte?

11.27. What are the different memory bank selection techniques?

11.28. Why we don't require to use seperate bank read strobe signals when the 8086 selects memory banks by the use of seperate bank write strobes?

11.29. Interface two 8 K x 8 SRAM chips to the 8086 using seperate write bank strobe signals?

●●●

Chapter-12

THE 8051 MICRO CONTROLLER

INTRODUCTION

In this chapter, we shall present a brief description and give a basic idea about microcontrollers. Microcontrollers are not as well known as their relatively popular cousin's the microprocessor. However, microcontrollers today are used almost everywhere, from burglar alarms to VCRs, in automobiles, videogames, copiers and a host of other machines that are just about called programmable machines. Automatic machines, microwaves ovens, robots and many more systems use microcontrollers. A lot of non-technical text use the terms microprocessors and microcontrollers as similar terms. However, you as an engineer must know the many differences between these two very similar chips and the different advantages of both. We shall discuss the intel 8051 microcontroller, give a brief idea about its architecture and its uses and give you a detailed understanding of the difference between the microprocessor and micro controller. We shall in this chapter however discuss nothing about microprocessors as we have done that in the preceding chapters.

12.1. MICROCONTROLLERS

Following the widespread popularity that microprocessors gained, it was found that using microprocessors for control applications for devices such as microwave ovens, automatic washing machines etc. was very expensive. The main reason was that the microprocessor required many components such as RAM, ROM etc. for a designer to be able to use them as a control device. Also this required a large amount of space which is generally not available in such devices. This led to the demand for small, dedicated processors, that required minimal components (like RAM, ROM etc.) for their working and had the advantage of being cheap. The microcontroller essentially was developed keeping these things in mind. A microcontroller incorporates all features of a microprocessor. It also has some added features which include the presence of RAM, ROM, I/O Ports, counters and a clock circuit, All these features are sufficient to make it a complete computer and the best part of all is that it is present on a single chip. This means that microcontrollers are essentially equivalent to a small computer built into a single chip requiring no other external devices such as RAM or ROM for its working. More simply we can say that microprocessors are intended to be used in general purpose digital computers while microcontrollers are special purpose digital computers.

Microprocessors	Microcontrollers
1. Microprocessors are designed to be used as the CPU of general purpose computers.	1. Microcontrollers are special purpose digital computers requiring few or no external component for their working.
2. Microprocessors contains an ALU, General purpose registers, control and interrupt circuits, in one chip.	2. Microcontrollers have all these as well as timers, 110 ports, RAM and ROM on the chip.
3. Microprocessor require many components such as RAM, ROM etc. for their workmg.	3. Many micro controllers require no external components for their workings as all these components are built onto the microcontroller chip.
4. Microprocessors have higher speeds, greater data handling capability and can address large amounts of memory.	4. Microcontrollers have RAM and ROM built on the same chip. Since the microcontroller are designed for use in machines such as a control device, they have slower speed and lesser address memory.

Figure 12.1. Key differences between Microprocessor and Microcontrollers

The general block diagram for a microcontroller is shown below in figure 12.2.

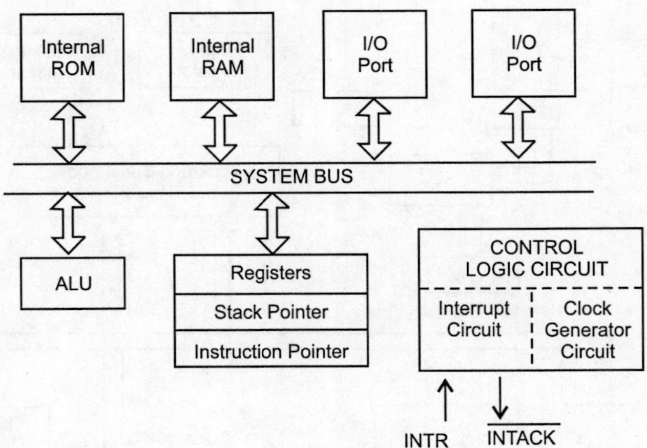

Figure 12.2. Basic Microcontroller block diagram

Microcontrollers are also known as dedicated or embedded controllers. Microcontrollers are designed for use as a dedicated computer which controls some device such as a microwave oven, automobiles etc. Microcontrollers are available as 4 bit, 8 bit, 16 bit, 32 bit and 64 bit versions. A 8 bit microcontroller implies that the microcontroller has an 8 bit CPU. Some common microcontrollers available in the market are Intel 8048, Intel 8031, Intel 8051, Motorola MC6801, Motorola MC 6805, Zilog Z8 and Texas Instrument TMS 7500. The 8051 microcontroller from Intel corporation is one of the most popular and widely used micro controller. For this reason we shall discuss the 8051 microcontroller in great detail.

12.2. 8051 INTERNAL ARCHITECTURE

The intel 8051 micro controller is a part of a family of microcontrollers ranging from 8031 to 8751 microcontrollers. The 8051 is fabricated by N channel Metal Oxide Silicon (NMOS) gate technology. It is also available in the complementary Metal Oxide Silicon (CMOS) gate technology version. The 8051 is an 8 bit micro controller and is available in a 40 pin DIP (Dual Inline Package). The 8051 microcontroller can be made to run at clock frequencies ranging from 1 MHz to 16 MHz.

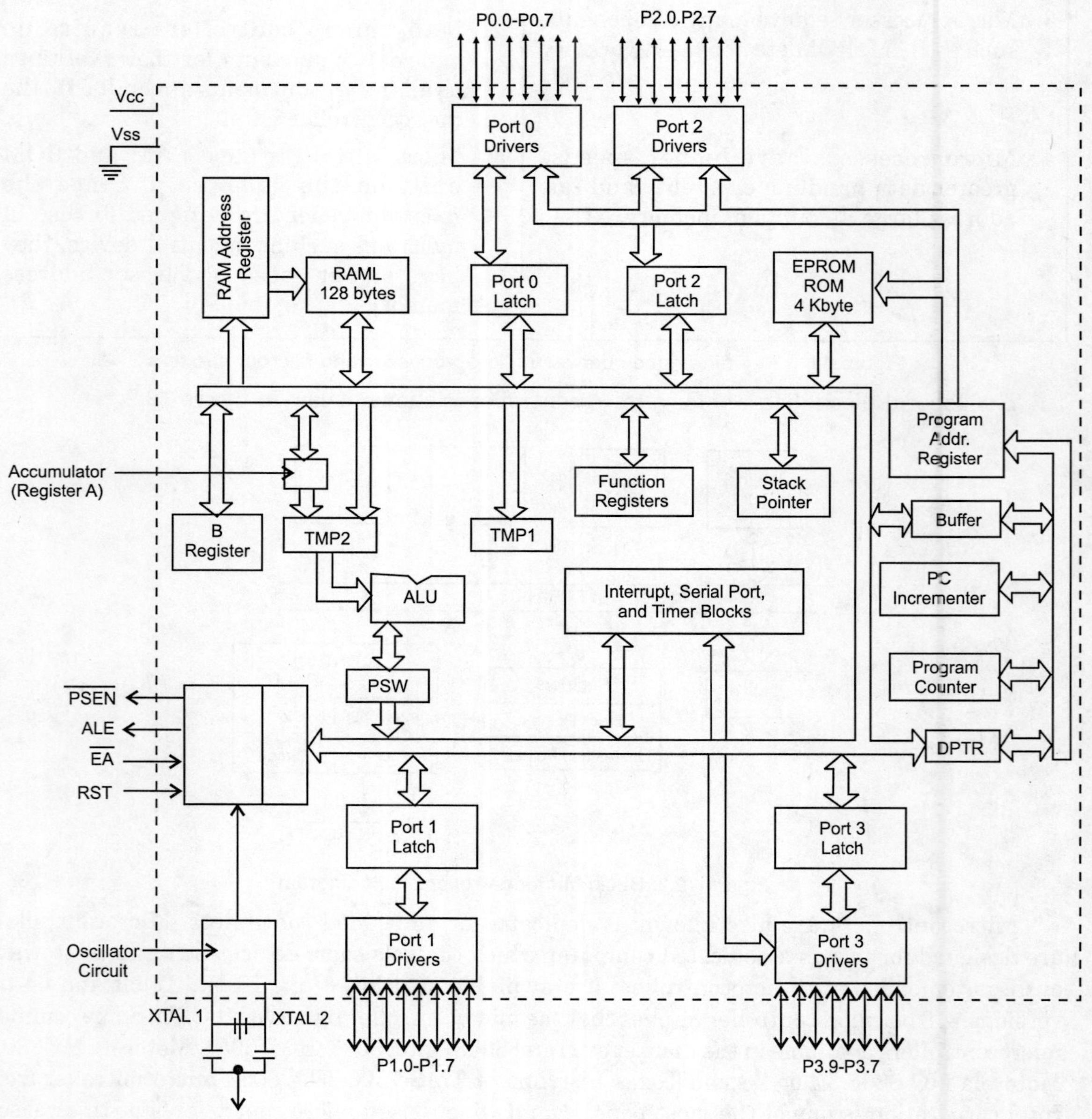

Figure 12.3. The 8051 Microcontroller Architectural Block Diagram

The 8051 microcontroller block diagram is shown in figure 12.3. The Block diagram clearly shows the difference between the microprocessor and microcontroller. The figure 12.3 also shows all the main internal features present in the 8051 microcontroller which include

1. INTERNAL RAM: 128 Bytes
2. INTERNAL ROM: 4 KBytes
3. I/O PORTS
4. TIMER CIRCUITRY
5. COUNTER CIRCUITRY MODULE
6. REGISTERS
7. INTERRUPTS MODULE
8. CLOCK CIRCUITRY MODULE
9. ALU
10. CONTROL MODULE

However, for certain applications the 8051 micro controllers internal memory may be insufficient. However special care has been taken by its designers enabling the 8051 microcontroller to be interfaced to external memory* (Maximum of 128 Kbytes). However, this external memory is not an essential component, i.e., if the application does not need external memory (i.e., the application program is small enough to fit in the 8051 internal memory) it does not have to be attached.

The figure 12.3. also shows the basic CPU components such as the ALU, Program counter, registers, interrupt circuits and clock circuits. Now let us discuss the 8051 microcontroller architecture in detail. The 8051 microcontroller contains the following internal hardware.

1. 8 bit ALU
2. 8 bit registers named A and B.
3. 16 bit registers PC and DPTR
4. 8 bit Program Status Word (PSW)
5. 8 bit Stack Pointer
6. Internal ROM 4 Kbytes
7. Internal RAM 128 bytes which can be subdivided as
   - 4 Registers banks where each bank is a group of eight 8-bit registers.
   - 16 bytes of bit addressable registers.
   - 80 bytes of general purpose memory
8. Two sixteen bit timers/counters (TO, Tl)
9. Full Duplex serial data receiver/transmitter

---

* since the 8051 is used as dedicated processors all programs are fed into the ROM. This allows a microcontroller to be used to run a single set of programs only for the task it is assigned for.

10. Interrupts
11. Clock generator circuits
12. Control Registers (Function registers)

Though the 8051 microcontroller architecture may seem a little complex at the first glance, it is not so. The basic complexity here is just that that all components that are required to make a working micro-controller have been fitted into one chip. We now explain the register structure of the 8051 in detail. The breakup of the registers as well as their functions are shown in the figure 12.4. We also show the RAM and ROM breakup. Figure 12.5. also shows the 8051 pin out of its 40 pin DIP package. Note that some pins serve more than one purpose.

Figure 12.4(a) and 12.4(b) shows the 8051 registers and internal memory organization. All 8051 registers has a one byte address (except the program counter). In the 8051 there are two types of register:

1. Bit addressable registers
2. Byte addressable registers

8 bit Registers

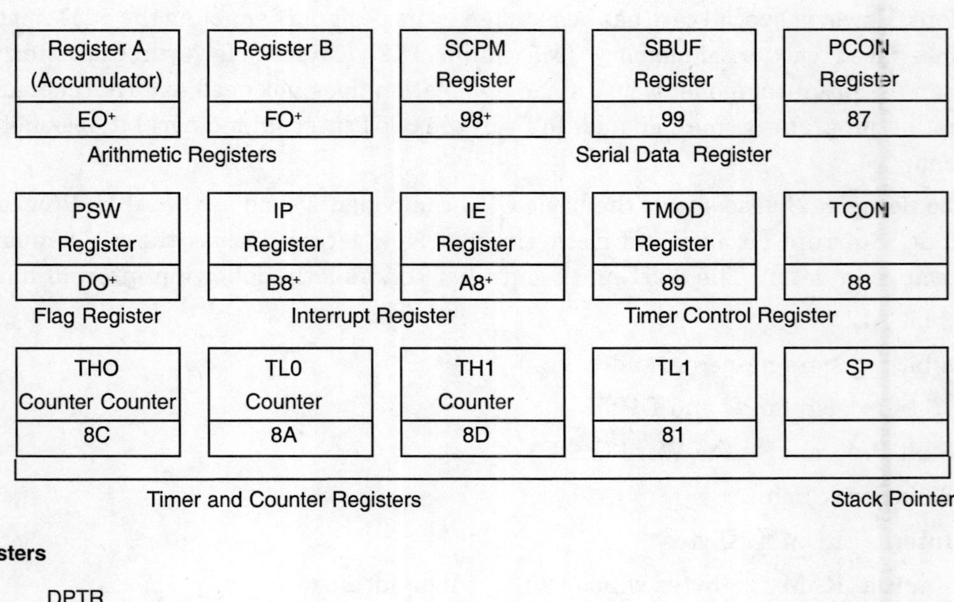

Port Latch

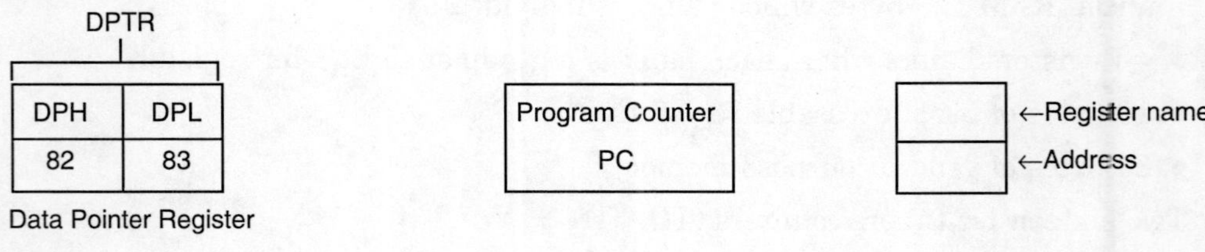

The byte addressable registers are similar to the previous 8 bit registers that you have studied about. These registers can stores 8 bits at a time. However, the bit addressable registers are significantly different. The bit addressable registers can be accessed bit by bit or byte by byte. These registers are included due to shortage of RAM space. When microcontrollers are used in control applications sometimes it becomes necessary to store the state (on or off) of some interfaced circuitry. The bit addressable registers are extremely useful in such situations as we can store the state of the interfaced circuitry in these registers. These registers thus are extremely helpful because (due to shortage of RAM) we can now only store the condition by changing a single bit of a bit addressable register (instead of a bit if only byte addressable registers were present).

We shall discuss the control and function registers in the following sections. In the pinout diagram each bit of the specific port is marked with all the functions that it is designed for e.g. the bit 0 of Port 3 is marked as P3.0 which is used as a general purpose I/O pin or as an input (receive/RXD) to SBUF register for serial communication. Most pins have dual functions which are shown in brackets. It is for you to decide the usage of the pin according to your applications and needs.

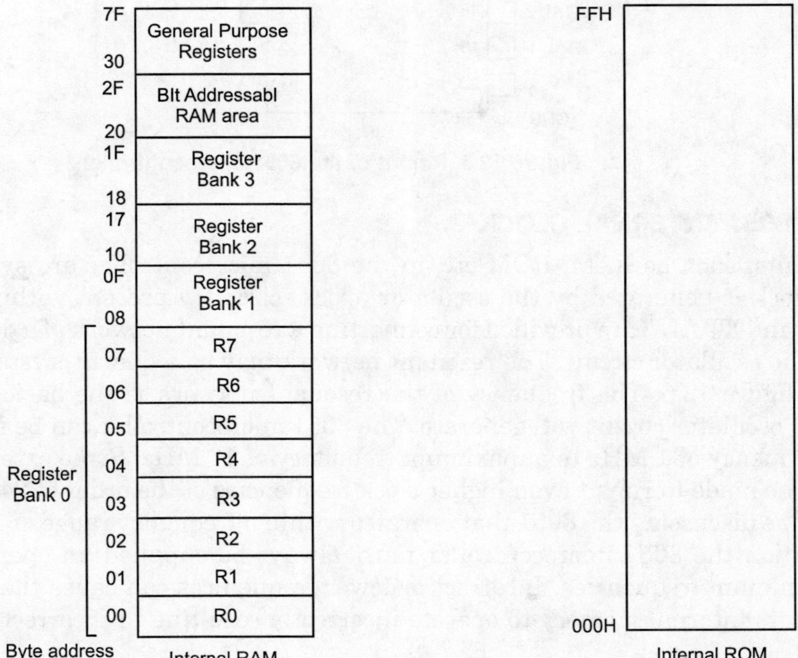

Figure 12.4.(b) 8051 Internal Memory Organization

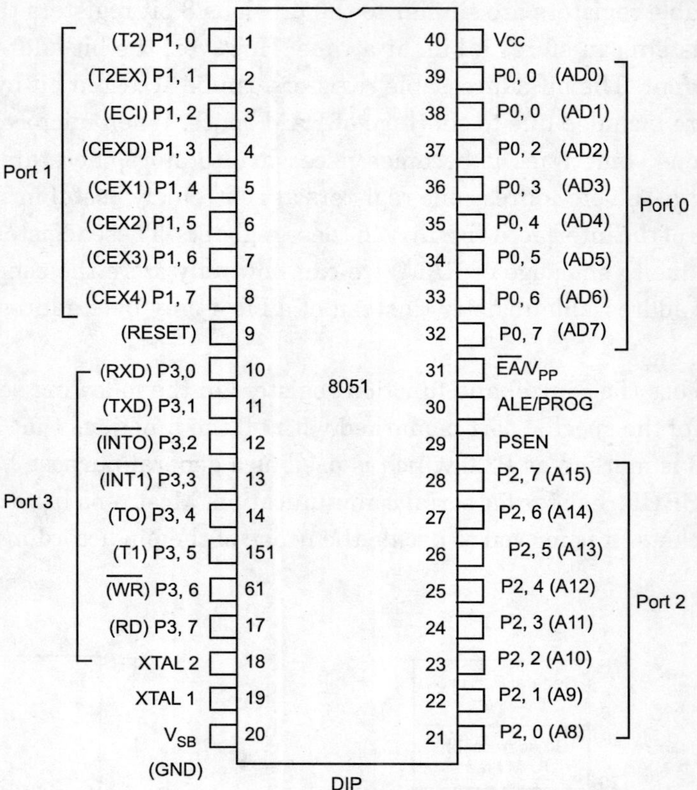

Figure 12.5. Pinout of the 8051 Microcontroller

12.2.1. THE 8051 INTERNAL CLOCK

Internal circuits, such as RAM, ROM etc, in the 8051 microcontroller are synchronized by a clock. This clock is generated by the oscillator (clock) circuitry present within the 805I. Two pins XT AL 1 and XT AL 2 are provided for connecting a resonant network of a desired frequency to complete the oscillator circuit. The resonant network may be a quartz crystal and capacitors as shown in figure 12.6. The frequency of the resonant network is the basic clock frequency that the 8051 oscillator circuit will generate. The 8051 microcontroller can be made run from a minimum frequency of 1 MHz to a maximum frequency of 16 MHz. Some extended versions of the 8051 can be made to run at even higher clock frequencies of the order of 24 MHz. However, we shall only be discussing the 8051 that operates within a frequency range of 1 to 16 MHz. For correct operation the 8051 microcontroller must always be supplied an operating frequency above the minimum frequencies. Incorrect or lower frequenices can cause the 8051 to become unstable and its internal memory to operate incorrectly resulting in incorrect outputs

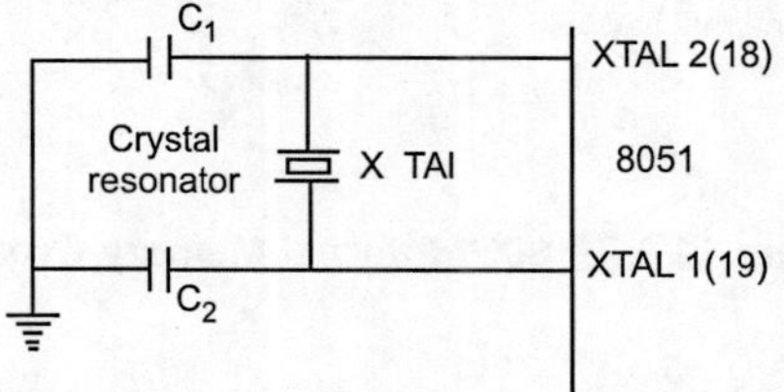

(a) Oscillator circuit connected to XTAL 1 and XTAL 2 pins

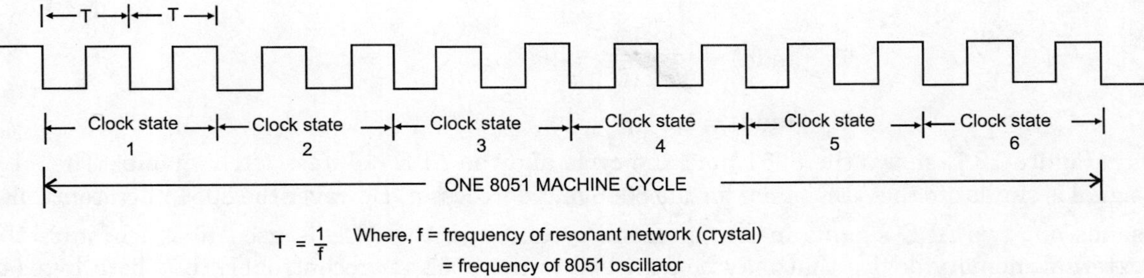

Figure 12.6. Crystal Resonant Network Connected to 8051 and the generated interal clock

The frequency of the resonant network as well as the type of resonant network must be decided depending on the usage environment of the 8051 microcontrollers*. Crystal and Ceramic resonator circuits are most commonly used. However, since the ceramic resonator circuits do not yield very accurate clock frequencies hence they should not be used when the 8051 microcontroller is being used in a system where high speed data transfer or precise timing is to be done. Crystal circuits though expensive are highly accurate and offer better frequency stability than other resonant circuits and should be preferably used. The oscillator circuit formed by the resonant network, the 8051 internal oscillator (amplifier) and an inverter generates a pulse train at the frequency of resonant circuit as shown in figure 12.6.

The clock frequency F decides the time period T of the clock signal. However, as in the 8086 processor we too define a machine cycle here. A machine cycle is the minimum duration required for processing instructions. A machine cycle in the 8051 microcontroller is made up of 6 time status. However, in the 8051 microcontroller each time state is made up of two clock pulses or time periods. Thus, we have

1 Machine Cycle = 6 time states

1 Time state = 2 Clock pulses (Time periods)

Now a machine cycle for the 8051 (as in the 8086) is the minimum time required to execute a simple instruction or the part of a complex instruction. The 8051 instructions may require one, two or four machine cycles. Thus the simplest instruction will be executed in 12 clock pulse (1 Machine cycle x no. of pulses per Machine cycle) while the most complex instruction will execute in 48 clock pulses (4 machine cycle × no. of pulse per machine cycle). Thus to calculate the time required to execute an instruction can be computed as

$$T_{instr} = \frac{MC \times 12}{\text{Crystal resonant frequency (Hz)}}$$

where, MC = Machine cycles required

Example 12.1. An 8051 add instruction requires one machine cycle. With a crystal resonant frequency of 12 MHz. Calculate the time required to execute an add instruction.

$$T_{instr} = \frac{MC \times 12}{\text{Crystal resonant frequency}}$$

* A faster execution time would require a higher clock frequencies. For example videogames might require higher clock frequencies while microwave oven would do with lower clocks.

MC = 1 for add instruction

$$\Rightarrow \qquad T_{instr}\,(\text{add}) = \frac{1\times12}{12\times10^6} = 1\mu s$$

$$T_{instr}\,(\text{add}) = 1\mu s$$

Figure 12.7. shows the 8051 machine cycle and the ALE (address latch enable). The ALE signal is similar to the ALE signal for the 8086 microprocessor. However the 8051 microcontroller sends out two ALE signals in one machine cycle. Since the ALE is used to synchronize the external memory device that may be attached to the 8051 microcontroller, thus both bytes of an instruction may be fetched and executed by the 8051 *. However, single byte instruction too execute in one machine cycle where the second byte is rejected as it is same as first byte.

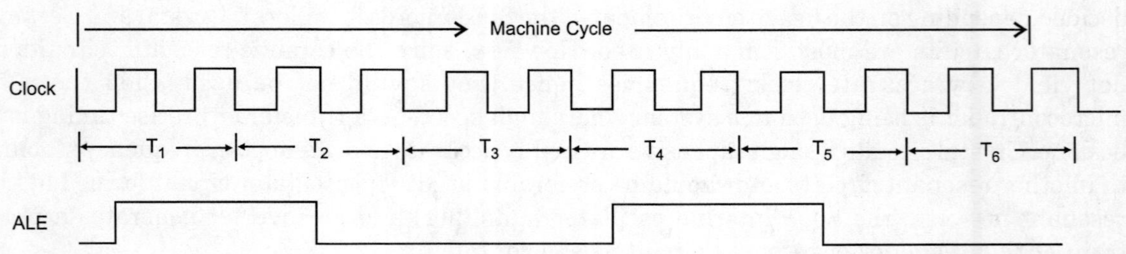

Figure 12.7. The 8051 Machine cycle and ALE

12.2.2. Program Counter And Data Pointer Registers

The most important registers of the 8051 microcbntroller are the program counter and data pointer register. Both of these registers are 16 bit registers and are used to hold the address of a byte in memory (internal or external).

The program counter is used to hold the address of instructions that are held in memory. The address can range from 0000H to FFFFH. If the value of the program counter is between 0000H to 0FFFH it assumes that the instruction bytes are stored in the internal ROM. The program counter is automatically incremented after execution of each instruction. The program counter may be altered by instructions such as jump instruction etc. This is the only register in the 8051 microcontroller that does not have an address.

The data pointer register or DPTR is a sixteen bit register which is further subdivided into the DPH (data pointer high byte) and DPL (Data Pointer low byte). These registers are used to access the internal RAM as well as other special register. Further these two may be combined to access data bytes from the external memory. However, the DPTR register does not have a single address. The DPH and DDL are both assigned their address. (83H and 82H respectively). The DPTR register is accessed by the 8051 microcontroller instructions. However. the 8051 microcontroller can access code or data by use of either the DPTR, DPH or DPL register depending on the instruction and address.

* Here the first byte refers to the instruction while the second byte may he data or operands.

12.2.3. Accumulator (Register A), Register B, And Other Available Registers

The 8051 contains 34 general purpose 8 bit registers. These are used to hold data which is to be operated on by other instructions such as ADD, SUB or are used to hold the results of instructions. The register A is also called the accumulator and is used with almost every 8051 instruction. The register A and register B are also called the arithmetic registers as arithmetic instructions generally use at least one of these registers to hold data. The multiplication and division instructions require the use of both these registers.

The other 32 registers available to the 8051 microcontroller are arranged as part of the 8051 microcontrollers internal RAM. These 32 registers are arranged in four banks where each bank contains a set of eight registers.

12.2.4. Program Status Word

Like other microprocessors the 8051 microcontroller is provided with flag bits which are used to store conditions that may be encountered during program execution. Some instructions of the 8051 can test the state of these flags and take a corresponding decision based on the state. The 8051 flag bits are grouped inside the PSW and PCON (Power Mode Control Register). The 8051 micro controller has four arithmetic flags that automatically set or reset depending on the result of an arithmetic operation. The arithmetic flags available are Carry Flag (CF), Auxiliary Carry Flag (ACF), Overflow flag (OF) and Parity Flag (PF).

Other flags that are included in the 8051 microcontroller include user flags. User flags are flag bits that can be programmed to store some condition during program execution. There are three user flags in the 8051, which are FO, GFO and GFl. Out of these three user flags, the FO flag is present in the PSW register. The program status word in addition to these three flags (i.e. CF, ACF, OF, PF. FO flag bits) also contains two register bank access flags (RSO and RS1). The state of these two flag bits decides the RAM register bank that will be accessed by the program. Figure 12.8 shows the flag bit positions in the 8051 PSW. The GFO and GF1 user flags are stored in the PCON register.

All flags in the 8051 can be set or reset at will. The PSW can be cleared or set to a specific value by the use of the data move instruction.

7	6	5	4	3	2	1	0
CF	ACF	FO	RS1	RS0	OF	R	PF

Bit	**Symbol**	**Function**	
0	PF	Parity flag bit shows the parity of the contents of accumulator. PF = 1 odd Parity, PF = 0 even parity	
1	Reserved		
2	OF	Overflow flag bit	
3	RSO	Register bank access bit 0	Defines which bank of RAM will
4	RS1	Register bank access bit 1	be used by the program
5	FO	User flag	
6	ACF	Auxiliary Carry flag	
7	CF	Carry flag	

Figure 12.8.(a) PSW Register

RS1	RSO	
0	0	Access Register bank 0
0	1	Access Register bank 1
1	0	Access Register bank 2
1	1	Access Register bank 3

Figure 12.8(b) Truth table for RSO, RS1 bits

The 8051 program status word is bit addressable and thus provides for a programmer to alter a specific flag bit. Each bit can be accessed as PSW.O to PSW.7.

12.2.5. Memory

Microcontroller are designed for specific applications and are a complete computer on a chip. A computer must have memory locations for storing programs as well as a variable memory area where data can be altered as the computer executes the program operating on this data. Since microcontrollers are designed for specific applications where the program size is relatively small compared to computers, their programs can be stored in ROM memory devices. (EPROM, EEPROM). They must also have a RAM so as to be able to store the data which is continuously altered during program execution. The 8051 microcontroller has a ROM and a RAM built in so as to be able to work as a complete computer. However, unlike other microprocessors (such as 8086) and microcontrollers which use the Von Nuemann architecture, the 8051 microcontroller uses a more advanced HARVARD architecture. Microporcessors/Microcontroller using Von Nuemann architecture uses different address for instructions and databytes. The Harvard architecture based microprocessors/microcontrollers use same address in different memories for instruction and data bytes. The 8051 internal circuitry automatically decides the correct memory that must be enabled depending upon the operation that is being executed.

Internal RAM

The 8051 microcontroller has a RAM (Random Access Memory) whose size is 128 bytes built in on the chip. The RAM is further subdivided into three main areas where each area serves different purposes. The RAM organization of the 8051 is shown in figure 12.9.

1. Register Banks : The 8051 has 4 register banks where each bank has eight byte-wide registers making a total of 32 bytes. The memory address of the register banks range from 00H to 1FH. The figure 12.9 shows the four register banks labelled B0 to B3. Each of the eight registers in a register bank can be accessed by its 8 bit address or by its name (when the bank is selected). The selection of the register bank is through the RS0 and RS1 bits of the PSW. The bank 0 (*i.e.* RS0 = 0, RS1 = 0) is selected when the microcontroller is reset.

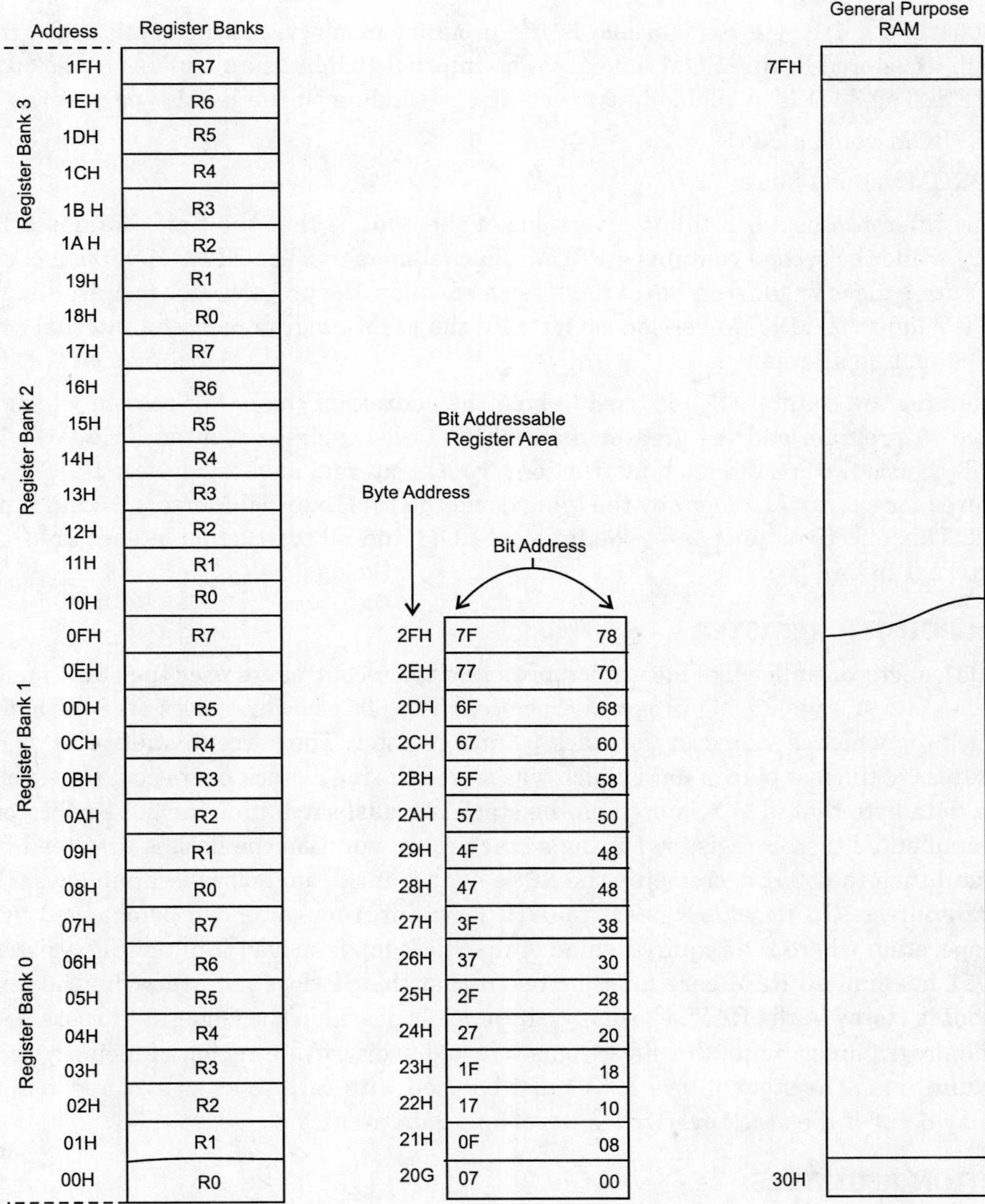

Figure. 12.9. 8051 RAM Organization

2. Bit Addressable Registers Area : There are a total of 16 bytes of bit addressable RAM whose address ranges from 2OH to 2FH. This area can be byte selected or bit selected depending on the address. These registers are especially useful when conditions (on/off) are to be stored.

3. General Purpose RAM : An 80 byte RAM area having an address ranging from 30H to 7FH is also present. This area can be used for any purpose such as data storage, storing results etc.

Internal ROM

The 8051 has a 4 Kbyte ROM which is the program memory. All instructions for the 8051 program are stored in this ROM memory. The internal ROM has an address range ofOOOOH to 0FFFH. The 8051 is available in two versions depending on the ROM type :

1. EPROM version 8051
2. PROM version 8051

The difference between the two versions of the 8051 is that the first contains a EPROM memory while the second contains a PROM internal memory. The EPROM version of the 8051 comes with a glass window on top of the IC so as to allow the programmer to erase the ROM by use of UV light. The PROM version comes with the ROM programmed and internal programs which cannot be altered.

The program counter (PC) is used to hold the address of the ROM location which is to be accessed. A program address greater than OFFFH (last address of internal ROM) will automatically cause the 8051 to fetch instructions from a external memory device. The 8051 can be configured for external memory by the tying down the EA (External Memory Access Pin) Pin to ground. This effectively by passes the internal ROM and all instruction fetches are from then the external ROM.

STACK POINTER REGISTER

The 8051 microcontroller like most microprocessor/microcontrollers uses the stack memory for program/state storage during program execution. The 8051 however uses an 8 bit address for the stack top which is stored in the stack pointer register. The current value of SP is actually the address of the last byte of data which was stored during a stack operation. However, in the 8051 a data byte that is to be stored on the stack is transferred after the 8051 microcontroller has incremented the SP register. During a stack read operation the data is first read from the stack and then the 8051 decrements the SP so as to point to a memory location lower than the current address. On Reset, SP is set to 07H. However, this value can be changed by a data move operation where the required value of the stack top is moved into the SP register. Since the 8051 has limited RAM care must be taken that the SP does not overwrite the important data that is stored in the RAM. Normally, the stack is placed in the General Purpose RAM area and the programmer limits the stack so as to avoid overwritting region of memory that holds important data. The stack in the 8051 must be used with care so as to avoid system crashes that may occur if the stack overwrites important data areas.

FUNCTION REGISTERS

Some 8051 instructions operate on data which is stored in special function registers. These registers are not a part of the 8051 internal 128 byte RAM (address from OOH to 7 FH) but can be accessed by their address to which range from 80H to FFH.

These registers however, are accessed in the same way that the 8051 internal RAM is accessed. Certain function registers are also bit addressable (Similar to bit addressable RAM area of 8086). This is provided to allow for programmers to alter the certain data bits of the 8051 function registers while leaving the remaining bits untouched. Figure 12.4(a) and 12.10. shows the different function registers, their addresses and functions.

Function registers are named directly in certain instructions by their functions name such as in the instruction ADD A, B* or accessed by their address. However all addresses must start with a number between 0 to 9 therefore, registers whose address begin with A to F will actually have a zero introduced before them, i.e., the Accumulator will have actually address OEOH (instead of EO H).

Name	Address (Hex)	Address Ability	Function
A	EO	Bit/Byte	Accumulator
B	FO	Bit/Byte	Arithmetic register
DPH	B3	Byte	Data pointer high byte
DPL	B2	Byte	Data pointer low byte
IE	A8	Bit/Byte	Interrupt Enable Register
IP	B8	Bit/Byte	Interrupt Priority Register
PO	80	Bit/Byte	Input Ouput Port 0 Latch
P1	90	Bit/Byte	Input Ouput Port 1 Latch
P2	AO	Bit/Byte	Input Ouput Port 2 Latch
P3	BO	Bit/Byte	Input Ouput Port 3 Latch
PCON	87	Byte	Power Control
PSW	DO	Bit/Byte	Program Status word
SCON	98	Byte	Serial port control
SBUF	99	Byte	Serial port data transfer
SP	81	Byte	Stack Pointer
TMOD	89	Byte	Timer mode control
TCON	88	Bit/Byte	Timer control
TLO	8A	Byte	Timer 0 low byte
THO	8B	Byte	Timer 0 high byte
TL1	8C	Byte	Timer 1 low byte
TLO	8D	Byte	Timer 1 high byte

Figure 12.10. Function Registers to 8051

12.2.6. IC Pins, Ports And Port Circuits

We have shown the 8051 microcontroller pin diagram in figure 12.5. However, now we shall discuss the 8051 pin diagram in more detail. Unlike the 8086 that we have discussed in earlier chapters, which required external components to be connected so as to interface with external circuitry, the 8051 microcontroller integrates all the circuitry on a single chip. This has a cumulative effect of reducing the total number of discrete components that are required to interface circuitry thereby reducing the cost of the overall 8051 circuit. However, increasing the functions that are done by single pins increases the complexity of a system. But the 8051 is designed so as to allow the program to physically control the function of these pins thereby

* When using assemblers

reducing complexity and increasing the flexibility of the 8051 system.

This flexibility that the intel designers has bestowed on the 8051 allows it to be introduced in a system where it acts as a complete component with Input/Output port only (i.e., the pins are used only for Input/Output purposes) or allows it to be expanded to interface additional memory and ports by using the alternate pin assignments. This flexibility of the pins is a result of the port circuitry which is present in the ports (figure 12.11).

As discussed earlier, the 8051 has a total of four ports where each port has eight pins. Each port pin has a clocked D-Latch which is used for data storage. Each port effectively thus has eight output latches which forms the function register of that port. Therefore, a port address addresses the port latches. For example a port address of 90H addresses (accesses) the eight latches of port 1 and the fourth pin 4 of port 1 is the bit 3 of the port 1 (PI) function register.

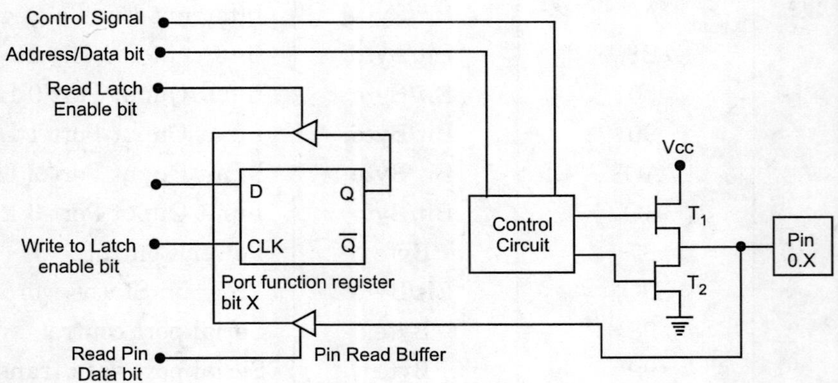

Figure 12.11. (a) Port 0 Latch and Pin Configuration

Each port (i.e. PO, PI, P2, P3) however, have different arrangements of these latches according to the functions that the port pins have to handle. Now let us discuss each of the ports and their latch arrangement.

PORT 0

Port 0 is an 8 bit open drain bidirectional I/O port. As an output port this port can drive several LS TTL gates. The pin configuration is shown below in figure 12.11(a).

A Port 0 pin that has the one written on them essentially causes both the transistors to turn off thereby causing them to float. Thus we would require external pull up resistors to supply a high output. When the two transistors turn off it causes the port 0 pin to float to high impedance state causing the pin to be connected to the pin read buffer. Thus the port 0 is now configured as an input port. When used as an output port, pin latches that are outputting zero will cause the lower FET to turn on, hence grounding the output pin (effectively thus at zero state). An output bit of one will cause the pin to float to high impedance state thus requiring external pull up resistors to output a data state of one.

Port zero is also used as the multiplexed lower order address and data bus to interface to external memory. When the address is to be transferred to the pin the control signals cause the control circuitry by pass the latch and automatically connect the address lines to the transistors T_1 and T_2 A logic 1 on an address bit will cause the upper FET T_1 to turn on and lower FET T_1 to turn off providing a logic high on the port pin. When the address bit is 0 the lower FET is turned on and upper FET T_1 turned off thereby grounding the output pin and providing logic zero at the output pin. During the end of the second time state the 8051 reconfigures the pin for receiving data by placing a logic 1 on the D-latch. This cause the FET T_1 and T_2 to float and connects the output pin to the pin read buffer. Thereby enabling the 8051 microcontroller to read the data sent out by the external memory.

PORT1

Port 1 is a bidirectional I/O port with internal pullups. The pin configuration of Port 1 is shown in figure 12.11(b).

The inverted output of the D latch is connected to the gate of the FET 1'1' This FET uses an internal FET pullup formed by the two FET's T_2 and T_3 T_2 is a depletion type NMOSFET while T_3 is a enhancement type NMOSFET.

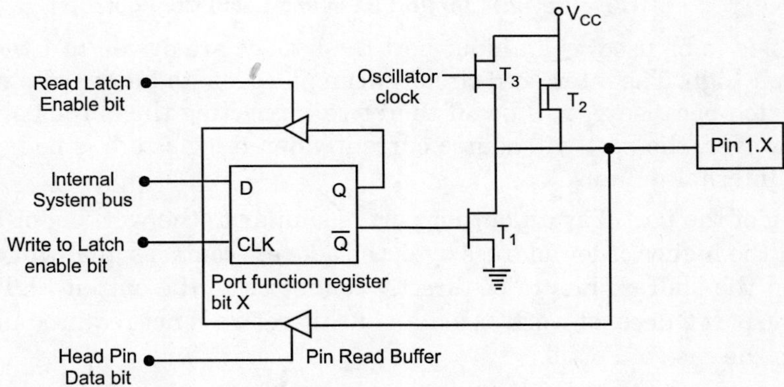

Figure 12.11.(b) Port 1 Latch and Pin Configuration

When Port 1 is used as an output port, the latch's containing a logic high (i.e. 1) turn off T_1 thus driving the pin high through the pullup resistor formed by T_2 and T_3 If the latch contains a zero bit then T_1 gets turned on thereby grounding the output pin (Pin l.X). Thus the pin effectively shows a logic level of zero to the circuitry connected.

When port 1 is used as an input port a logic level of 1 is written to the latch. This effectively turns the FET T_1 off (i.e., high impedance state). Therefore, the pin is directly connected to the pin Read buffer. However, at this time the input to the pin read buffer and the port 1 pin is high. An external device can place a zero by driving the pin to ground or place a one on the port by leaving the input pin at logic 1 state.

For high speed operation the internal pullup resistor uses two FETs T_2 and T_3 FET T_3 is turned on for two clock pulses (i.e. one time state) which provides a low impedance path to the supply voltage helping to reduce transients in the external circuitry.

PORT 2

Port 2 can be used as an I/O port and operates similar to the port l. The Port 2 latch and pin configuration is shown below in figure 12.11(c). Port 2 also emits the higher order address byte during program fetches from external memory.

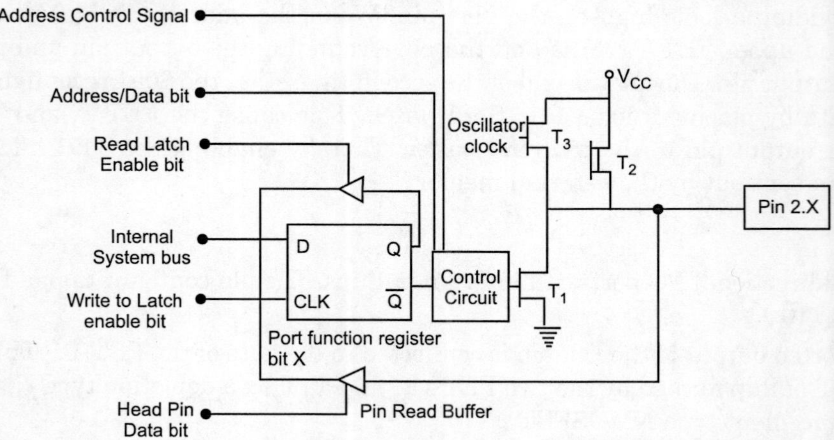

Figure 12.11 (c) Port 2 Pin and Latch Configuration

When the Port 2 is to be used as an input port the latches are driven to 1 thereby pulling the output pin to logic high. The interfaced circuitry can place a data bit of 1 or o which will change the state of D latch because T_1 will be off therefore connecting the output pin to the read pin data buffer. However, the port will source current when a logic zero is being inputted on the port pin due to internal pullups.

The working of the port 2 as an output port is similar to the working of Port 1. When the Port 2 is to emit the higher order address byte, the address control signal will cause the latch to be bypassed and the address bits to be directly connected to the output FET stage. However, unlike port 0, the port 2 doesnot work as a data bus therefore is not required to be configured as a data bus as in the case of Port 0.

PORT3

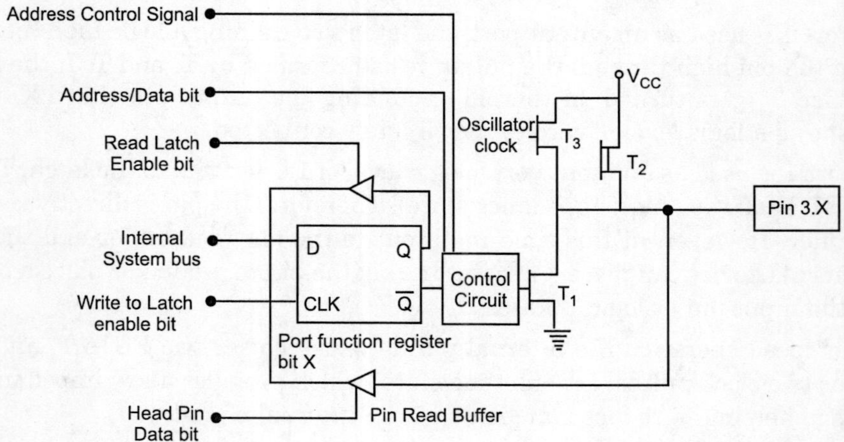

Figure 12.11.(d) Port 3 Port and Pin Configuration

The Port 3 is an 8 bit bidirectional I/O port with internal pullup. The port configuration is shown in figure 12.11(d) The Port 3 pins also serve dual functions as listed in figure 12.12. The input and output function of the port 3 pins-is under control of P3 latches or under control of certain function registers. (Refer figure 12.12). However, Port 3 pins are individually programmable and are unlike the other ports which change function of all port pins when in alternate use. However, to be used as an input port, port 3 latches must be raised to logic one so as to turn off transistor T1.

Pin Name	Pin Number	Other Use	Function Register
P3.0	10	Serial data Input (RXD)	SBUF
P3.1	11	Serial data Output (TXD)	SBUF
P3.2	12	External interrupt (INTO)	TCON.1
P3.3.	13	External interrupt 1 (INT1)	TCON.3
P3.4.	14	External times output (TO)	TMOD
P3.5.	15	External times 1 input (T1)	TMOD
P3.6	16	External Memory write pulse (WR)	–
P3.7	17	External memory read pulse (RD)	–

Figure 12.12. Alternate Function of Port 3

12.3. INTERFACING THE 8051 TO EXTERNAL MEMORY

For some applications of the 8051 microcontroller, the internal memory is insufficient. However, the 8051 has the ability to access external ROM and RAM by the use of addresses stored in 16 bit PC and 16 bit DPTR register. However, the 8051 has different control signals to access RAM and ROM unlike other microprocessor chips which use theasame signals for both. Also a ROM access will use the value stored in the program counter (PC) as the address whilea RAM access will use the value stored in DPTR register as the address.

The external memory chips that are connected to the 8051 are limited to 64 Kbytes as the 8051 uses a 16 bit address for memory access. Figure 12.13 shows an 8051 connected to a 32 Kbyte EPROM chip and a 8 Kbyte SRAM chip. The 8051 accesses external ROM whenever. the address exceed FFFH or when the EA pin is tied low. Thus, the 8051 can use a combination of internal and external ROM. However, as in figure 12.12 since the EA pin is grounded low the 8051 will only access the external ROM. The external RAM is accessed by special instructions present in the 8051 instruction set. Figure 12.14(0) shows the timing diagram for external ROM read cycle while figure 12.14(b) and (c) show the external RAM read and write cycles.

During, a memory access port 0 is the multiplexed address data bus. During the first time state the port 0 holds the lower byte of the sixteen bit memory address. After the address has been outputted the port 0 then acts as a bidirectional data bus for data input or output. Port 2 holds the upper 8 bits throughout the read/write cycle.

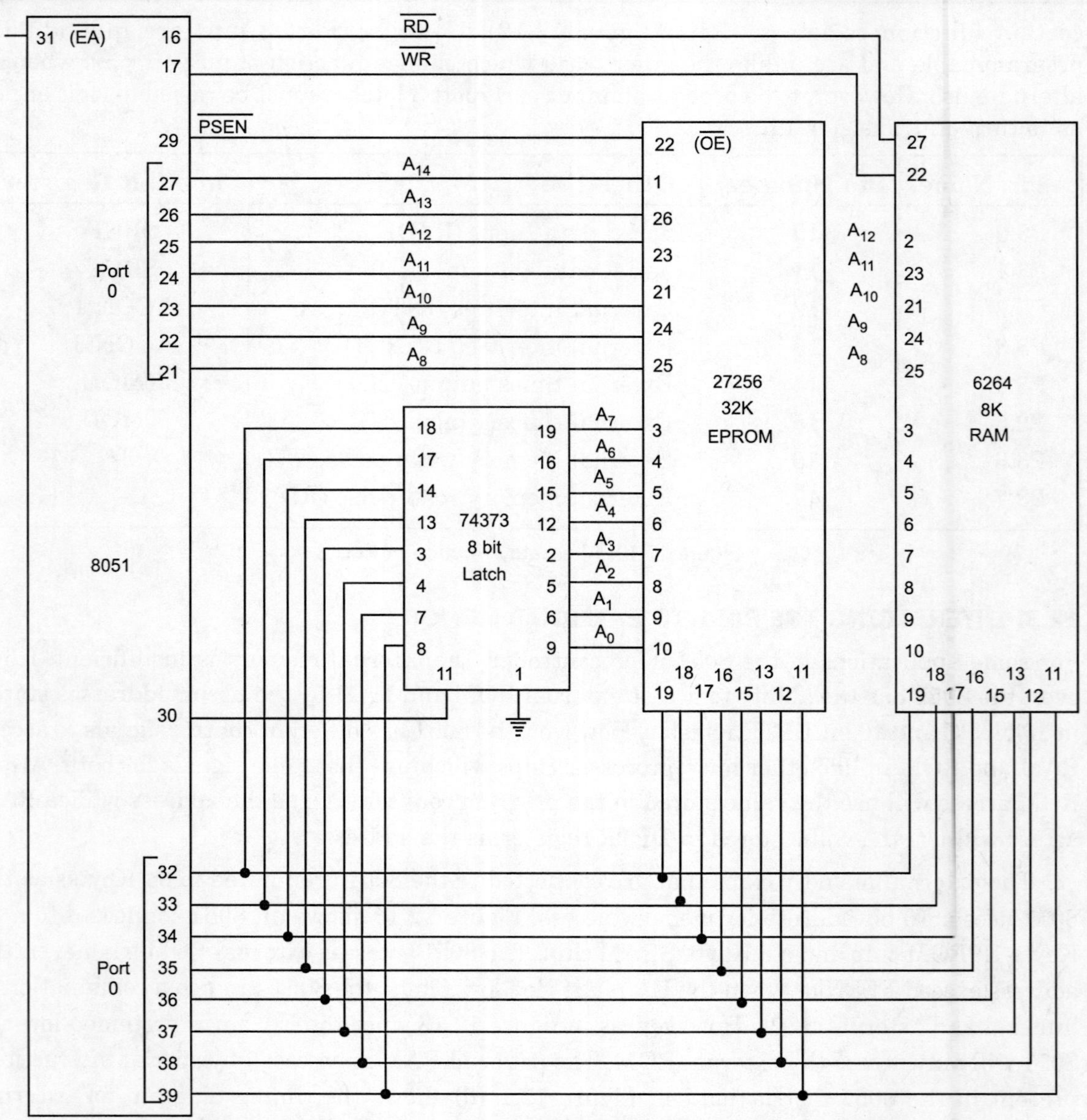

Figure 12.13. An 8051 interfaced to 32 K byte EPROM and a 8 K byte SRAM chip

The lower order address must be stored as this bus is time multiplexed and after certain time will behave as the data bus. In case the external device has an access time greater than the time the address stays on this bus, then the memory device will transmit incorrect data. To bypass this problem we use a 8 bit latch to hold the address. The latch is enabled by the ALE signal, which is forced high as long as thelower order bus is serving as an address bus.

When the 8051 accesses a program code byte in the ROM, the PSEN (Program Store Enable) pin will go low thereby enabling the ROM to place a code byte on the data bus. A memory access requiring bytes stored in RAM will cause the RD or WR pins to go low depending on the type of memory access thereby enabling data to be written or read from the RAM.

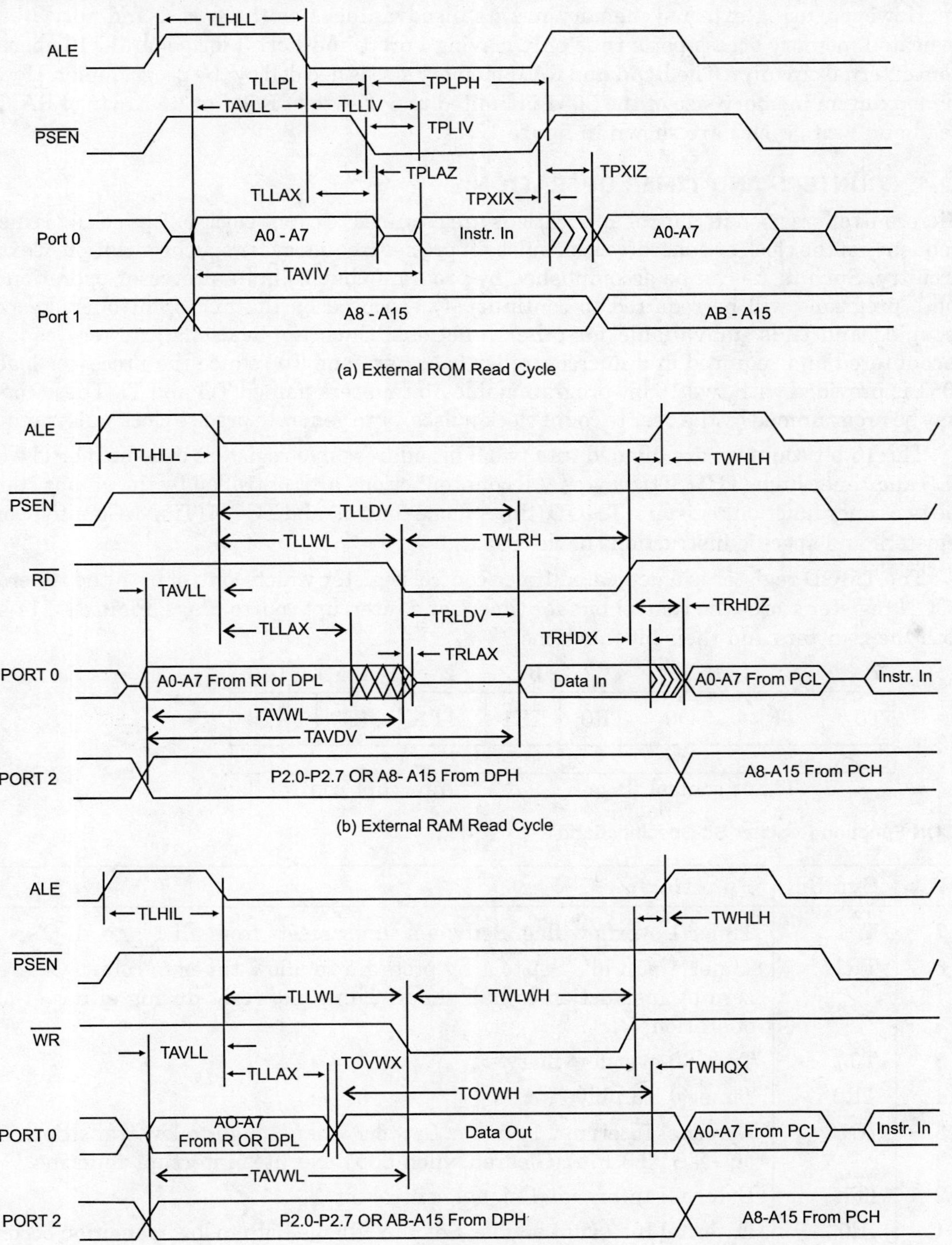

(a) External ROM Read Cycle

(b) External RAM Read Cycle

(c) External RAM Write Cycle

Figure 12.14 ROM and RAM access cycles

However, using external memory has its disadvantages as the port 1 and port 2 become dedicated memory access ports thus only leaving Port 1 and Port 3 for general 110 operations from external circuitry. The ROM and RAM memory size is decided by the programmer. However, the maximum memory size of the 8051 is limited to 64Kbyte of ROM or 64 Kbyte of RAM. The Read and write cycles are shown in figure 12.14.

12.4. COUNTERS AND TIMER OPERATION

Microcontrollers used in control applications are required to count clock pulses so as to calculate frequency of the clock or generate clock pulses of precise frequency for synchronization of external circuitry. Such tasks can be accomplished by programs but it limits processor operation since these programs will be required to continuously executed by the microcontroller, keeping it occupied and thus unavailable for other functions. Such applications are very commonly encountered and required in a micrcontroller based system. To reduce the processor load, the 8051 is provided with two 16 bit programmable up counters named TO and Tl. These counters may be programmed by the user to count clock pulses, or to generate precise clock pulses required.

The 16 bit counters are divided into two 8 bit addressable registers called timer low (TLO, TLl) and timer high (THO, THl) bytes. All counter actions are controlled by the values stored in the two timer function registers: TMOD (timer mode control) and TCON (Timer/Counter Control) registers and specific instructions of the 8051.

The TMOD register is a dedicated timer control register which control both the timers. The TCON register stores the control bits for timers and interrupt control flags. Figure 12.14 shows both the registers and their bit functions.

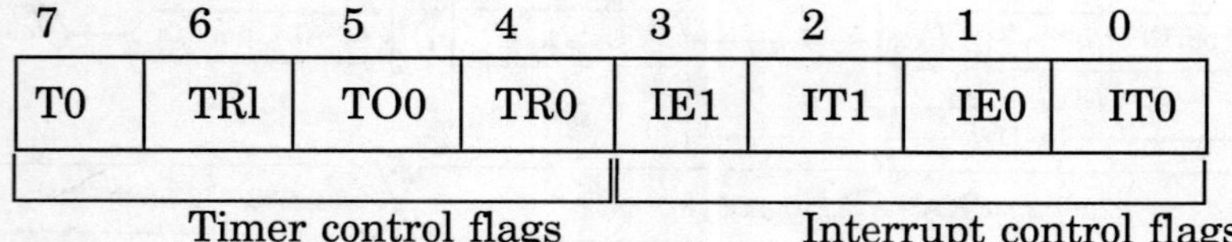

TCON Function Register Bit Specifications

Bit	Symbol	Function
7	T01	Timer 1 overflow flag. Set when timer resets from all 1 's to all O's.
6	TR1	Timer 1 run bit: Set to 1 by program to allow timer to count. Cleared to zero by instruction to half timer. Timer no, reset during either of above operation.
5	T00	Timer 0 overflow flag
4	TR0	Timer 0 Run bit
3	IE1	External Interrupt 1 enable: Set to 1 when a high to low transition occurs on INT(P3.3 Pin). Cleared when 8051 executes interrupt rountine.
2	IT1	External Interrupt 1 signal control bit
1	IE0	External Interrupt enable: Set 1 to when a high to low transition occurs on INT0 (P3.2 pin) cleared when 8051 executes interrupt service rountine.
0	IT0	External Interrupt 0 signal control bit.

(a) TCON Register

The value of ITO and IT1 sets the signal type that will cause on interrupt. If thus value is 0 then a low level signal on the INTO (or INT) will cause a interrupt. If 1 a high to low transition on INTO (or INT) will cause an interrupt.

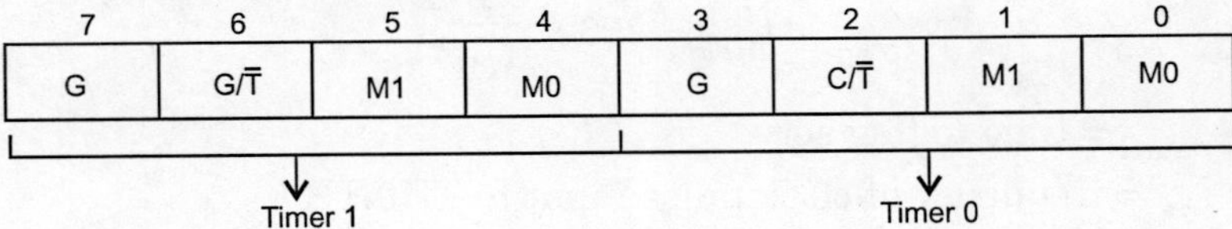

TMOD Function Register Bit Specifications

Bit	Symbol	Function
7/3	G	Gate enable bit: This bit controls the start/stop of Timer 1/0. Set to 1 by 8051 instruction enable timer to run if TR1/TRO in TCON is set and signal on INT1/INTO is high (i.e. no interrupt caused). Cleared to 0 by instruction enabling timer 1/0 to run if TR1/TRO is set irrespective of INT1/INTO state.
6/2	CIT	Set to 1 by instruction to cause timer 1/0 to act as counter by counting pulse from external input caused on 8051 pins P3.5(for T1) and P3.4 (for TO). Cleared to 0 by instruction causing timer I/O to act as timer by connecting 8051 clock pulses.
5/1	M1	Timer mode select bit 1
4/0	MO	Timer mode select bit 0

Time Mode Select Truth Table

M1	MO	Mode Selected
0	0	ModeO
0	1	Mode 1
1	0	Mode2
1	1	Mode3

(b) TMOD Function Register

Figure 12.14. Timer Function Registers TCON and TMOD

TIMER MODES

The 8051 timer may operate in four modes. The mode in which the 8051 microcontroller times is operating is defined by the user by the states of the mode control bits MO and M1 of the TMOD function register.

TIMER MODE 0

Setting the MO and M1 bits of timer 0/1 to 00 causes the timer to operate in mode O. This is the 13 bit count mode where the TLO/TL1 acts as a 5 bit counter and THO/TH1 acts as a eight bit counter. The pulse input is inputted to the timer every machine cycle (i.e after 12 clock pulse of 8051 internal clock) and is used to increment the TLO/TL1 register. The TLO/TL1 register counts from 0 to 32 and restarts. Before restarting it increments the THO/TH1 register. Thus TH counts the pulse frequency divided by $\frac{1}{32 \times 12} = \frac{1}{384}$. Therefore, a 8 MHz refrom the value frequency would result in a pulse whose frequency is 20833 Hz which is counted by THO/TH 1. The timer overflow flag (TO) is set every time the THO/TH 1 goes from FFH to OOH in

$$t_{Toset} = 256 \times \frac{1}{f_{TH}}$$

where. $t_{To\ set}$ = time to flag set

f_{TH} = frequency of clock pulse input to TH0/TH 1

i.e., for above all the To flag is set every 0.0122 seconds.

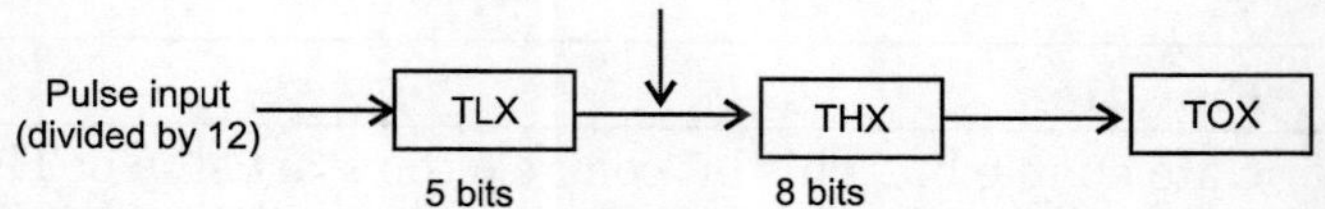

Figure 12.15. Time Mode 0 (X = 0 or 1)

TIMER MODE 1

Mode 1 is similar to Mode 0 except that TLO/TL1 is an 8 bit count register. Mode 1 is set when MO = 1 and Ml = 0 in TMOD register. Here the frequency of pulse that is input to TH0/TH 1 is input to THO/TH 1 is

f_{TH} = frequency of pulse input to TH

f_{TL} = frequency of pulse input to TL

$$f_{TL} = \frac{f}{12}, f_{TH} = \frac{f_{TL}}{236} = \frac{f}{256 \times 12} = \frac{f}{3072}$$

f = frequency of clock pulse to be counted.

where t_{Toset} = time after which timer overflow flag is set

TIMER MODE 2

When the M1 = 1 and MO = 0 the timer operates in mode 2. In mode 2 operation, the TH1/THO register operates as a 8 bit counter. THO/TH1 is used to held a eight bit value which is loaded into TLO/TL1 every time the timer overflows. On overflow the timer overflow flag T00/T01 is set and the timer register TLO/TL1 is loaded with the value stored in THO/TH1 This mode can be used for generating precise time delays as timer will count up from the value stored in THO/TH1 overflows (and set Timer overflow flag) and again restart counting from the value stored in THO/TH1 The delay can be calculated as

f = frequency of internal clock = 10 MHz

d = delay required = 0.0003 see

= time required after which overflow flag should be set

x = value stored in TH i.e., from where counties will start committing

y = final count before setting Timer overflow flag

= 2^8 (8 bit counts) = 256

c = total increment period = 12 (y – x) = (256 – x) 12

[As count will increment every machine cycle

Now $$d = \frac{c}{f} = CT \text{ where T= time period of interrial clock pulse}$$

$$\Rightarrow \quad c = \frac{d}{t}$$

$$\Rightarrow \quad (256 - X) = \frac{df}{12}$$

$$\boxed{X = 256 - \frac{df}{12}}$$

$$\Rightarrow \quad X = 256 - \frac{0.003 \times 10 \times 10^6}{12}$$

$$X = 256 - 250 = 6$$

$$\boxed{X = 6}$$

∴ The TH0/TH1 should be loaded with a value of 6H to produce a delay of 0.0003 Sf However this is the delay time after which the overflow flag shall be set. Programs can the overflow flag state and take a decision accordingly.

TIMER MODE 3

Both timers 0 and timer 1 can independently operate in mode 0,1 or 2. Both timers can be operating in same modes or separate modes. However both timers cannot, at the same operate in mode 3. If mode 3 is chosen for Timer 1 it will cause it to stop counting and hold current value. Timer 0 operating in mode 3 essentially behaves as two 8 bit counters where TLO is controlled by TRO (timer 0 Run bit) in TCON register and THO is controlled by 1. (timer tram bit). The TLO register count from OOH to FFH. On counting till FFH it then the timer 0 overflow flag and restarts counting from OOH. The THO receives timer coder puls only if the TR1 bit is set and sets the Timer 1 overflow flag (TF1) when it overflow from FF to OOH.

Timer 1 however can be used in mode 0, 1 and 2 but will not cause interrupts as the time overflow flag is now under control of Timer O.

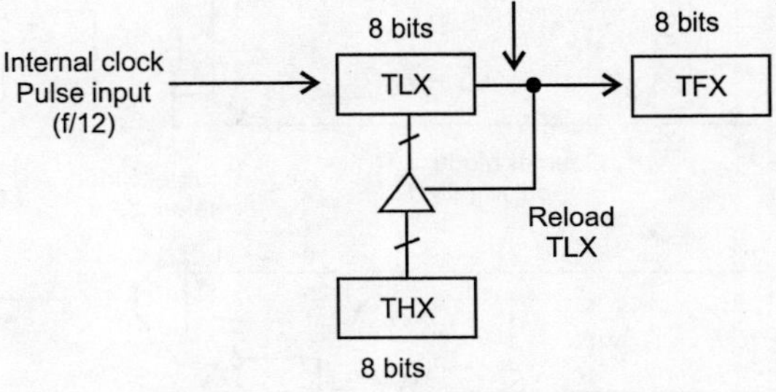

Figure 12.16. Mode 2 Operation

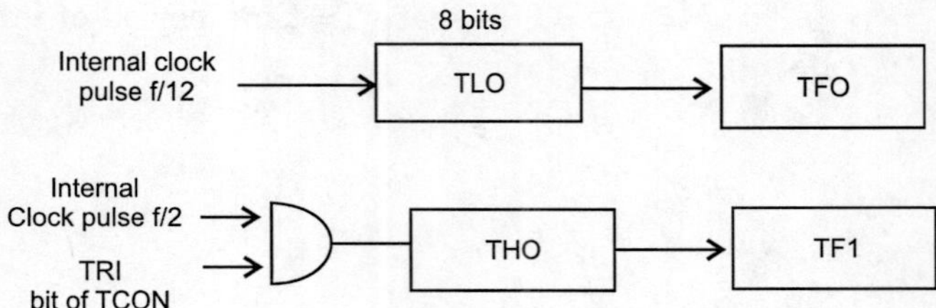

Fig. 12.17 Mode 3 Operation

TIMING AND COUNTING

Since we have talked about the timer modes let us understand the basic difference between the timer registers that are programmed to operate in timer mode (by the $C/\overline{T}$ bit in TMOD register) it will be transmitting the 8051 internal clock as the count source. That is it will count the internal frequency of the 8051 oscillator divided by 12. That is if the 8051 crystal frequency is 12 MHz the timer clock will have a frequency of 1 MHz.

When the circuit is used as a counter the source of clock pulses is from the counts pins T0 (P 3.4) or Tl (P 3.5). The counter pins TO supplies clock pulses as input to Timer 0 while pin Tl supplies pulses as input to Timer 1. The $C/\overline{T}$ bit must bet set to 1 to enable the T0/T1 pin to be directly connected to the timer circuitry. The pulse on T0/T1 is sampled in the second pulse of time state 5 of every machine cycle. A change of high to low between two samples causes an increment in the counter. Each high and low state must be held low for at least one machine cycle so as to generate the correct samples of the clock pulse. Thus the maximum frequency that can be accurately counted is the oscillator frequency divided by 24 (2 machine cycles). Thus for a 12 MHz crystal the counter can correctly count a signal of maximum external frequency 0.5 MHz.

The timer/counter logic diagram is shown below. Observe that the $C/\overline{T}$ bit state routes the corresponding input to the timer. An intermediate gate controls the timer counting depending on the states of bits in the TMOD and TCON registers.

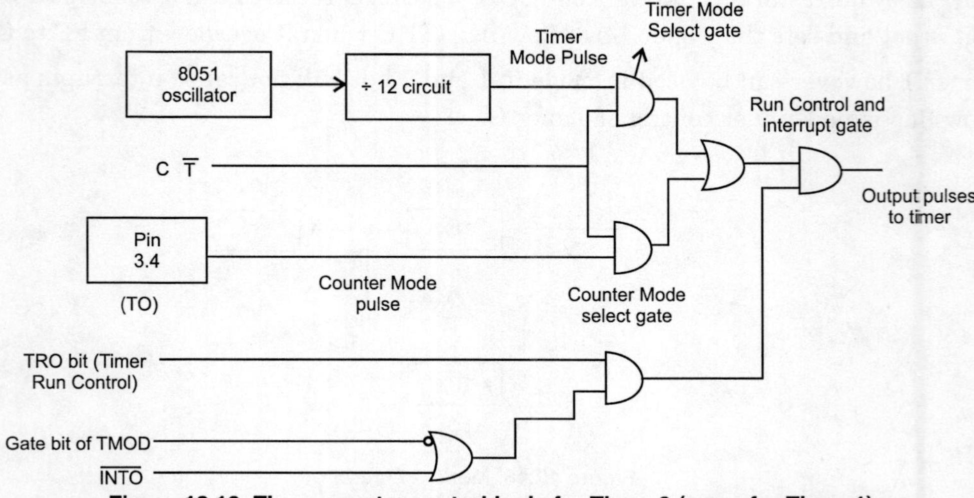

Figure 12.18. Timer counter control logic for Timer 0 (same for Timer 1)

TIMER INTERRUPTS FOR TIMES DELAY

The timer that have been discussed of the 8051 microcontroller sets the timer overflow flag whenever the timer overflows. Timer overflow occurs when the timer increments the register which is already at its maximum value. For example an increment to an 8 bit register holding the value of FFH will cause it to increment to 100H. This 1 sets the overflow flag (figure 12.19)

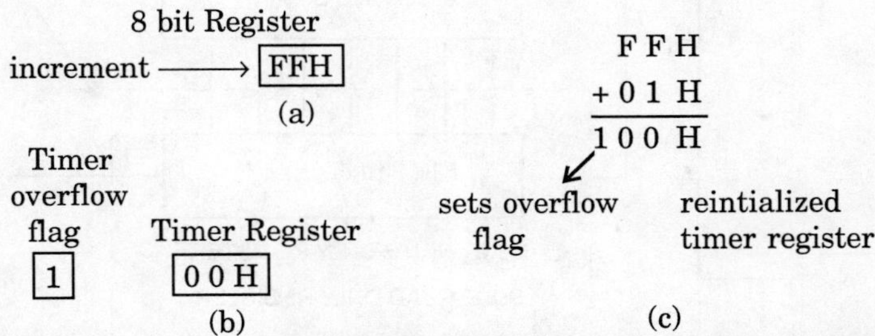

Figure 12.19(a) increment to a register holding max value (b) causing an overflow (c) Mathematical logic of timer overflow

This set overflow flag can be tested by instructions to tell a program that the required count has been accomplished or the flag may be used to interrupt the program. For time delays or counting of specific number of pulses a count number can be placed in one of timer registers. This count value should be equal to 256 – desired count. The timer will then increment the count value to the final value (255) and then roll over to zero. At that time the timer overflow flag shall be set.

12.5. SERIAL DATA COMMUNICATION

In any microprocessor/microcontroller based system it becomes important for data communication from one device to the microcontroller. There are two methods of data communication: serial and parallel. However serial data communication is preferred as it is cheaper and easier to handle. (Transferring 8 bits at one time would require 8 communication lines and circuitry to be precisely synchronized there by increasing the hardware cost).

The 8051 microcontroller supports serial data communication. The 8051 part pin P3.0 (R × D) is used to receive data from an external device while the pin P 3.1 (T × D) is used to transmit data from the 8051microcontroller to the external device. The SBUF (serial Data Communication Buffer) register is used to hold data, while the overall serial communication circuitry is controlled by the PCON (Power Control) and SCON (Serial Communications Control) function register bits.

The SBUF function register is physically conferred of two 8 bit registers. One register is write only and used to store data that is to be transmitted serially from the 8051 P3.1 (T × D) pin. The second register is read only and is used to store the data that is received from external devices over the P3.0 (R × D) pin. Both register have the address 99H. The control logic register have the address 99H. The control logic for reading and writing to SBUF function register is shown in figure 12.20.

The 8051 programmer can use four modes for serial data communication by setting the corresponding communication of SMO and SM 1 bits in SCON register (figure 12.20).

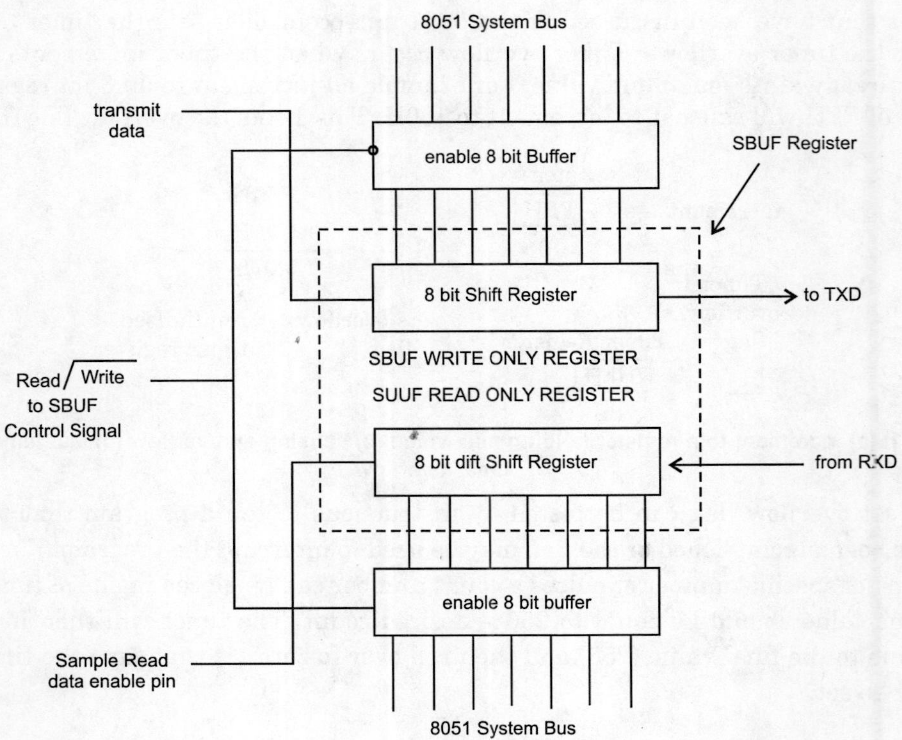

Figure 12.20 SBUF Read and Write Only Register

7	6	5	4	3	2	1	0
SMO	SM1	SM2	REN	TB	RB	R1	T1

Bit	Symbol	Function
7	SMO	Serial port mode control bit. Set/Reset by
6	SMI	program to select mode
5	SM 2	Multiprocessor Communication Bit
4	REN	Data Reception enable bit. If 1 then reception enabled, 0 reception disabled
3	TB	transmitted bit 8
2	RB	Received bit 8
1	T1	Transmit Interrupt flag. Set to one at end of bit 7 transmission in mode 0, and before transmission of stop bit in other modes.
0	R1	Receive Interrupt flag. Set to one after reception of bit 7 in mode 0 and during reception of stop bit for other modes

Figure 12.21. (a) SCON Function Registers Bit Configuration

SMO	SMl	Mode Selected	Transmissron Type and band rate
0	0	0	Shift Register. band rate = f/12 8 bit UART, band rate
0	1	1	= variable bit UART.baul rate = f/32 or f/64 9 bit UART.
1	0	2	baud rate = variable
1	1	3	

Figure 12.21. (b) Mode Select truth table

SERIAL COMMUNICATION INTERRUPT PROCESS

Serial data communication is relatively slower than parallel data transfer. Data bytes may take several milliseconds per 8 bits of data that is to be transferred. After data has been transmitted the TI flag of SCON register is set. Similarly when a data byte is received through the RXD pin the RI flag is set. The 8051 internal circuitry uses an OR gate to generate a 8051 interrupt. However programs must access the SCON register to see which of these two causes the 8051 interrupt. However unlike the timer flag which is cleared automatically (when causing an interrupt) these flags must be cleared by the programmer.

TRANSMISSION AND RECEPTION

Serial data transmission requires the program to first load SBUF register with eight bit data. After data has been transferred the 8051 can begin data transmission. After the data byte has been transmitted the T1 flag is set. This effectively cumulates to that the data byte has been transmitted and the SBUF register is empty. A program should therefore wait for the TI flag to be set before loading the SBUF register with the next data byte. Serial data reception mode is enabled when the REN bit (Data reception enable) is set to l. For mode 0 the RI bit also must be cleared (not required for other modes). The RI bit is set after the data byte is received via the RXD pin of the 8051 and should be used to read the last data byte from SBUF register and store it in RAM while a data byte is being received through the RXD pin.

DATA TRANSFER MODES

1. Mode 0 – Shift Register Mode

The data transfer mode 0 or shift register mode is set when SM1 and SMO is set to OOH. This mode uses RXD pin for both data transmission and reception while the TXD pin supplies a clock of frequency f12 (f = internal clock frequency of 8051). The TXD pin is connected to external devices to synchronize them to the 805l.

7	6	5	4	3	2	1	0
SMOD	R	R	R	GF1	GF0	PD	IDL

Bit	Symbol	Function
7	SMOD	Serial baud rate modify bit: 1 causes double baud rate using timer l. 0 causes baud rate equal to timer 1 band rate.
4,5,6	R	Reserved
3. 2	GFlIGFO	User flag bit.
1	PD	Power down control bit for CHMOS 805l.
0	IDL	Idle mode control bit for CHMOS 8051

Figure 12.22. PCON Function Register

12.6. 8051 INTERRUPT

The 8051 like other processors has interrupts for interrupting the flow of a program. Interrupts are hardware signals that may cause the 8051 microcontroller to jump and execute an interrupt service routine that is located on a predetermined address in the code memory. There are a total of five interrupts in the 8051. Three of these interrupts are generated by internal operations. Two interrupts are provided which respond to external signal changes on the 8051 pins INTO and INT 1.

Internal Interrupts: There are three internal interrupts which are caused when the TOO (Timer 0 overflow flag) TOI (Timer 1 overflow flag) or the RI (Receive Interrupt) and TI (Transmit interrupt) are enabled. The timer overflow flag as discussed earlier is set when a timer/counter overflows from FFH to OOH. This sets TOO/TOI depending on the timer which overflows. The RI is set when a data byte is received in the SBUF register while Tl is set when a data byte is transmitted from the SBUF register. Both these flags are OR-ed and the output is used to cause a interrupt to the 8051.

EXTERNAL INTERRUPTS

External circuitry can use the 8051 pins INTO or INT 1 to cause an interrupt. A signal on either of these can cause the interrupt flag IEO or TE 1 to be set in the TCON register. However the 8051 system designer has a choice in deciding what type of signal will cause an interrupt. Bits ITO and ITI in TCON register are used for this. IflTO/ITI is reset, a signal that is low on the INTO/INT 1 pins can cause the interrupt enable bit to be set. If ITO/ITI is set, a high to low transition on INTO/I NT 1 will cause the interrupt enable bit (lEO/IE 1) to be set.

INTERRUPT CONTROL

However the set state of an interrupt flag is not a sufficient criteria for the 8051 to enter an interrupt state. The interrupt control register is shown in figure 12.18 (c) has specific bits which must be set so as for an interrupt to occur. This is sometime necessary when a critical system is running an important task and therefore 111 must not be interrupted.

7	6	5	4	3	2	1	0
EX	R	R	ES	ET1	EXL	ETO	EXO

Interrupt enable (IE) Function Register

Bit	Symbol	Function
7	EA	Enable interrupt mode bit. Cleared to one to disable all interrupts. Set to 1 to per unit individual interrupts to be enabled by respective bit.
6, 5	R	Reserved
4	ES	Enable Serial Port interrupt
3	ETI	Enable Timer 1 overflow interrupt
2	EX	Enable external interrupt 1
1	ET0	Enable Timer 0 overflow interrupt
0	EX0	Enable External interrupt 0

Bit 4 to Bit 0 : If set, enable that interrupt. If reset, disables respective interrupt.

Figure 12.23(a) IE function register bit specifications

If however an interrupt is enabled and its control bit is also set, an interrupt occurs. At this point the program counter is loaded onto the stack memory and the program counter is reloaded with a predefined address (corresponding to the interrupt that occurs). The microcontroller now accesses code bytes from this location. The interrupt service routine (or the task. that the microcontroller has to perform for that interrupt) is written at this address. The interrupt service routine always ends with a RET (Return from Interrupt Routine) instruction. This causes the microcontroller to reload the program counter from stack (i.e., the address when it was interrupted) and restore program execution.

INTERRUPT PRIORITY

The 8051 interrupts can be given a priority level. Setting a priority level causes a higher priority interrupt to be serviced before an interrupt of lower priority. The IP register (Interrupt Priority Function Register) bits are used to decide the interrupt Priority of the five interrupts and can be changed according to a designer's need. A IP register bit that is set causes the corresponding interrupt to be given higher priority level. A IP register bit that is reset causes the corresponding interrupt to be given lower priority. However if two or more interrupts have the same priority then their priority levels are set by the 8051 designers as :

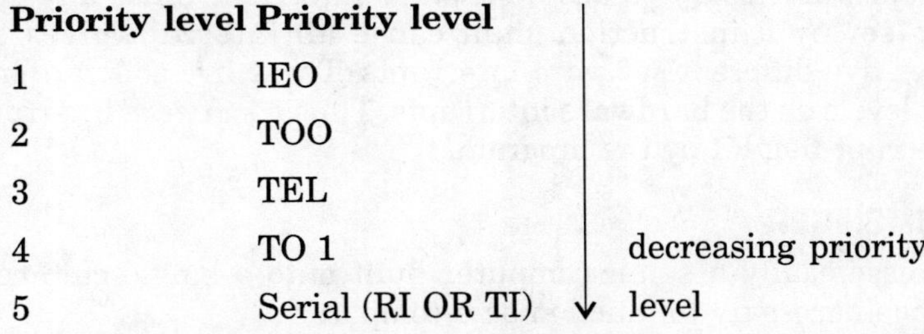

Priority level	Priority level
1	lEO
2	TOO
3	TEL
4	TO 1
5	Serial (RI OR TI)

The IP register bit configuration is shown below in figure 12.19.

7	6	5	4	3	2	1	0
R	R	R	P_S	P_{TI}	R_{XI}	P_{TO}	P_{XO}

Interrupt Priority (IP) function register

Bit	Symbol	Function
7,6,5	R	Reserved
4	PS	Serial interrupt priority level
3	PTI	Timer 1 overflow interrupt priority level
2	PXl	External interrupt 1 priority level
1	PTO	Timer 0 overflow interrupt priority level
0	PXO	External interrupt 0 priority level

Note: Each bit if 1 : higher priority if 0: low priority

Figure 12.24. In Interrupt Priority Function Register Bit Configuration

INTERRUPT SERVICE ADDRESSES

Each of the 8051's five interrupts causes the microcontroller to do a hardware call to predefined addresses. However the responsibility of placing the interrupt service routine at this address is the job of the 8051 programmer. However an interrupt call saves the program counties (PC) onto the stack memory located on the 8051 internal RAM. This is done so that when the 8051 encounters a RETI instruction at the end of the interrupt service routine it can restore the program execution at the point (address) where the interrupt was caused. The predefined. addresses for each of the five interrupts is listed below.

Interrupt	Service Address
IEO	0003H
TOO	000BH
lEI	0013H
T01	001BH
Serial	0023H

PROGRAMMABLE INTERRUPTS

However sometimes it becomes necessary to use a software interrupt. Since any of the five interrupt bits even when set by a instruction shall cause an interrupt gives the 8051 microcontroller programmer five different software interrupts. These five software interrupts will have the same priority levels as the hardware interrupts. Thus a software interrupt can be initiated by setting an interrupt flag bit by the program.

12.7. KEY POINTS TO REMEMBER

1. A microcontroller is essentially a small computer built onto a single chip capable of performing all functions necessary to function as a computer.
2. Microcontrollers are dedicated or embedded controllers and are used in a wide number of applications ranging from automatic ovens to air craft and missile guidance systems.
3. The 8051 is an 8 bit microcontroller available in 40 pin DIP package. It can be made to run on frequencies ranging from 1 MHz to 16 MHz.
4. The 8051 contains the following internal hardware
   - 8 bit ALU
   - 16 bit registers PC and DPTR
   - 8 bit Program Status word
   - 8 bit stack points
   - Internal ROM 4 K bytes
   - Internal RAM 128 bytes
   - Two sixteen bit counters
   - Full Duplex serial data transmitter/receiver
   - Interrupts
   - Check generator
   - Control registers
   - function registers
5. An 8051 machine cycle has 6 time states where each time state is equal to two clock periods

6. The ALE signal is sent twice in one machine cycle allowing both bytes of instructions to be fetched by the microcontroller .

REVIEW QUESTIONS

12.1. What is the total size of the 8051 internal RAM?

12.2. Calculate the execution time of a single machine cycle instruction for the following clock frequencies.

(a) 5MHz

(b) 6 MHz

(c) 9 MHz

12.3. Explain the flag register (PSW) ofthe 8051 microcontroller.

12.4. What is the total size of a program that can be stored on the 8051 internal ROM.

12.5. Explain the 8086 clock generator circuit? Also calculate the instruction execution time for the 8051?

12.6. Explain the 8051 machine cycle?

12.7. List the difference between microprocessors and microcontroller ?

12.8. Explain the port circuitry of the following ports and the functioning

(a) port 0

(b) port 1

(c) port 2

(d) port 3

12.9. Explain the 8051 RAM structure?

12.10. What is the difference between bit and byte addressable registers? How do the b it addressable register and the 8051 architecture?

12.11. Explain the structure of the bit addressable registers of the internal RAM.

12.12. Explain the method employed for storing a data byte onto the stack? What is the value of SP register on a 8051 reset?

12.13. Interface an 8051 to external memory where the RAM size is 16 K and the ROM size is 64K.

12.14. Explain the TCON and TMOD function registers?

12.15. Explain the operation of the different timer mode? What is the maximum pulse rate that can be counted on pin Tl if the oscillator frequency is 5MHz.

12.16. Explain the timer and counter control logic circuitry?

12.17. Explain the different bits and their use in the SBUF and SCON registers?

12.18. What would be the baud rate if the Mode 0 of serial data transmission when the clock frequency is 8 MHz.

12.19. Explain the PCON function register? What is the use of GF1 and GFO bits?

12.20. Explain the 8051 interrupts and timer operation?

12.21. What are the various internal interrupts available in the 8051 ?

12.22. Flow does the IE function register control the interrupts of the 8051 ?

12.23. How is the 8051 interrupt priority set?

12.24. Which signals are used to access the 8051 external ROM and RAM chips?

12.25. Explain the technique used to select the timer modes of the 8051microcontroller ?

Chapter-13

D/A AND A/D CONVERTERS

INTRODUCTION

In electronics, a digital-to-analog converter (DAC or D-to-A) is a device that converts a digital (usually binary) code to an analog signal (current, voltage, or electric charge). An analog-to-digital converter (ADC) performs the reverse operction. Signals are easily stored and transmitted in digital form, but a DAC is needed for the signal to be recognized by human senses or other non-digital systems. A common use of digital-to-analog converters is generation of audio signals from digital information in music players. Digital video signals are converted to analog in televisions and cell phones to display colors and shades. Digital-to-analog conversion can degrade a Signal, so conversion details are normally chosen so that the errors are negligible.

13.1 DIGITAL TO ANALOG CONVERSION

Digital to analog conversion is the process of taking a value represented in digital code and converting it to voltage or current which is proportional to the digital value. The simplified block diagram for Digital to analog converter is shown in figure 13 .1. The D/A converter have n-bit digital data as its input and one analog data as its output.

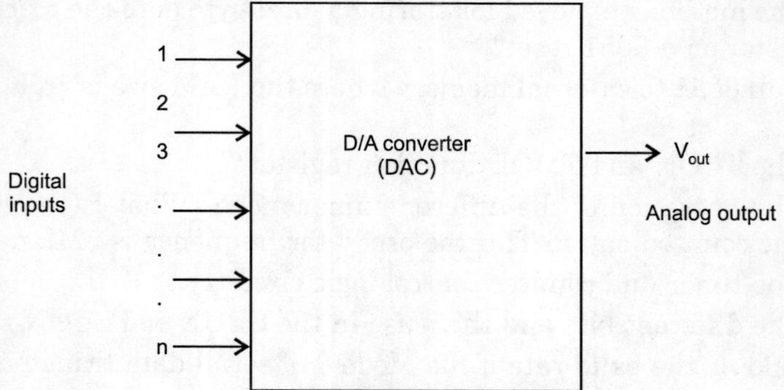

Fig 13.1 Simplified diagram of a Digital to Analog converter.

There are two types of D/A converter

(i) Weighted Resister Type D/A Converter

(ii) R-2R Ladder type D/A converter

13.2 WEIGHTED RESISTER TYPE D/A CONVERTER

The weighted resister type D/A converter can convert each digital level into equivalent analog voltage or current. Figure 13.2 shows the basic circuit for weighted resister type D/A converter in which the operational amplifier is used as a summing amplifier. The summing amplifier produces the weighted sum of the input voltages. As we know summing amplifier multiplies each input voltage by the ratio of the feedback resister R_f to the corresponding input resister R_{IN}. By applying the Kirchhoff's current law at point A in the figure 11.2, we get, current coming to A = current going away from A

$$I_1 + I_2 + I_3 + I_4 = I_F$$

$$\frac{V_1}{R} + \frac{V_2}{2R} + \frac{V_3}{4R} + \frac{V_4}{8R} = \frac{V_0}{R_F}$$

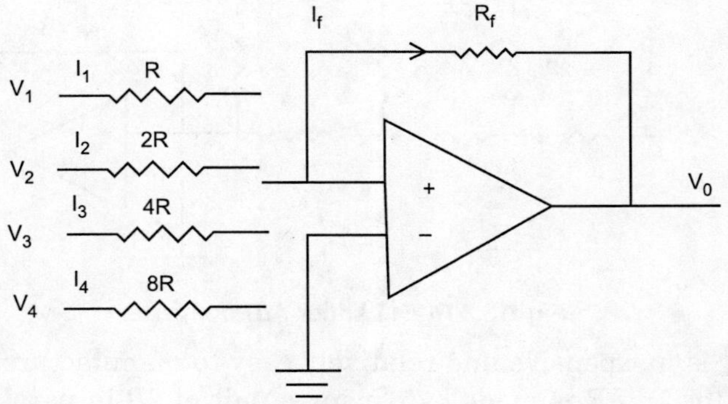

Fig. 13.2 Weighted resistor ND converter

If we select, R f = 8 R. Then, we get

$$-V_0 = 8\,V_1 + 4\,V_2 + 2\,V_3 + V_4$$

Let us now consider a binary input 1110 is applied at the input of D/A converter. If we assume 1 V for logic 1 and OV for logic 0, Then we obtain

$$-V_0 = 8 + 4 + 2 + 0 = 14\text{ V}$$

But 14 is the decimal equivalent of the binary number 1110. It is clearly concluded that the circuit shown in figure 11.2 performing as digital to analog converter.

13.3 R-2r Ladder Type D/A Converter

The previous D/A converter use binary weighted resisters to produce the proper weighting of each bit. In this type of D/A converter, there is large difference in the resister values between MSB and LSB and further in IC technology, it is very difficult to fabricate resister range that maintain an accurate ratio especially with variation in temperature. Hence, the binary weighted resister type of D/A has above practical limitations. For this reason, it is preferable to have a circuit that uses resistances that are fairly close in value. The figure 13.3 shows, a circuit of R to 2R ladder type of D/A converter which uses only 2 to 1 resistance value. In this circuit, only two different values resistances are used, Rand 2R. The I_{OUT} current depends on the positions of the four switches and the binary inputs B_3 B_2 B_1 Bo control the states of the switches. This current is flow into an op- amp, current- to- voltage converter, to produce V_{out}. The value of V_{out} can be written as

$$V_{OUT} = \frac{V_{ref}}{8} \times B$$

Where, B is the value of the binary input, which can range from 0000 to 1111.

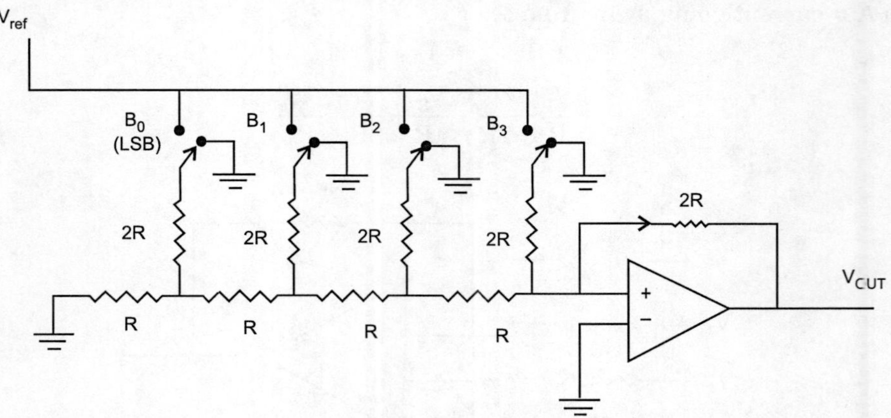

Fig. 13.3 R-2R Ladder A/D converter

The R-2R ladder is inexpensive and relatively easy to manufacture since only two resistor values are required (or 1, if R is made by placing a pair of 2R in parallel, or if 2R is made by placing a pair of R in series). It is fast and has fixed output impedance R. The R-2R ladder operates as a string of current dividers whose output accuracy is solely dependent on how well each resistor is matched to the others. Small inaccuracies in the higher significant bit resistors can entirely overwhelm the contribution of the less significant bits. This may result in non-monotonic behavior at major crossings, such as from 01111 to 10000. Depending on the type of logic gates used and design of the logic circuits, there may be transitional voltage spikes at such major crossings even with perfect resistor values. These can be filtered, with capacitance at the output node for instance (the consequent reduction in bandwidth may be significant in some applications). Finally, the 2R resistance is in series with the digital output impedance.

High output impedance gates (e.g., LVDS) may be unsuitable in some cases. For all of the above reasons, this type of DAC tends to be restricted to a relatively small number of bits, although integrated circuits may push the number of bits to 14 or even more, 8 bits or fewer is more typical.

13.4 SPECIFICATION OF D/A CONVERTER

The specification of D/A converters are linearity, resolution, accuracy, settling time and temperature sensitivity.

13.4.1 Resolution

The output of a D/A converter is not truly continuous; it produces a finite set of output values. Therefore, resolution of a D/A converter is defined as the smallest change that can occur in the analog output as a result of a change in digital input. The resolution is always equal to the weight of the LSB and is also referred as the step size. Since it is the amount that Vout will change as the digital input value is changed from one step to the next. The resolution (step size) is a proportionality factor for input/output relationship of D/A converter.

$$\text{Analog output} = K \times \text{digital input}$$

Where, K is resolution or step size and can be calculated as follows

$$\text{resolution} = K = \frac{A_{fs}}{2^n - 1}$$

Where A_{fs} is the analog full scale output and n is number of bits. The resolution as a percentage of full scale output can be defined as

$$\% \text{ resolution} = \frac{\text{Step size}}{\text{Full scale}} \times 100\%$$

$$\% \text{ resolution} = \frac{1}{\text{number of steps}} \times 100$$

For an N- bit binary input code the total number of step is $2^N - 1$. Thus % resolution can be written as

$$\% \text{ resolution} = \frac{1}{2^N - 1} \times 100\%$$

This means that it is only the number of bits that determines the percentage resolution. Therefore, most D/A converter manufactures specify resolution as the number of bits. The system designer must decide what resolution is needed on the basis of the required system performance.

13.4.2. Accuracy

Accuracy means how the measured value is close to true value. The D/A converter manufacturers have many ways to specifying accuracy. The two main ways are full scale error and linearly error which are normally expressed as a percentage of the converter's full scale output. Full scale error is the maximum deviation of the D/A's output from its expected (ideal) value expressed as a percentage of full scale.

13.4.3. Linearity error

Linearity error is the maximum deviation in step size from the ideal step size. it may also be noted that accuracy and resolution of a D/A converter must be compatible. It would be illogical to have a resolution of 1 percent and accuracy of 0.1 percentage or vice versa. For example, a D/A converter with a resolution of 1 percentage and an F. S. output of 10 V can produce an output analog voltage within 0.1V of any desired value, assuming perfect accuracy. It makes no sense to have a costly accuracy of 0.01 % F. S. (or 1 mv) if the resolution already limits the closeness of the desired value to 0.1V. The same can be said for having a resolution that is very small (many bits) while accuracy in poor it is a waste of input bits.

13.4.4 Offset error

Ideally, the output of a D/A converter will be zero volt when the binary inputs are all 0. However, practically there will be a very small output voltage called offset voltage or offset error.

13.4.5 Settling Time

The time required for the output of the DAC to settle to within (±1/2) LSB of the final value for a given digital input is referred as settling time. Thus, it specifies the operating speed of the DAC.

13.4.6 Monotonicity

A DAC is monotonic if its output increases with the binary input increment from one value to the next. In other words, this is that the staircase output will have no downward step as the binary input is incremented from zero to full scale.

13.4.7 Sensitivity

DAC are sensitivity to temperature change. This is expressed as parts per million (ppm)/°C. Typical value of temperature sensitivity may be taken as 10 ppm/°C

13.5 ANALOG-TO-DIGITAL (ADC) CONVERTER

An analog-to-digital converter (abbreviated ADC, A/D or A to D) is a device that converts analog quantity to its equivalent digital signal. Typically, an ADC is an electronic device that converts an input analog voltage or current to a digital number proportional to the magnitude of the voltage or current. However, some non-electronic or only partially electronic devices, such as rotary encoders, can also be considered ADCs. A/D conversion is usually more complex and time consuming than D/A conversion process.

The most common ADCs are:

1. Digital Ramp ADC
2. Successive approximation ADC
3. Flash ADC
4. Single slope ADC
5. Dual slope ADC

13.6 DIGITAL RAMP ADC

A ramp A/D converter is simplest A/D converter as shown in figure 11.4, it has a control AND gate, an counter that feeds a digital to analog converter (DAC) and analog comparator. The input signal and the DAC signal both go to a comparator. The comparator controls the counter. The circuit uses negative feedback from the comparator to adjust the counter until the DAC's output is close enough to the input signal. The number is read from the counter. Ramp converters have very wide ranges and high resolution, but the conversion time is dependent on the input signal level, though it will always have a guaranteed worst-case. Delta converters are often very good choices to read real-world signals. Most signals from physical systems do not change abruptly.

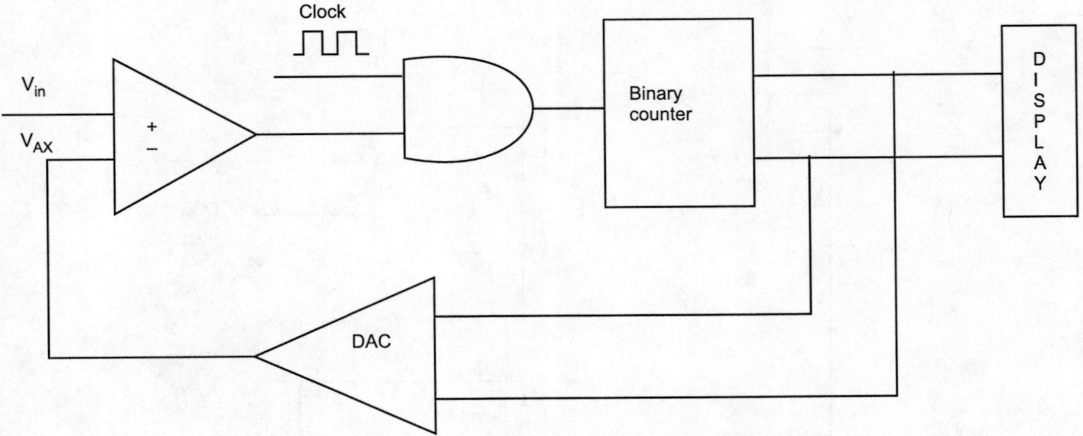

Fig. 13.4 Ramp Type A/D converter

The operation of the ramp type ADC is very simple. Initially all inputs to DAC is zero, thus V_{in} is greater than V_{AX} the output of comparator will be high. As the clock goes high the counter will enable and start counting. This binary output is fed to DAC that converts these digital bits back to corresponding analog data V_{AX} for comparison with input analog signal Vin. The DAC output starts increasing and reaches a value equal to or greater V_T (typically 10 to 1000 µV). At this point output of comparator goes low and it disables the counter and counter stop counting. The conversion process is now completed as the comparator output goes high-to-low transition and contents of counter are the digital representation of input voltage V_{in}. The counter will hold the digital values until the next conversion is initiated.

13.7 SUCCESSIVE-APPROXIMATION ADC

A successive-approximation ADC is most widely used AD converter. It has more complex circuitry that the ramp ADC. But it has much shorter cc..version time usually fixed value that is not dependent on the value of the analog input. The Figure 11.5 shows the basic arrangement of successive-approximation ADC similar to ramp ADC but the counter is replaced by a register in this case. A successive approximation ADC uses a comparator to successively narrow a range that contains the input voltage. At each successive step, the converter compares the input voltage to the output of an internal digital to analog converter which might represent the midpoint of a selected voltage range. At each step in this process, the approximation is stored in a successive approximation register (SAR).

For example, consider an input voltage of 6.3 V and the initial range is 0 to 16 V. For the first step, the input 6.3 V is compared to 8 V (the midpoint of the 0-16 V range). The comparator reports that the input voltage is less than 8 V, so the SAR is updated to narrow the range to 0-8 V. For the second step, the input voltage is compared to 4 V (midpoint of 0-8). The comparator reports the input voltage is above 4 V, so the SAR is updated to reflect the input voltage is in the range 4-8 V. For the third step, the input voltage is compared with 6 V (halfway between 4

V and 8 V); the comparator reports the input voltage is greater than 6 volts, and search range becomes 6-8 V. The steps are continued until the desired resolution is reached.

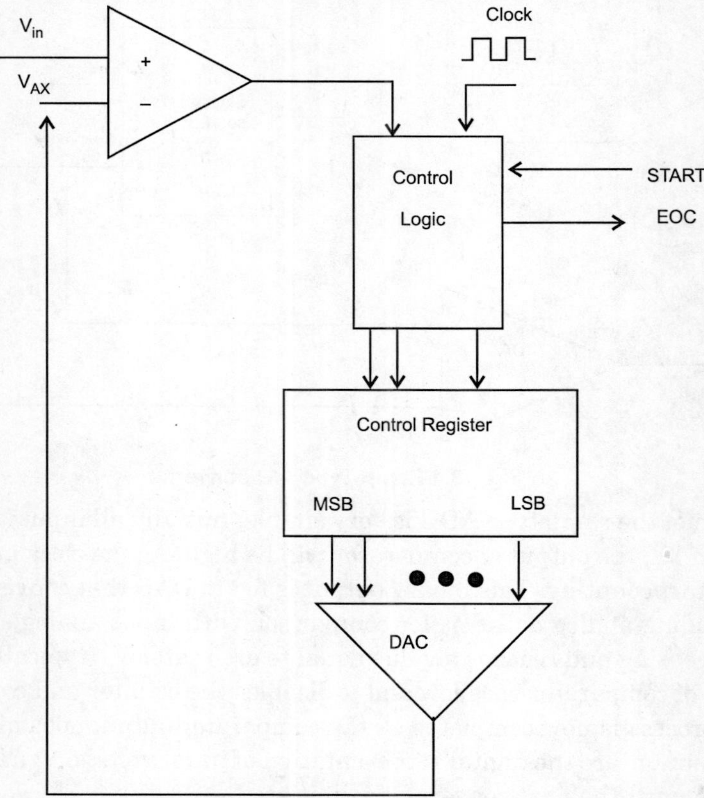

Fig. 13.5 Successive-approximation ADC

13.8 FLASH AID CONVERTER

The flash converter is the fastest ADC available now but it requires much more circuitry than other types. A direct-conversion ADC or flash ADC has a bank of comparators sampling the input signal in parallel, each firing for their decoded voltage range. The comparator bank feeds a logic circuit that generates a code for each voltage range. Direct conversion is very fast, capable of gigahertz sampling rates, but usually has only 8 bits of resolution or fewer, since the number of comparators needed, $2^N - 1$, doubles with each additional bit, requiring a large, expensive circuit. ADCs of this type have a large die size, a high input capacitance, high power dissipation, and are prone to produce glitches at the output (by outputting an out-of-sequence code).

The flash converter in fig 13.6 has a three-bit resolution and a step size of 1 V. The voltage divider sets up references levels for each comparator so that there are seven levels corresponding to 1V (weight of LSB), 2V, 3V and 7V (full scale). The other input of each comparator is connected to analog input Vin. All or any comparator output will be high if input V_{in} is less than 1V and low if greater than 1V. The output of each comparator is fed to an active LOW priority encoder that generates a binary output corresponding to the highest-number comparator output that is LOW. For example, when V_{in} is between 3 and 4, output of first, second and third comparator will be low and all other will be high. The priority encoder will responds only to the LOW at third comparator output and will produce a binary output 011, which represents the digital equivalent of V_{in}, within the resolution of 1V. The encoder will produced 111 when V_{in} is greater than 7V, output of all comparators will be low.

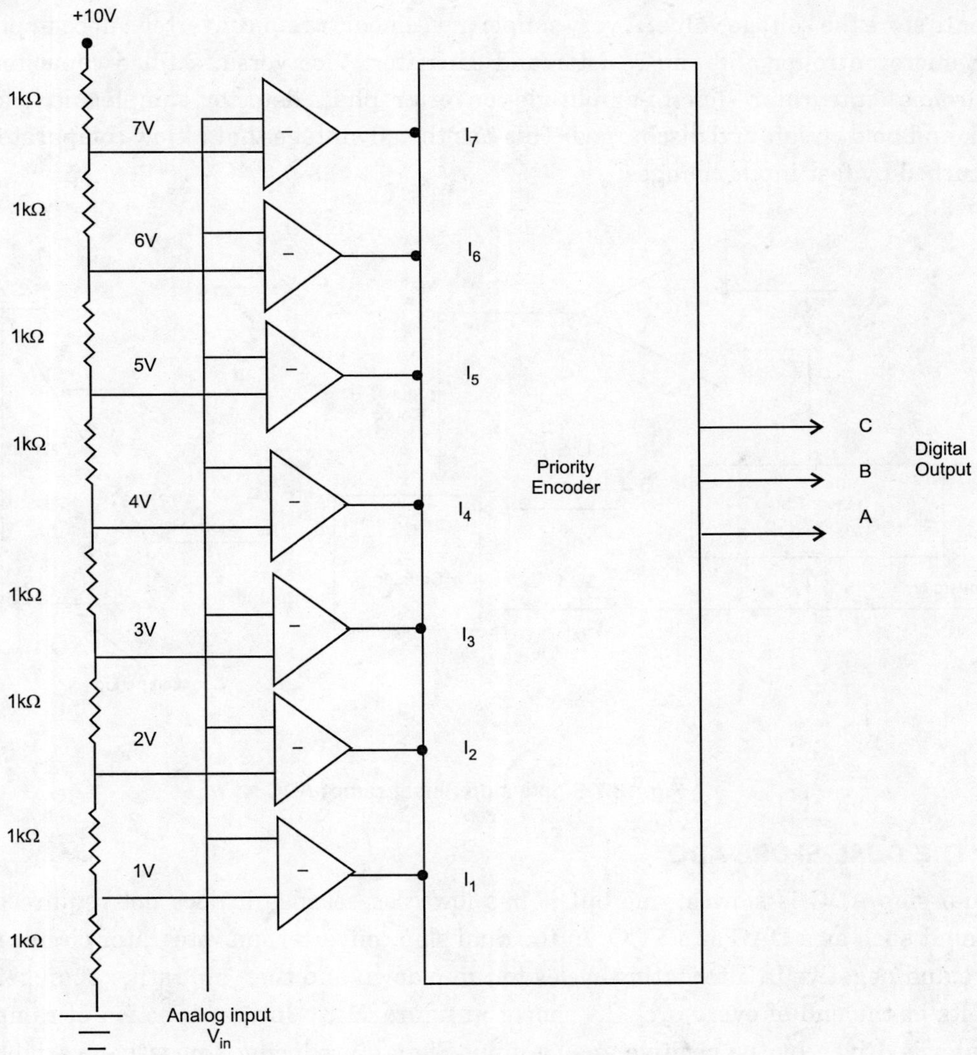

Fig. 13.6 Three-bit flash A/D Converter

13.9 THE SINGLE-SLOPE ADC

A single slope ADC produces a saw-tooth signal that ramps up or down then quickly returns to zero. Figure 11.7 shows a relatively simple ADC that uses a ramp generator as comparison. When the ramp starts, a timer starts counting. When the ramp voltage matches the input, a comparator fires, and the timer's value is recorded. Timed ramp converters require the least number of transistors. The ramp time is sensitive to temperature because the circuit generating the ramp is often just some simple oscillator.

There are two solutions: use a clocked counter driving a DAC and then use the comparator to preserve the counter's value, or calibrate the timed ramp. A special advantage of the ramp-compare system is that comparing a second signal just requires another comparator, and another register to store the voltage value. A very simple (non-linear) ramp-converter can be implemented with a microcontroller and one resistor and capacitor. Vice versa, a filled capacitor can be taken from an integrator, time-to-amplitude converter, phase detector, sample and hold circuit, or peak and hold circuit and discharged. This has the advantage that a slow comparator cannot be disturbed by fast input changes

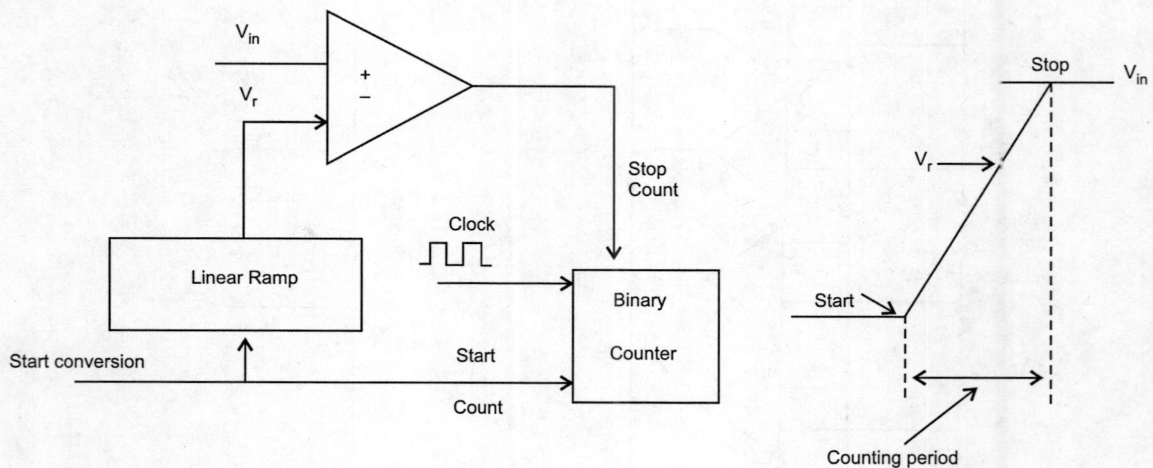

Fig. 13.7 Single slop (linear ramp) ADC

13.10 THE DUAL-SLOPE ADC

The dual slop ADC is slowest one but it has low cost because it does not require precision component such as a DAC or a VCO. In the dual slop converter, an integrator circuit is driven positive and negative in alternating cycles to ramp down and then up, rather than being reset to 0 volts at the end of every cycle as shown in figure 11.8 . In one direction of ramping, the integrator is driven by the positive analog input Signal (producing a negative, variable rate of output voltage change, or output slope) for a fixed amount of time, as measured by a counter with a precision frequency clock. Then, in the other direction, with a fixed reference voltage (producing a fixed rate of output voltage change) with time measured by the same counter. The counter stops counting when the integrator's output reaches the same voltage as it was when it started the fixed-time portion of the cycle. The amount of time it takes for the integrator's capacitor to discharge back to its original output voltage, as measured by the magnitude accrued by the counter, becomes the digital output of the ADC circuit.

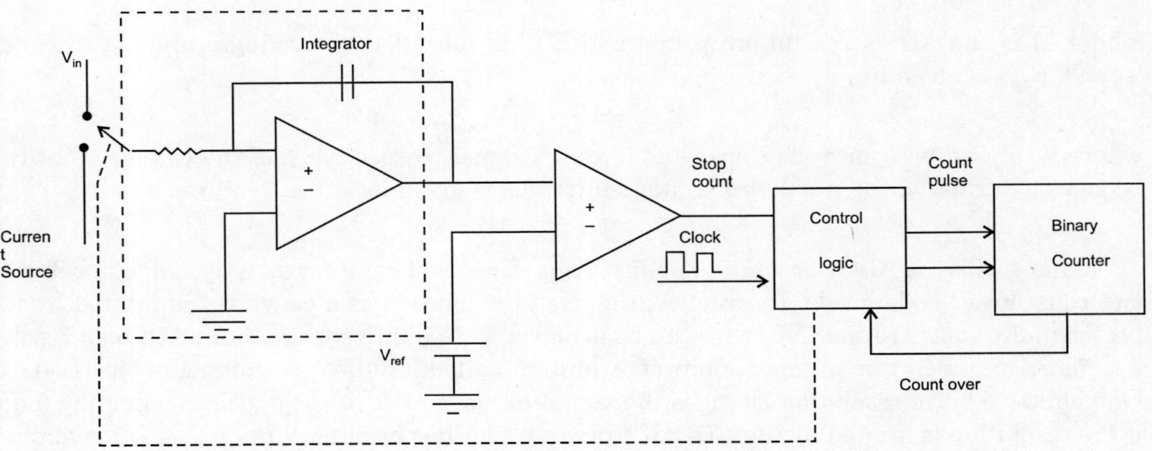

Fig. 13.8 Dual slope (linear ramp) ADC

13.11 SPECIFICATION OF ADC CONVERTER

Some of the specifications of ADC are accuracy, quantization error, conversion time and resolution as discussed below:

13.11.1 Accuracy

Since analog and digital, both systems are used to design ADC, so, the overall accuracy of ADC converters depends on the errors from both analog and digital systems. An ADC has several sources of errors, Quantization error and (assuming the ADC is intended to be linear) non-linearity are intrinsic to any analog-to-digital conversion. There is also a so-called aperture error which is due to a clock jitter and is revealed when digitizing a time-variant signal (not a constant value).

13.11.2 Resolution

The resolution of the converter indicates the number of discrete values it can produce over the range of analog values. The values are usually stored electronically in binary form, so the resolution is usually expressed in bits. In consequence, the number of discrete values available, or "levels", is a power of two. For example, an ADC with a resolution of 8 bits can encode an analog input to one in 256 different levels, since $2^8 = 256$. The values can represent the ranges from 0 to 255 (i.e. unsigned integer) or from -128 to 127 (i.e. signed integer), depending on the application.

Resolution can also be defined electrically, and expressed in volts. The minimum change in voltage required to guarantee a change in the output code level is called the least significant bit (LSB) voltage.

The resolution Q of the ADC is equal to the LSB voltage. The voltage resolution of an ADC is equal to its overall voltage measurement range divided by the number of discrete values:

$$Q = \frac{E_{FSR}}{2^M}$$

where M is the ADC's resolution in bits and E_{FSR} is the full scale voltage range (also called 'span'). E_{FSR} is given by

$$E_{FSR} = V_{RefHi} - V_{Reflow}$$

where V_{RefHi} and V_{Reflow} are the upper and lower extremes, respectively, of the voltages that can be coded. Normally, the number of voltage intervals is given by

$$N = 2^M - 1$$

Where, M is the ADC's resolution in bits. That is, one voltage interval is assigned in between two consecutive code levels. In practice, the useful resolution of a converter is limited by the best signal-to-noise ratio (SNR) that can be achieved for a digitized signal. An ADC can resolve a signal to only a certain number of bits of resolution, called the effective number of bits (ENOB). One effective bit of resolution changes the signal-to-noise ratio of the digitized signal by 6 dB, if the resolution is limited by the ADC. If a preamplifier has been used prior to A/D conversion, the noise introduced by the amplifier can be an important contributing factor towards the overall SNR.

13.11.3 Response type

Most ADCs are linear types. The term linear implies that the range of input values has a linear relationship with the output value.

13.11.4 Quantization error

Quantization error (or quantization noise) is the difference between the original signal and the digitized signal. Hence, the magnitude of the quantization error at the sampling instant is between zero and half of one LSB. Quantization error is due to the finite resolution of the digital representation of the signal, and is an unavoidable imperfection in all types of ADCs.

13.11.5 Non-linearity

All ADCs suffer from non-linearity errors caused by their physical imperfections, causing their output to deviate from a linear function (or some other function, in the case of a deliberately non-linear ADC) of their input. These errors can sometimes be mitigated by calibration, or prevented by testing.

Example 13.1: An 8-bit DAC has a full scale output of 2mA and a full scale error of ±0.5% F.S. What is the range of possible output for an input of 10000000?

Solution: The step size is 2mA/255= 7.84 μA. Since 10000000 = 128_{10}, the ideal output should be 128 × 7.84 μA = 1004 μA. The error can be as much as

$$\pm 0.5\% \times 2\ mA = \pm 10\ 10\mu A$$

Thus, the actual output can deviate by this amount from ideal 1004 μA , so the actual output can be anywhere from 994 to 1004 μA.

Example 13.2: A 12-bit DAC has a full scale output of 9.99V.

(a) Determine the percentage resolution.

(b) Determine the converter's step size.

Solution: (a) Twelve bits correspond to three decimal digits, that is, decimal numbers from 000 to 999 Therefore, the output of this DAC has 999 possible steps from 0 to 9.99 V. Thus, we have

$$\% \text{ resolution} = \frac{1}{\text{number of steps}} \times 100\%$$

$$= \frac{1}{999} \times 100\%$$

$$\approx 0.1\%$$

(b)

$$\text{step size} = \frac{\text{F.S.}}{\text{number of steps}}$$

$$= \frac{9.99\text{V}}{999}$$

$$= 0.01\text{V}$$

Example 13.3 A 5-bit DAC has a current output. For a digital input of 10100, an output current of 10 mA is produced. What will I_{OUT} be for a digital input of 11101?

Solution: The digital output 10100_2 is equal to decimal 20. Since I_{OUT} = 10 mA for this case, the proportionality factor must be 0.5 mA. Thus, we can find I_{OUT} for any digital input such as $11101_2 = 29_2$ as follows:

$$I_{OUT} = 0.5 \text{ mA} \times 29$$

$$I_{OUT} = 14.5 \text{ mA}$$

Example 13.4: Assume that V_{ref} = 10 V and R = 10 kΩ. Determine the resolution and the full-scale output for this DAC. Assume that R_L is much smaller than R.

Solution:

$$I_{OUT} = \frac{V_{ref}}{R} = 1\text{ma}$$

This is the weight of the MSB. The other three current will be 0.5, 0.25, and 0.125 mA. The LSB is 0.125 mA, which is also the resolution. The full scale output will occur when the binary inputs are all HIGH so that each current switch is closed and

$$I_{OUT} = 1 + 0.5 + 0.25 + 0.125 = 1.875 \text{ mA}$$

Example13.5: Assume that V_{ref} = 5 V for R-2R DAC. What are the resolution and the full-scale output for this DAC?

Solution: the resolution is equal to the weight of LSB, which we can determine by setting B = 0001 = 1 in the output equation of R-2R ladder DAC

$$\text{resolution} = \frac{-5\text{V} \times 1}{8}$$

$$= -0.625 \text{ V}$$

The full scale output occurs for B=1111 = 15_{10}. Again using the output equation

$$\text{full} - \text{scale} = \frac{-5V \times 15}{8}$$

$$= -9.375$$

Example13.6: Assume the following values for the digital ramp ADC: clock frequency = 1 MHz; V_T = 0.1 mV; DAC has F.S. output = 10.23 V and 10-bit input. Determine the following values.

(a) The digital equivalent obtained for Vin = 3.728 V.

(b) The conversion time

(c) The resolution of this converter.

Solution:

(a) The DAC has a 10-bit input and a 10.23 V F. S. output. Thus, the number of total possible steps is $2^{10} - 1$ = 1023, and so the step size is 10.23/1023 = 10 mV

This means that V_{AX} increases in steps of 10 mV as the counter counts up from O. Since Vin = 3.728 V and V_T = 0.1 mV, V_{AX} must reach 3.7281 V or more before the comparator switches Low. This will require

$$\frac{3.7281}{10mV} = 372.81 = 373 \text{ steps}$$

Then at the end of the conversion, the converter will hold the binary equivalent of 373, which is 0101110101. This is the desire digital equivalent of Vin = 3.728 V, as produced by this ADC.

(b) 373 steps were required to complete the conversion. Thus, 373 clock pulses occurred at the rate of one per microsecond. This gives a total conversion time 373 μs.

(c) The resolution of this converter is equal to the step size of the DAC, which is 10 mV. In percent it is 1/1023 × 100% ≃ 0.1 %.

Example13.7: Compare the maximum conversion times of a 10-bit digital-ramp ADC and a 10-bit successive approximation ADC if both utilize a 500-kHz clock frequency.

Solution: for digital-ramp converter, the maximum conversion time is $(2^N - 1) \times$ (1 clock cycle)

$$= 1023 \times 2 \ \mu s$$

$$= 2046 \ \mu s$$

For a 10-bit successive-approximation converter, the conversion time is always 10 clock periods or

$$10 \times 2 \ \mu s = 20 \mu s$$

Thus, it is about 100 times faster than the digital-ramp converter.

Example13.8: As eight-bit successive approximation ADC has a resolution of 20 mY. What will it digital output is for an analog input of 2.17 V?

Solution:

$$\frac{2.17V}{20mV} = 108.5$$

So that step 108 would produce V_{AX} = 2.16 V and step 109 would produce 2.18 V. The SAC always produces a final V_{AX} that is at the step below Vin. Therefore, for the case of Vin = 2.17 V, the digit result would be $108_{10} = 01101100_2$.

REVIEW QUESTIONS

1. Digital to analog conversion is the process of taking a value represented in digital code and converting it to voltage or current which is proportional to the digital value.
2. The R-2 ladder is inexpensive and relatively easy to manufacture since only two resistor values are required (or 1, if R is made by placing a pair of 2R in parallel, or if 2R is made by placing a pair of R in series).
3. Accuracy means how the measured value is close to true value.
4. The resolution as a percentage of full scale output can be defined as

$$\% \text{ resolution} = \frac{\text{Step size}}{\text{Full scale}} \times 100\%$$

5. Linearity error is the maximum deviation in step size from the ideal step size.
6. DAC are sensitivity to temperature change. This is expressed as parts per million (ppm)/ °C. Typical value of temperature sensitivity may be taken as 10 ppm / °C
7. An analog-to-digital converter (abbreviated ADC, A/D or A to D) is a device that converts analog quantity to its equivalent digital signal.
8. Ramp converters have very wide ranges and high resolution, but the conversion time is dependent on the input signal level, though it will always have a guaranteed worst-case. Delta converters are often very good choices to read real-world signals. Most signals from physical systems do not change abruptly.
9. A successive-approximation ADC is most widely used AD converter. It has more complex circuitry than the ramp ADC. But it has much shorter conversion time usually fixed value that is not dependent on the value of the analog input.
10. The flash converter is the fastest ADC available it now but it requires much more circuitry than other types.
11. The dual slop ADC is slowest one but it has low cost because it does not require precision component such as a DAC or a VCO.
12. Since analog and digital both systems are used to design ADC, so, the overall accuracy of ADC converters depends on the errors from both analog and digital systems.
13. Most ADCs are linear types. The term linear implies that the range of input values has a linear relationship with the output value.
14. Quantization error (or quantizatior noise) is the difference between the original signal and the digitized signal.
15. All ADCs suffer from non-linearity errors caused by their physical imperfections, causing their output to deviate from a linear function (or some other function, in the case of a deliberately non-linear ADC) of their input.

PROBLEM

1. What is the function of an ADC?
2. What is the function of a DAC?
3. An 8-bit DAC has an output of 3.92 mA for an input of 01100010. What are the DAC's resolution and full-scale output?
4. What is the percentage resolution of an eight-bit DAC?
5. What is the advantage of R-2R ladder DACs over those that use binary-weighted resistors?
6. Define full scale error.
7. What is settling time?
8. What is the function of the comparator in the ADC?
9. Describe the basic operation of the digital ramp ADC.
10. Explain quantization error.
11. Describe the basic operation of the successive approximation ADC.
12. Describe the basic operation of the digital ramp ADC.
13. Describe the basic operation of the flash ADC.
14. How many comparators would a 12-bit flash converter require? How rnary resistors?
15. Name three types of ADCs that do not use a DAC.
16. Describe the basic operation of the R-2R ladder DAC.
17. Describe the basic operation of the binary weighted resistors ADC.
18. Describe the basic operation of the single slop ADC.
19. Describe the basic operation of the dual slop ADC.
20. Determine the weight of each bit for the how a flash ADC with a conversion time of 1μs would work for the situation.